Lecture Notes in Physics

Bisher erschienen/Already published

Vol. 1: J. C. Erdmann. Wärmeleitung in Kristallen, theoretische Grundlagen und fortgeschrittene experimentelle Methoden. II, 283 Seiten. 1969.

Vol. 2: K. Hepp, Théorie de la renormalisation. III, 215 pages. 1969.

Vol. 3: A. Martin, Scattering Theory: Unitarity, Analyticity and Crossing. IV, 125 pages. 1969.

Vol. 4: G. Ludwig, Deutung des Begriffs „physikalische Theorie" und axiomatische Grundlegung der Hilbertraumstruktur der Quantenmechanik durch Hauptsätze des Messens. 1970. Vergriffen.

Vol. 5: Schaaf, The Reduction of the Product of Two Irreducible Unitary Representations of the Proper Orthochronous Quantummechanical Poincare Group. IV, 120 pages. 1970.

Vol. 6: Group Representations in Mathematics and Physics. Edited by V. Bargmann. V, 340 pages. 1970.

Vol. 7: R. Balescu, J. L. Lebowitz, I. Prigogine, P. Résibois, Z. W. Salsburg, Lectures in Statistical Physics. V, 181 pages. 1971.

Vol. 8: Proceedings of the Second International Conference on Numerical Methods in Fluid Dynamics. Edited by M. Holt. 1971. Out of print.

Vol. 9: D. W. Robinson, The Thermodynamic Pressure in Quantum Statistical Mechanics. V, 115 pages. 1971.

Vol. 10: J. M. Stewart, Non-Equilibrium-Relativistic Kinetic Theory. III, 113 pages. 1971.

Vol. 11: O. Steinmann, Pertubation Expansions in Axiomatic Field Theory. III, 126 pages. 1976.

Vol. 12: Statistical Models and Turbulence. Edited by C. Van Atta and M. Rosenblatt. Reprint of the First Edition. VIII, 492 pages. 1975.

Vol. 13: M. Ryan, Hamiltonian Cosmology. VII, 169 pages. 1972.

Vol. 14: Methods of Local and Global Differential Geometry in General Relativity. Edited by D. Farnsworth, J. Fink, J. Porter, and A. Thompson. V, 188 pages.

Vol. 15: M. Fierz, Vorlesungen zur Entwicklungsgeschichte der Mechanik. V, 97 Seiten. 1972.

Vol. 16: H.-O. Georgii, Phasenübergang 1. Art bei Gittergasmodellen. IX, 167 Seiten. 1972.

Vol. 17: Strong Interaction Physics. Edited by W. Rühl and A. Vancura. V, 405 pages. 1973.

Vol. 18: Proceedings of the Third International Conference on Numerical Methods in Fluid Mechanics, Vol. I. Edited by H. Cabannes and R. Temam. VII, 186 pages. 1973.

Vol. 19: Proceedings of the Third International Conference on Numerical Methods in Fluid Mechanics, Vol. II. Edited by H. Cabannes and R. Temam. VII, 275 pages. 1973.

Vol. 20: Statistical Mechanics and Mathematical Problems. Edited by A. Lenard. VIII, 247 pages. 1973.

Vol. 21: Optimization and Stability Problems in Continuum Mechanics. Edited by P. K. C. Wang. V, 94 pages. 1973.

Vol. 22: Proceedings of the Europhysics Study Conference on Intermediate Processes in Nuclear Reactions. Edited by N. Cindro, P. Kulišic and Th. Mayer-Kuckuk. XIV, 329 pages. 1973.

Vol. 23: Nuclear Structure Physics. Proceedings 1973. Edited by U. Smilansky, I. Talmi, and H. A. Weidenmüller. XII, 296 pages. 1973.

Vol. 24: R. F. Snipes, Statistical Mechanical Theory of the Electrolytic Transport of Nonelectrolytes. V, 210 pages. 1973.

Vol. 25: Constructive Quantum Field Theory. The 1973 "Ettore Majorana" International School of Mathematical Physics. Edited by G. Velo and A. Wightman. III, 331 pages. 1973.

Vol. 26: A. Hubert, Theorie der Domänenwände in geordneten Medien. XII, 377 Seiten. 1974.

Vol. 27: R. K. Zeytounian, Notes sur les Ecoulements Rotationnels de Fluides Parfaits. XIII, 407 pages. 1974.

Vol. 28: Lectures in Statistical Physics. Edited by W. C. Schieve and J. S. Turner. V, 342 pages. 1974.

Vol. 29: Foundations of Quantum Mechanics and Ordered Linear Spaces. Advanced Study Institute, Marburg 1973. Edited by A. Hartkämper and H. Neumann. VI, 355 pages. 1974.

Vol. 30: Polarization Nuclear Physics. Proceedings 1973. Edited by D. Fick. IX, 292 pages. 1974.

Vol. 31: Transport Phenomena. Sitges International Schools of Statistical Mechanics, June 1974. Edited by G. Kirczenow and J. Marro. XIV, 517 pages. 1974.

Vol. 32: Particles, Quantum Fields and Statistical Mechanics. Proceedings 1973. Edited by M. Alexanian and A. Zepeda. V, 132 pages. 1975.

Vol. 33: Classical and Quantum Mechanical Aspects of Heavy Ion Collisions. Proceedings 1974. Edited by H. L. Harney, P. Braun-Munzinger, and C. K. Gelbke. VII, 311 pages. 1975.

Vol. 34: One-Dimensional Conductors GPS Summer School Proceedings, 1974. Edited by H. G. Schuster. VII, 371 pages. 1975.

Vol. 35: Proceedings of the Fourth International Conference on Numerical Methods in Fluid Dynamics, 1974. Edited by R. D. Richtmyer. V, 457 pages. 1975.

Vol. 36: R. Gatignol, Théorie Cinétique des Gaz à Répartition Discrète de Vitesses. II, 219 pages. 1975.

Vol. 37: Trends in Elementary Particle Theory. Proceedings 1974. Edited by H. Rollnik and K. Dietz. V, 472 pages. 1975.

Vol. 38: Dynamical Systems, Theory and Applications. Proceedings 1974. Edited by J. Moser. VI, 624 pages. 1975.

Vol. 39: International Symposium on Mathematical Problems in Theoretical Physics. Proceedings 1975. Edited by H. Araki. XII, 562 pages. 1975.

Vol. 40: Effective Interactions and Operators in Nuclei. Proceedings 1975. Edited by B. R. Barrett. XII, 339 pages. 1975.

Vol. 41: Progress in Numerical Fluid Dynamics. Proceedings 1974. Edited by H. J. Wirz. V, 471 pages. 1975.

Vol. 42: H II Regions and Related Topics. Proceedings 1975. Edited by D. Downes and T. L. Wilson. XII, 488 pages. 1975.

Vol. 43: Laser Spectroscopy. Proceedings 1975. Edited by S. Haroche, J. C. Pebay-Peyroula, T. W. Hänsch, and S. E. Harris. X, 466 pages. 1975.

Lecture Notes in Physics

Edited by J. Ehlers, München, K. Hepp, Zürich
R. Kippenhahn, München, H. A. Weidenmüller, Heidelberg
and J. Zittartz, Köln
Managing Editor: W. Beiglböck, Heidelberg

80

Mathematical Problems in Theoretical Physics

International Conference
Held in Rome, June 6–15, 1977

Edited by
G. Dell'Antonio, S. Doplicher and G. Jona-Lasinio

Springer-Verlag Berlin Heidelberg GmbH

Editors
G. Dell'Antonio
S. Doplicher
Istituto Matematico
„G. Castelnuovo"
Università di Roma
Citta Universitaria
Piazzale delle Scienze, 5
I-00100 Roma/Italy

G. Jona-Lasinio
Istituto di Fisica „G. Marconi"
Università di Roma
Città Universitaria
Piazzale delle Scienze, 5
I-00100 Roma/Italy

Library of Congress Cataloging in Publication Data

International Conference on the Mathematical
Problems in Theoretical Physics, University of
Rome, 1977.
Mathematical problems in theoretical physics.

(Lecture notes in physics ; 80)
Sponsored by the International Mathematical
Union.
Bibliography: p.
Includes index.
1. Mathematical physics--Congresses.
I. Dell'Antonio, G., 1933- II. Doplicher, S.
III. Jona-Lasinio, G. IV. International Mathematical Union (Founded 1950) V. Title.

ISBN 978-3-540-08853-0 ISBN 978-3-540-35811-4 (eBook)
DOI 10.1007/978-3-540-35811-4

2153/3140-543210

FOREWORD

The present volume contains the Proceedings of the International Conference on the Mathematical Problems in Theoretical Physics held at the University of Rome from 6 to 15 June 1977. This Conference continued the tradition of the Moscow 1972 and Kyoto 1975 Conferences. All the Conferences in this serie have been sponsored by the International Mathematical Union.

According to a tradition by now established, these Conferences are not only an occasion to communicate results but also a means to provide prespectives and outlook relevant to scientists with different specializations, and should be helpful for the orientation of our younger colleagues. To this purpose, and in the spirit of experimenting a new formula, we avoided fractioning the Conference in more technical parallel sessions, planning instead only review talks on Mathematics and Physics. In fact, the general subject of such a Conference is not only the intersection of Mathematics and Physics but anything in Mathematics and Physics which brings to evidence their mutual motivations and relevance.

Accordingly we have chosen to include besides the more standard and quickly expanding fields also subjects such as fluid dynamics, turbulence and dynamical systems in the large; solitons and non linear evolution problems; algebraic topology, whose relevance to Theoretical Physics may grow at various levels.

In this connection we believe that research fields which are bordeline between different areas or new viewpoints on classical problems are often especially rich of conceptual structure and stimulating of independent thinking.

It is in this spirit that we asked the speakers to provide motivations and a view on the landscape where their subject is located as well as relations to neighbouring subjects.

We thank all the contributors for their collaboration to this purpose, which we think would best serve the unity of knowledge and provide stimulation and perspectives to people engaged in research in Mathematics and Physics.

The International Advisory Board of the Conference has been consulted at several stages and was composed by: H.Araki, F.Calogero, L.Faddeev, F.Guerra, D.Kastler, I.M.Khalatnikov, O.E.Lanford III, J.Moser, L.Salvadori, Ya.G.Sinai, V.S.Vladimirov.

We would like to thank each of these persons for the collaboration given to us.

We are pleased to express our special thanks to Giovanni Gallavotti for his deep engagement as a coorganizer in an early stage of the preparation of the Conference.

The content of this volume covers almost completely the material presented at the Conference, and in our opinion provides a significant perspective on those fields of Mathematical Physics we intended to cover.

In addition we present the contribution of some invited lecturers who were unable to attend but were kind enough to send a manuscript.

We tried to reflect affinity and connections between the contributions through the order and grouping in which they appear in this volume; the essential continuity of the subject led us to avoid any division into chapters.

At the end of the volume one finds also some of the discussion contributions and of the short communications which were organized on the spot in connection with last minute changes of the schedule.

At the beginning of the Conference we learned of the untimely death of Professor Seymour Sherman; we want to pay here a tribute to his memory, in these proceedings which cover a field to which he devoted much of his scientific activity.

In organizing the Conference we had to operate under rather adverse conditions.

It was necessary to set a comparatively high participation fee, to face the most urgent and undelayable part of the organization expenses. We are very grateful to the participants who, with their personal contribution, made the Conference possible.

Contributions from the International Mathematical Union, the National Institute for Nuclear Physics (INFN), the Ministry of Foreign Affairs, the Mayor of Rome, were equally important in meeting the most immediate financial obligations.

Subsequently, support from the University of Rome and from the Italian National Research Council (CNR) both through its National Committee for Mathematics and through the Committee for International Affairs, made it possible to complete the financing of the Conference.

We would like to thank all the Organizations and Institutions mentioned above for having contributed to the success of the Conference.

We are pleased to take this opportunity to express our warm thanks to Professor Giorgio Tecce, Dean of the Faculty of Sciences of the University of Rome, for the continuous, precious aid in overcoming all kinds of serious difficulties.

Finally, we want to express here our thanks and appreciation to our secretaries Ms. Elisabetta Di Silvestro and Ms. Lidia Paoluzi for their invaluable help in the preparation and organization of the Conference.

Gianfausto Dell'Antonio

Sergio Doplicher

Giovanni Jona-Lasinio

CONTENTS

Main Lectures

Short Communications

MAIN LECTURES

LATTICE INSTANTONS: WHAT THEY ARE AND WHY THEY ARE IMPORTANT

Arthur Jaffe*

Harvard University
Cambridge, Massachusetts 02138/USA

1. Introduction

The central theme of this lecture is that gauge fields provide an interesting direction for mathematical physics. While gauge invariance is far from new — it appears centrally in the work of H. Weyl — gauge theories have recently had a resurgence in particle physics. This new interest stems, on the one hand, from the acceptance of the Salam-Weinberg-Glashow unified theory of weak, strong and electromagnetic interactions. This theory involves the coupling of Yang-Mills gauge field with a Higgs boson and with fermions (quarks). Experimental evidence for this model has come from many directions the last few years, especially with the discovery of charmed (and other) particles which fit into the framework. Furthermore, the theoretical property of "asymptotic freedom" allows one to expect that nonabelian gauge models may be renormalized, and that perturbation calculations are applicable. All these features have lead to a renaissance of gauge theories, and these avenues encompass much physics: the physics of particles, of phase transitions, of critical phenomena, and of Debye screening.

But, gauge theories also have an intrinsic mathematical interest. The classical gauge potential is a connection form to geometers, and the gauge field is the curvature form. The differential geometry of classical Yang-Mills fields has been pursued by Polyakov, Schwarz, Atiyah, Penrose, Singer, Ward and others, as we will hear at this conference. Study of special solutions of the Yang-Mills equations has become fashionable: On the one hand, it is a mathematically beautiful subject. On the other hand, the classical solutions provide insight as classical limits of quantum fields; they reveal new physics. There are certainly many years work left to uncover the mathematical structure of gauge theories and to understand their quantization. In fact, because of "asymptotic freedom", it may actually be easier to construct an example of interacting Yang-Mills fields in four space-time dimensions,

*Supported in part by the National Science Foundation under Grant PHY-77-18762.

rather than to construct a scalar field theory or quantum electrodynamics.

The quantization of (Euclidean) gauge fields can be considered statistical mechanics of gauge potentials, i.e. the statistical mechanics of geometry. When the statistical mechanics on the space of connections has been defined, the Osterwalder-Schrader axioms will provide an analytic continuation to an appropriate quantum field theory in Minkowski space. The construction of the measure, of course, is a difficult unsolved problem.

The first step in understanding the statistical mechanics of gauge potentials is to understand an approximate theory, namely, one in which the usual divergences of quantum fields do not occur. K. Wilson proposed a lattice cutoff which preserves gauge invariance; this approximation was studied by Osterwalder and Seiler, who have established the basic property of reflection positivity (Osterwalder-Schrader positivity). This property is preserved under limits, so the basic existence theorem for quantum gauge fields is reduced to establishing Euclidean invariance (and regularity) in the limit of zero lattice spacing.

Aside from being an intermediate tool, the lattice theories themselves are interesting. As approximations to their presumed continuum limits, they contain interesting physics. At this conference, de Angelis, de Falco and Guerra report on correlation inequalities for lattice gauge fields. Related to this is recent work of Fröhlich and Park. Osterwalder and Seiler report on the Higgs mechanism for lattice gauge fields.

We remark, however, there are many elementary unsolved problems. For example, what is the lattice analogs of the SU(2) Polyakov instanton? How should one approximate on a lattice the differential geometry analysis of classical, continuum lattice gauge fields? Perhaps we will hear these results at the next conference!

In this lecture we investigate instantons in the U(1) lattice theory. Here the instantons are Coulomb dipoles, and the quantization of lattice U(1) (Maxwell) fields leads to an analysis of the statistical mechanics of a Coulomb dipole gas, i.e. to the theory of dielectrics. In fact, the presence of a phase transition in the dipole gas indicates that a nonabelian gauge theory is necessary to describe the confinement of quarks.

In summary, it is clear that $M \cap \Phi$ is very large; the more we learn the harder it becomes to distinguish $M \cap \Phi$ from $M \cup \Phi$!

2. Lattice Gauge Theory [1]

2.1 Definitions.

In order to avoid the short distance (ultra-violet) problems of field theory we introduce an ultraviolet cutoff, namely, a cubic lattice with distance ε between adjacent sites. Let (ij) denote a directed bond from lattice site i to a nearest neighbor j, and let Γ denote the set of all such bonds. Let G denote a unitary matrix representation of a compact group, the gauge group.

Definition. A lattice gauge field γ is a map

$$\gamma:\Gamma \to G ,$$

with the restriction that

$$\gamma(ij) = \gamma(ji)^{-1} .$$

Definition. The Yang-Mills action associated with a subset Λ of the lattice is defined by

$$\mathcal{A}(\Lambda) = \sum_{P\subset\Lambda} \mathcal{A}(P) , \qquad (1)$$

where $\mathcal{A}(P)$ is the action for a plaquette P bounded by lattice sites 1, 2, 3, and 4.

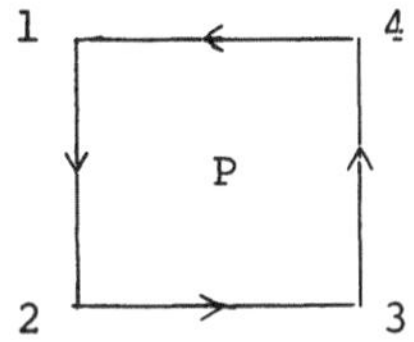

Here

$$\mathcal{A}(P) = -\beta \operatorname{Re}\left(\operatorname{tr} \gamma(12)\gamma(23)\gamma(34)\gamma(41)\right), \quad \beta = \varepsilon^{d-4}g^{-2} .$$

Definition. The lattice pure gauge field theory corresponding to G is the statistical mechanics defined by the expectation

$$d\mu_{\Lambda} = \frac{1}{N(\Lambda)} e^{-\mathcal{A}(\Lambda)} d\nu(\Lambda)$$

where $N(\Lambda)$ is a normalizing factor and the measure $d\nu(\Lambda)$ is

$$d\nu(\Lambda) = \prod_{(ij)\subset\Gamma_{\Lambda}} d\gamma(ij) .$$

Here Γ_{Λ} is the subset of bonds Γ inside Λ.

Here $d\gamma(ij)$ is the Haar Measure for the group G and $d\nu$ is a Cartesian product of Haar Measure over the bonds in Λ.

The continuum limit $\varepsilon \to 0$ leads to the usual Yang-Mills action

$$\mathfrak{a} = \frac{1}{2} \text{ trace} \int \sum_{\mu,\nu} F_{\mu\nu}{}^2(x)\, dx \, .$$

Here the vector potential $A_\mu(x) = A(jk)$ is defined as the infinitesimal generator of the gauge transformation $\gamma(jk)$,

$$\gamma(jk) = e^{i\varepsilon g A_\mu(j)} = e^{i\varepsilon g A(jk)}$$

where μ denotes the direction from j to k. With this definition, the formal $\varepsilon \to 0$ limit is

$$F_{\mu\nu}(x) = \partial_\mu A_\nu(x) - \partial_\nu A_\mu(x) + ig\left[A_\mu(x),\, A_\nu(x)\right] \, ,$$

The Euler-Lagrange variational equations arising from $\mathfrak{a}$ are

$$\partial_\mu F_{\mu\nu}(x) + ig\left[A_\mu(x),\, F_{\mu\nu}(x)\right] = 0 \, ,$$

or in invariant notation

$$D^* F = 0 \, ,$$

where D denotes the covariant derivative and * the adjoint in the Hodge inner product. The Bianchi identities for the curvature can be written $DF = 0$.

2.2 Gauge Invariance.

Let $G(\cdot)$ denote a map from vertices in the lattice to G. A <u>local gauge transformation</u> is defined by

$$\gamma(ij) \to \gamma'(ij) = G(i)\, \gamma(ij)\, G(j)^{-1} \, .$$

Clearly $\mathfrak{a}(\Lambda) = \mathfrak{a}'(\Lambda)$ since

$$\text{tr}\left(\gamma(12)\gamma(23)\gamma(34)\gamma(41)\right) = \text{tr}\left(\gamma'(12)\gamma'(23)\gamma'(34)\gamma'(41)\right) \, .$$

Furthermore, Haar Measure $d\nu$ is gauge invariant, so $d\mu_\Lambda$ is invariant under any local gauge transformation. If F is any gauge invariant function of the lattice gauge field, then the expectation

$$\langle F \rangle = \int F \, d\mu_\Lambda$$

is also gauge invariant.

2.3 The Wilson Loop Integral. A common example of a gauge invariant function F occurs as follows: Let C denote a closed curve in Λ, made up of bonds (12),(23),...(n-1,n),(n,1), and let

$$F(C) = \mathrm{tr}\left(\gamma(12)\gamma(23)\ldots\gamma(n,1)\right) .$$

In case G is an abelian group,

$$F(C) = \mathrm{tr}\; e^{i\varepsilon g \oint A(ij)} ,$$

where the sum $\oint$ ranges over C. We can write

$$F(C) = \mathrm{tr}\,\exp\left(ig \oint_C A_\mu dx^\mu\right) . \qquad (2)$$

In the nonabelian case this same formula is often used, with the convention that the integral is ordered, i.e. the exponential is written as a product of factors ordered by their order along C.

The loop integral is defined as

$$\langle F(C) \rangle = \int F(C) \, d\mu_\Lambda = e^{-f(C)} . \qquad (3)$$

One basic question is the asymptotic behavior of f(C) as a function of the size of C. Wilson has interpreted f(C), when C is an L x T rectangle, in terms of the potential energy E(L) between two quarks separated by distance L. He proposes [1]

$$E(L) = \lim_{T\to\infty} \frac{1}{T} f(C) . \qquad (4)$$

If $E(L) \to \infty$ as $L \to \infty$, then we expect the quarks to always be bound, i.e. confined. We can ask in lattice gauge models how E(L) depends on L, and whether $E(L) \not\to \infty$. The answer will depend on the dimension of space d, the gauge group G and the parameter g.

The two simplest behaviors for f(C) are the

Area Law:

$$f(C) \sim \alpha|C| + \beta\,|\partial C| + \ldots. ,\qquad (5)$$

where $|C|$ denotes the area of C and $|\partial C|$ denotes the length of the boundary. If the area law is obeyed, then $E(L) \sim \alpha L$, and confinement occurs for $\alpha > 0$.

Length Law: The length law is the special case $\alpha = 0$, $\beta > 0$ of (5), yielding $E(L) \sim \beta$. In this case the potential $E(L)$ does not increase with L and confinement does not occur.

Some expected behavior is summarized in the following chart:

	Group	Area Law	Length Law
"Ising" Gauge Model	Z_2	$d \leq 2$, all g $d \geq 3$, $g > g_c(d)$	$d \geq 3$, $g < g_c(d)$
Electrodynamics	U(1)	$d \leq 3$, all g $d \geq 4$, $g > g_c$	$d \geq 4$, $g < g_c$
SU(2) Yang-Mills	SU(2)	$d \leq 3$, all g $d = 4$?	$d \geq 5$, $g < g_c$ $d = 4$?

The transition from area decay to length decay is a phase transition. This phase transition may not be associated with a change in the number of ground states, but rather with nonanalyticity in g and the disappearance of a mass gap.

Very little of the picture in this table has been established. We do know that for g sufficiently large (depending on d and G) the length decay is generic. This has been established by Osterwalder and Seiler, with minor restrictions on the representation of G (e.g., the identity representation and certain others are excluded) c.f. [1]. Furthermore, for the U(1) case, length decay can not occur for $d \leq 3$ [2]. Our bound allows a possible logarithmic (Coulomb) force law. Polyakov has argued that the $d = 3$ U(1) problem is equivalent to a dilute Coulomb gas. For this problem one expects Debye screening to produce a linear potential [3]. The problem of Debye screening has been analyzed recently on a lattice by Brydges [4], who used our cluster expansions developed to study $P(\phi)$ quantum fields. It would be of interest to show that Brydges' work could be used to justify Polyakov's picture for $d = 3$. In the Ising case, the study of $g << 1$ can be studied by a

$g >> 1$ expansion by utilizing a duality transformation [5]. Any mathematical work for the SU(2) case with $g << 1$ would be of interest.

3. The Instanton Gas

We are faced with the problem of understanding a function space measure of the form

$$d\mu = \frac{1}{Z} e^{-\mathfrak{a}(A)} [\mathscr{D}A]$$

where $\mathfrak{a}$ is the action. Suppose A_{cl} is a particular field configuration. We can expand about $A = A_{cl}$ by defining a new variable

$$\psi = A - A_{cl} ,$$

in terms of which

$$\mathfrak{a}(A) = \mathfrak{a}_{cl}(A_{cl}) + \mathfrak{a}_0(\psi) + R(\psi, A_{cl}) .$$

Here $\mathfrak{a}_{cl}$ is the classical approximation and $\mathfrak{a}_0$ is quadratic. We choose A_{cl} so the remainder term R is small (at least for ψ small) and if A_{cl} is a classical extremum, then no linear term in ψ occurs.

More generally, if we expand about different classical points A^i_{cl}, we can write

$$d\mu = \sum_i e^{-\mathfrak{a}_{cl}\left(A^i_{cl}\right)} \frac{1}{Z} e^{-\mathfrak{a}_0(\psi) - R_i\left(\psi, A^i_{cl}\right)} \chi_i [\mathscr{D}\psi]$$

where χ_i is a partition of unity, $\sum \chi_i = 1$. If we neglect R_i, then we see that the integration $d\mu$ approximately factors into

$$d\mu = \left(\sum_i e^{-\mathfrak{a}_{cl}\left(A^i_{cl}\right)}\right) d\mu_0$$

$$= (\text{Instanton gas}) \times (\text{Gaussian}) ,$$

where the instanton gas is given by the classical approximation

$$d\mu_{\text{Instanton}} = d\mu_{\text{classical}} = \sum_i e^{-\mathfrak{a}_{cl}\left(A^i_{cl}\right)} . \qquad (6)$$

This decomposition of the functional integral can give insight into the phase transitions in the model $d\mu$. For instance, we can often identify

$\sum \exp(-\mathcal{G}_{cl})$ as the partition function for a system of particles (instantons) interacting with a pair potential V. Whether V is short range or long range will determine the qualitative properties of the system. There is a duality between instanton variables and the variable $d\mu$. A phase transition in $d\mu$ is associated with a short range potential and independence of the instantons. On the other hand, a lack of a phase transition in $d\mu$ is associated with a phase transition in the instanton variables. Such a picture has been used to generate a convergent expansion for the study of quantum field theories in a mathematically rigorous and quantitative way [6]. As a byproduct the expansion can be used to show the existence of phase transitions for quantum fields.

4. The d = 4 Maxwell Field [7]

To illustrate these ideas in more detail, we consider the case G = U(1), i.e. the lattice Maxwell field, for d = 4. In this case the instanton gas turns out to have the properties of a Coulomb dipole gas i.e. a dielectric. To derive this result, we note that the action (1) is

$$\mathcal{G} = g^{-2} \sum_{P} \left(1-\cos\left(\varepsilon^2 gF(P)\right)\right) , \qquad (7)$$

where

$$F(P) = \varepsilon^{-1} \sum_{\ell \in \partial P} A(\ell)$$

is the field. Geometrically

$$F = \partial^* A \qquad (8)$$

i.e. F is the coboundary of A. We wish to decompose (7),

$$\mathcal{G} = \mathcal{G}_{cl} + \mathcal{G}_0 + R . \qquad (9)$$

This requires a choice of the classical expansion points A^i_{cl}. We label the expansion points by the minima of the cosine in (7). Thus we define an "integer part" [F] of F(P) by

$$\varepsilon^2 g[F] = 2\pi j \quad \text{if} \quad |\varepsilon^2 gF - 2\pi j| < \pi . \qquad (10)$$

We then define the charge Q by

$$Q = *\partial^{*}[F] \qquad (11)$$

where ∂^{*} is the coboundary operator and * denotes Hodge duality. The identities

$$Q = \partial *[F] = \partial S \qquad (12)$$

and

$$0 = \partial Q = \partial\partial S \qquad (13)$$

have the geometric interpretation that the charge is the $(d-3) = 1$ dimensional boundary of a two dimensional surface, the Dirac surface $S = *[F]$. Furthermore, (13) expresses the Kirchoff law that the charge flowing into each vertex must equal the charge flowing out. Hence, Q is a superposition of constant line charges on closed loops. Since the definition (11) of Q involves duality, we can think of Q as a magnetic charge.

In order to obtain a gauge invariant classical configuration, we define

$$F_{cl} = [F] + \partial\Delta^{-1}*Q \ , \qquad (14)$$

where Δ is the lattice Laplacian, yielding

$$\partial^{*}F_{cl} = \partial^{*}[F] - *Q = 0 \ . \qquad (15)$$

We then choose A_{cl} so that

$$F_{cl} = \partial^{*}A_{cl} \qquad (16)$$

Inserting our choice of A_{cl} into (9) we find (for details see Ref. 7) the instanton gas to be

$$\mathfrak{a}_{cl} = \mathfrak{a}_{cl}(Q) = \frac{1}{2}\sum \varepsilon^4\vec{q}_i(-\Delta)^{-1}\vec{q}_i \ , \qquad (17)$$

while $\mathfrak{a}_0$ is a massless free field. In other words, $\mathfrak{a}_0$ determines the oscillation within one well of the cosine function (7), while $\mathfrak{a}_{cl}(Q)$ parameterizes the interaction due to transitions between wells.

Since (17) includes self-action terms, the partition function (6) for the instanton gas tends to suppress all but the smallest configurations Q. The elementary configuration is

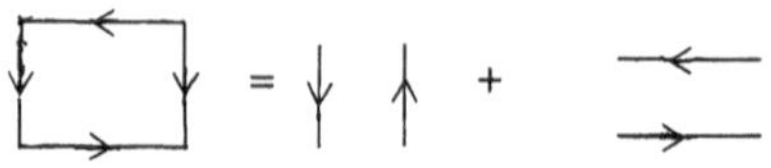

a single charge loop. Because of the Kirchoff law and the orthogonality of the vector charges, the loop charge is composed of two orthogonal dipoles. (Vector charges in different directions do not interact to reduce the dipole to a quadrapole.) The self-action of each charge loop grows at least linearly with the length of the loop. For this reason the action (17) is dominated by the effects of small loops, i.e. by the dipoles. The dominant action is the self-action plus a Coulomb dipole gas. The kernel of $(-\Delta)^{-1}$ is just the Coulomb Green's function which is $O(r^{-2})$ for large r (in d = 4).

We have established one estimate, namely, the free energy of the dilute instanton gas equals (up to $O(g^{\gamma})$) the activity of a small charge loop (presumably the elementary 2-dipole loop above) [7]. The method of proof uses block charges to isolate the self-action, and a sequence of increasing length scales to analyze dipole fluctuations (In large blocks, the probability of a large dipole moment is small.) We hope that these renormalization group methods can be used to determine the leading long range behavior in the dilute dipole gas, and to establish the existence of a phase transition for $U(1)_4$ gauge fields. If these methods are successful, they may also provide insight into the phase transition in the d = 2, X - Y model and its presumed line of critical points, see [8]. In fact, a detailed analysis of dipoles appears useful for numerous infrared analyses.

References

1. There is much literature on lattice gauge theories, following the basic paper by K. Wilson, Phys. Rev. D10, 2245 (1974). We refer the reader to K. Osterwalder and E. Seiler, gauge field theories on a lattice, Annals of Physics (to appear) which contains an extenxive bibliography. See also the contributions of E. Seiler, J. Fröhlich and F. Guerra contained here and the contribution of K. Osterwalder to the 1976 Cargese Summer School. The material in this section is based on these references, and we use the notation of Osterwalder and Seiler. A good review of the physics of lattice gauge fields in L. Kadanoff, Rev. Mod. Phys. 49, 267 (1977). See also A. A. Migdal, J.E.T.P. 69, 810, 1457 (1975) and A. M. Polyakov, Phys. Lett. 59B, 82 (1975).
2. J. Glimm and A. Jaffe, Phys. Lett. 66B, 67 (1977).
3. A. M. Polyakov, Nuclear Physics B120, 429 (1977).

4. D. Brydges, A rigorous approach to Debye screening in dilute classical Coulomb systems, to appear.
5. F. Wegner, J. of Math. Phys., 12, 2259 (1971). See also G. Gallavotti and F. Guerra, these proceedings.
6. J. Glimm, A. Jaffe and T. Spencer, Ann. Phys. (N.Y.), 101, 610 (1976).
7. J. Glimm and A. Jaffe, Instantons in a U(1) Lattice Gauge Theory, Commun. Math. Phys., to appear. See also A. M. Polyakov, Phys. Lett. 59B, 79 (1975).
8. J. V. José, L. Kadanoff, S. Kirkpatrick, and D. R. Nelson, Phys. Rev. B16, 1217 (1977).

GAUGE FIELDS ON THE LATTICE

by

Gian Fabrizio De Angelis
Diego de Falco
Francesco Guerra
Institute of Physics, University of Salerno, Via Vernieri 42,
84100 Salerno, Italy

Ideas and techniques of classical statistical mechanics have been made available for the rigorous study of quantum gauge fields by an ultraviolet cutoff and quantization procedure on a Euclidean space-time lattice proposed by Wilson. Here we focus our attention on Scalar Quantum Electrodynamics and the Abelian Higgs Model and point out some consequences of Griffiths-Kelly-Sherman inequalities, which are easily shown to hold for the model.
For the pure U(1) model these inequalities, combined with a suitable high temperature expansion, lead to definite statements about a long range order parameter the change in whose qualitative behavior with the coupling constant is expected, on heuristic grounds, to indicate the presence of some kind of phase transition.
For the complete model GKS inequalities, combined with exponential bounds, not only show the existence and non triviality of the infinite volume limit but also give a quantitative estimate of the decoupling effect of the local gauge symmetry providing a definite support to the expectation that if there is a phase transition it is not marked by the appearence of any non trivial local order parameter. Some conjectures about the nature of this phase transition are formulated.

§ 1 . The Model: Definitions and Notations.

Consider a two component scalar field $\vec{\Phi} = (\Phi_1, \Phi_2)$ in d space-time dimensions described by the Euclidean action

$$S[\vec{\Phi}] = \int_{R^d} \{\tfrac{1}{2}[\,\underline{\nabla}\vec{\Phi} \cdot \underline{\nabla}\vec{\Phi} + \mu^2\vec{\Phi} \cdot \vec{\Phi}\,] + \lambda(\vec{\Phi} \cdot \vec{\Phi})^2\}\, d^d x \;,\; \lambda > 0 \;,$$

whose lattice version on $\varepsilon\mathbb{Z}^d$, in a given cutoff volume Λ, with a definite choice of boundary conditions, is given by

$$S_{\varepsilon,\Lambda}[\vec{\Phi}] = -\varepsilon^{d-2} \sum_{\substack{(n,n')\subset\Lambda \\ n < n'}} \rho(n)\rho(n')\cos(\Theta(n) - \Theta(n')) + \frac{\varepsilon^d}{2}(\mu^2 + 2d\varepsilon^{-2}) \sum_{n\in\Lambda} \rho(n)^2 +$$

$$+ \lambda \varepsilon^d \sum_{n \in \Lambda} \rho(n)^4,$$

where we have chosen to describe the field $\vec{\Phi}$ at site n in terms of its polar coordinates by $\vec{\Phi}(n) = (\rho(n)\cos\Theta(n), \rho(n)\sin\Theta(n))$ and where the first summation extend to the set of ordered pairs of nearest neighbor sites ("links") in Λ.

The requirement that the global U(1) invariance of this action, namely its invariance under the transformations

$\Theta(n) \to \Theta(n) + g\alpha$,

for fixed real g and arbitrary real α , be promoted to a local gauge invariance admitting also site dependent α's, is satisfied by associating to each link (n,n'), $n < n'$, a new variable $\mathcal{A}(n,n')$ with $-\pi \le \varepsilon g \mathcal{A}(n,n') \le \pi$, and changing the nearest neighbor coupling terms in the action into

$\rho(n)\rho(n')\cos(\Theta(n) - \Theta(n') - \varepsilon g \mathcal{A}(n,n'))$.

A self interaction term for the $\mathcal{A}$ field is then introduced subject to the conditions that it preserves the gauge invariance of the action, namely its invariance under the transformations

$\Theta(n) \to \Theta(n) + g\alpha(n)$

$$\mathcal{A}(n,n') \to \mathcal{A}(n,n') + \frac{\alpha(n) - \alpha(n')}{\varepsilon} ,$$

and its periodicity in the $\mathcal{A}$ variables and that the complete lattice action reduces, in the formal classical continuum limit, to the action $S[\vec{\Phi}]$ minimally coupled to the Maxwell field.

This term is best described in the following notations suited to the developments of the subsequent sections: we define first of all $\mathcal{A}(n,n') = -\mathcal{A}(n,n')$; we call then $\mathfrak{M}$ the family of divergenceless vector fields of compact support with values in $\mathbb{Z}$ namely the set of functions m on the links with

$m(n,n') = -m(n',n)$

$$(\operatorname{div} m)(n) = \sum_{|n'-n|=\varepsilon} m(n,n') = 0$$

$m(n,n') = 0$ for n and n' outside some bounded region and call $\mathcal{A}(m) = \sum_{n<n'} \mathcal{A}(n,n') \cdot m(n,n')$ the field $\mathcal{A}$ smeared with the test function $m \in \mathfrak{M}$. To every closed path C on $\varepsilon\mathbb{Z}^d$ $C = \{n_1, \ldots\ldots, n_k = n_1\}$, $|n_{s+1} - n_s| = \varepsilon$ we associate a field $m_C \in \mathfrak{M}$ defined by $m(n,n') = -m(n',n) = k$ if the elementary step $n \to n'$ appears k times in C . If now p is a plaquette, namely an elementary lattice square with a

definite orientation, we call m_p the field associated to it.
The self interaction term for the $\mathfrak{A}$ field is written, in these notation, as

$$\frac{-\varepsilon^{d-4}}{2g^2} \sum_{p \subset \Lambda} \cos(\varepsilon g\, \mathfrak{A}(m_p))$$

where for the parity of the cosine function we need not specify the orientation of each p .

Having thus constructed the complete action, which we prefer to rewrite in terms of rescaled gauge field variables $A = \varepsilon g\, \mathfrak{A}$ as

$$S_\Lambda[A,\vec{\Phi}] \equiv S_\Lambda[A,\Theta,\rho] = \sum_{n \in \Lambda} P(\rho(n)) - \beta_\ell \sum_{(n,n') \subset \Lambda} \rho(n)\rho(n')\cos(\Theta(n) - \Theta(n') - A(n,n')) -$$
$$- \beta_p \sum_{p \subset \Lambda} \cos(A(m_p)),$$

with obviously redefined coupling polynomial P , bounded below, and non negative coupling constants β_ℓ, β_p, the corresponding ultraviolet and volume cutoff quantized Euclidean system will have the variables ρ, Θ, A as random variables distributed according to a probability measure having $\exp(-S_\Lambda(\rho,\Theta,A))$ as density with respect to the natural (Lebesgue and normalized Haar) a priori measure.

Thus, Wilson's (1) requirement of preserving manifest gauge invariance at each stage of the cutoff and quantization procedure leads to a perfectly well defined classical statistical mechanical system.

In the last two years this model and similar non Abelian ones have been the subject of an extensive exploration on a rigorous level.

Osterwalder and Seiler (2,3) have performed, also in the non Abelian case, a fairly complete analysis of the high temperature (strong coupling) behavior of these systems by cluster expansion techniques; Glimm and Jaffe, focusing their attention on the pure U(1) model, have given important upper bounds holding at every temperature (4) and have set up an interesting framework for the study of the instantonic configurations (5) which are expected to be important in the weak coupling regime (6); Lüscher (7,8) has analyzed the transfer matrix formalism and, in the Hamiltonian framework, the gauge invariance of the ground state.

Much less is known about the possibility of performing the continuum limit: for the many interesting possibilities suggested by renormalization group ideas we refer to the important review articles by Kadanoff (8,9) and to the many references quoted there.

For the Abelian model just described, the infinite volume limit can be easily constructed for any value of the coupling parameters due to the fact that Griffiths-Kelly-

Sherman inequalities (11) hold (12).
Due to the unbounded character of the spins $\vec{\Phi}(n)$ some care is necessary in order to prove the non triviality of such a limit: exponential bounds of Fröhlich's type (13, 14) are easily shown to complete the monotonicity statements following from GKS inequalities into convergence statements (15).

§ 2 . Some Consequences of Gauge Invariance.

We have stressed at the end of the previous section that the classical statistical mechanical models motivated by Wilson's lattice approximation fit very nicely into the general program of (Euclidean) Quantum Field Theory as Classical Statistical Mechanics (16,17,18) in the definite operative sense that most of the relevant techniques acquired in the development of that program for the $P(\phi)$ models (correlation inequalities, cluster expansion, transfer matrix formalism,...) have shown to be relevant also in the study of theories, such as gauge theories, which are strongly hoped to give a very deep insight into the structure of the fundamental interactions (for an updated review on this last point we refer to (19)).
In this section we wish to point out some peculiar properties of the models just introduced referring to their most striking feature, namely their high degree of (local) symmetry.
All the previous experience on the critical behavior of the $P(\phi)$ models (20,21) shows that their phase transition correspond to the spontaneous breaking of some internal symmetry and that a detailed analysis of their dynamical properties (say their mass spectrum) is strongly dependent on the appearence of local order parameters marking this symmetry breaking.
To be definite, consider the effect of coupling the globally invariant action $S_\Lambda(\vec{\Phi})$ of Section 1 to an external field $\vec{\lambda}$ through an additional term $-\vec{\lambda}\cdot\vec{\Phi}(x)$ in the Lagrangian.
Call $\langle\cdot\rangle^{\vec{\lambda}}$ the thermodynamic limit of the Gibbs state constructed from this modified action (which can be proven to be a pure state).
As a remnant of the rotational invariance of the original action, if A is any observable, R any rotation in "spin space", RA the correspondingly rotated observable, one will have $\langle A\rangle^{\vec{\lambda}} = \langle RA\rangle^{R\vec{\lambda}}$, so that, in particular the one and two point functions, setting $\hat{\lambda} = \frac{\vec{\lambda}}{|\lambda|} = (\hat{\lambda}_1,\hat{\lambda}_2)$, will have the form

$$\langle \Phi_i(x)\rangle^{\vec{\lambda}} = \lambda_i M(|\lambda|)$$

$$\langle \Phi_i(x)\Phi_j(y)\rangle^{\vec{\lambda}} = \hat{\lambda}_i\hat{\lambda}_j M^2(|\lambda|) + (\delta_{ij} - \hat{\lambda}_i\hat{\lambda}_j)D_\perp(x,y) + \hat{\lambda}_i\hat{\lambda}_j D_\parallel(x,y)$$

where $D_\perp$ and $D_{/\!/}$, depending on $\vec{\lambda}$ only through $|\vec{\lambda}|$, define, respectively, the transverse and longitudinal part, of the two point function.

The presence of a spontaneous magnetization, namely of the non vanishing of $M \equiv \equiv \lim_{|\lambda|\to 0} M(|\lambda|)$ which can be shown to occur for suitable values of the coupling parameters (21) has definite consequences on the mass spectrum of the theory (Goldstone's theorem) which can be easily described in terms of the zero momentum contributions to $D_\perp$ and $D_{/\!/}$, namely the "susceptibilities"

$$\chi_\perp(|\lambda|) = \int D_\perp(x,y)dy$$
$$\chi_{/\!/}(|\lambda|) = \int D_{/\!/}(x,y)dy$$

As one can easily compute them to be

$$\chi_{/\!/} = \frac{\partial M(|\lambda|)}{\partial|\lambda|} \quad , \quad |\lambda|\,\chi_\perp(|\lambda|) = M(|\lambda|)$$

one immediately sees that the $\lim_{|\lambda|\to 0} M(|\lambda|)$ can be different from zero only if there appear long range excitations in $D_\perp$ marked by the divergence of $\chi_\perp$ as $|\lambda| \to 0$.

Now, there is widespread expectation, started by interesting conjectures of Ref.(1), that the changes with the coupling parameters of the gauge invariant models described before, such as the disappearence of the mass gap known to be present in the strong coupling regime (2,3), or the related onset of the Schwinger-Wilson confinement mechanism (22,1), might be marked by some kind of phase transition; beyond that, one expects (9,10,23,24) that working close to a critical point will make it easier to recover the Euclidean space-time symmetry of the continuum model.

Preliminary mean field explorations of this problem (25) suggest, indeed,very interesting phase diagrams.

What we wish to point out is that these different phases cannot be singled out by the appearence of any non trivial local order parameter. Our analysis, which sharpens the one of Ref.(26) in order to deal with the specific difficulties introduced by the couplings $\rho(n)\rho(n')$ between the unbounded variables $\rho(n)$ is based on the following considerations (27): let f be any bounded continuous function of the field configurations localized in a given bound region M ; the combined effect on the expectation of f of any gauge transformation α , under whose natural actions f goes into f_α and $\vec{\Phi}(n) = (\Phi_1(n),\Phi_2(n))$ goes into $\vec{\Phi}(n) + \delta\vec{\Phi}(n)$, and of an additional external field term $-\vec{\lambda}\cdot\sum_{n\in\Lambda}\vec{\Phi}(n)$ in the action, can be easily estimated in the form

$$|\langle f_\alpha\rangle^{\vec{\lambda}}_\Lambda - \langle f\rangle^{\vec{\lambda}}_\Lambda| \le c \langle |f| [\exp(|\vec{\lambda}| \sum_{n\in M} |\delta\vec{\Phi}(n)|) - 1] R_M(\rho)\rangle_M$$

where we have called $< \cdot >_{\Lambda}^{\vec{\lambda}}$ the Gibbs state generated by the action

$S_{\Lambda}(A,\vec{\Phi}) - \vec{\lambda}\cdot\sum_{n\in\Lambda} \vec{\Phi}(n)$,

C is a non negative constant and R_M is a non negative function of $\rho(n)$, $n \in M$, both of them independent of Λ and of $\vec{h}$ for $\vec{h}$ in any bounded disk.

It is then an immediate application of Lebesgue's dominated convergence theorem to conclude that the state $\lim_{|\lambda|\to 0} \lim_{\Lambda\to\mathbf{Z}^d} < \cdot >_{\Lambda}^{|\lambda|\hat{\lambda}}$ is gauge invariant, in particular independent of $\hat{\lambda}$, so that in this state the expectation of $\vec{\Phi}(n)$ must vanish.

We are <u>not</u> claiming, of course, that there may not be a Higg's phenomenon, but are just pointing out that if there is any dynamical mass generation (22,28,3) it must show up, at least in Wilson's explicitly gauge invariant formulation, with much more subtle phenomena than the appearence of a non trivial expectation of the scalar field.

§ 3 . High Temperature Expansion and the "Confinement bound".

Restricting, for the moment, our considerations just to the pure U(1) model, described by the a priori measure

$$dA_{\Lambda} = \prod_{\substack{(n,n')\subset\Lambda \\ n<n'}} \frac{dA(n,n')}{2\pi}$$

and by the action

$$S_{\Lambda}(A) = -\beta_p \sum_{p\subset\Lambda} \cos(A(m_p)),$$

what is expected as to its critical behavior (1) can be described in terms of the expectations $< \exp iA(m_C)>$, where C is some lattice loop (beyond the physical interpretation of these expectations as "Bohm-Aharanov factors" (1), we want to stress that, from a mathematical point of view, they are the building blocks, by convolutions, of the Fourier transform $< \exp iA(m) >$, as, by gauge invariance, this is different from zero only for $m \in \mathfrak{M}$ and therefore for support of m made of unions of closed loops).

More precisely, taking for C a square loop C_L of side L , and defining the two long range order parameters

$$c_1(\beta) = -\lim_{L\to\infty} \frac{\log< \exp iA(m_{C_L})>}{L^2}$$

$$c_2(\beta) = -\lim_{L\to\infty} \frac{\log< \exp iA(m_{C_L})>}{L},$$

the heuristic expectations for the pure U(1) model for $d > 3$ (for $d = 2$ the model is trivial, for $d = 3$ see (4)) can be described by the following picture:

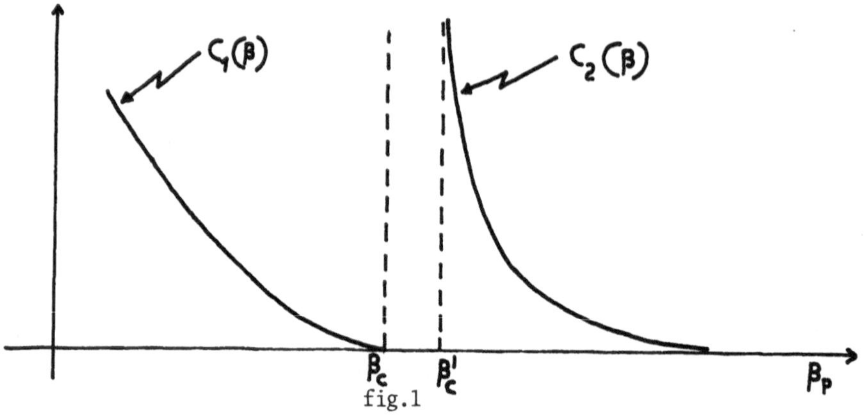

fig.1

For β_p small $<\exp iA(m_{C_L})>$ is expected to be $o(e^{-L^2})$ so that c_1 is expected to be finite and non zero (c_2 is then obviously infinite). At some $\beta_c > 0$ c_1 is supposed to vanish allowing then c_2 to become finite for β_p greater than some $\beta_c' \geq \beta_c$ so that for β_p large enough $< \exp iA(m_{C_L}) >$ is expected to be $o(e^{-L})$. GKS inequalities give a fairly good <u>qualitative</u> control on the guesses described by fig.1; they establish that:

i) as a consequence of the monotonic increase of $<\exp iA(m_c)>$ with β_p, c_1 and c_2, provided the limits defining them exist, are monotone decreasing functions of β_p. In particular if c_1 becomes zero at some β_c, it stays such for $\beta_p > \beta_c$

ii) c_1 does exist. To see this observe first of all that if a rectangle C is the union of two equal and equally oriented rectangles A and B as in fig.2

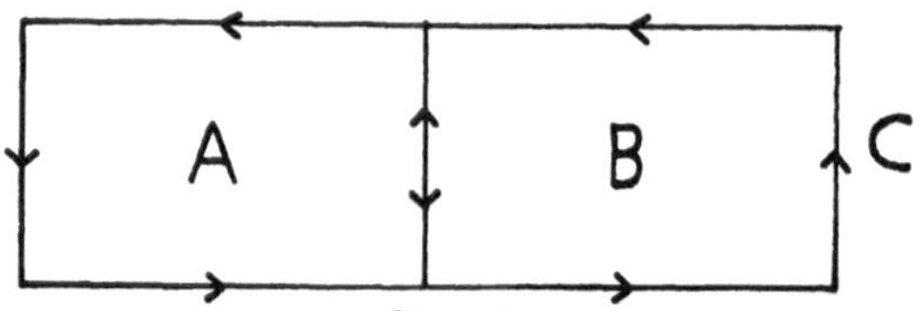

fig. 2

$$<\exp iA(m_c)> = <\cos(A(m_A) + A(m_B))> \geq <\cos A(m_A)\cos A(m_B)> \geq <\cos A(m_A)><\cos A(m_B)>$$

where the last inequality is just Griffiths' second inequality while the first one requires the observation that

$$<\sin A(m_A)\sin A(-m_B)> \geq 0$$

as in any cutoff volume Λ symmetric with respect to $A\cap B$ it is the integral with respect to the link variables on $A \cap B$ of a square. It immediately follows that, if $\{C_n\}$ is a sequence of square of sides $L_n = 2^n$

$$<\cos A(m_{C_{n+1}})> \geq <\cos A(m_{C_n})>^4$$

so that

$- \frac{\log\langle\cos A(m_{c_n})\rangle}{L_n^2}$ is a monotone decreasing function of n, whose limit as $n \to \infty$

obviously exists as it is bounded below by zero.

iii) The explicitly known expression of c_1 for $d = 2$ (23)

$$c_1(\beta_p)\Big|_{d=2} = - \lg \frac{\int_{-\pi}^{\pi} e^{\beta\cos x}\cos x\,dx}{\int_{-\pi}^{\pi} e^{\beta\cos x}dx}$$

is an upper bound for $c_1(\beta_p)$ for every $d > 2$ (12).

In fact we need only observe that deleting from the action all the terms connecting one lattice plane to the others causes the expectations of functions satisfying GKS inequalities to decrease.

Further information on the system can be obtained from a set of recurrent equations for the generating functional

$u(m) = \langle\exp iA(m)\rangle \qquad m \in \mathfrak{M}$

which we describe below (29).

Given $m \in \mathfrak{M}$, $m \neq 0$, call $\ell_1 = (n_1, n_2)$, $n_1 < n_2$ the first (in lexicographic order) link for which $\bar{m} \equiv m(n_1, n_2) > 0$.

Call P_i, $1 \leq i \leq 2(d-1)$, the plaquettes having ℓ_1 as one side, oriented in such a way that $n_1 \to n_2$ is an elementary step, and call m_{P_i} the field in $\mathfrak{M}$ associated to P_i. Having chosen a volume cutoff Λ large enough with respect to the support of m, and performing by parts the integration with respect to $A(n_1, n_2)$ in the integral defining

$u_\Lambda(m) = \langle\exp iA(m)\rangle_\Lambda$

we find

$$u_\Lambda(m) = \frac{\beta}{2\bar{m}} \sum_i \left[u_\Lambda(m-m_i) - u_\Lambda(m+m_i)\right]$$

The same equation holds for the thermodynamic limit $u(m)$ and can be conveniently rewritten in the form

$u = u_0 + \beta S u$

where u_0 is the function

$$u_0 : m \in \mathfrak{M} \to u_0(m) = \begin{cases} 1 & \text{if } m=0 \\ 0 & \text{otherwise} \end{cases}$$

and S is the linear operator on the functions on $\mathfrak{M}$ defined by

$$(S\nu)(m) = \begin{cases} 0 & \text{if } m = 0 \\ \frac{1}{2\bar{m}} \sum_i |\nu(m-m_i) - \nu(m-m_i)| & \text{if } m \neq 0 \end{cases}$$

In the Banach space B of bounded functions on $\mathfrak{M}$ with the sup norm, S is a bounded operator with $||S|| \leq 2(d-1)$.

Therefore the solution of the previous equation is surely unique and given by

$u = (1 + \beta S + \beta^2 S^2 + \ldots)u_\circ$

for $\beta < \frac{1}{2(d-1)}$.

Notice that, as $u(m + m_i) \geq 0$ by Griffiths' first inequality, we also get the recurrent inequalities:

$u \leq u_\circ + \beta Tu$

where T is the linear operator

$$(T\nu)(m) = \begin{cases} 0 & \text{if } m = 0 \\ \frac{1}{2\bar{m}} \sum_i \nu(m-m_i) \end{cases}$$

having in B the norm $||T|| = d-1$ because $|(T\nu)(m)| \leq (d-1) \sup_{m'} |\nu(m')|$ and $(T\nu)(m) = (d-1)\nu(m)$ for ν identically equal to 1 and $m=m_p$.

As T is positivity preserving we can iterate the previous inequality to obtain:

$u(m) \leq u_\circ(m) + (\beta Tu_\circ)(m) + \ldots + [(\beta T)^{k-1}u_\circ](m) + [(\beta T)^k u](m)$.

In particular, given m, if k is the minimum number of plaquette fields m_p to be added to m in order to make it vanish

$u(m) \leq [(\beta T)^k u](m)$

which leads to the "confinement bound" (see also (1,2,3):

$u(m) \leq (\beta(d-1))^k$

and, in particular, $h_\circ$:

$c_1(\beta) \geq - \log \beta(d-1)$.

We can thus conclude that the smallest value β_c of β_p at which c_1 may vanish satisfies the inequality

$\beta_c \geq \frac{1}{d-1}$.

Obtaining a complete control of fig.1, in particular the existence of such a β_c and the finiteness of c_2 above β_c will require further work and new ideas.

We refer to (6) for an approach to the high β_p regime, and to (2) for a promising transformation (the construction of the dual model) connecting the high and low β_p regimes.

As a concluding remark we want to emphasize that c_1 and c_2, as indicated by work

in progress (30) on the "gauge invariant xy model" (25), might not be the natural order parameters for the full model of Section 1. The point is that, taking into account both sets of angular variables A and Θ, the Fourier transform $\langle \exp i(A(m) + \Theta(q)) \rangle$ is allowed to take non vanishing values also from $m \notin \mathcal{M}$.
A natural long range order parameter might then be constructed from "string", as opposed to "Bohm-Aharanov" variables corresponding to situations in which m is different from zero on an open line with the integer valued field q having non vanishing values at the end points of such a line.
A rich source of conjectures about how the complete phase structure of the gauge invariant models might appear is given by the "gauge invariant Ising model", whose group of symmetry is the discrete group Z_2 (31).
In terms of suitable reduced variables it can be described in terms of a collection of random variables $\sigma(n,n')$ sitting at each link (n,n') and taking the values ± 1.
The action is given by

$$- \beta_\ell \sum_{(n,n')\subset\Lambda} \sigma(n,n') - \beta_p \sum_{p\subset\Lambda} \sigma_p \quad ,$$

where β_ℓ and β_p are non negative constants, and σ_p is the product of σ's corresponding to the sides of the plaquette P. The duality transform [31,32] is very useful in the study of this model. For the sake of definiteness let us consider the case of a three-dimensional lattice. Then duality relates the model (β_ℓ, β_p) with the model $(\beta_\ell^*, \beta_p^*)$ where

$$\beta_\ell^* = - \frac{1}{2} \log t g h \beta_p \ ,$$

$$\beta_p^* = - \frac{1}{2} \log t g h \beta_\ell \ .$$

If $\beta_\ell = 0$ then the "pure gauge" model with some coupling β_p is related to the Ising model in three dimensions with coupling $\beta^* = - \frac{1}{2} \log t g h \beta_p$, therefore it has a phase transition for a definite value $\beta_{p,c}$ of the coupling. As for the long range order parameters it is easy to show that the average $\langle \sigma_\Gamma \rangle$, where σ_Γ is the product of σ's associate to links forming a closed durve Γ, has an exponential damping ruled by the area of Γ below the critical point and by the perimeter of Γ beyond the critical point. When $\beta_\ell > 0$ the behavior of $\langle\sigma_\Gamma\rangle$ is always ruled by the perimeter. Since $(\beta_\ell = 0, \beta_{p,c})$ is a critical point, then, by duality, also the point $(\beta_\ell^*, \beta_p = \infty)$ is critical. It is an attractive conjecture, supported in part by mean field calculations, that the two critical points are joined by a whole critical line which divides the plane (β_ℓ, β_p) in two regions, characterized by two different regimes, as far as the "confinement" of quasi-particle excitations is concerned. A natural candidate as long range order parameter characterizing the two regions is given by

$$c = \lim_{L\to\infty} - \log \langle\sigma_L\rangle / |L| ,$$

where L is a line of length $|L|$ and σ_L is the product of σ's associate to the bonds of L, (the existence of C follows easily from GKS correlation inequalities).
A detailed study of the Z_2 model and its critical behavior can be found in a forthcoming paper [33].

REFERENCES

1) K.G. Wilson, Phys. Rev. D10, 2445 (1974).
2) K. Osterwalder, E. Seiler, Cargèse Lectures 1976.
3) ______________, Gauge Field Theories on the Lattice, preprint (1977).
4) J. Glimm, A. Jaffe, Phys. Letters 66B, 67 (1977).
5) A.M. Polyakov: Nucl. Phys. B120, 429 (1977).
6) J. Glimm, A. Jaffe, Instantons in a U(1) Lattice Gauge Thoery: A Coulomb Dipole Gas, preprint (1977).
7) M. Lüscher, Construction of a Self-adjoint, Strictly Positive Transfer Matrix for Euclidean Lattice Gauge Theory, preprint (1976).
8) ______________, Absence of Spontaneous Gauge Symmetry Breaking in Hamiltonian Lattice Gauge Theories, preprint (1977).
9) L.P. Kadanoff, in Lecture Notes in Physics, Vol. 54.
10) __________, Reviews of Modern Physics 49, 267 (1977).
11) J. Ginibre, Comm. Math. Phys. 16, 310 (1970).
12) G.F. De Angelis, D. de Falco, Lettere al Nuovo Cimento 18, 536 (1977).
13) F. Guerra, in Mathematical Methods of Quantum Field Theory, CNRS Marseille (1976).
14) J. Fröhlich, Proceedings of the Internationale Universitätswochen für Kernphysik, Schladmig (1976).
15) G.F. De Angelis, D. de Falco, F. Guerra: Scalar Quantum Electrodynamics on the Lattice as Classical Statistical Mechanics, preprint (1977).
16) E. Nelson, Probability Theory and Euclidean Field Theory, in "Constructive Field Theory", G. Velo, A. S. Wightman Editors, Springer- Verlag (1973).
17) F. Guerra, L. Rosen, B. Simon, Ann. Math. 101, 111 (1975).
18) J. Glimm, A. Jaffe, T. Spencer, The Particle Structure of the Weakly Coupled

$P(\phi)_2$ Model and other Applications of High Temperature Expansions, in "Constructive Quantum Field Theory".

19) L. Maiani, An Elementary Introduction to Yang-Mills Theories and to their Applications to the Weak and Electromagnetic Interactions, Proceedings of the 1976 CERN School of Physics, Wépion, June 1976.

20) J. Glimm, A. Jaffe, T. Spencer, Comm. Math. Phys. 45, 203 (1975).

21) J. Fröhlich, B. Simon, T. Spencer, Comm. Math. Phys. 50, 79 (1976).

22) J. Schwinger, Phys. Rev. 125, 397 (1962) and 128, 2425 (1962).

23) J. Glimm, A. Jaffe, Comm. Math. Phys. 51, 1 (1976).

24) R. Schrader, Comm. Math. Phys. 49, 131 (1976), 50, 27 (1976).

25) R. Balian, J.M. Drouffe, C. Itzykson, Phys. Rev. D10, 3376 (1974).

26) S. Elitzur, Phys. Rev. D12, 3978 (1975).

27) G.F. De Angelis, D. de Falco, F. Guerra, A note on the Abelian Higgs-Kibble Model on the Lattice: Absence of Spontaneous Magnetization, preprint (1977).

28) P.W. Higgs, Phys. Letters 12, 132 (1964) and Phys. Rev. 145, 1156 (1966).

29) G.F. De Angelis, D. de Falco, F. Guerra, Lettere al Nuovo Cimento 19, 55 (1977).

30) S. De Martino, S. De Siena, In preparation.

31) R. Balian, J.M. Drouffe, C. Itzykson, Phys. Rev. D10, 2098 (1975).

32) F.J. Wegner, J. Math. Phys. 12, 2259 (1971).

33) S. De Martino, S. De Siena, R. Marra, in preparation.

LATTICE GAUGE THEORIES

K. Osterwalder*†
Lyman Laboratory of Physics
Harvard University
Cambridge, MA
02138

E. Seiler**
Joseph Henry Laboratories of Physics
Princeton University
Princeton, N.J.
08540

(Talk given by E. Seiler)

I. Introduction

Gauge field theories are attractive for the constructive field theorist for two reasons: First they are believed to play an important role in the description of the real world and second their ultraviolet divergencies seem to be milder than one would expect for renormalizable theories ("asymptotic freedom" [GW], [P]).

Unfortunately, one encounters the first difficulty right away: The usual cutoffs destroy the gauge invariance and nobody knows how to use the gauge invariant renormalization schemes of perturbation theory in a nonperturbative treatment. Therefore, we take up the idea first advanced by Wilson [W1] (see also [BDI]) of a lattice cutoff that preserves this invariance and also, as we will see, physical positivity.

So far the problem of the continuum limit has not been resolved (I will make some short remarks about it at the end). But it is both interesting and necessary to obtain information on the mathematical structure of the theory on the lattice before attacking this certainly very hard problem. We present here an overview of the results we have obtained so far (cf. also the contributions of Guerra and Jaffe to this conference). First results were reported almost a year ago [O]; a detailed account is given in [O Se].

II. The Formalism of Lattice Gauge Theories

We work on a cubic d-dimensional ($d \geq 2$) lattice $\Lambda \subset a\mathbf{Z}^d + (\frac{a}{2}, \ldots, \frac{a}{2})$ where a is the lattice constant; i.e. the points in Λ are of the form $(a(\frac{1}{2} + n_1), \ldots, a(\frac{1}{2} + n_d))$ ($n_1, \ldots, n_d$ integers). Whenever possible we will replace a by 1. The reason for the added $\frac{1}{2}$ should become clear in Section III.

Basic objects in the lattice are the sites, the bonds (i.e. pairs of nearest neighbors) and the so-called plaquettes (closed loops formed by four bonds).

*A. Sloan Foundation Fellow

†Supported in part by NSF Grant PHY 75-21212

**Supported in part by NSF Grant MPS 74-22844

We assume that a compact Lie group G (the "gauge group") is given.

Next we introduce fields:

1. Yang-Mills Field [W,BDI]

A "field configuration" is a map $g_{\bullet}$ from the oriented bonds in Λ into G such that

$$g_{xy} = g_{yx}^{-1}$$

This introduces in a natural way a map from the oriented plaquettes into the conjugacy classes of G: If $P = (xy,yz,zu,ux)$, we set $g_P = g_{xy}\, g_{yz}\, g_{zu}\, g_{ux}$ and the class of g_P is independent of the starting point.

We define an action $A_{YM} = \frac{1}{2g_0^2}\, \mathrm{Re} \sum_{\text{plaq.}} \chi(g_P)$ where χ is a character on G. (This corresponds to $A_{YM} = \frac{1}{2}\, \mathrm{tr} \int F^2$ in the continuum.)

2. Higgs Field

Here we have a map $\phi(\cdot)$ from the sites into $\mathbb{C}^k$ where $\mathbb{C}^k$ carries a unitary representation $U(\cdot)$ of G (typically, though not necessarily the one of which $\chi(\cdot)$ is the character).

Action: $A = A_{YM} + A_H$

$$A_H = -\frac{1}{2} \sum_{\text{bonds}} |\phi(x) - U(g_{xy})\phi(y)|^2 - \sum_{\text{sites}} V(|\phi(x)|)$$

(in the continuum this would be $A_H = -\frac{1}{2} \int |D_\mu \phi|^2 - \int V(|\phi|)$)

Special cases:

(a) $G = U(1)$

$$\phi(x) = e^{i\theta(x)}\, R(x) \equiv U(g_x) R(x)$$

(b) $G = SU(2)$

$\phi(x) \quad \mathbb{C}^2$; $U(\cdot)$ spinor representation

$$\phi(x) = U(g_x) \begin{pmatrix} 1 \\ 0 \end{pmatrix} R(x)$$

In these cases we can rewrite the action as follows:

$$A_H = \lambda\, \mathrm{Re} \sum_{\text{bonds}} R(x)R(y)\chi(g_x^{-1} g_{xy} g_y) - \sum_{\text{sites}} \left(V(R(x)) + \frac{1}{2} d(x) DR(x)^2\right)$$

where $\lambda = 1/\chi(e)$ (e the unit element of G) and $d(x)$ equal to d except on a boundary layer.

3. Fermions (cf. [B], [OS 2])

To each site we associate a finite number of generators of a Grassmann algebra: $\{\psi^i_{\alpha\ell}(x)\}$ where $i = 1,2$ is a charge degree of freedom (in some sense ψ^1 corresponds to the relativistic $\bar{\psi}$ and ψ^2 to ψ); the first lower index refers to "spin", the second one to internal degrees of freedom ($\ell = 1,\dots,k$, so that a k-dimensional unitary representation $U(\cdot)$ of G can operate on this index).

Action: $A = A_{YM} + A_F$

$$A_F = \frac{1}{2} \sum_{\text{bonds}} \psi^1(x)^T \gamma_{xy} U(g_{xy})\psi^2(y)$$

$$+ m \sum_{\text{sites}} \psi^1(x)^T \psi^2(x)$$

where

$$\gamma_{xy} = \begin{cases} \gamma^\mu & \text{if} \quad y = x + e_\mu \\ -\gamma^\mu & \text{if} \quad y = x - e_\mu \end{cases}$$

(e_μ a unit lattice vector) and $\gamma^\mu\gamma^\nu + \gamma^\nu\gamma^\mu = 2\delta^{\mu\nu}$ $(\mu,\nu = 0,1,\dots,d-1)$.

There are reasons to consider more complicated possibilities, for instance

$$A_F' = A_F + \frac{1}{2} \sum_{\text{bonds}} \psi^1(x)^T U(g_{xy}) i\gamma_5 \psi^2(y)$$

$$- \sum \psi^1(x)^T i\gamma_5 \psi^2(x)$$

where

$$\gamma_5{}^* = \gamma_5 \ ; \ \gamma_5\gamma^\mu + \gamma^\mu\gamma_5 = 0 \qquad (\mu = 0,1,\dots,d-1).$$

The formal continuum limit in both cases is

$$A_F = \int \psi^1 \not{\partial} \psi^2 + m \int \psi^1\psi^2$$

but only the second choice seems to give the right continuum limit for the interacting theory (cf. [W2]); at least this is true if one studies the Y_2 theory instead of a gauge theory. The first case contains additional, possibly unwanted degrees of freedom (for a discussion of the continuum limit for Y_2 see [MD]).

Expectations:

Let F be a "local observable", that is a function of the basic fields (say, continuous; depending on only finitely many). We define

$$\langle F\rangle_\Lambda = Z_\Lambda^{-1} \int F e^A$$

$$Z_\Lambda = \int e^A$$

Here the symbol $\int \dots$ means integration with Haar measure over all bond variables, integration over the Higgs fields (if present) with Lebesgue measure and with respect to the fermions it means Berezin's "anticommuting integration" [B] defined by

$$\int \bigwedge_{\alpha,k,x} \psi^1_{\alpha,k}(x) \quad \psi^2_{\alpha,k}(x) = 1$$

if all generators appear, $\int \bigwedge_{\alpha,k,x,i} \psi^i_{\alpha,k}(x) = 0$

if not all generators appear, and extended by linearity.

III. Basic Properties

1. Gauge Invariance

A map $\gamma_\cdot$ from the sites into G - such that $\gamma_x = e$ for all but finitely many sites x - induces a local gauge transformation as follows:

$$g_{xy} \rightarrow \gamma_x \, g_{xy} \, \gamma_y^{-1}$$

$$\phi(x) \rightarrow U(\gamma_x)\phi(x)$$

$$\psi^1(x) \rightarrow U(\gamma_x^{-1})\psi^1(x)$$

$$\psi^2(x) \rightarrow U(\gamma_x)\psi^2(x)$$

Both the action and the "measure" are invariant under these gauge transformations. The observables can therefore be replaced by averages over gauge transforms.

The gauge invariance can be used to eliminate redundant degrees of freedom by restricting the integrations (here we only mean the true integrations) to a "hypersurface" intersecting each orbit of the gauge group G^Λ at least once. In general this will introduce a weight factor (Jacobian) leading to the so-called Faddeev-Popov ghosts.

2. A Special Gauge ("Radiation Gauge")

Let e_0 be a unit vector in "time" direction. By a gauge transformation and a subsequent change of integration variables we can in effect fix all bond variables $g_{x,x\pm e_0}$ to the unit element $e \in G$. Because of the invariance of Haar measure no Jacobian appears in this case.

3. Physical Positivity

All the models described in II have the physical positivity property which is fundamental for the eventual reconstruction of a relativistic field theory (see [OS 1]). M. Lüscher has proved this positivity in a somewhat different setting and under certain restrictions on the parameters [[L1]; see also [W]). I will describe here a very simple general scheme for the proof of this property.

Let Λ be symmetric under "time" reflections, $\mathfrak{A}_\Lambda$ the corresponding algebra of observables, $\mathfrak{A}_+$ ($\mathfrak{A}_-$) the subalgebra of functions of the fields at positive (negative) times.

We define a map

$$\Theta : \mathfrak{A}_+ \to \mathfrak{A}_-$$

as an antilinear antiisomorphism that reflects the time coordinates and takes complex conjugates; furthermore

$$\Theta\psi^2(x) = \gamma_0\, \psi^1(\theta x)$$

$$\Theta\psi^1(x) = \gamma_0\, \psi^2(\theta x)$$

$$\left(\text{if } x = (x^0, x^1, \ldots, x^{d-1})\ ,\quad \theta x = (-x^0, x^1, \ldots, x^{d-1})\right)$$

<u>Theorem:</u> $\langle(\Theta F)F\rangle_\Lambda \geqslant 0$ if $F \in \mathfrak{A}_+$ and F gauge invariant.

The proof uses the radiation gauge described above. We denote by $\mathfrak{A}_+^{rad}$ the algebra at positive times in this gauge and define

$$\mathcal{P} \equiv \{F \in \mathfrak{A} \,|\, F = \sum_{i=1}^{N} (\Theta G_i) G_i\ ;\quad G_i \in \mathfrak{A}_+^{rad} \quad \text{for all } i\}\ .$$

It is not hard to see that $\mathcal{P}$ is a multiplicative convex cone.

Next one checks that for $F \in \mathcal{P}$ $\quad \int F > 0$:

If $F = (\Theta G)G$

$$\int F = \int_{\mathfrak{A}_-} \Theta G \int_{\mathfrak{A}_+} G = \left|\int_{\mathfrak{A}_+} G\right|^2 \geqslant 0$$

($\int_{\mathfrak{A}_\pm}$ defined in the obvious way).

The third step is to prove that

$$F\, e^A \in \mathcal{P} \quad \text{if} \quad F \in \mathcal{P}\ .$$

This follows from the fact that

$$A = \underbrace{A_+}_{\in \mathfrak{A}_+} + \underbrace{\Theta A_+}_{\in \mathfrak{A}_-} + \underbrace{W}_{\in \mathcal{P}}$$

and the multiplicativity of $\mathcal{P}$.

Standard Consequences:

(a) Physical Hilbert space

$$\mathcal{H} = \overline{\mathfrak{A}_+^{\mathrm{rad}}/\mathcal{N}}$$

where $\mathcal{N}$ is the null space of the scalar product

$$\langle(\Theta F)F\rangle_\Lambda \equiv (F,F)$$

(cf. [OS 1,2])

(b) There is a positive self-adjoint transfer matrix corresponding to a shift by two lattice units (this requires either periodic boundary conditions or taking the infinite volume limit in time direction; see [OS 1], [Se S]).

(c) Existence of the pressure $\lim \frac{1}{|\Lambda|} \log Z_\Lambda$ ("Guerra's theorem", see [G1]), bounds on exponentials of the fields, so-called chessboard estimates (cf. [G2], [SeS], [FS]).

IV. Strong Coupling Results

1. For large coupling constant g_0 the pure Yang-Mills theory can be treated by an expansion closely related to well known high temperature expansions of statistical mechanics (see for instance [GK]). The form of the expansion we are using is the Glimm-Jaffe-Spencer cluster expansion [GJS] which becomes extremely simple in our case. For the special case of an abelian gauge theory De Angelis, de Falco, and Guerra [DA DF G1] have obtained similar results (cf. Guerra's contribution to this conference). Our result, which was reported already in [O], is the following

Theorem: For $|g_0|$ large enough the correlation functions for the pure Yang-Mills theory have a unique thermodynamic limit, are analytic in g_0^{-2} and cluster exponentially with a mass gap m that is asymptotically $4 \ln g_0^2$; they have the physical positivity property and are invariant under lattice symmetries.

The crucial fact for the proof is that

$$\rho_P \equiv \exp\left(\frac{1}{2g_0^2}\chi(g_P)\right) - 1$$

is (uniformly) small for large $|g_0|$.

The expectation of an observable F can be written

$$\langle F\rangle_\Lambda = \sum_{Q\subset\Lambda} Z_\Lambda^{-1} \int F \prod_{P\in Q} \rho_P$$

($Q \subset \Lambda$ means that Q is a set of plaquettes lying in Λ) . Resummation over the components of Q disconnected from $P_0 \equiv \mathrm{supp}\, F$ gives

$$(*)\qquad \langle F\rangle_\Lambda = \sum_{Q'} \int F \prod_{P\in Q'} \rho_P \, (Z_{\Lambda\setminus\overline{P_0\cup Q'}}/Z_\Lambda)$$

where the sum is now over all sets of plaquettes Q' such that $Q'\cup P_0$ is connected; $\overline{P_0\cup Q'}$ is the union of $P_0\cup Q'$ with all plaquettes that have a bond in common with it. (*) is the GJS form of the cluster expansion; it is not very hard to show convergence uniformly in Λ for large $|g_0|$ (see [O], [O Se]); the consequences listed in the theorem are standard.

2. The cluster expansion (*) can also be used to prove Wilson's "confinement bound" for strong coupling (see also [GJ 2]). We limit ourselves to $G = U(1)$ or $SU(n)$. Let σ be an irreducible representation of G and define

a) for $G = SU(n)$:

$n_\sigma = \frac{1}{n}$ (number of boxes in the Young tableau for σ , mod n)

b) for $G = U(1)$:

$n_\sigma = \sigma$

where $U(e^{i\phi}) = e^{i\sigma\phi}$ $(\sigma\in\mathbb{Z})$

Furthermore, let $C = (x_1x_2, x_2x_3, \ldots, x_nx_1)$ be a closed (planar) loop in Λ (x_1x_2 etc. bonds in Λ) and define $g_C = g_{x_1x_2}g_{x_2x_3}\cdots g_{x_nx_1}$; let $A(C)$ be the area enclosed by C . Then we have the

<u>Theorem:</u> If $n_\sigma \neq 0$, then $|\langle\chi_\sigma(g_C)\rangle_\Lambda| \leq a_1 e^{-a_2A(C)}$ uniformly in Λ $(a_1, a_2 > 0)$.

The proof is based on the fact that in the cluster expansion (*) for $\langle\chi_\sigma(g_C)\rangle_\Lambda$ the only nonvanishing terms have the loop C totally filled with plaquettes and therefore a common factor $(\sup|\rho_P|)^{A(C)}$ can be extracted. The fact that all terms in which C is not totally filled have to vanish is most easily understood in two dimensions: Integration over the bond variables projects onto the trivial representation for each bond variable (by the Peter-Weyl theorem). This forces the sum of all n_τ (taken with the appropriate sign) for each bond to vanish; therefore all plaquettes inside C must come with the same n_τ , and all plaquettes outside C must come with the same $n_{\tau'}$. But becaue only a finite number of plaquettes is involved, $n_{\tau'} = 0$ (the argument has to be modified slightly for periodic boundary conditions); across C there is a jump by n_σ, so $n_\tau = n_{\tau'} + n_\sigma = n_\sigma \neq 0$, which means that all plaquettes inside C have to be "filled".

The general (d-dimensional) case can be reduced to two dimensions by projection onto the plane of C (see [O Se]; the mathematically inclined reader might want to use cohomology theory instead, which would also cover the nonplanar case).

V. Mass Generation Through the Higgs Mechanism

The situation becomes more interesting when we couple the Yang-Mills field to a Higgs field (cf. [EB], [H]). Let us concentrate on the situation of "totally broken symmetry", that is the Higgs action

$$A_H = \frac{1}{2} \mathrm{Re} \sum_{\text{bonds}} R(x)R(y)\, \chi\, (g_x^{-1} g_{xy} g_y)$$

$$- \sum_{\text{sites}} (V(R(x)) + \frac{1}{2} d(x) D\, R(x)^2)$$

where V has a sharp minimum at $\bar{R} > 0$, more precisely

$$V(R) \geqslant c_1 (R-\bar{R})^2 \qquad \text{for} \qquad R \geqslant 0$$

$$V(R) \leqslant c_2 (R-\bar{R})^2 \qquad \text{for} \qquad |R-\bar{R}| \leqslant \frac{1}{2} \bar{R}$$

$(c_1, c_2$ large)

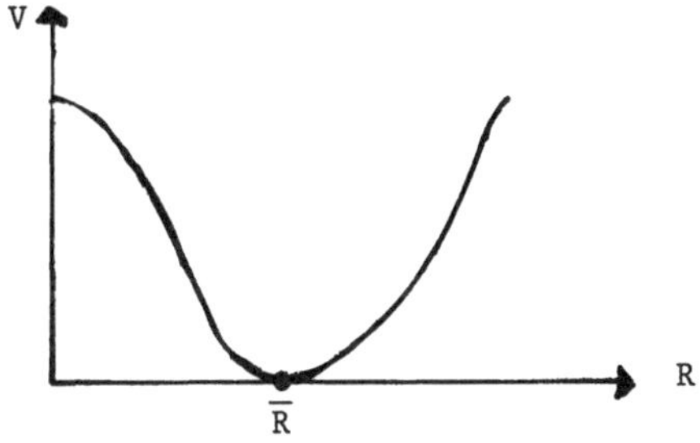

In this case we have the

<u>Theorem:</u> For <u>any</u> g_0 $\mathbb{R}$ there are $\bar{R}$, c_1, c_2 such that there is a convergent cluster expansion and therefore a unique infinite volume limit and a positive mass gap (exponential clustering).

<u>Idea of proof</u>: First change variables:

$$h_{xy} \equiv g_x^{-1}\, g_{xy}\, g_y$$

The g_x then disappear from the action and the corresponding integrations can be done trivially (this is reminiscent of Higgs' original treatment [H]). The remaining variables $\{R(x)\}$, $\{h_{xy}\}$ are gauge invariant; the action becomes

$$A = A_{YM} - \sum_{\text{sites}} (V(R(x)) + \frac{1}{2} d(x) D R(x)^2)$$

$$+ \frac{1}{2} \mathrm{Re} \sum R(x)R(y)\, \chi\, (h_{xy}) \quad .$$

Part of the action is used to modify the "single spin measure" to

$$d\mu = Z_0^{-1} \prod_{\text{bonds}} e^{\overline{R}^2(\chi(h_{xy}) - D)} \, dh_{xy} \times$$

$$\times \prod_{\text{sites}} R(x)^k \, e^{-V(R(x))} \, dR(x)$$

($D = \chi(e)$; Z_0 normalizing factor to make $\int d\mu = 1$)

There are three types of remaining coupling terms:

(a) $\frac{1}{2g_0^2} (\chi(h_p) - D)$

(b) $(\overline{R}^2 - R(x)R(y))(D - \chi(h_{xy}))$

(c) $-\frac{1}{2} D \; (R(x) - R(y))^2$

Now it is no longer true that $|e^{\text{coupling terms}} - 1|$ is uniformly small; but $d\mu$ favors values of R near $\overline{R}$ and of h_{xy} near $e \in G$. So by choosing $\overline{R}, c_1, c_2$ large enough we can still make $\|e^{\text{coupling terms}} - 1\|_p$ very small (p- norm with respect to $d\mu$) which is sufficient for convergence of the cluster expansion. A Gaussian approximation for $d\mu$ can give the right qualitative idea of the size of the coupling terms:

(a) $|\frac{1}{2g_0^2} (\chi(h_p) - D)| \sim \frac{1}{2g_0^2\overline{R}^2}$

(b) $|\overline{R}^2 - R(x)R(y)| \sim c^{-1} + 2\overline{R}c^{-\frac{1}{2}}$

where c is somewhere between c_1 and c_2

(c) $\frac{1}{2} D(R(x) - R(y))^2 \sim \frac{1}{2} Dc^{-1}$

So choosing $\overline{R}$ large and c large compared to $\overline{R}^2$ makes the coupling terms small.

These heuristic considerations can be converted into precise estimates [O Se].

We could rephrase the result in the language of statistical mechanics; there it would give an expansion converging for all positive temperatures ($\beta \sim \frac{1}{2g_0^2}$; $V \sim \beta W$) somewhat similar to strong field expansions in ferromagnets.

<u>Note:</u> The expansion works exactly in the region where - without the presence of the gauge field - there would be spontaneous symmetry breaking and massless Goldstone excitations in $d \geqslant 3$ (see [FSS]); so the mechanism can be interpreted as "mass generation". It works, however, in the same way for $d = 2$ where no spontaneous symmetry breaking exists [[M],[BFL]). The conventional explanation of the Higgs mechanism which starts with spontaneous symmetry breaking should therefore not be taken too literally.

VI. Additional Remarks

The study of the continuum limit is intimately connected with the study of critical behavior, as has been noted by various authors (see e.g. [GJ 1], [Sch]) : To let the lattice constant a go to zero while keeping the physical masses fixed is (by scaling) equivalent to keeping the lattice spacing fixed while letting the masses go to zero (and rescaling the observables appropriately).

One therefore has to understand the map from the ("bare") parameters of the theory to physical quantities like various masses. This would be a "renormalization map" as discussed in [Sch]. For the pure Yang-Mills theory the situation is much simpler than for ϕ^4 since there is only one parameter (g_0) ; one has to study the map $g_0 \to m$ (mass gap). The belief is that in the nonabelian case for $d < 4$ the only critical point is at $g_0 = 0$, whereas the U(1) theory has a finite critical point in $d = 4$ (cf. Jaffe's contribution to this conference and [GJ 3]).

In order to control the continuum limit one has to invert the renormalization map and rescale the observables. This can be done explicitly in the admittedly trivial case of two-dimensional pure Yang-Mills theory and gives the expected result that g_0 has to go to zero proportionally to a (which coincides with the naive continuum limit). In four dimensions there will be logarithmic corrections (predicted by "asymptotic freedom" [P], [GW]) to the naive behavior $g_0 \sim a^{2-\frac{1}{2}d}$.

Let me close with a remark on "spontaneous breakdown of gauge invariance" since this is a point that seems to cause some confusion: Any infinite volume state constructed as a limit of finite volume Gibbs states, or more generally, any state that fulfills the DLR equations [La] is locally gauge invariant (it is here that the local character of gauge invariance is crucial); so there can be no spontaneous breaking of gauge invariance (boundary conditions do not propagate; cf. also [L2],[DA DF G 2]). The situation is quite different, however, when a "gauge is fixed" (see Sec. III). Then gauge invariance is always broken; this happens for instance in the usual continuum QED as has been noted by Strocchi [St].

References

[BDI] R. Balian, J.M. Drouffe, C. Itzykson, Phys. Rev. D10, 3376 (1974); D11, 2098 (1975); D11, 2104 (1975)

[B] F.A. Berezin, The Method of Second Quantization, Academic Press, New York 1966

[BFL] J. Bricmont, J.R. Fontaine, L.J. Landau, On the Uniqueness of the Equilibrium State in Plane Rotators, Louvain la Neuve preprint UCL-1PT-77/03

[DA DF G 1] G.F. DeAngelis, D. deFalco, F. Guerra, Lattice Gauge Models in the Strong Coupling Regime, Salerno preprint 1977

[DA DF G 2] ________ ________ ________, A Note on the Abelian Higgs-Kibble Model on a Lattice: Absence of Spontaneous Magnetization, Salerno preprint 1977

[EB] F. Englert, R. Brout, Phys. Rev. Lett. 13, 321 (1964)

[FS] J. Fröhlich, B. Simon, Ann. Math. 105 (1977)

[FSS] J. Fröhlich, B. Simon, T. Spencer, Commun. Math. Phys. 50 79 (1976)

[GJ 1] J. Glimm, A. Jaffe, Commun. Math. Phys. 51, 1 (1976)

[GJ 2] ________ ________ Phys. Lett. 66B, 67 (1977)

[GJ 3] ________ ________ Instantons in a U(1) Lattice Gauge Theory: A Coulomb Dipole Gas, Harvard preprint 1977

[GJS] J. Glimm, A. Jaffe, T. Spencer, in : Constructive Quantum Field Theory, G. Velo and A.S. Wightman eds., Spinger Lecture Notes in Physics 25 (1973)

[GW] D. Gross, F. Wilczek, Phys. Rev. Lett. 26, 1343 (1973)

[GK] C. Gruber, H. Kunz, Commun. Math. Phys. 22, 133 (1971)

[G1] F. Guerra, Phys. Rev. Lett. 28, 1213 (1972)

[G2] ________ in Mathematical Methods of Quantum Field Theory, CNRS Marseille 1976

[H] P. Higgs, Phys. Lett. 12, 132 (1964); Phys. Rev. 145, 1156 (1966)

[La] O. Lanford, in Statistical Mechanics and Mathematical Problems, A. Lenard ed., Springer Lecture Notes in Physics 20 (1973)

[L1] M. Lüscher, Construction of a Self-Adjoint, Strictly Positive Transfer Matrix for Euclidean Lattice Gauge Theory, DESY preprint 1976

[L2] ________ Absence of Spontaneous Symmetry Breaking in Lattice Gauge Theories, DESY preprint 1977

[MD] A. MacDermot, Ph.D. Thesis, Cornell University 1976

[M] N.D. Mermin, J. Math. Phys. 8, 1061 (1967)

[O] K. Osterwalder, Yang-Mills Fields on the Lattice, lecture delivered at the 1976 Cargèse summer school, Harvard preprint 1976

[OS 1] K. Osterwalder, R. Schrader, Commun. Math. Phys. 31, 83 (1973); 42, 281 (1975)

[OS 2] ________ ________ Helv. Phys. Acta 46, 277 (1973)

[O Se] K. Osterwalder, E. Seiler, Gauge Field Theories on the Lattice, Harvard preprint 1977 (subm. to Ann. Phys.)

[P] H.D. Politzer, Phys. Rev. Lett. 26, 1346 (1973)

[Sch] R. Schrader, Commun. Math. Phys. 49, 131 (1976); 50, 97 (1976)

[Se S] E. Seiler, B. Simon, Ann. Phys. 97, 470 (1976)

[W 1] K.G. Wilson, Phys. Rev. D10, 2445 (1975)

[W 2] ________ 1976 Cargèse lecture notes, to appear

Some Frontiers in Constructive Quantum Field Theory and Equilibrium Statistical Mechanics[1)]

Jürg Fröhlich[2)]
Department of Mathematics
Princeton University
Princeton, N. J. 08540

Abstract:
We present and discuss a list of important, mostly open problems in constructive quantum field theory and equilibrium statistical mechanics the solution of which requires (in rare cases : required) new ideas going beyond high - and low - temperature expansions guided by standard (super-renormalizable and infrared finite) perturbation theory about the critical points of some action or Hamilton function, beyond Peierls-type arguments and their variants and beyond spin wave theory and its rigorous counterparts. This list of problems includes higher order phase transitions, critical phenomena, long range forces, gauge theories, quantum solitons, etc.

[1)] Supported in part by the U. S. National Science Foundation under grant #MPS 75-11864
[2)] A. Sloan Foundation Fellow.

I. Introduction:

A list of important problems and table of contents.

I.1 Personal problems and acknowlegements.

"Die Phantasie wird nur von dem erregt, was man noch nicht oder nicht mehr besitzt;..." (R. Musil, Der Mann ohne Eigenschaften).

A free translation of this quote might read as follows: Our imagination gets only excited (inspired) by what we do not possess yet, or not possess anymore.

When I recently learned this quote I felt it would be the right motto for these notes to two lectures I presented at $M \cap \phi$ in Rome. First reason: In these notes I try to speak about some problems in theoretical and mathematical physics whose solutions we do not possess, yet. At best we have some vague ideas of how to approach them or some preliminary results. My hope is that stating those problems in a precise way will stimulate our imagination and eventually lead to significant progress. Second reason: I found those ten days in Rome very exciting not only because of the interesting news about $M \cap \phi$ I learned, but at least as much because Rome is a place that inspires our imagination by showing us witnesses of some wealth we do not possess anymore : an overwhelming variety of past culture and civilization sunken into history; (and it excites our imagination by its wide variety of future possiblities).

Visitors of Rome face a serious problem. Unless they have a vast amount of time available they have to make a choice:

1) They might just enjoy themselves, relax and have Frascati, Espresso and good meals.
2) They might concentrate on seeing only some of the antique, or the Renaissance or the modern sites.
3) They might rush through all or most of Rome and then try to look at this or that in more detail.

When preparing my lectures and writing these notes I was facing a similar problem: Should I relax and just write a few pages of trivialities, should I concentrate on one specific problem and try to discuss it carefully, or should I rush through many cf the problems that excite me and look only at a few in some more detail?

These notes are probably a bad compromise of alternatives 2) and 3). It might well be that they show nothing more than the author's ignorance, somewhat contrary to his intention and presumably the one of all those people from whom he has profited in innumerable discussions (or through correspondence): E. H. Lieb, O. McBryan, Y. M. Park, E. Seiler, B. Simon, T. Spencer and others. They should have written these notes. Apart from those people I wish to thank the organizers of $M \cap \phi$ for their great work and for giving me the opportunity to present ideas that are in part, to say the least, doubtful.

I.2 The main theme and table of contents.

In these notes we are concerned with problems in constructive quantum field theory and equilibrium statistical mechanics a complete solution of which requires to go beyond - standard (super-renormalizable and infrared finite) perturbation theory about finitely many isolated (constant) degenerate minimas of some classical action or Hamilton function and its convergent versions: High and low temperature expansions, Peierls-type contour arguments, etc.; - super-renormalizable perturbation expansions or approximations (e.g. spin wave theory) about infinitely many, non-isolated (but constant) degenerate minimas of some classical action or Hamilton function and its rigorous versions: Spin wave analysis and Infrared (Gaussian) domination, the Goldstone theorem (and scattering theory for zero mass particles or excitations).

Among such problems there are

A. Rigorous treatment of non-super-renormalizable ultraviolet divergences, field strength - and charge renormalization.

B. Gauge theories in general, (meaning of gauge invariance in the presence of instantons, infrared divergences, confinement, lattice approximation, etc.); super-renormalizable gauge theories, such as the abelian Higgs model in two space-time dimensions (which has instantons) or QED in three dimensions (existence, physical positivity, phase transitions, etc.).

C. The theory of (topological) charges and super-selection sectors; quantum solitons.

D. Higher order phase transitions, critical phenomena and infrared divergences, the theory of critical points, interactions of very long range.

E. Scattering of charged particles interacting with the radiation field.

None of the problems A.-E. has so far been understood-not to mention solved-in a mathematically rigorous way. (The great importance of these problems for theoretical physics need not be explained here).

To make it clear at the beginning: I have nothing interesting to say about A. Although mathematical physicists (Schrader[1] and Glimm-Jaffe[2], see also [3])have tried to formulate this problem in a precise way and developed some preliminary ideas, one is far from knowing what the main difficulties are and one could view it as a scandal that we still do not have any concrete ideas about how the predictions of the renormalization group (e.g. asymptotic freedom and its converse;one may also think of supersymmetry) can be made into precise hints to the constructivists or, more ambitiously, into provable results.

I shall not say much of interest about problem B. either. (A preliminary outline of a program towards constructing continuum gauge quantum field theories and some rigorous results for simple models in two space-time dimensions were first given in [4]). What I could say about B. may well not be of much interest and, furthermore, it would require much more space. It is limited to some partly rigorous [4,5,6] and partly semi-rigorous [7] results on two dimensional gauge theories and some comments on lattice theories and on the meaning of gauge invarinace in theories with instantons [28]. However, I do want to recommend the following references to the reader's attention: [8,9,10,11,12] and [13,14,15,16,27,28]. In Section III a few results are sketched. In particular, we find phase transitions and a breakdown of the Higgs mechanism in approximate models of gauge theories with instantons; (for the $\theta = \pi$ vacuum): A new result that might be interesting for particle physics.

For reasons of page limitation I cannot describe the recent rigorous work concerning C. (quantum solitons) either; see [17,18,19]. But I want to emphasize that in these references a point of view has been developed which I feel is the correct one and will survive (e.g. because of its mathematical precision, which has not yet been widely appreciated, though). A rather general theory of Poincaré covariant superselection sectors with non-trivial (e.g. topological) charges is now available [20,17,7,21], and for a large class of two dimensional models with non-trivial superselection (soliton) sectors a quantum field theory of solitons has been developed [17], and it has been proven that, to leading order, the mass of the quantum soliton is given by the rest energy of the classical soliton [19]. The question of whether an expansion in $\hbar$ of all interesting quantum soliton effects about classical soliton solutions is asymptotic at $\hbar = 0$ can now be posed in a precise fashion and is presently studied; see also [22,23,24,25,7].

A discussion of D. (higher order phase transitions, critical phenomena,...) is the main part of these notes. As to the methods available for proving rigorous results in the field of critical phenomena one is still almost entirely limited to using correlation inequalities, infrared domination (and reflection positivity) - see Sections II, III - and some special inequalities (e.g. for Coulomb systems) - or else rely on exactly solvable models [26] about which I have nothing to say. Such methods are insufficient and may not lend themselves to much hard analysis. What is missing is a constructive version of the renormalization group (or other methods for setting up expansions about zero mass situations) applicable to physically interesting models and amenable to rigorous mathematics. An exception is the very recent work of Glimm and Jaffe [27] concerning the U(1) lattice gauge theory in four dimensions which may turn out to be interesting for statistical mechanics, too. In Section II we give a new derivation of their approximation and in Section III we present some results complementary to theirs. Our methods also apply to the abelian Higgs model on the lattice [28].

As to problem E: The reader is advised to consult the contribution of D. Buchholz to these proceedings and refs. [29,30,31]. Buchholz' results [29] and earlier proposals and results of the author [30] may supply a suitable axiomatic framework for understanding the scattering of charged particles and photons. This framework has been tested and partially confirmed in a simple model of non-relativistic electrons interacting with massless, scalar photons which has infrared divergences typical of QED, [31].

II. Models, mathematical structures, inequalities.

II.1) Lattice spin systems and - gauge theories.

Let $\mathbb{Z}^\nu$ be the simple, cubic lattice in ν dimensions. At each site $i \in \mathbb{Z}^\nu$ there is a random variable (classical spin) $\vec{S}_i \in \mathbb{R}^N$ distributed according to a (generally, but not always finite) measure $d\lambda(\vec{S})$ on $\mathbb{R}^N$. With a bounded cube $\Lambda \subset \mathbb{Z}^\nu$ we associate a Hamilton function

$$H(\{\vec{S}\}_\Lambda) = -\sum_{i,j\in\Lambda} J(i-j)\vec{S}_i \cdot \vec{S}_j + \vec{h}\cdot\left(\sum_{i\in\Lambda} \vec{S}_i\right) . \qquad \text{(II.1)}$$

We usually impose periodic (Λ viewed as a torus) or **free** ($\vec{S}_i = 0$, for all $i \notin \Lambda$) boundary conditions. The couplings $J(m)$ are assumed to be non-negative for $m \neq 0$ (ferromagnetic), of exponential decrease in $|m|$, reflection positive [32] (which is equivalent to the existence of a selfadjoint transfer matrix [32])and isotropic (w.r. to interchanging lattice axes). Finally $\vec{h}$ is a fixed external field which we assume, from now on, to point in the 1-direction : $\vec{h} = h\cdot e_1$.

We let $\langle - \rangle_\Lambda(\beta, h)$ denote the Gibbs equilibrium expectation of the system so defined. We set $\langle - \rangle_\Lambda(\beta) = \langle - \rangle_\Lambda(\beta, 0)$. Here β is the inverse temperature.

For measures $d\lambda$ of compact support a standard compactness argument gives existence of at least one limiting Gibbs expectation, $\langle - \rangle(\beta, h)$, as $\Lambda \uparrow \mathbb{Z}^\nu$, and periodic boundary conditions (or correlation inequalities) guarantee translation invariance. The susceptibility χ is defined by

$$\chi(\beta, h) = \sum_j \langle \vec{S}_o \cdot \vec{S}_j \rangle(\beta, h) , \qquad \text{(II.2)}$$

and the inverse correlation length (mass) by

$$m(\beta, h) = -\lim_{n\to\infty} \frac{1}{n} \log\langle \vec{S}_o \cdot \vec{S}_{ne_\alpha} \rangle(\beta, h) , \qquad \text{(II.3)}$$

where e_α is the unit vector in the positive α-direction. Note that $m(\beta, h) > 0$ implies $\chi(\beta, h) < \infty$.

We now consider some examples:

II.1) N-vector models : $N = 1,2,3,\ldots,$

$$d\lambda(\vec{S}) = \delta(|\vec{S}| - 1)d^N S .$$

For $N = 1$ this is the Ising model, for $N = 2$ the rotator and for $N = 3$ the classical Heisenberg model. The rotator model can be rewritten in terms of angle variables:

$$\vec{S}_i = (\cos\theta_i, \sin\theta_i),\ \theta_i \in [0, 2\pi]$$

$$d\lambda(\vec{S}) \longmapsto \frac{d\theta}{2\pi} ,$$

$$H(\{\theta\}_\Lambda) = -\sum_{i,j\in\Lambda} J(i-j)[\cos(\theta_i - \theta_j) - 1] + h\sum_{i\in\Lambda}\cos\theta_i \tag{II.4}$$

For the Ising and the rotator model it is known that the equilibrium expectation in the limit $\Lambda = \mathbb{Z}^\nu$ is <u>unique</u> for $h \neq 0$ [33,34] and for $h = 0$ in the absence of spontaneous magnetization (i. e. $\langle S_i^1\rangle(\beta, 0_\pm) = 0$), [33,35].

II.2) <u>Dual Villain - (or V-) model</u>:

This is the model with $N = 1$, $h = 0$, $d\lambda(S) = (\sum_{m\in\mathbb{Z}} \delta(S-m))dS$ and <u>free</u> boundary conditions. If $J(m)$ is the kernel of the finite difference Laplacean on $\mathbb{Z}^\nu$ (nearest neighbor coupling) and $\nu = 2$ (in which case the moments $\langle\pi_{i\in A} S_i\rangle(\beta)$ do presumably not exist for <u>small</u> β !) this model is an approximation to the dual of the nearest neighbor rotator model at $h = 0$, provided one sets

$$\beta_V \propto \beta_{rot}^{-1} . \tag{II.5}$$

This follows from replacing $\exp \beta[\cos\theta - 1]$ by

$$\sum_{n\in\mathbb{Z}} \exp[-\frac{\beta}{2}(\theta + 2\pi\, n)^2] \tag{II.6}$$

in the definition of the partition function and the Gibbs expectation of the rotator model, then taking the Fourier transform with respect to $\{\theta_i\}$ and making a change of variables; see e.g. [36] and refs. given there. Next, we introduce <u>abelian lattice gauge theories</u> [14,15,16,27]. For this purpose we consider "p-form valued random fields on $\mathbb{Z}^\nu$". Such a p-form valued r.f. ω is of the form

$$\omega_i = \sum_{\alpha_1<\cdots<\alpha_p} \Omega_i^{\alpha_1\cdots\alpha_p}\, e_{\alpha_1}\wedge\cdots\wedge e_{\alpha_p} , \tag{II.7}$$

where, for each i and given $\alpha_1,\ldots,\alpha_p$ $\Omega_i^{\alpha_1\cdots\alpha_p}$ is a real random variable which is totally antisymmetric in $\alpha_1,\ldots,\alpha_p$. For such p-forms one can define the usual duality map $*$ so that $*\omega$ is a $(\nu-p)$-form when ω is a p-form. Furthermore, we define

$$d\omega_i = \sum_{\alpha,\alpha_1<\cdots<\alpha_p} \partial^\alpha \Omega_i^{\alpha_1\cdots\alpha_p}\, e_\alpha\wedge e_{\alpha_1}\wedge\cdots\wedge e_{\alpha_p} , \tag{II.8}$$

with

$$\partial^\alpha \Omega_i^\# = \Omega_{i+e_\alpha}^\# - \Omega_i^\# \tag{II.9}$$

II.3) <u>The U(1) lattice gauge theory</u>:

At each site $i \in \mathbb{Z}^\nu$ we are given a 1-form

$$\theta_i = \sum_\alpha \Theta_i^\alpha e_\alpha \ , \quad 0 \leqq \Theta_i^\alpha \leqq 2\pi \ . \tag{II.10}$$

We write $d\ \theta_i$ as

$$\phi_i = \sum_{\alpha_1<\alpha_2} \Phi_i^{\alpha_1\alpha_2} e_{\alpha_1} \wedge e_{\alpha_2}$$

with
$$\Phi_i^{\alpha_1\alpha_2} = \partial^{\alpha_1}\Theta_i^{\alpha_2} - \partial^{\alpha_2}\Theta_i^{\alpha_1} \ . \tag{II.11}$$

The single spin distribution $d\lambda$ of the $U(1)$ lattice gauge theory is given by $\prod_{\alpha=1}^{\nu} (2\pi)^{-1}\, d\Theta^\alpha$, and the Hamilton function (which should be called action in this context) by

$$H(\{\Theta^\alpha\}_\Lambda) = -\sum_{i\in\Lambda}\ \sum_{\alpha_1<\alpha_2} [\cos(\Phi_i^{\alpha_1\alpha_2}) - 1] \tag{II.12}$$

These definitions also determine the partition function and the Gibbs equilibrium expectation of this model which is non-trivial only for $\nu \geqq 3$. Similar expressions define the abelian Higgs model on the lattice which describes an additional pair of random fields (ρ_i, χ_i), $\rho_i \in \mathbb{R}^+$, $\chi_i \in [0, 2\pi]$, [37].

Suppose we now replace, in the definition of the partition function and the Gibbs expectations, the factors

$$\exp \beta[\cos(\Phi^{\alpha_1\alpha_2}) - 1] \quad \text{by}$$

$$\sum_{n_i^{\alpha_1\alpha_2}\in\mathbb{Z}} \exp[-\frac{\beta}{2}(\Phi_i^{\alpha_1\alpha_2} + 2\pi\, n_i^{\alpha_1\alpha_2})^2] \tag{II.13}$$

and take the Fourier transform with respect to $\{\Theta_i^\alpha\}_{i\in\Lambda}$.

Then we obtain, after some straight-forward calculations, rewriting a 2-form as the * of a $(\nu-2)$-form and using Poincaré's lemma

$$d\omega = 0 \quad \Longrightarrow \quad \omega = d\gamma \tag{II.14}$$

(always valid on the lattice) : [27,28]

II.4) For $\nu = 3$ the nearest neighbor Villain model with $\beta_V \propto \beta_{U(1)}^{-1}$;

II.5) For $\nu = 4$ a model (which we call the vector Villain - or VV-model, [28])for a 1-form lattice random field a_i^α with single spin distribution

$$\left.\begin{aligned} d\lambda(a_i) &= \prod_{\alpha=1}^{4} d\bar{\lambda}(a_i^{\alpha}) \ , \\ d\bar{\lambda}(a) &= \Big(\sum_{m\in\mathbb{Z}} \delta(a-m)\Big)da \ , \end{aligned}\right\} \tag{II.15}$$

Hamilton function (or action)

$$H(\{a^{\alpha}\}_{\Lambda}) = \frac{1}{2} \sum_{i\in\Lambda} (da_i)^2 \tag{II.16}$$

and with

$$\beta_{VV} \propto \beta_{U(1)}^{-1} \ . \tag{II.17}$$

In both cases we impose <u>free</u> boundary conditions at the boundary $\partial\Lambda$ of Λ. Similar <u>gauge-invariant</u> approximations can be made for the abelian Higgs model [28]. Finally we briefly discuss an isomorphism of the V- and the VV-model onto Coulomb-type models:

Given a V-model with couplings $J(m)$, let $\hat{J}(m)$ be - the (convolution) inverse of $J(m)$ and let $d\lambda(\frac{q}{2\pi})$ be as in defintion II.2) of the V-model (with S replaced by $^q/_{2\pi}$), but replace the Hamilton function H by

$$\hat{H}(\{q\}_{\Lambda}) = \frac{1}{2} \sum_{i,j\in\Lambda} \hat{J}(i-j)q_i q_j \tag{II.18}$$

The model so obtained is called the $\hat{V}$-model.

If $\nu = 2$ and J is the kernel of the finite difference Laplacean $\hat{J}$ is the two dimensional lattice Coulomb potential (which is only conditionally positive definite; see II.2) and we replace $\prod_{i\in\Lambda} d\lambda(q_i)$ by $\delta(0, \sum_{i\in\Lambda} q_i) \prod_{i\in\Lambda} d\lambda(q_i)$, where $\delta(m, n)$ is the Kronecker δ. (II.19)

Finally we introduce a $\hat{VV}$-model: $d\lambda(\frac{q_i}{2\pi})$ is as in (II.15) ($^{q}i/2\pi$ replacing a_i). We choose as an a priori measure on $\mathbb{R}^{4|\Lambda|}$ the measure

$$\prod_{i\in\Lambda} d\lambda(q_i)\delta(0, {*d*}\, q_i) \tag{II.20}$$

and as Hamilton function (action)

$$H(\{q\}_{\Lambda}) = \frac{1}{2} \sum_{i,j\in\Lambda} W_{\Lambda}(i-j)q_i q_j \ , \tag{II.21}$$

where W_{Λ} is the Green's function of the finite difference Laplacean on $\mathbb{Z}^4$ (the lattice Coulomb potential) with free boundary conditions at $\partial\Lambda$, q_i is a conserved vector charge. Using generalizations of the simple identities

$$\int e^{-\frac{1}{2}ax^2} e^{ibx} dx = \sqrt{\frac{2\pi}{a}}\, e^{-\frac{b^2}{2a}} \ , \tag{II.22}$$

and

$$\sum_{n\in\mathbb{Z}} e^{2\pi inx} = \sum_{m\in\mathbb{Z}} \delta(x-m) \tag{II.23}$$

we obtain

Theorem II.1:

1)[36] The V-model at inverse temperature β_V is isomorphic to the $\hat{V}$-model defined in (II.18) (II.19) at $\beta_{\hat{V}} = \beta_V^{-1}$; in particular the nearest neighbor V-model is mapped onto the $\hat{V}$-model with the Coulomb potential as couplings.

2)[27,28] The VV-model at inverse temperature β_{VV} is isomorphic to the $\hat{VV}$-model defined in (II.20) (II.21) at $\beta_{\hat{VV}} = \beta_{VV}^{-1}$.

In both cases, the partition function of the V-, resp. VV-model is the product of the partition function of the $\hat{V}$-, resp. $\hat{VV}$-model and a spin wave partition function $\det(-\sqrt{2\pi}\, J_\Lambda)^{-1}$. The $\hat{VV}$-model was first discussed in [67].

Remark: In Theorem II.1, 2) we recover (in a novel way) the Glimm-Jaffe approximation to the U(1)-model, [27]. See [27,28] for details.

II.2 Classical gases

Let x be a position vector in a configuration space $C = \mathbb{R}^\nu$ or $\mathbb{Z}^\nu$, and let q be a generalized charge, a vector in some topological vector space Q .

The potential between a particle with charge q at position x and one with charge q' at position x' is given by a function $V(q,x;\, q',x')$ on $(Q \times C)^2$ of positive type, satisfying translation invariance and

$$V(q,x;\, q',x') = -V(-q,x;\, q',x') = -V(q,x;\, -q',x') \quad \text{(II.24)}$$

and $V(q,x;\, q,x) \leq v < \infty$; see also [38].

The potential for n particles with parameters $W_i = (q_i,x_i)$, $i=1,\ldots,n$ is given by

$$U((W)_n) = \sum_{1\leq i<j\leq n} V(W_i;\, W_j) . \quad \text{(II.25)}$$

We let $d\lambda$ be a finite, positive measure on Q with $d\lambda(q) = d\lambda(-q)$. (II.26)

Set $d\lambda(q)_n = \prod_{i=1}^{n} d\lambda(q_i)$, $d(x)_n = \prod_{i=1}^{n} d^\nu x_i$.

The grand canonical partition function for the system in a bounded region $\Lambda \subset C$ is given by

$$\Xi_\Lambda(\beta,z) = \sum_{n=0}^{\infty} \frac{z^n}{n!} \int_{Q^n} d\lambda(q)_n \int_{\Lambda^n} d(x)_n \, e^{-\beta U((W)_n)}$$

where z is the activity (and $\int_{\Lambda^n} d(x)_n$ denotes the sum over all sites in $\Lambda^n = \Lambda^{xn}$, when $C = \mathbb{Z}^\nu$). Pressure $p_\Lambda(\beta,z)$ and correlation functions $\rho_\Lambda(\beta,z;\, W_1,\ldots,W_n)$ are defined in the usual way; see [39].

Examples:

II.6) Coulomb-type potentials

$Q = \mathbb{R}$ and $d\lambda(q) \overset{e.g.}{=} (\delta(q-1) + \delta(q+1))dq$, $V(q,x;\, q',x') = q\cdot q'\, W(x-x')$, where W is a potential on C of positive type and arbitrarily long range. The following cases are of special interest:

6.a) $C = \mathbb{Z}^\nu$, $\nu \geq 2$, W the Green's function of the finite difference Laplacean (i.e. the lattice Coulomb potential). When $\nu = 2$ W is not of positive type, but it is conditionally of positive type, i.e.

$$\sum \bar{c}_i\, c_j\, W(x_i - x_j) \geq 0 \, , \tag{II.28}$$

for arbitrary complex numbers $c_1, c_2, \ldots$ with $\sum_i c_i = 0$. This will suffice for a study of the neutral Coulomb gas; (see also (II.19)). As $z \to \infty$ this model converges to the nearest neighbor V-model at $\beta_V = \beta^{-1}_{\text{Coulomb}}$, [38].

6.b) $C = \mathbb{R}^\nu$, $\nu \geq 2$, W a regularized version of the Coulomb potential; ("ultraviolet cutoff").

6.c) $C = \mathbb{R}^2$, W the two dimensional Coulomb potential; see [40].

In cases 6.b), $\nu = 2$, and 6.c) the same comment as in 6.a), i.e. (II.28), applies.

II.7) Dipole potentials

$Q = \mathbb{R}^\nu$, $d\lambda(q) \overset{e.g.}{=} \delta(|q|-1)d^\nu q$, $V(q,x;\, q',x') = (q\cdot\nabla_x)(q'\cdot\nabla_{x'})W(x-x')$, with W a potential of positive type on $\mathbb{R}^\nu$ (or $\mathbb{Z}^\nu$), e.g. a (regularized) Coulomb potential such that $V(q,x;\, q,x) < \infty$.

II.3 Functional integrals

Let $\mathcal{H}_V$ be the Hilbert space of real functions $f, g, h, \ldots$ on $Q \times C$ with scalar product

$$\langle f,g\rangle_{\beta V} = \int d^\nu x\, d^\nu x'\, d\lambda(q) d\lambda(q') \overline{f(q,x)} \times \beta V(q,x;\, q',x') g(q',x') \tag{II.29}$$

Let ϕ be the Gaussian process with mean 0 and covariance βV, indexed by $\mathcal{H}_V$, and let $\langle - \rangle_{\beta V}$ denote the corresponding Gaussian expectation (given by a Gaussian measure $d\mu_{\beta V}$ with mean 0, covariance βV, defined on a suitable measure space; see [41]).

One defines Wick ordering with respect to $\langle - \rangle_{\beta V}$ by

$$:e^{i\phi(f)}: = e^{i\phi(f)} \langle e^{i\phi(f)} \rangle^{-1}_{\beta V} \tag{II.30}$$

We set

$$C_\Lambda \equiv -S_\Lambda \equiv \int_Q d\lambda(q) \int_\Lambda d^\nu x \; : \cos\phi : (q,x) \tag{II.31}$$

The following gives a connection between the quantities Ξ_Λ and ρ_Λ introduced in II.2 and (Gaussian) functional integrals.

Theorem II.2, [42,40,38]:

$$\Xi_\Lambda(\beta,z) = \langle \exp z\, C_\Lambda \rangle_{\beta V} \;, \quad \text{and}$$

$$\rho_\Lambda(\beta,z;\, W_1,\ldots,W_n) = \Xi_\Lambda(\beta,z)^{-1}\, z^n \times \langle \prod_{i=1}^{n} :e^{i\phi(W_i)}:\, e^{zC_\Lambda} \rangle_{\beta V} \qquad \text{(II.32)}$$

$$\equiv z^n \langle \prod_{i=1}^{n} :e^{i\phi(W_i)}: \rangle_\Lambda(\beta,z) \;.$$

Next we discuss a general inequality extending Ginibre's inequality for the rotator [43] and some recent inequalities of Park and the author [44,38].

Let ϕ and ϕ' be two isomorphic random fields indexed by a real Hilbert space $\mathcal{H}$ and distributed according to some measure $d\mu$. Let $\langle - \rangle_\mu$ denote expectation w.r. to $d\mu$. Let X be some measure space and $d\rho$ a finite measure on X. Let $x \longmapsto \ell_x$ be a measurable mapping $: X \longrightarrow \mathcal{H}_1 \subseteqq \mathcal{H}$. Define

$$C(\rho) = \int_X d\rho(x) \cos\phi(\ell_x) \;,$$

$$\langle - \rangle(\mu,\rho) \equiv \langle e^{C(\rho)} \rangle_\mu^{-1} \langle - \, e^{C(\rho)} \rangle_\mu \;, \text{ and}$$

$$\langle A;\, B\rangle(\mu,\rho) \equiv \langle AB\rangle(\mu,\rho) - \langle A\rangle(\mu,\rho)\cdot\langle B\rangle(\mu,\rho) \;.$$

Theorem II.3, [38]:

Let S be the class of random variables of the form $\prod_i [\cos\phi(f_i) \pm \cos\phi'(f_i)]$, $f_i \in \mathcal{H}_1$. Suppose that

$$d\mu(\phi)d\mu(\phi') \restriction S = d\nu\left(\frac{\phi+\phi'}{\sqrt{2}}\right) d\nu\left(\frac{\phi-\phi'}{\sqrt{2}}\right) \restriction S \;,$$

for some finite measure $d\nu$.

Then, for arbitrary f and g in $\mathcal{H}_1$,

1) $\langle \cos\phi(f)\rangle(\mu,\rho) \geqq 0$, (provided μ is of positive type)

2) $\langle \cos\phi(f);\, \cos\phi(g)\rangle(\mu,\rho) \geqq 0$

Remark: 2) implies e.g. that $\langle \cos\phi(f)\rangle(\mu,\rho)$ is monotone increasing in ρ. (II.33)

Application 1: Ginibre's inequalities [43] for the rotator, the $U(1)$ lattice gauge theory and the abelian Higgs model; (the latter two cases have been noticed in [45,37]. They are straightforward consequences of [43], resp. Theorem II.3).

Application 2: The inequalities of Park and the author [38] for the correlation functions of classical gases, e.g., for all f, g in $\mathcal{H}_V$,

$$\left.\begin{array}{l} \langle :e^{i\phi(f)}: \rangle_\Lambda(\beta,z) \geq 0 \\ \langle :e^{i\phi(f)}: \,;\, :\cos\phi(g): \rangle_\Lambda(\beta,z) \geq 0 \\ \langle :e^{i\phi(f)}: \rangle_\Lambda(\beta,z) \uparrow \text{ in } \Lambda \text{ and } z \end{array}\right\} \qquad (II.34)$$

(They follow from (II.30) and Thm. II.3:1), 2), (II.33), resp., by noting that $:e^{i\phi(f)}: = :\cos\phi(f): + i :\sin\phi(f):$ and that $\langle - \rangle_\Lambda(\beta,z)$ is invariant under $\phi \mapsto -\phi$).

II.4 Functional integrals and quantum field theory

Functional integrals are also used to construct relativistic quantum field models such as the $\phi^4_{2,3}$ - [46,47,48], $(\vec{\phi}\cdot\vec{\phi})^2_{2,3}$ - [49,50,51] or the sine-Gordon model [4,40,5,44], etc.

In the context of Euclidean field theory q labels different fields in the theory, x is a Euclidean space-time point; $\phi(q,x)$ is a generalized stochastic process with expectation $\langle - \rangle$ given by some probability measure on a space of generalized functions. The moments $\langle \prod_{i=1}^{n} \phi(q_i,x_i) \rangle$ are tentatively interpreted as the Euclidean Green (or Schwinger) functions of a relativistic quantum field theory satisfying all the Wightman axioms (except possibly uniqueness of the vacuum). Sufficient conditions for this interpretation to be correct have been given in a basic paper of Osterwalder and Schrader [52]. For detailed, rigorous information on the Euclidean description of relativistic quantum field theory and functional integrals see also [47,48] and refs. given there, and [31,53]. A formal version of the conditions of Osterwalder and Schrader [52,31] is as follows: Let $S(\phi) \equiv \int d^\nu x \; S(\phi(\cdot,x))$ be the classical Euclidean action of some field theory. Formally, the expectation $\langle - \rangle := \langle - \rangle_S$ is given by the Euclidean Gell'Mann-Low formula

$$"\langle - \rangle_S = \left[\int e^{-\hbar^{-1} :S_{ren.}(\hbar^{1/2}\phi):} \Pi \mathcal{D}\phi(q,x)\right]^{-1} \times \int - \; e^{-\hbar^{-1} :S_{ren.}(\hbar^{1/2}\phi):} \Pi \mathcal{D}\phi(q,x)" , \qquad (II.35)$$

where $\hbar$ is (proportional to) Planck's constant, the double colons denote some Wick order (depending on the curvature of the classical action at some of its absolute minimas), and $S_{ren.}$ is the renormalized action (possibly including ∞ (for $\hbar > 0$) counterterms).

Quasi-Theorem II.4:

Suppose (II.35) can be given a rigorous meaning, and

1) $:S_{ren.}(\hbar^{1/2}\phi):$ is Euclidean invariant;

2) if θ represents Euclidean time reflection on random variables

(with expectation $<\!-\!>_S$) then $\overline{\theta :S_{ren.}(\hbar^{1/2}\phi(\vec{x},t)): = :S_{ren.}(\hbar^{1/2}\phi(\vec{x},-t)):}$ with $(\vec{x},t) \equiv x$.

3) $<e^{\phi(f)}>_S$ exists for a suitable class of test functions. Then the moments of $<\!-\!>_S$ exist and are the Schwinger functions of a unique relativistic quantum field theory.

For
$$S(\phi) = S_o(\phi) \equiv \frac{1}{2}\left[(\nabla\phi)^2 + m^2\phi^2\right], \qquad \text{(II.36)}$$

$<\!-\!>_S$ ("given" by (II.35)) is the Gaussian expectation with mean 0 and covariance $(-\Delta+m^2)^{-1}$, the kernel of which is the Yukawa-($m > 0$), resp. the Coulomb potential ($m=0$). (II.37)

Its moments are the Schwinger functions of the free field. For $m = 0$ and

$$S(\phi) = S_o(\phi) + \lambda \cos\phi \qquad \text{(II.38)}$$

we obtain the sine-Gordon theory. If we compare (II.35) - (II.38) with (II.30)-(II.32) and with model II.6.c) we obtain

Theorem II.5:

1) For $\hbar < 4\pi$ ("no ultraviolet divergences") the sine-Gordon theory is isomorphic to the two dimensional, two component Coulomb gas II.6.c) [40], and

2) The inequalitites of Theorem II.3 apply to the Schwinger functions of the fields $:e^{i\phi}:$ of the sine-Gordon theory, [44].

Rigorous connections between the standard ϕ^4_ν - and $(\vec{\phi}\cdot\vec{\phi})^2_\nu$-models $(\vec{\phi} = (\phi_1,\phi_2))$ and the classical N-vector models, example II.1), due to [48,50,54], are by now well known.

$$\left.\begin{aligned} &\text{The case } \nu = 2,\ S_o(\phi) = \sum_{i,j\in\mathbb{Z}^2} J(i-j)\phi_i\phi_j \\ &S(\phi) = S_o(\phi) + \lambda\sum_i \cos(\phi_i - \theta),\ \theta \in [0,2\pi], \end{aligned}\right\} \qquad \text{(II.39)}$$

with $J^{-1} = q^2V_C + W$, where V_C is the two dim. lattice Coulomb potential and W is a positive type potential of very short range, gives an approximate description of the Euclidean vortex (magnetic) field in the abelian Higgs model in two space-time dimensions if q is chosen to be the ratio of the electric charge of a massless fermion Dirac field and the Higgs scalar [7]. The interaction term $\lambda\sum_i\cos(\phi_i - \theta)$ comes from the instantons of this model: the Nielsen-Olesen vortices[55]. For $q = 0$ the angle θ parametrizes the θ-vacua first described in [11,12]. Using Theorem II.2 one sees that our approximation is obviously a version of the dilute gas approximation of Polyakov [9] and others [10,13]. In this form it is proposed and discussed in [7]. The lattice Higgs model in the Villain approximation leads to a similar effective field theory, but there are some important differences.

III Critical points, critical phenomena, long range forces

III.1 Critical point in the N-vector and $|\vec{\phi}|^4$-models

First we consider the N-vector models defined in Section II.1, for $N = 1, 2, 3$, i.e. the Ising, the rotator and the classical Heisenberg model. They are known to satisfy the Lee-Yang theorem of [56,50]. From this we get

Theorem III.1:

1) For $\mathrm{Re}\, h \neq 0$ the correlation functions $< \prod_{i \in A} S_i^{\alpha_i} > (\beta,h)$ are analytic in h and real analytic in β in an h-dependent neighborhood of $(0,\infty)$. Periodic and free boundary conditions coincide in the thermodynamic limit.

2) For real $h \neq 0$, $m(\beta,h) > 0$, and $m(\beta,h) = O(h)$ when $m(\beta,0) = 0$. For real $h \neq 0$ the exponential decay rate of all truncated correlations is uniformly bounded away from 0.

This theorem has been derived in [49] as a consequence of the Lee-Yang theorem. It holds for the $(\vec{\phi}\cdot\vec{\phi})^2_{2,3}$-models $(\vec{\phi}=(\phi^1,\ldots,\phi^N)$, $N = 1, 2, 3)$, too [49]. Part 2) extends earlier results of [57,58].

Open problem: Do Lee-Yang theorem and Theorem III.1 remain true for $N > 3$?

Theorem III.2:

1) For $N = 1$, $\nu \geq 2$, and for $N = 2, 3$, $\nu \geq 3$, there exists $\bar{\beta}_c < \infty$ such that, for $\beta > \bar{\beta}_c$, $\lim_{h\to 0} <S^1> (\beta,h) \neq 0$ (spontaneous magnetization).

2) For $h = 0$, $N = 1, 2, 3, 4$ there exists $\underline{\beta}_c \leq \bar{\beta}_c$ (with $\bar{\beta}_c < \infty$ for $N = 1$, $\nu \geq 2$ and $N = 2, 3, 4$, $\nu \geq 3$) such that $\lim_{\beta\uparrow\underline{\beta}_c} m(\beta) = 0$, $\lim_{\beta\uparrow\underline{\beta}_c} \chi(\beta) = \infty$.

For $\nu \geq 3$ the expectation $<\!-\!>(\underline{\beta}_c) \equiv \lim_{\beta\uparrow\underline{\beta}_c} <\!-\!>(\beta)$ is clustering. I.e. there exists a critcal model with mass = 0, infinite susceptibility, but no long range order.

3) If $<\vec{S}_0\cdot\vec{S}_j> (\underline{\beta}_c) \approx |j|^{-(d-2+\eta)}$, as $|j| \to \infty$, then $0 \leq \eta \leq 2$.

Remarks and comments: 1) For $N = 1, 2$ $m^{-1}(\beta)$ and $\chi(\beta)$ are monotone increasing in β (for $h=0$): A consequence of Theorem II.3, resp. [43]. 2) For $N \geq 2$, part 1) follows from infrared domination [59]; for more details concerning this and related results for these and a class of quantum models see [60,61] and Lieb's contribution to these proceedings. Part 2) is based on infrared bounds [59] and the "Lebowitz inequalities" $<S_i^\alpha;\, S_j^\alpha;\, S_\ell^\alpha;\, S_m^\alpha> (\beta) \leq 0$, and

$$<S_i^\alpha\, S_j^\alpha;\, S_\ell^\gamma\, S_m^\gamma> (\beta) \leq 0$$

proven in [62,63]. See [64,65,21] for $N = 1$, and [28] for $N = 2, 3, 4$. Finally 3) follows from infrared domination [59] and 2); see [59, 21].

3) Clearly $\overline{\beta}_C = \infty$ for $N \geq 2$ and $\nu = 2$.

4) Theorem III.2 extends to $(\vec{\phi}\cdot\vec{\phi})^2_{2,3}$(N=1,2,3) for which the following additional results hold: Absence of two particle bound states [66,63] and existence of N degenerate elementary particles [21] (for almost every physical mass > 0) at $h = 0$, in the one phase region.

Open problems:

1) Show (or disprove that $\underline{\beta}_C = \overline{\beta}_C$ for $\nu \geq 3$, (or $\nu = 2$ and $N \geq 3$, [67]) and that $\lim_{\beta \downarrow \overline{\beta}_C} \langle S_i^1 \rangle (\beta, 0+) = 0$; (for $N = 1$ this would imply $m(\beta, 0+) \downarrow 0$, as $\beta \downarrow \overline{\beta}_C$, [31]).

2) Prove (or disprove) the existence of a Euclidean invariant scaling limit (and hence of an associated relativistic quantum field theory, e.g. [32,53]) for <—> $(\underline{\beta}_C)$. For the $\nu = 2$ Ising model this problem has been partially solved in [69] by rather direct, very difficult calculations. A proper, general and rigorous understanding of the scaling limit is however still missing; see e.g. [53].

3) Does the scaling limit in $\phi^4_{2,3}$ [23] teach us something about non-super renormalizable ultraviolet divergences and triviality or non-triviality of ϕ^4_4 [1,2,53] ?

Theorem III.3:

Let $N = 2, 3, \ldots$ and $\nu = 2$.

1) [70] For arbitrary $\varepsilon > 0$ there exists $\beta_o(\varepsilon) < \infty$ such that for all $\beta > \beta_o(\varepsilon)$

$$\langle \vec{S}_o \cdot \vec{S}_j \rangle (\beta) \leq \text{const.} \cdot |j|^{-[(2\pi+\varepsilon)\beta]^{-1}}$$

2) [51] $m(\beta) \leq \text{const. } e^{-\text{const.}\beta N^{-1}}$, for $\beta \gg 1$.

Remark: This result has been extended to all truncated correlations in [63]. A new proof of 1) has been found in [71]. Part 2) also holds for the field theory case [72], and it seems to us that the methods of [71] presumably give 1) for the field theory case, too.

Conjecture III.4, [67,36,73]:

For $N \geq 3$, $\nu = 2$ $\underline{\beta}_C$ is infinite, for $N = 2$, $\nu = 2$ $\underline{\beta}_C$ is finite, more specifically, the bound of Theorem III.3.1) is saturated for $N = 2$, and the factor "N^{-1}" in the exponent on the r.h.s. of 2) can be replaced by $(N-2)^{-1}$, provided β is very large; see also [68].

Obvious open problem: Prove Conjecture III.4.

A proof would be an impressive and promising beginning in our understanding of higher order phase transitions and critical phenomena.

III.2 Gases with long range forces

Theorem III.5, [38]:

In the notations and under the hypotheses of Section II.2

$$\lim_{\Lambda\uparrow C} p_\Lambda(\beta, z) \equiv p(\beta, z) \quad \text{and, for all } n, \qquad \text{(III.1)}$$

$$\lim_{\Lambda\uparrow C} \rho_\Lambda(\beta,z; W_1,\dots,W_n) \equiv \rho(\beta,z; W_1,\dots,W_n) \qquad \text{(III.2)}$$

exist and are independent of $\{\Lambda\}$. Moreover, the correlation functions $\rho_{(\Lambda)}(\beta,z;\cdots)$ are monotone increasing in z, the Fourier transform of the "effective potential function" $\langle\hat\phi(q, k)\hat\phi(q,-k)\rangle\ (\beta, z)$ is monotone decreasing in z and bounded above by $\beta\hat V(q,k; q,-k)$, (its value for $z = 0$!). (III.3)

This Theorem is a direct consequence of Theorems II.2 and II.3. It is proven in [38], where it has also been shown to be true for the corresponding quantum gases with "Boltzmann statistics". To our knowledge this is the first existence theorem for thermodynamic and correlation functions valid for potentials of arbitrarily long range and for all positive β, z. Part (III.1) extends to certain gases with statistics and to potentials that include hard cores ; (III.3) is an infrared bound to be compared with the one of [59]. Under various additional assumptions it implies clustering of the correlation functions in the thermodynamic limit [38].

Corollary III.6:

The thermodynamic limit of the pressure (resp. vacuum energy density) and all correlation (resp. Schwinger) functions of the following models exists and is shape independent:

1) The two dimensional Coulomb gas-example II.6.c)-above collapse temperature [40], equivalently (see Theorem II.5), the $\nu = 2$ sine-Gordon theory for $\hbar < 4\pi$; see [44]. The "bosonized" [74,5] $\nu = 2$ Yukawa - and a model for $\nu = 2$ QED of massive fermions and massive photons [4]; see [6].

2) The classical gases of examples II.6.a) - c) and II.7), the V- and VV-models (examples II.2), II.6.a), II.5)), and the isomorphic $\hat V$- and $\hat{VV}$-models; (see Theorem II.1).

3) The rotator [43], the U(1) lattice gauge theory [45,37] and a class of abelian Higgs models on the lattice [37], resp. their Villain approximation [28].

Remarks:

1) Cor. III.6 is a direct consequence of Thms. II.2, II.3 [38,28]. It would be of considerable interest to prove Thm. II.3 resp. Cor. III.6.1) for the Yukawa model in the Matthews-Salam-Seiler [75] representation.

2) Cor. III.6.2) adds information to the study of the VV-resp. the $\hat{VV}$-approximation

to the U(1) (resp. the abelian Higgs [28]) lattice gauge theory complementary to the (deeper!) one of Glimm and Jaffe [27].

Next we state a beautiful result due to Brydges [76].

Theorem III.7:

For the lattice Coulomb gas, example II.6.a) in ν dimensions exponential Debye screening is valid in a region of high enough temperature and activity approximately given by the scaling properties of the corresponding continuum Coulomb gases.

Remarks: 1) This result is a lot more difficult to prove than a corresponding result that affirms exponential Debye screening for the $\hat{V}$- and $\hat{V}V$-models which are isomorphic to special types of Coulomb gases (Thm. II.1). Brydges' methods [76] are based on the difficult "expansion in phase boundaries" due to Glimm, Jaffe and Spencer [77] (which can be applied to this problem thanks to Thms. II.2, II.5 [40], whereas in the case of $\hat{V}$-resp. $\hat{V}V$-models standard Peierls arguments (convertable into high temperature expansions) suffice.

2) Brydges' methods apply to a larger class of lattice gases than the one he considers: If $W^{-1}(m)$ is of exponential decrease in $|m|$ one always gets screening in some range of (high) temperatures and activities. This may show that screening (in particular Debye screening) may not really depend too much on special properties of the Coulomb potential (such as Newton's theorem [78]).

3) Applied to the $\nu = 2$ lattice Coulomb gas (example II.6.a)) Brydges' results give Debye screening only for high temperatures. The obvious open problem is thus: Is there a $\beta_C < \infty$ such that for $\beta \geq \beta_C$ Debye screening disappears, e.g. in the sense that the susceptibility (defined in terms of the effective potential function) is infinite? The following inequalities are relevant to this problem.

Theorem III.8 [38,28]:

Let $\langle\hat{S}(k)\hat{S}(-k)\rangle_V(\beta)$ be the two point function of the nearest neighbor V-model and $\langle\hat{\phi}(k)\hat{\phi}(-k)\rangle\ (\beta, z)$ the effective potential function of the lattice Coulomb gas, in ν-dim. momentum space. Then

1)
$$\langle\hat{S}(k)\hat{S}(-k)\rangle_V(\beta^{-1}) = \langle\hat{\phi}(k)\hat{\phi}(-k)\rangle\ (\beta, z = \infty)$$
$$\leq \langle\hat{\phi}(k)\hat{\phi}(-k)\rangle\ (\beta, z) \leq O(\beta k^{-2}).$$

2) In any V-model $\langle\hat{S}(k)\hat{S}(-k)\rangle_V(\beta^{-1})$ is monotone increasing in β.

3) in the VV-model $\langle|da(k)|^2\rangle_{VV}(\beta^{-1})$ is monotone increasing in β and $\leq O(\beta)$.

Remark: The V- and the VV-model satisfy "inverse" Lebowitz inequalities [38]. This theorem follows from Theorems II.2, II.3, III.5.

Some consequences of it are:

If the susceptibilities $\chi_V(\beta)$ and $\chi_{VV}(\beta)(= [k^{-2}<|da(k)|^2>_{VV}(\beta)]_{k=0})$ of the V-resp. VV-model are infinite for some $\beta = \beta_C$ they are infinite for all $\beta \leq \beta_C$, and screening in the $\hat{V}$- resp. $\hat{VV}$-model disappears. Furthermore

$$<\hat{\phi}(0)\hat{\phi}(0)>(\beta,z) \geq \chi_V(\beta^{-1}), \forall (\beta,z),$$

so that the ν dim. lattice Coulomb gas II.6.a) has a higher order phase transition (break down of Debye screening) if the V-(resp. $\hat{V}$-) model have one.

Furthermore, the connection between the $\nu = 2$ V-model and the $\nu = 2$ rotator suggests [36,73] that the $\nu = 2$ rotator has a higher order phase transition, provided the $\nu = 2$ V-model has one. Motivated by the results of Section III, the renormalization group [36] and [67,27] we make the following

Conjecture III.9, [36,67,73,27]:

1) The $\nu = 2$ rotator, V-model and lattice Coulomb gas II.6.a) have a critical interval (β_C, ∞), $\beta_C < \infty$ on which $m(\beta) = 0$ and the susceptibility is infinite; (see also Thm. III.2.2)).

2) The $\nu = 3$ U(1) lattice gauge theory [27], the $\nu = 3$ V-model and, presumably, the $\nu = 3$ lattice Coulomb gas have some form of screening for all $\beta < \infty$. However, the $\nu = 3$ VV-model has a phase transition [28].

3) The $\nu = 4$ U(1) lattice gauge theory and the $\nu = 4$ VV-model [27] have a phase transition of the form described in 1).

Remarks: 1) From Theorem III.8 (see also Thm. III.2.2)) we know that it suffices to exhibit one β_C such that $m(\beta_C) = 0$, $(\chi(\beta_C) = \infty)$. 2) Results of Glimm and Jaffe [27,79] will probably soon provide a proof of 3) and possibly of 1), too. See also [67].

An important open problem is to rigorously investigate the scaling properties of these models for $\beta \geq \beta_C$ (once $\beta_C < \infty$ is established) and their continuum limits.

Finally we come to some comments on the $\nu = 2$ abelian Higgs model in the approximation of Section II.4, (II.39). (We refer the reader to II.4 for the definition of the approximate, effective field theory for the vorticity, the angle θ and the charge ratio q). In the approximation (II.39) to the $\nu = 2$ continuum abelian Higgs model the following results are rigorous [7,28,6]:

1) For $q = 0$ and $\theta = 0$ the expectation $<\text{—}>_S$ has cluster properties (follows from (III.3) and [80]), i.e. the vacuum is unique [52], for all $\hbar$ and λ. For small $\hbar$ and all λ or small λ and all $\hbar$ they are exponential.

2) For $q = 0$ and $\theta = \pi$ we find, for small $\hbar$ and large λ, a first order phase transition in the vorticity and two different vacua (corresp. to $\pi+0$, $\pi-0$) with opposite spontaneous vorticity; (this follows from a Peierls argument [81,18,61]).

We note that 1) also holds for an abelian Higgs model on the lattice [16], resp. its

Villain approximation [28] (which we call H-V model).

The standard lattice Higgs, resp. H-V model only gives an analogue of the $\theta = 0$ vacuum. However a modified lattice model [28] gives the θ-vacua; it has a first order phase transition at $\theta = \pi$, and the existence of two different Gibbs expectations (with opposite spontaneous "vorticity") can be proven rigorously in some range of coupling constants [28]. In particular, the usual Higgs mechanism can be proven to occur for arbitrary coupling constants only for the $\theta = 0$ H-V model. At the critical point of the $\theta = \pi$ theory the Higgs mechanism breaks down. A heuristic approximation to a non-abelian Higgs model with instantons in four space-time dimensions (similar in spirit to the approximation (II.39) of Section II.4 with $q = 0$ - one may think of an SU(2) Higgs model [8] without fermions) suggests that a first order phase transition accompanied by spontaneous instanton density and the break down of the Higgs mechanism at the critical point may be typical features of the $\theta = \pi$ theory; see [28]. However, for the three - or more dimensional (abelian) H-V models there is only one vacuum (equilibrium state), and the Higgs mechanism occurs for arbitrary values of the coupling constants [28].

3) For the two dimensional, abelian Higgs model coupled to massless fermions in the approximation II.4, (II.39), i.e. for $q > 0$, and for $0 < \hbar q^2 << 1$ and large enough λ a slight variation of Brydges' results [76] (see Remark 2 following Theorem III.7) gives exponential screening which can be interpreted as dynamical mass generation for the fermions [13,7]. If Conjecture III.9.1) is true this dynamical mass generation disappears at large value of $\hbar q^2$ (which plays the role of β in the $\nu = 2$ lattice Coulomb gas II.6.a));the collapse phenomenon encountered in model II.6.c) [40] and the renormalization group [36] predict $\hbar q^2 \gtrsim 4\pi$ as the critical value).

4) For arbitrary $q > 0$ chiral invariance is broken, and fermionic charges which are not an integer multiple of the charge of the Higgs scalar are confined. See [13,7].

Remark: The resulting picture for the $\nu = 2$ continuum abelian Higgs model has similarities to the one found for $\nu = 2$ massive QED in [82,5].

Open problems:

1) Construct the $\nu = 2$ continuum abelian Higgs model rigorously and show the approximation II.4, (II.39) discussed above gives qualitatively correct results.

2) Which version of gauge-invariant lattice approximation to this model has a continuum limit equal to it?

3) Does the $\nu = 2$ Higgs model teach us something about more interesting gauge theories with instantons?

References

General:

I. Constructive Quantum Field Theory, G. Velo and A. S. Wightman, eds., Lecture Notes in Physics 25, Springer-Verlag, Berlin-Heidelberg-New York, 1973.

II. Les Méthodes Mathématiques de la Théorie Quantique des Champs, F. Guerra, D. W. Robinson, R. Stora, eds., éditions du C.N.R.S., Paris, 1976.

III. Proceedings of the conference on "Quantum Dynamics: Models and Mathematics", Acta Phys. Austriaca Suppl. XVI, (1975).

IV. Proceedings of the Cargèse summer school in theoretical physics, 1976; to appear.

Special:

1. R. Schrader, in II; Commun. math. Phys. 49, 131, (1976) and 50, 97 (1976); Phys. Rev. B14, 172, (1976).
2. J. Glimm and A. Jaffe, 1. Commun math. Phys. 52, 203, (1977), 2. Commun. math. Phys. 51, 1, (1976), 3. Phys. Rev. D11, 2816, (1975), 4. Phys. Rev. Letters 33, 440, (1974).
3. 1. J. Fröhlich, Ann. Phys. 97, 1, (1976).
   2. J. Glimm and A. Jaffe, contributions to II-IV.
4. J. Fröhlich, in "Renormalization Theory", G. Velo and A. S. Wightman, eds., Nato Adv. St. Inst. Series C, Reidel, Dordrecht-Boston, 1976.
5. J. Fröhlich and E. Seiler, Helv. Phys. Acta 49, 889, (1976).
6. J. Fröhlich and Y. M. Park, in preparation.
7. J. Fröhlich, Erice lectures 1977, to appear (G. Velo and A. S. Wightman, eds.).
8. S. Coleman, proceedings of the 1973 Int. School on Subnuclear Physics "Ettore Majorana", A. Zichichi (ed.), Academic Press, New York, 1975.
9. A. M. Polyakov, Nuclear Physics B 20, 429, (1977), and refs. given there.
10. G. 't Hooft, Phys. Rev. D14, 3432, (1976), and refs. given there.
11. C. Callan, R. Dashen and D. Gross, Physics Letters 63B, 334,(1976).
12. R. Jackiw and C. Rebbi, Phys. Rev. Letters 37, 172, (1976).
13. C. Callan, R. Dashen and D. Gross, Princeton preprint COO-2220-115, (1977).
14. K. Wilson, Phys. Rev. D10, 2445, (1974).
15. R. Balian, J. M. Drouffé and C. Itzykson, Phys. Rev. D10, 3376, (1974), D11, 2098, (1975), D11, 2104, (1975).
16. K. Osterwalder and E. Seiler, "Gauge Field Theories on the Lattice", Harvard preprint, 1977. See also E. Seiler's contribution to these proceedings.
17. J. Fröhlich, Commun. math Phys. 47, 269, (1976).
18. J. Fröhlich, Acta Phys. Austriaca, Suppl. XV, 133, (1976).
19. J. Béllissard, J. Fröhlich and V. Gidas, Phys. Rev. Letters 38, 619, (1977); paper in preparation.
20. R. Haag and D. Kastler, J. Math. Phys. 5, 848, (1964). S. Doplicher, R. Haag and J. Roberts, Commun. math. Phys. 23, 199, (1971), 35, 49, (1974), and refs. given there.
21. J. Roberts, Commun. math. Phys. 51, 107, (1976).
22. S. Coleman, to appear in the proceedings of the 1975 Int. School on Subnuclear Physics "Ettore Majorana", A. Zichichi (ed.).
23. J. Gervais, proceedings of the XVIth Schladming winter school, to appear in Acta Phys. Austriaca, Suppl.
24. K. Hepp, Commun. math. Phys. 35, 265, (1974).

25. J.-P. Eckmann, "Remarks on the Classical Limit of Quantum Field Theories", Geneva, preprint, 1976.

26. E. H. Lieb and F. Y. Wu, in "Phase Transitions and Critical Phenomena", C. Domb and M. S. Green (eds.), Academic Press, London-New York, 1972; (p.p. 332-490).
R. J. Baxter, "Solvable Eight Vertex Model on an Arbitrary Planar Lattice", to appear in Proc. Roy. Soc., and refs. given there.
E. H. Lieb, Physica 73, 226, (1974).
K. Hepp and E. H. Lieb, in I., and refs. given there.

27. J. Glimm and A. Jaffe, Physics Letters 66B, 67, (1977) and "Instantons in a U(1) Lattice Gauge Theory : A Coulomb Dipole Gas", Harvard preprint, 1977.

28. J. Fröhlich, paper on critical phenomena, in preparation.

29. D. Buchholz, Commun. math. Phys. 52, 147, (1977).

30. J. Fröhlich, "Asymptotic Condition and Structure of Scattering States in ···", unpubl. report, 1974.

31. J. Fröhlich, Ann. Inst. H. Poincaré 19, 1, (1973).

32. J. Fröhlich, R. Israel, E. H. Lieb and B. Simon, to appear; see also J. Fröhlich, Bulletin A.M.S., to appear; B. Simon, to appear in the proceedings of the 1977 Haifa conference on Stat. Physics; E. H. Lieb, these proceedings.

33. J. L. Lebowitz and A. Martin-Löf, Commun. math Phys. 25, 276, (1972); J. L. Lebowitz, these proceedings, and refs. given there.

34. A. Messager, S. Miracle-Sole, C. Pfister, private communication by S. Miracle-Sole.

35. J. Bricmont, J. R. Fontaine and L. J. Landau, Louvain, preprint 1977, to appear in Commun. math. Phys.

36. J. V. José, L. P. Kadanoff, S. Kirkpatrick and D. R. Nelson, IBM Research Report R C 6428 (27401), 1977, and refs. given there.
J. M. Kosterlitz and D. J. Thouless, J. Phys. C6, 1181, (1973). J. M. Kosterlitz, J. Phys. C7, 1046, (1974). J. Zittartz, Z. Phys. B23, 55, (1976), B23, 63, (1976). See also refs. 67, IV and E. Brézin and J. Zinn-Justin, Phys. Rev. B14, 3110, (1976).

37. G. F. DeAngelis, D. de Falco and F. Guerra, Physics Letters 68B, 255, (1977); Lettere al Nuovo Cimento, to appear. Also F. Guerra's contribution to these proceedings.

38. J. Fröhlich and Y. M. Park, Princeton preprint, 1977.

39. D. Ruelle, "Statistical Mechanics", W. A. Benjamin, Reading-London-Amsterdam-Don Mills-Sydney-Tokyo, 1969.

40. J. Fröhlich, Commun. math. Phys. 47, 233, (1976).

41. E. Nelson, in ref. I.

42. S. Edwards and A. Lenard, J. Math. Phys. 3, 778, (1962).
S. Albeverio and R. Høegh-Krohn, Commun. math. Phys. 30, 171, (1973).

43. J. Ginibre, Commun. math. Phys. 16, 310, (1970).

44. Y. M. Park, Seoul, preprint 1977, to appear in J. Math. Phys.

45. E. Seiler, private communication and these proceedings.

46. See the contributions of Nelson, Glimm and Jaffe, Glimm, Jaffe and Spencer, Guerra, Rosen and Simon, Osterwalder to refs. I-IV and their refs. to the original articles. See also J. Glimm and A. Jaffe, in "Statistical Mechanics and Quantum Field Theory", Les Houches 1970, C. de Witt and R. Stora (eds.), Gordon and Breach, New York 1971. In addition: J. Magnen and R. Sénéor, Ann. Inst. H. Poincaré 24, 95, (1976), and Harvard preprint, 1977. J. Feldman and K. Osterwalder, Ann. Phys. 97, 80, (1976). Refs. 81 and 58, 59.

47. See ref. I.

48. B. Simon, The $P(\phi)_2$ Euclidean (Quantum) Field Theory, Princeton University Press, Princeton, N. J., 1974.

49. J. Fröhlich, in ref. II and 17.

50. F. Dunlop and C. M. Newman, Commun. math. Phys. 44, 223, (1975).

51. J. Fröhlich and T. Spencer, in ref. IV.

52. K. Osterwalder and R. Schrader, Commun. math. Phys. 31, 83, (1973) and 42, 281, (1975). (See also V. Glaser, Commun. math. Phys. 37, 257, (1974)).

53. J. Glimm and A. Jaffe, in ref. IV.

54. B. Simon and R. Griffiths, Commun. math. Phys. 33, 145, (1973).

55. H. Nielsen and P. Olesen, Nuclear Physics B61, 45, (1973).

56. T. D. Lee and C. N. Yang, Phys. Rev. 87. 410, (1952).

57. J. L. Lebowitz and O. Penrose, Commun. math. Phys. 39, 165, (1974).

58. F. Guerra, L. Rosen and B. Simon, Commun. math Phys. 41, 19, (1975).

59. J. Fröhlich, B. Simon and T. Spencer, Commun. math Phys. 50, 79, (1976).

60. F. Dyson, E. H. Lieb and B. Simon, Phys. Rev. Letters 37, 120 (1976), and paper in preparation.

61. J. Fröhlich and E. H. Lieb, Phys. Rev. Letters 38, 440, (1977); and in preparation.

62. J. L. Lebowitz, Commun. math. Phys. 35, 87, (1974).

63. J. Bricmont, Louvain, preprint 1977.

64. J. Rosen, Adv. Math., to appear.

65. O. McBryan and J. Rosen, Commun. math. Phys. 51, 97, (1976).

66. T. Spencer, Commun. math. Phys. 39. 77, (1974).

67. A. M. Polyakov, Physics Letters 59B, 79, (1975) and 59B, 82, (1975).

68. E. Brézin and J. Zinn-Justin, in IV; see also ref. 36.

69. B. M. McCoy, C. A. Tracy and T. T. Wu, Phys. Rev. Letters 38, 793, (1977). See also: D. Abrahams, Physics Letters (1977).

70. O. McBryan and T. Spencer, Commun math. Phys. 53, 299, (1977).

71. H. Kunz (private communication through B. Simon).

72. A. Klein and L. J. Landau, in ref. II.

73. O. McBryan and T. Spencer, private communication.

74. S. Coleman, Phys. Rev. D11, 2088, (1975).

75. E. Seiler, Commun. math. Phys. 42, 163, (1975).

76. D. Brydges, Rockefeller preprint, 1977.

77. J. Glimm, A. Jaffe and T. Spencer, Ann. Physics 101, 610 and 631, (1976).

78. E. H. Lieb and J. L. Lebowitz, Adv. Math. 9, 316, (1972); see also : E. H. Lieb, Rev. Mod. Physics 48, 553, (1976).

79. J. Glimm and A. Jaffe, private communication.

80. B. Simon, Commun math Phys. 31, 127, (1973), and ref.(to J.L.Lebowitz) given there.

81. J. Glimm, A. Jaffe and T. Spencer, Commun. math. Phys. 45, 203, (1975).

82. S. Coleman, R. Jackiw and L. Susskind,Ann. Phys 93, 267, (1975); S. Coleman Ann. Phys. 101, 239, (1976)

NEW PROOFS OF LONG RANGE ORDER

Elliott H. Lieb*
Departments of Mathematics and Physics
Princeton University
Princeton, N.J. 08540, USA

I. Introduction

I will try to sketch some ideas and techniques (reflection positivity and infrared bounds) that have been developed recently to prove the existence of phase transitions in lattice systems, both classical and quantum and with discrete or continuous symmetry. Specifically I have in mind the work of Fröhlich, Simon and Spencer (FSS)[1] on the ≥ 3 dimensional classical Heisenberg model, Dyson, Lieb and Simon (DLS)[2] on the ≥ 3 dimensional quantum Heisenberg model, Fröhlich, Israel, Lieb and Simon (FILS)[3] on classical models (all dimensions) with long range interactions, Fröhlich and Lieb (FL)[4] on the anisotropic 2 dimensional quantum and classical Heisenberg models.

In addition there is some new work (after the M∩Ø meeting) by Heilmann and Lieb (HL)[5] on models of a liquid crystal. Along these lines I will also discuss how the new technology greatly simplifies the beautiful, but complicated, proofs of Heilmann(H)[6] of the phase transition in the 2 dimensional, nearest neighbor, hard core model on the hexagonal and triangular lattices.

By "proof of phase transition" is meant a proof that long range order (LRO) exists at sufficiently low temperatures (or high density in the case of hard core models). In all the models to be discussed, it is easy to prove (by means of high temperature or virial expansions) that such order does not exist at high temperature (or low density).

These notes are not meant to be complete in any sense. They are meant to be light reading, and to give the reader a glimpse of the past and present status of the field.

II. Summary of Results for Spin Systems

To give a flavor of the sort of results that can be obtained, we consider here Heisenberg type models with Hamiltonian

$$H = \mp\ |S|^{-2} \sum_{x,y} J(x-y)\ S_x \cdot S_y$$

$J(x) \geq 0$, - is the ferromagnet, + is the antiferromagnet, and S is an N (=1,2,3) component spin. $|S|$ is the magnitude of the individual spins.

One Dimension: For short range interactions there is no LRO. This can be proved by the Mermin, Wagner, Hohenberg (MWH)[7,8] method or, in the classical case, by the transfer matrix method of Van Hove[9]. For long range interactions with $J(x) \sim |x|^{-\alpha}$, $1 < \alpha < 2$ (one needs $\alpha > 1$ for the thermodynamic limit) Dyson[10] invented the heierarchical model and used Griffiths inequalities to prove LRO for the ferromagnetic Ising model. FILS extend this to any J such that $k(x,y) = J(x+y-1)$, for $x,y = 1,2,\ldots,\infty$ is a positive definite matrix. $|x|^{-\alpha}$, $1 < \alpha < 2$, is included. The FILS result also includes the N vector case (classical). In fact the FILS proof extends to > 1 dimensions if $J(x_1+y_1-1, x_2-y_2, \ldots, x_n-y_n)$ is positive definite on $x_1, y_1 \geq 1$, $-\infty < x_j, y_j < \infty$, and similarly for the other n-1 coordinates. The FILS proof also extends to some quantum cases.

Two Dimensions: If $N \geq 2$, classical or quantum, there is no LRO by the MWH argument. If N=1 (Ising) there is LRO by the Peierls argument[11]. In the N=3 anisotropic case (in which $S \cdot T$ is replaced by $\alpha(S^x T^x + S^y T^y) + S^3 T^3$, $\alpha < 1$) one expects LRO. Malyshev[12] proved this in the classical case by an extension of the original Peierls argument (earlier, Bortz and Griffiths[13] proved it for small α). In the quantum case, Ginibre[14] and Robinson[15] proved it for small α.

FL use the Peierls argument and reflection positivity (RP) (cf. section VI) to prove:

Antiferromagnet: For each $|S|$ (in the quantum case) there is an $\alpha_c(|S|)$ such that the model with $\alpha < \alpha_c(|S|)$ has LRO for low temperature. Furthermore, $\alpha_c(|S|) \to 1$ as $S \to \infty$. (Note: $|S| = \infty$ is the same as the classical model[16].) An open problem is to extend this to all $\alpha < 1$ for all $|S|$.

Ferromagnet: FL claimed there is LRO at low temperature for all $\alpha < 1$. This is not justified because it was assumed that RP holds for the ferromagnet. If RP holds then the rest of the FL proof is valid. It is not known whether RP (in the weak sense that it is used by FL) holds; the assumption was based on the DLS work, which also has the same error.

Three Dimensions: In the N=3, nearest neighbor isotropic case, LRO is expected on the basis of spin wave theory (cf. section IV). FSS proved this in the classical case and DLS proved it for the quantum antiferromagnet when $|S| \geq 1$ (in three dimensions) and $|S| \geq 1/2$ (in sufficiently large dimension). Both use the infrared bounds (section VII). DLS also claimed a proof of LRO for the ferromagnet in ≥ 3 dimensions and all $|S| \geq 1/2$. This claim is unjustified because the proof of the infrared bounds (and reflection positivity) was erroneous. However the DLS infrared bounds and RP is correct for the XY model; thus there is LRO in this case in ≥ 3 dimensions.

III. The Peierls Argument

This is one of the most beautiful ideas in mathematical physics. Suppose 2 spins, S_x and S_y, far apart, are oppositely oriented. Then, in any configuration, there must either be a contour γ surrounding S_x or one surrounding S_y. A contour is a closed polygonal curve in the lattice such that just on the inside all spins are in one direction, and just on the outside they are in the opposite direction. Suppose one can show that $P(\gamma)$, the probability of γ occurring, satisfies $P(\gamma) \le \exp[-c(\beta)|\gamma|]$, where $|\gamma|$ is the length of γ and $c(\beta) \to \infty$ as $\beta \to \infty$. Since the number of contours of length $|\gamma|$ is (essentially) bounded by $3^{|\gamma|}$ (in ≥ 2 dimensions) and $|\gamma| \ge 4$, one has that the probability of opposite spins, P_{+-}, satisfies

$$P_{+-} \le \sum_{|\gamma|=4}^{\infty} \exp[(-c(\beta) + \ln 3)|\gamma|]$$

and this goes to zero as $\beta \to \infty$.

So far, what we have said applies to any case, classical or quantum. The difficulty is to obtain the estimate on $P(\gamma)$. For the classical nearest neighbor Ising model, the original Peierls estimate proceeds as follows: $P(\gamma) = Z^{-1} \Sigma_\gamma \exp(-\beta H)$, where the sum is on all spin configurations that contain γ. For each term in the numerator, consider the corresponding term in the denominator (i.e. Z) in which <u>all</u> the spins inside γ are reversed. This gives $c(\beta) \le \beta J$. Obviously this method does not work in the quantum case, or with continuous symmetry, or with long range interactions. It can be made to work, with difficulty, in the anisotropic classical case[12]. In hard core models, great ingenuity is required to find a term in the denominator to compare with a term in the numerator. Dobrushin[17] did this for the nearest neighbor exclusion on the square lattice, and Heilmann[6] did it for the triangular and hexagonal lattices.

It is here that reflection positivity (RP) comes in. Using it, one can reduce the estimate of $P(\gamma)$, which is the probability of a local quantity, to an estimate for $P(\tilde{\gamma})$, where $\tilde{\gamma}$ is a "universal contour" that covers the whole lattice. In fact $P(\gamma) \le P(\tilde{\gamma})^{a|\gamma|/|\Lambda|}$ with $a \sim 1$. Here, Λ is the lattice. The idea of combining reflection positivity (which is due to Osterwalder and Schrader[18] and Nelson[19]) with the Peierls argument is due to Glimm, Jaffe and Spencer[20] in quantum field theory. The first application in statistical mechanics (which requires some additional ideas in the quantum case) is due to FL.

When RP holds, it makes the proof comparatively simple, especially for the hard core models. But to have it, one needs <u>perfect</u> translation invariance - hence the need to use periodic boundary conditions. This is a great defect of the method because translation invariance was not needed in the original Peierls estimate - a positive lower bound on J would have sufficed. This is discussed more fully in sections V and VI.

IV. Spin Wave Theory

This is a theory of small fluctuations about the completely ordered state for a system with continuous symmetry (e.g. the isotropic Heisenberg ferromagnet). An important open problem is to make it precise.

Apart from irrelevant factors, the energy of a spin wave (SW) of momentum p is

$$E_p = \sum_{j=1}^{\nu} 1 \mp \cos p_j,$$

ν = dimension, - for the ferromagnet and + for the antiferromagnet. $E_p \sim p^2$ (ferro) for small p, and p runs over the Brillouin zone (e.g. $|p_j| < \pi$). If

$$\hat{S}_p = |\Lambda|^{-1/2} \sum_{x\in\Lambda} S_x \exp(ip\cdot x),$$

then SW theory yields

$$A_p \equiv < \hat{S}_p \cdot \hat{S}_{-p} > \approx 1/\beta E_p, \ \beta \text{ large.} \tag{1}$$

LRO is equivalent to $|\Lambda|^{-1} <\hat{S}_o \cdot \hat{S}_o> \neq 0$ as $\Lambda \to \infty$. But

$$|\Lambda|^{-1} \sum_p < \hat{S}_p \cdot \hat{S}_{-p} > = < S_o \cdot S_o > = S^2 \text{ or } S(S+1)$$

in the classical or quantum case. Thus, we want $|\Lambda|^{-1} \sum_{p\neq 0} < \hat{S}_p \cdot \hat{S}_{-p} >$ to be small. As $\Lambda \to \infty$, this latter quantity becomes (using (1)) $\beta^{-1} \int d^\nu p(E_p)^{-1}$. The integral is finite for $\nu \geq 3$, but infinite for $\nu = 1,2$. Thus, for $\nu \geq 3$, $|\Lambda|^{-1} < \hat{S}_o \cdot \hat{S}_o > \neq 0$ when β is large enough.

The MWH proof shows that $1/\beta E_p$ is essentially (apart from factors) a lower bound for A_p. Hence there is no LRO for $\nu = 1,2$. The infrared bounds of FSS and DLS show that $1/\beta E_p$ is essentially also an upper bound. Hence there is LRO for $\nu \geq 3$.

Two remarks are in order: (i) It should be possible to show that SW theory is asymptotically exact as $\beta \to \infty$. (ii) There should be some connection between SW theory and the Peierls argument. One tenuous connection is this: Put the spins at the boundary up and the spin at the origin down. Arrange the other spins to give minimum energy. Then the size of the ball of down spins goes to ∞ as $\Lambda \to \infty$, for $\nu = 1,2$, whereas it remains finite for $\nu \geq 3$.

V. Reflection Positivity

The simplest example of RP is the Coulomb potential. Let $\rho(w)$, $w=(x,y,z)$, be *any* charge density in $\mathbb{R}^3$ supported in $x \geq 0$, and let $\theta\rho(w)=\rho(-x,y,z)$ be its mirror image. Then the interaction $\iint \theta\rho(w')\bar{\rho}(w)|w-w'|^{-1}d^3w\,d^3w'$ is positive. This is *not* the same as the ordinary positivity of $|w-w'|^{-1}$. Proof: Write $[(x+x')^2+(y-y')^2+(z-z')^2]^{-1/2}$ as the p, q, r integral of $\exp[ip(x+x')+iq(y-y')+ir(z-z')](p^2+q^2+r^2)^{-1}$. Do the p integration to obtain $(q^2+r^2)^{-1/2}\exp[-(q^2+r^2)^{1/2}|x+x'|]$. But since x and $x' \geq 0$, $|x+x'| = x+x'$, and we see at once that the kernel is positive definite.

An analogous result holds for *some* statistical mechanical models. Consider the Ising ferromagnet, for instance. Let λ be a vertical line (plane) that passes through the midpoints of a column of horizontal bonds. (Since the lattice is on a torus, λ really cuts the lattice in 2 lines). λ divides the lattice into 2 symmetric pieces. Let F be any function of the spins lying to the left of λ, and θF the reflected function (i.e. $\theta F(S_1,S_2,\ldots) = F(\theta S_1,\theta S_2,\ldots)$ with θS_i being the mirror image of S_i through λ. Then,

$$<\bar{F}\,\theta\,F> \geq 0$$

This implies, by the Schwarz inequality proof, that

$$|<F\,\theta\,G>|^2 \leq <\bar{F}\,\theta\,F><\bar{G}\,\theta\,G> \qquad (2)$$

In particular, if $G \equiv 1$, then $|<F>|^2 \leq <\bar{F}\,\theta\,F>$.

The proof is, briefly, the following: Write $\exp(-\beta H)=\exp(-\beta H_L-\beta H_R+\beta\Sigma S_i\theta S_i)$, where H_L (resp. H_R) is the Hamiltonian of the spins to the left (resp. right) of λ, and the last sum is the interaction across λ. The plus sign is crucial! Next, expand $\exp(\beta\Sigma S_i\theta S_i)$ in a power series. Then one easily sees that every term in this series gives a positive contribution to $\sum_S \bar{F}(S)\theta F(S)\exp(-\beta H)$.

There are three ways in which the foregoing can be altered or generalized: (i) The notion of θF can be changed. For the antiferromagnet the appropriate choice is $\theta F(S_1,S_2,\ldots) = F(\pm\theta S_1,\pm\theta S_2,\ldots)$ where + holds for the A sublattice and - holds for the B sublattice. (ii) By using the Trotter product formula one can sometimes get RP for quantum systems. This work, e.g., for the Heisenberg antiferromagnet and the **XY** model. It does *not* work for the Heisenberg ferromagnet (this was the oversight in the DLS "proof"). (iii) λ can be chosen to be a line through a column of vertices, or through a diagonal of vertices. In this case, F is allowed to depend on the spins to the left of λ *and* the spins on λ. Of course, λ can also be a horizontal line.

The interaction J does not have to be short range. In one dimension, the requirement is that the matrix $J(i+j-1)\big|_{i,j=1}^{\infty}$ be positive semidefinite. Alternatively, $J(x) = \int_{-1}^{1} t^{|x|-1}\,d\rho(t)$, $d\rho \geq 0$. $|x|^{-\alpha}$, $1<\alpha<2$, is satisfactory.

For the nearest neighbor exclusion hard core models (square, triangular or hexagonal lattices), it is easy to check that RP holds for reflections through the (extended) lines which define the squares, triangles or hexagons.

Another example is the HL[5] model of a liquid crystal. The particles are hard core dimers on a square lattice with the same chemical potential for vertical and horizontal dimers (there are also 3 dimensional versions of the model). There is an interaction energy $-J$ for every pair of colinear, adjacent dimers, i.e. —···— or ⋮ . RP holds for reflections through the centers of horizontal or vertical bonds in the lattice (as in our first example). The problem posed by this model is to show: (a) at low temperature and not too negative chemical potential there is long range orientational order, i.e. the dimers are mostly either horizontal or vertical. (b) There is never any long range translational order, i.e. every two point function clusters. HL can prove (a) but not (b), even though (b) is "obvious" because at low temperatures the system is essentially a product of independent one dimensional systems.

VI. Reflection Positivity and the Peierls Argument

We will show, by the example of nearest neighbor exclusion on the square lattice, how RP yields a simple estimate for $P(\gamma)$. The reader is urged to try this for himself in the following cases: (i) The Ising ferromagnet. (ii) nearest neighbor exclusion on the triangular and hexagonal lattices[6]. (iii) The HL liquid crystal model to show that there is orientational ordering at large β, i.e. that the probability of finding a horizontal and a vertical dimer at two sites arbitrarily far apart is small. The case we consider, and (i), can be done by the conventional Peierls method[17], but (ii) and (iii) are very difficult to do that way[6].

Let $x \in A$=A sublattice, $y \in B$=B sublattice. Particles at x and $y \Rightarrow$ either x is surrounded by an A-contour, γ, or y is surrounded by a B-contour. An A-contour, γ, runs through B points, and is characterized by the A sites inside γ being occupied and the A sites outside γ being empty. Associated with each segment, i, of γ is a projector Q_i, $i \in B$, which we may indicate schematically as $\times$-o. $\times$ stands for the inner A site being occupied and o stands for the outer A site being empty. There are 4 kinds of Q's, namely $\times$-o, o-$\times$ and the 2 vertical pairs. Assuming the contour in question is an A type, $P(\gamma) = \langle Q \rangle$ where $Q = \prod_{i \in \gamma} Q_i$. The number of Q_i's is $|\gamma|$. We write $Q = Q^h Q^v$, where $Q^h = \prod_{i \in h} Q_i$ is the product of the horizontal Q's, $\times$-o or o-$\times$. Suppose that $|h|$ = (number of horizontal Q_i's) $\geq |\gamma|/2$. Then $P(\gamma)^{1/|\gamma|} \leq \langle Q^h \rangle^{1/2|h|}$. Define the projector $R_i = (\times\text{-o}) + (\text{o-}\times)$. Then $P(\gamma) \leq \langle R \rangle^{1/2|h|}$, where $R = \prod_{i \in h} R_i$. Now define

$$\xi = \max \left\langle \prod_{i=1}^{n} R_i \right\rangle^{1/n}$$

where the maximum is over all $n \geq 1$ and, for each n, over all choices of n distinct sites $R_1, \ldots, R_n \in B$. Then $P(\gamma) \leq \xi^{|\gamma|/2}$ by definition. This is our desired goal provided we can show that $\xi \to 0$ as μ = (chemical potential) goes to infinity.

Suppose $R = \pi R_i$ is a maximizing choice for ξ. Take a vertical reflection line λ (one of the lines in the lattice). Write $R = R^L R^\lambda R^R$ where R^L is the product of the R_i's to the left of λ, R^λ is the product of the Ri's which straddle λ. Then, by RP

$$< R >^2 \le < R^L R^\lambda \theta R^L > \quad < \theta R^R R^\lambda R^R >. \qquad (3)$$

Note that (the total number of R_i's on the right side of (3)) = 2n = (the number of R_i's on the left side of (3)). Assume $R^L R^\lambda \neq 1$. Thus, if R maximizes, so does $R^L R^\lambda \theta R^L$. Proceeding in this way, one can construct a maximizing R with the following property: on one row all the B sites have an R_i. Repeating the argument with horizontal reflection lines one concludes that there is a maximizing R of the following type: On even numbered rows every B site, i, has an R_i; the odd numbered rows are identical and are either full of R_i's or empty of R_i's.

Now we can easily bound $< R >^{1/n}$. If $n = |\Lambda|/2$ (odd rows full) the numerator is essentially $\exp(\beta\mu|\Lambda|/4)$. For the denominator we take only a term with the maximum number of particles; thus $Z \ge \exp(\beta\mu|\Lambda|/2)$. This gives $\xi = \exp(-\beta\mu/2)$. In the other case ($n = |\Lambda|/4$, odd rows empty) a similar, but slightly different result holds. Thus, ξ (and $P(\gamma)^{1/|\gamma|}) \to 0$ exponentially fast as $\beta \to \infty$.

It is hoped that this example, briefly presented, illustrates the method. Note that the last step was to estimate $< R >$, where R is a "universal" projector. This is easy to do in the classical case. In quantum systems, e.g. the Heisenberg antiferromagnet treated by FL, it is much harder to do. Additional tricks are needed, e.g. "the principle of exponential localization" used by FL.

VII. Infrared Bounds

These bounds come from a variation on the theme of RP. They are central to the FSS, DLS and FILS proofs of phase transitions in ≥ 3 dimensional systems with continuous symmetry. (See section IV.)

We shall illustrate the argument for the classical Heisenberg model on the 3-dimensional square lattice with Hamiltonian $H = -\Sigma J(x-y)\, S_x \cdot S_y$. J is assumed to satisfy the RP condition: $J(x^1 + y^1 - 1, x^2 - y^2, x^3 - y^3)$ is a positive definite matrix for $1 \le x^1, y^1 < \infty$, $-\infty < x^2, y^2, x^3, y^3 < +\infty$; a similar condition holds for the second and third coordinates. Let $\hat{J}(k)$ be the (periodic lattice) Fourier transform of $J(x)$. $\hat{J}$ is assumed to have its maximum at k=0. Therefore $K(x-y) = \hat{J}(0)\delta(x-y) - J(x-y)$ is positive semidefinite, and $K\cdot 1=0$ where 1 is the constant vector. Since $S_x^2 = 1$, all x, we can replace H, for the purpose of calculating expectation values, by H = SKS (dot product being understood).

Now let $\underline{h} = \{h_x | x\epsilon\Lambda\}$ be a set of (3-dimensional) vectors, one for each point of Λ. Let $H(\underline{h}) = (S-\underline{h})K(S-\underline{h}) \ge 0$, and consider $Z(\underline{h}) = T_r \exp(-\beta H(\underline{h}))$. The claim is that $Z(\underline{h})$ has its maximum at $\underline{h} = 0$. Simply mimic the proof of RP in section V and its application in section VI. By reflecting in lines which are perpendicular to the midpoints of bonds, one finds that a maximizing $\underline{h}$ for $Z(\underline{h})$ must be symmetric about every such line. Hence h_x is independent of x. But since $K\cdot 1=0$, we can take $\underline{h}=0$.

To apply this fact, fix $\underline{h}$ and let $f(t) = Z(t\underline{h})$. Then $f'' \leq 0$ at $t=0$. Hence, we reach the important conclusion:

$$2\beta\, h\cdot K\, M\, K\cdot h \leq h\cdot K\cdot h \qquad (4)$$

where $M(x-y) = \langle S_x\, S_y \rangle$.

Now let $h_x = (\cos p^1x^1 \cos p^2x^2 \cos p^3x^3, 0, 0)$. Then from (4)

$$\langle \hat{S}^1_p\, \hat{S}^1_{-p} \rangle \leq 1/2\beta\, E_p \qquad (5)$$

and similarly for components 2 and 3 of $\hat{S}_p$. The spin wave discussion of section IV can now be carried to completion to show that there is LRO for large β.

The quantum case is a bit more subtle. Assuming that RP holds (as it does for the XY model or the Heisenberg antiferromagnet), the analogue of (5) that is obtained is the following:

$$(\hat{S}^j_p, S^j_{-p}) \leq 1/2\beta E_p$$

where (A,B) is the <u>Duhamel</u> 2-point function:

$$(A,B) = \int_0^1 dt\ \mathrm{Tr}\ \exp(-\beta tH) A \exp(-\beta(1-t)H) B$$

A useful lower bound for (A^*,A) in terms of the ordinary $\langle A^*A \rangle$ is needed. This is provided by the inequality of Falk and Bruch[21]:

$$(A^*,A) \geq g\ f\ (c/4g)$$

$$g = (\langle A^*A \rangle + \langle A\, A^* \rangle)/2$$

$$c = \beta \langle [A^*, [H,A]] \rangle \geq 0$$

and $f(x)$, $x \geq 0$, is the convex function given by

$$f(x \tanh x) = (\tanh x)/x$$

This inequality is sufficient for carrying through the spin wave argument.

*Work partially supported by U. S. National Science Foundation grant MCS 75-21684 A01.

<u>References</u>

[1] J. Fröhlich, B. Simon and T. Spencer, Phys. Rev. Lett. <u>36</u>, 804 (1976); Commun. Math. Phys. <u>50</u>, 79 (1976).

[2] F. Dyson, E. Lieb and B. Simon, Phys. Rev. Lett. <u>37</u>, 120 (1976) and J. Stat. Phys. (to appear).

[3] J. Fröhlich, R. Israel, E. Lieb and B. Simon, in preparation.

[4] J. Fröhlich and E. Lieb, Phys. Rev. Lett. <u>38</u>, 440 (1977), and paper in preparation.

[5] O. Heilmann and E. Lieb, Lattice Models for Liquid Crystals, in preparation.
[6] O. Heilmann, Commun. Math. Phys. 36, 91 (1974); Lett. Nuovo Cim. 3, 95 (1972).
[7] N. Mermin and H. Wagner, Phys. Rev. Lett. 17, 113 (1966); N. Mermin, J. Math. Phys. 8, 1061 (1967).
[8] P. Hohenberg, Phys. Rev. 158, 383 (1967).
[9] L. Van Hove, Physica 16, 137 (1950).
[10] F. Dyson, Commun. Math. Phys. 12, 91 (1969) and 21, 269 (1971).
[11] R. Peierls, Proc. Camb. Phil Soc. 32, 477 (1936).
[12] S. Malyshev, Commun. Math. Phys. 40, 75 (1975).
[13] A. Bortz and R. Griffiths, Commun. Math. Phys. 26, 102 (1972).
[14] J. Ginibre, Commun. Math. Phys. 14, 205 (1969).
[15] D. Robinson, Commun. Math. Phys. 14, 195 (1969).
[16] E. Lieb, Commun. Math. Phys. 31, 327 (1973).
[17] R. Dobrushin, Funct. Anal. Appl. 2, 44 (1968); Engl. Trans. 2, 302 (1968).
[18] K. Osterwalder and R. Schrader, Helv. Phys. Acta 46, 277 (1973); Commun. Math. Phys. 31, 83 (1973).
[19] E. Nelson, in Constructive Quantum Field Theory, G. Velo and A. Wightman eds., Springer (1973).
[20] J. Glimm, A. Jaffe and T. Spencer, Commun. Math. Phys. 45, 203 (1975).
[21] H. Falk and D. W. Bruch, Phys. Rev. 180, 442 (1969).

NUMBER OF PHASES IN ONE COMPONENT FERROMAGNETS*+Δ

Joel L. Lebowitz
Department of Mathematics
Rutgers University
New Brunswick, New Jersey 08903

Abstract

Using a new inequality, derived here, we obtain information about the number of pure phases which can coexist in one component spin system with (many body) ferromagnetic interactions. This extends previous results [1] for spin-$\frac{1}{2}$ Ising systems to continuous spin systems.

1. Introduction

As is well known it follows from the general formalism of statistical mechanics that phase transitions, e.g. the coexistence of two phases in equilibrium or the non-analytic behavior of the free energy as a function of temperature or magnetic field, can occur strictly only in infinite systems - the proper mathematical idealization of macroscopic systems which are described thermodynamically by intensive variables [2,3]. The microscopic correlations in such a system are described by Gibbs states which are probability measures on the phase space of the system satisfying the DLR equations [3,4,5]. These states are the appropriate limits of finite volume Gibbs ensembles. Equivalently one may describe the state of the infinite system by means of correlation functions. The latter are obtained as infinite volume limits of the equilibrium correlations in a finite system with specified "boundary conditions". A pure thermodynamic phase then corresponds (loosely speaking) to a translation invariant Gibbs state μ $(\mu \varepsilon I)$ with correlation functions which "cluster" at infinity, i.e. correlations between different local regions of the system decay (however weakly) as the distance between these regions becomes larger and larger [3]. The latter condition is equivalent to the requirement that intensive variables be well definded, i.e. that fluctuations in "all"

*Based on lectures given at the Rencontres Physique Mathematique held in Strasbourg in May 1977 and at the International Conference on the Mathematical Problems in Theoretical Physics held in Rome in June 1977.

+Part of this work was done while the author was a visitor at IHES in Bures-sur-Yvette and in the Department Physique Theorique, CEN, Saclay, France, as a John Guggenheim Fellow on sabbatical leave from Yeshiva University, N.Y. .

ΔWork supported by NSF Grant #MPS 75-20638.

intensive variables, local functions averaged over the volume of the system, vanish as the volume tends to infinity. The coexistence of several phases then corresponds to the existence, for a given interaction, temperature and magnetic field, of more than one translation invariant solution of the DLR equations. This is the same as the possibility of obtaining different translation invariant infinite volume limits of the Gibbs measure (or the correlation functions) from different boundary conditions. These states have also been shown to be (in many cases) the solution of a variational principle minimizing the infinite volume free energy density [3]. The latter states are sometimes called "equilibrium states" E ($E \subset I$).

By a very general theory [3-5] it is always possible to decompose any Gibbs state <u>uniquely</u> into "extremal" Gibbs states; the translation invariant extremal states corresponding to the pure phases. This means the following: given any "observable" f then its expectation value $\langle f\rangle$ in any I. equilibrium state can be written in the form $\langle f\rangle = \sum_{k=1}^{n} \alpha_k \langle f\rangle_k$ where $\langle f\rangle_k$ is the expectation value of $\langle f\rangle$ in the k<u>th</u> pure phase, $0 \leq \alpha_k \leq 1$, and $\sum_{k=1}^{n} \alpha_k = 1$, i.e. α_k measures the fraction of volume occupied by the k<u>th</u> phase. The crucial point here is that the α_k are independent of the observable f: n thus clearly represents the total number of phases which can coexist (at a given temperature and magnetic field) and the question then is to determine n. (The Gibbs phase rule states that for an m-component fluid $n \leq m+2$, but this is far from proven and does not apply to spin systems with general interactions [3,6].)

This lecture is devoted mainly to the description of some new results regarding the number of possible phases in one component spin system with ferromagnetic interactions. We consider first the case of spin $\frac{1}{2}$ Ising systems. These are the simplest non-trivial systems for which such results can be derived in a mathematically rigorous way. The main new result is that for such a system with even spin interactions (païr, quadruple,etc.) there can coexist, at zero magnetic field, only two phases (up and down magnetization) at all temperatures at which the energy is continuous in the temperature In particular, there are no intervals of temperature, below the critical temperature T_c, at which three or more phases can coexist. This extends results previously known only for the two dimensional spin $\frac{1}{2}$ Ising system with nearest neighbor pair interactions [7] and for higher dimension spin $\frac{1}{2}$ Ising systems only at low temperatures [8]. We then indicate how similar results can be obtained also for general, bounded and unbounded, one component spin systems. For the unbounded case there are still some gaps in the argument relating invariant Gibbs states to solutions of the variational principle, e.g. for what class of states are the two equivalent. It appears however that this is a soluble technical problem and that our results may be extended also to the field theory case.

The main results are derived in section 3. They are based on a new inequality for ferromagnetic systems which is derived for spin $\frac{1}{2}$ Ising systems in section 2. Section 4 is devoted to proving a similar inequality for general spin systems.

2. Inequality

Let Λ be a finite set of $|\Lambda|$ sites, which for later applications we shall think of as a subset of a regular ν-dimensional lattice, say $\mathbb{Z}^\nu$. Call $s_i \varepsilon R$, $i\varepsilon\Lambda$, the spin variable at the site i and define, $s_A = \prod_{i\varepsilon A} s_i^{\ell_i}$, for $A \subset \Lambda$ (with the index i repeated ℓ_i times) $\ell_i \varepsilon \mathbb{Z}_+$. We let $d\rho_i(s_i)$ be the free measure of the spin at the site i and $\beta H = -\sum_{K\subset\Lambda} J_K s_K$ the energy (times the reciprocal temperature) of a spin configuration in Λ. The Gibbs measure $d\mu(\underline{S}_\Lambda)$, $\underline{S}_\Lambda = \{s_i\}$, $i\varepsilon\Lambda$, has the expectation values for $F(\underline{S}_\Lambda)$.

$$<F>_\mu = Z^{-1} \int F(\underline{S}_\Lambda) \exp [\Sigma J_K s_K] \prod_{i\varepsilon\Lambda} d\rho_i(s_i) \tag{2.1}$$

We assume that the free measures, ρ_i, are even and have a sufficiently strong decay as $|s_i| \to \infty$ for all the moments of μ to exist.

We wish to compare these expectations $<\ >_\mu$ with those obtained from the Gibbs measure $d\mu'(\underline{S}_\Lambda)$ for a different spin system in Λ - one having free measures $d\rho_i(s_i)$ and energy $\beta' H' = -\Sigma J'_K s_K$.

<u>Lemma 1</u>. Let $f_\alpha(s)$, $\alpha=1,\dots,n$ be odd monotone non-decreasing functions of $s\varepsilon\mathbb{R}$ and let $Q(s,s')$ be a symmetric, even, non-negative function of s and s', $s'\varepsilon R$; $Q(s,s') = Q(s',s) = Q(-s,-s') \geq 0$. Then

$$M_i \equiv \int \int \prod_{\alpha=1}^{n} [f_\alpha(s) - f_\alpha(s')]^{k_\alpha} [f_\alpha(s) + f_\alpha(s')]^{\ell_\alpha} Q(s,s') d\rho_i(s) d\rho_i(s') \geq 0 \tag{2.2}$$

<u>Proof</u>: Letting $s \leftrightarrow s'$ and $s \leftrightarrow -s$, $s' \leftrightarrow -s'$, shows (remembering that ρ_i is an even measure) that $M_i = 0$ unless $h = \Sigma k_\alpha$ and $\ell = \Sigma \ell_\alpha$ are even integers in which case the integrand is non-negative. This is very similar to Ginibre's proof of the GKS inequalities [9,10].

<u>Lemma 2</u>. Let $J_K \geq |J'_K|$ and let $f_i(s_i)$ be odd monotone and $g_i(s_i)$ be

either an odd or even bounded functions of s_i, $|g_i(s_i)|\leq\lambda_i$. Define $f_A(s_A)=\prod_{i\varepsilon A} f_i(s_i)$, $g_A(s_A)=\prod_{i\varepsilon A}[g_i(s_i)/\lambda_i]$. Then

$$I\equiv\int[1\pm g_B(s_B)g_B(s_B')][f_A(s_A)-f_A(s_A')]d\mu(\underline{S}_\Lambda)d\mu'(\underline{S}_\Lambda') \tag{2.4}$$
$$\equiv<f_A>-<f_A>'\pm[<g_Bf_A><g_B>'-<g_Bf_A>'<g_B>]\geq 0.$$

Proof: Noting that $d\mu(\underline{S}_\Lambda)d\mu(\underline{S}_\Lambda')=\exp\,[\Sigma(J_K s_K+J_K' s_K')]\Pi d\rho_i(s_i)d\rho_i(s_i')/ZZ'$ we put $J_K s_K+J_K' s_K'=\frac{1}{2}[(J_K+J_K')(s_K+s_K')+(J_K-J_K')(s_K-s_K')]$ and expand the exponential. We then factorize $S_K\pm S_K'$, $f_A(s_A)-f_A(s_A')$, and $[1-g_B(s_B)g_B(s_B')]$ into products of terms of the form $(s_i\pm s_i')$, $(f_i(s_i)\pm f_i(s_i'))$, $[1\pm g_i(s_i)g_i(s_i')]$; e.g. $f_i(s_i)f_i(s_j)-f_i(s_i')f_i(s_j')=\frac{1}{2}\{[f_i(s_i)+f_i(s_i')][f_j(s_j)-f_j(s_j')]+[f_i(s_i)-f_i(s_i')][f_j(s_j)+f_j(s_j')]\}$. The final result is that I can be written as a sum of products of terms of the form M_i in (2.3). By our assumption, $J_K\geq|J_K'|$, all these terms have positive coefficients. Hence the lemma is proven.

We can rewrite (2.4) in the form

$$<f_A>-<f_A>'\geq|<g_Bf_A><g_B>'-<g_Bf_A>'<g_B>|\geq 0. \tag{2.5}$$

It now follows from (2.5) that

Corollary 3: Let $J_K\geq|J_K'|$, $<f_A>=<f_A>'$ and $<g_B>=<g_B>'\neq 0$, then $<f_Ag_B>=<f_Ag_B>'$.

Corollary 3 is particularly useful for the case of spin $\frac{1}{2}$ Ising systems which correspond to having $d\rho_i(s_i)=\frac{1}{2}\delta(|s_i|-1)$. Setting $f_i(s_i)=g_i(s_i)=s_i$ (and writing $s_i\equiv\sigma_i=\pm 1$ to emphasize that we are dealing with a special case) we may use the following basic group property for the $\sigma_A=\prod_{i\varepsilon A}\sigma_i$, (this is just like s_A with $\ell_i=1$, $V_i\varepsilon A$, since $\sigma_i^2=1$) $\sigma_A\sigma_B=\sigma_C$ with $C=A\,\Delta\,B$, $A\,\Delta\,B$ the symmetric difference between A, $B\subset\Lambda$. This yields the additional results.

Corollary 4: Let $J_K\geq|J_K'|$. Then $<\sigma_A>=<\sigma_A>'$ and $<\sigma_B>=<\sigma_B>'\neq 0$ imply $<\sigma_A\sigma_B>=<\sigma_A\sigma_B>'$ for all $A,B\subset\Lambda$.

Corollary 5: Let $J_K\geq|J_K'|$. Then: (1) $<\sigma_i>=<\sigma_i>'\neq 0$ for all the one site sets $i\varepsilon\Lambda$ implies $<\sigma_A>=<\sigma_A>'$ for all $A\subset\Lambda$.
(2) $<\sigma_i\sigma_j>=<\sigma_i\sigma_j>'\neq 0$ for all $i,j\varepsilon\Lambda$ implies $<\sigma_E>=<\sigma_E>'$ for all sets E containing an even number of sites, $|E|$ even.

Proof: By Corollary 3 $<\sigma_i>=<\sigma_i>'\neq 0$ and $<\sigma_j>=<\sigma_j>'\neq 0$ implies $<\sigma_i\sigma_j>=<\sigma_i\sigma_j>'$. Furthermore since $J_K\geq 0$ it follows from the GKS inequalities that $<\sigma_A\sigma_B>\geq<\sigma_A><\sigma_B>\geq 0$. Hence $<\sigma_i\sigma_j>\geq<\sigma_i><\sigma_j>>0$. The rest follows by induction. The proof of (2) is similar since $\langle\sigma_B\rangle=\langle\sigma_B\sigma_k\sigma_\ell\sigma_k\sigma_\ell\rangle$ for all $B\subset\Lambda$.

The proof of Corollary's 4 and 5 for spins with general measures ρ_i is a bit more complicated. It is postponed to section 4 following the discussion in the next sections of some consequences of these inequalities.

3. Equilibrium States for Spin $\frac{1}{2}$ Systems

We shall now use the inequalities derived in the last section to obtain information about the number of equilibrium states for infinite Ising systems. To do this we assume that the interactions are translation invariant $J_A=\beta\Phi_{A+x}$ where $A+x$ is the set A translated by a lattice vector x. In particular for the one point sets $A=i\varepsilon\mathbb{Z}^\nu$, $\beta\Phi_i=h$, the magnetic field (times β) and for $|A|=2$, $J_{\{i,j\}}=\beta\phi(i-j)$, etc... The energy of a spin configuration σ_Λ in $\Lambda\subset\mathbb{Z}^\nu$ will depend on the specified values of the spins outside Λ, i.e. we consider the spins outside Λ to be fixed and act as boundary conditions for the spins in Λ. A particular boundary condition "b" then corresponds to a lattice spin configuration σ^b such that $\sigma_i=\sigma_i^b$ for $i\varepsilon\Lambda_c$. (Generally $\sigma_i^b=\pm1$; $\sigma_i^o=0$ correspond to zero b.c.). We then have, corresponding to Eq. (1),

$$H(\underline{\sigma}_\Lambda;b)=-\sum_{B\ni\{0\}}\sum_x \Phi_B\sigma_{B+x}, \quad \Phi_B\geq0 \tag{3.1}$$

where $\{0\}$ designates the origin and the sum over x goes over all x such that $\{B+x\}\cap\Lambda$ is not empty, i.e., at least some of the sites in $B+x$ are in Λ. We assume from now on that $\Phi_B\geq0$, i.e. positive ferromagnetic interactions. It is then clear that μ_+ corresponding to plus b.c., $\sigma_i^+=1$, 'dominates' all other b.c.. Hence defining $\langle\sigma_A\rangle(\beta,h;b,\Lambda)$ as the expectation value of σ_A, $A\subset\Lambda$, for the Hamiltonian (3.1) at reciprocal temperature β and magnetic field h we can identify $\langle\sigma_A\rangle$ of Sec. 2 with $\langle\sigma_A\rangle(\beta,h;+,\Lambda)$ and $\langle\sigma_A\rangle'$ with $\langle\sigma_A\rangle(\beta,h;b,\Lambda)$ for any other boundary condition. (Our notation implies the "physicist" point of view where β and $h=\{\beta\Phi_i\}$ are independent "externally controlled" variables while Φ_K, $|K|\geq2$, are "given" interactions).

It follows from the GKS inequalities [10,11] that

$$\lim_{\Lambda\uparrow\mathbb{Z}}\langle\sigma_A\rangle(\beta,h;+,\Lambda)=\langle\sigma_A\rangle(\beta,h;+) \tag{3.2}$$

exist and are translation invariant

$$\langle\sigma_{A+x}\rangle(\beta,h;+)=\langle\sigma_A\rangle(\beta,h;+). \tag{3.3}$$

To avoid unnecessary complications we assume that the interactions are of "finite range", $\Phi_B=0$ unless $B\subset N$, N bounded. The thermodynamic free energy per site, $\psi(\beta,h)=\lim\{|\Lambda|^{-1}\ln \mathrm{Tr}(\exp[-\beta H(\underline{\sigma}_\Lambda;b)])\}$ then exists and is independent of b.

We shall write $\langle\sigma_i\rangle(\beta,h;+)=m(\beta,h;+)$, the magnetization per site with $+$ b.c.. For more general boundary conditions, (including a superposition, with specified weights, of different $\underline{\sigma}^b$) the limit $\Lambda\uparrow\mathbb{Z}^\nu$ might have to be taken along subsequences to obtain infinite volume correlation functions $\langle\sigma_A\rangle(\beta,h;b)$ which need not, in general, be translation invariant [12]. It is however always possible to average over translations to obtain translation invariant correlation functions. The set of correlations, $\langle\sigma_A\rangle(\beta,h;b)$, $A\subset\mathbb{Z}^\nu$, obtained from $\langle\sigma_A\rangle(\beta,h;b,\Lambda)$ as $\Lambda\uparrow\mathbb{Z}$ define an infinite volume Gibbs measure. These measures are identical to the ones which satisfy the DLR equations and the translation invariant ones are identical to the solutions of a variational principle (minimizing the free energy per unit volume) [3-5]. We shall sometimes write $\langle\sigma_A\rangle_+$ for $\int\sigma_A\mu_+(d\underline{\sigma})$, $\mu_+\varepsilon I$ being the measure obtained with $+$ b.c..

These considerations also lead to an identification of the $\langle\sigma_A\rangle_\mu$, $\mu\varepsilon I$, with derivatives of the free energy density $\Psi(\underline{J})$ with respect to $J_A(=\beta\Phi_A)$ [3-5].

$$\Psi(\underline{J})=\lim_{\Lambda\uparrow\mathbb{Z}^\nu}|\Lambda|^{-1}\ln Z(\underline{J}:b,\Lambda), \tag{3.4}$$

and we have used $\underline{J}$ for the argument of Ψ to emphasize that Ψ can be thought of as a function of "all possible" potentials J_K. $\Psi(\underline{J})$, being a convex function of each J_A, will be differentiable for almost all values of J_A (keeping the other interactions fixed).

We are now ready to state our first theorem about the number of possible equilibrium states.

Theorem 6. Let $\Psi(\beta,h)$ be the infinite volume free energy per site of an Ising spin system with translation invariant interactions; $\Phi_K=\Phi_{K+x}\geq 0$, $x\varepsilon\mathbb{Z}^\nu$, $\beta\Phi_{\{0\}}=h$. If the derivative of Ψ with respect to h exists (is continuous) and is positive, $\frac{\partial\Psi(\beta,h)}{\partial h}>0$, then there is a unique translation invariant Gibbs state. In particular $\langle\sigma_A\rangle(\beta,h;b)=\langle\sigma_A\rangle(\beta,h:+)=\partial\Psi/\partial J_A$ for all boundary conditions b.

Proof: Given any $\mu\epsilon I$, $\langle\sigma_A\rangle_\mu=\partial\Psi/\partial J_A$, when the latter exists [3-5], and the theorem then follows from Corollary 5 with $\langle\sigma_i\rangle=\partial\Psi/\partial h$.

Remark: Theorem 6 states that differentiability of Ψ with respect to h implies differentiability of Ψ with respect to all interactions. It thus generalizes to ferromagnetic many spin interactions the results of Lebowitz and Martin-Löf [11] for the case when the interactions are such that the Fortuin, Kasteleyn and Ginibre, (FKG) inequalities hold, e.g. when only pair interactions are present, $\Phi_K=0$, $|K|>2$ [13]. In that case however the results are stronger; there is a unique Gibbs state, (and so $I=G$) whenever $\partial\Psi(\beta,h)/\partial h$ exists. For pair interactions this is true for all $h\neq0$, and is always true at sufficiently high temperatures [2,3].

The positivity requirement on $\partial\Psi/\partial h$ is however not as restrictive as it might appear. First, by GKS, $\langle\sigma_i\rangle(\beta,h;+)>0$ if $h>0$ and hence $\frac{\partial\Psi(\beta,h)}{\partial h}=0\Longrightarrow h=0$. Second, if the interactions are such that $\langle\sigma_E\rangle(\beta,h=0;+)>0$ for $|E|$ even, e.g. when the nearest neighbor pair interactions is positive, then it is easy to show [14] that $\langle\sigma_i\rangle(\beta,0;+)=0\Longrightarrow\langle\sigma_Q\rangle(\beta,0;+)=0$ for all $|Q|$ odd. This implies, by GKS, that $\Phi_K=0$ for all $|K|$ odd. These facts in turn imply that the odd correlations vanish for all b.c. since, for $|Q|$ odd,

$$0=\langle\sigma_Q\rangle(\beta,0;+)\geq\langle\sigma_Q\rangle(\beta,0;b)=-\langle\sigma_Q\rangle(\beta,0;-b) \qquad (3.5)$$

where $-b$ is the b.c. obtained from b by reflection; $\sigma_i^{-b}=-\sigma_i^b$. We are therefore left, when $\partial\Psi(\beta,h)/\partial h=0$ at $h=0$, only with the possible nonuniqueness of the even correlation functions. We shall now consider this problem which is also, as we shall see, the central problem when $\partial\Psi(\beta,h)/\partial h$ is discontinuous at $h=0$ and there are only even interactions, e.g. in the Ising model with ferromagnetic pair interactions.

Definition: We call a (finite) collection of bounded sets $\{K_\alpha\}$, $K_\alpha\ni\{0\}$ all α, generating for the even sets, $\{K_\alpha\}=\bar{G}$ iff; given any bounded set $E\subset\mathbb{Z}^\nu$, $|E|$ even, we can write $\sigma_E=\prod_{n=1}^{m}\sigma_{\{K_{\alpha_i}+x_n\}}$, m finite, with $K_{\alpha_i}\epsilon G$, and x_n a lattice vector (we may have $K_{\alpha_i}=K_{\alpha_j}$).

By the proof of part (2) of Corollary 5, $\bar{G}$ will be generating iff it generates all the sets consisting of pairs of sites $\{i,j\}$. Letting e_α be the unit vector in the αth direction it is now easy to see that the ν nearest neighbor sets, $K_\alpha=\{0,e_\alpha\}$, $\alpha=1,\dots,\nu$ are generating, e.g. the product $(\sigma_0\sigma_{e_1})(\sigma_{e_1}\sigma_{e_1+e_2})=\sigma_0\sigma_{e_1+e_2}$ where e_1+e_2

is one of the next nearest neighbor sites of the origin, etc. .

It follows from part (2) of Corollary 5 that if the expectation values of σ_{K_α} in a translation invariant state μ are positive and equal to $\langle\sigma_{K_\alpha}\rangle(\beta,h;+)>0$, for all $K_\alpha\varepsilon\bar{G}$, then all the even correlation functions of μ are the same as in the + state. This will be the case for all translation invariant μ whenever Ψ is differentiable with respect to J_{K_α}, for all $K_\alpha\varepsilon K$ and $\partial\Psi/\partial J_{K_\alpha}>0$. We now show that this is equivalent to having $\Psi(\beta,h)$ differentiable with respect to β.

<u>Theorem 7</u>: Let the conditions of theorem 6 hold and let $\Phi_{K_\alpha}>0$ for all $K_\alpha\varepsilon\bar{G}$. If $\partial\Psi(\beta,h)/\partial\beta$ exists, i.e. the energy per site (apart from the magnetic field contribution) is continuous in β, then the expectation value of σ_E, $|E|$ even, is the same in all translation invariant states: $\langle\sigma_E\rangle_\mu=\langle\sigma_E\rangle_+$ for $\mu\varepsilon I$.

<u>Proof</u>: By the general arguments [3-5] mentioned earlier $\partial\Psi(\beta,h)/\partial\beta$ continuous implies that for every $\mu\varepsilon I$, $\sum_{K\ni\{0\}}\Phi_K\langle\sigma_K\rangle_+=\sum_{K\ni\{0\}}\Phi_K\langle\sigma_K\rangle_\mu$. By (7) $\langle\sigma_K\rangle_+\geq\langle\sigma_K\rangle_\mu$, hence the continuity of $\frac{\partial\Psi(\beta,h)}{\partial\beta}$ implies that $\langle\sigma_K\rangle_+=\langle\sigma_K\rangle_\mu$ for all $\mu\varepsilon I$ and all K such that $\Phi_K>0$. In particular $\Phi_{K_\alpha}>0$ for all $K_\alpha\varepsilon\bar{G}$ and by GKS $\langle\sigma_{K_\alpha}\rangle_+>0$ so part (2) of Corollary 5 implies that $\langle\sigma_E\rangle_+=\langle\sigma_E\rangle_\mu$ for all $|E|$ even.

The interest of theorem 4 lies primiarly in what it tells us about the number of extremal translation invariant Gibbs states for a system with even ferromagnetic interactions, when $h=0$, and $\Psi(\beta,h)$ not differentiable at $h=0$. Since $\Psi(\beta,h)$ is now symmetric (and convex) in h the non differentiability of Ψ at $h=0$ corresponds to the existence of a spontaneous magnetization with [11]

$$m^*(\beta) = \lim_{h\downarrow 0}\frac{\partial\Psi(\beta,h)}{\partial h} = -\lim_{h\uparrow 0}\frac{\partial\Psi(\beta,h)}{\partial h} = \langle\sigma_i\rangle(\beta,h=0;+)= = -\langle\sigma_i\rangle(\beta,h=0;-) \tag{3.6}$$

Here $\langle\sigma_A\rangle(\beta,h;-)=(-1)^{|A|}\langle\sigma_A\rangle(\beta,h;+)$ is the expectation of σ_A in the infinite volume Gibbs state μ_- obtained, as $\Lambda\uparrow\mathbb{Z}^\nu$, with "minus" boundary conditions (translation invariance is assured if $h\geq 0$). As already mentioned there are cases, i.e. only pair interactions (ferromagnetic), when $h=0$ is the only place where a phase transition is possible. With more general even interactions only the symmetry $h\to -h$, is known a priori. In a recent paper [14] we were able, using the GKS inequalities, to obtain some information about the Gibbs states of such a system at $h=0$. The following theorem greatly extends those results.

<u>Theorem 8</u>: Let the condition of theorem 6 hold and let $\Phi_K=0$ for all $|K|$ odd and $\Phi_{K_\alpha}>0$ for all $K_\alpha \varepsilon \bar{G}$. If $\partial\Psi(\beta,h=0)/\partial\beta$ exists then there are at most two extremal translation invariant Gibbs states, μ_+ and μ_-. These states coincide if $\partial\Psi(\beta,h)/\partial h$ exists at $h=0$.

<u>Proof</u>: By theorem 7 the differentiability of $\Psi(\beta,0)$ implies that the $\langle\sigma_E\rangle_\mu$, $|E|$ even, are the same in all $\mu\varepsilon T$. If furthermore $\partial\Psi(\beta,h)/\partial h=0$, at $h=0$, then by the remarks following theorem 6 the odd correlations vanish for all $\mu\varepsilon G$ and the state $\mu\varepsilon I$ is then unique. (When the FKG inequalities hold differentiability with respect to h implies differentiability with respect to β). When Ψ is not differentiable at $h=0$, $m^*(\beta)>0$, there are at least two extremal translation invariant Gibbs states, μ_+ and μ_-, [11]. Let $\mu(\beta,0;b)$ be an invariant state then $\bar{\mu}(\beta,0;b)\equiv\frac{1}{2}[\mu(\beta,0;b)+\mu(\beta,0;-b)]$ is an invariant state in which all the odd correlations vanish by symmetry. Hence $\bar{\mu}(\beta,0;b)=\frac{1}{2}[\mu_++\mu_-]$ which, since invariant Gibbs states form a simplex, i.e. each state has a unique decomposition into extremal states, implies that $\mu(\beta,0;b)=\gamma\mu_++(1-\gamma)\mu_-$, $0\leq\gamma\leq1$. This completes the proof. (The last part of the argument, which is also used in refs. [7] and [8], I heard originally from Ruelle).

<u>Remarks</u>: i) It follows [2] from GKS that there exists a unique β_c such that

$$m^*(\beta) = \begin{cases} 0, & \beta<\beta_c \\ >0, & \beta>\beta_c \end{cases} .$$

We always have [2,15] $\beta_c\geq\beta_0>0$ and for $\nu\geq2$ (with non-vanishing Φ_K), $\beta_c\leq\beta_p<\infty$ by the Peierls argument (or the more recent method of Frohlich, Simon and Spencer [16] for $\nu\geq3$). Using the convexity of $\Psi(\beta,0)$ it follows from theorem 8 that with the possible exception at a countable number of values of β, there is a unique $\mu\varepsilon I$ for $\beta<\beta_c$ and two extremal states $\mu\varepsilon I$ for $\beta>\beta_c$. In particular there are no triple or higher order points at $h=0$ when the energy is continuous in β.

ii) The state, at $h=0$, obtained with "zero" (or periodic) b.c. $\mu_0(\mu_p)$ is translation invariant and has vanishing odd correlations

[2,15]. Hence $\mu_o=\mu_p=\frac{1}{2}[\mu_+ + \mu_-]$. This implies in particular the existence of "long range order" in these states for $\beta>\beta_c$, i.e. $\langle\sigma_i\sigma_j\rangle_{\mu_0} \xrightarrow[|i-j|\to\infty]{} [m^*(\beta)]^2>0$, for $\beta>\beta_c$. (The converse of this statement, long range order $\Longrightarrow m^*(\beta)>0$, is also true [2]).

iii) For the two dimensional Ising system with nearest neighbor pair interactions the continuity of $\partial\Psi(\beta,h=0)/\partial\beta$ follows from Onsager's [1,17] exact computation of $\Psi(\beta,0)$. Hence theorem 6 establishes the existence of exactly two extremal states for all $\beta>\beta_c$, (β_c being here the place where the second derivative of $\Psi(\beta,0)$ diverges logarithmically [18]). This result for the square lattice was proven earlier, using duality, by Messager and Miracle-Sole [7]. For more general Ising systems with even ferromagnetic interactions this result is known at low temperatures (not all the way to T_c) from the work of Gallavotti and Miracle-Sole and of Slawny [8]. Gallavotti and Miracle-Sole used (for nearest neighbor interactions) a beautiful version of the Peierls argument while Slawny uses the Asano-Ruelle method of locating zeros of the partition function to prove analyticity of $\psi(\underline{J})$ in the even interactions at sufficiently large β. Using the above theorem it is sufficient to establish that $\psi(\beta,0)$ is C^1. This can be done readily if the correlation function in the + state cluster sufficiently well for $\sum_x [\langle\sigma_A\sigma_{B+x}\rangle(\beta,0;+)-\langle\sigma_A\rangle(\beta,0;+)\langle\sigma_{B+x}\rangle(\beta,0;+)]<\infty$ [18]. The latter can be easily proven for large β by a Peierls type argument [19] which actually establishes exponential clustering.

iv) Theorem 8 can be generalized, in a fairly direct way, using the ideas of Slawny, Gruber and their coworkers to non-even interactions. One then gets a larger number of extremal states: these are related to the group, acting on the spins, which leaves the Hamiltonian invariant.

4. General One Component Spin Systems

In order to generalize theorems 5-8 to arbitrary (rapidly decaying) free measures $\rho_i(ds_i)$ we first need the analog of corollarys 4 and 5 for such ρ_i's. This can be achieved by defining

$$\sigma(s;\lambda) = \begin{cases} s, & |s|\leq\lambda \\ \lambda, & |s|\geq\lambda, \qquad 0<\lambda<\infty \end{cases} \tag{4.1}$$

and choosing in Lemma 2, $f_i(s_i)=g_i(s_i)=\sigma(s_i,\lambda_i)$. Using (2.5) the analog of Corollary 4 would now be:

Corollary 4': Let $J_K \geq |J'_K|$. Then $\langle\sigma_A(\underline{s}_A;\lambda_A)\rangle = \langle\sigma_A(\underline{s}_A;\lambda_A)\rangle'$ and $\langle\sigma_B(\underline{s}_B;\lambda_B)\rangle = \langle\sigma_B(\underline{s}_B;\lambda_B)\rangle'$ implies $\langle\sigma_A(\underline{s}_A;\lambda_A)\sigma_B(\underline{s}_B;\lambda_B)\rangle = \langle\sigma_A(\underline{s}_A;\lambda_A)\sigma_B(\underline{s}_B;\lambda_B)\rangle'$ where $\sigma_C(\underline{s}_C;\lambda_C) = \prod_{i\in C}[\sigma(s_i;\lambda_i)]^{\ell_i}$, $\ell_i = 1,2,\ldots.$

To obtain the analog of Corollary 5 we make two remarks: (a) since both $\sigma(s;\lambda)$ and $s-\sigma(s;\lambda)$ are odd monotone increasing functions of s it follows from Lemma 2 that $\langle\sigma(s;\lambda)\rangle \geq \langle\sigma(s;\lambda)\rangle'$ and $\langle s-\sigma(s;\lambda)\rangle \geq \langle s-\sigma(s;\lambda)\rangle'$. Hence $\langle s_i\rangle = \langle s_i\rangle'$ implies $\langle\sigma(s_i;\lambda_i)\rangle = \langle\sigma(s_i;\lambda_i)\rangle'$ for all $\lambda_i \geq 0$. Similarly $\langle s_i s_j\rangle = \langle s_i s_j\rangle'$ implies $\langle\sigma(s_i;\lambda_i)\sigma(s_j;\lambda_j)\rangle = \langle\sigma(s_i;\lambda_i)\sigma(s_j;\lambda_j)\rangle'$, etc. Since $[\sigma(s_i;\lambda_i)/\lambda_i]^2 \to 1$ as $\lambda_i \to 0$, and $s_i \neq 0$, $\langle\sigma(s_1;\lambda_1)[\sigma(s_2;\lambda_2)]^2\sigma(s_3;\lambda_3)\rangle = \langle\sigma(s_1;\lambda_1)[\sigma(s_2;\lambda_2)]^2\sigma(s_3;\lambda_3)\rangle'$ for all $\lambda_2 \Longrightarrow \langle\sigma(s_1;\lambda_1)\sigma(s_3;\lambda_3)\rangle = \langle\sigma(s_1;\lambda_1)\sigma(s_3;\lambda_3)\rangle'$ (if $\rho_2(ds_2)$ is not concentrated on $s_2=0$, which is an irrelevant case).

Permitting now the $\{\lambda_i\}$ in Corollary 4' to vary arbitrarily we obtain

Corollary 5': Let $J_K \geq |J'_K|$. Then: (1) $\langle s_i\rangle = \langle s_i\rangle'$ $\forall i\in\Lambda$ implies $\mu(d\underline{S}_\Lambda) = \mu'(d\underline{S}_\Lambda)$. (2) $\langle s_i s_j\rangle = \langle s_i s_j\rangle'$ for all $i,j\in\Lambda$ implies that all even moments of μ and μ' are equal.

Let us consider first the case when $\rho(dS)$ has compact support in some interval $[-R,R]$. Identifying ±b.c. with $s_i^b = \pm R$, for $i\in\Lambda_c$, Theorems 6-8, remain unchanged. In the case of unbounded spins however the situation is more complicated. It was shown by Lebowitz and Presulti [20] for the case of pair interactions (but this restriction is unessential) that if one restricts oneself to "regular" Gibbs states - those for which $\mu(d\underline{S}_\Omega)$, the projection of μ on the region $\Omega \subset \mathbb{Z}^\nu$, does not grow "too fast" - then the role of $\mu_\pm$ will be played by states obtained as the infinite volume limit of Gibbs states with b.c. $s_j^b = \pm a \ln |j|$. It seems likely that the translation invariant regular Gibbs states satisfy the variational principle (indeed all the solutions of variational principle may be regular) and theorems 6-8 would then apply to these states. (Indeed, using methods similar to these, Theorem 6 has been extended recently to classical rotators [21]).

Acknowledgements: I would like to thank F. Dunlop and E. Presutti for many valuable discussions.

REFERENCES

1. J.L. Lebowitz, Coexistence of Phases in Ising Ferromagnets, Jour. Stat. Phys. 16, 463 (1977).

2. c.f. R.B. Griffiths, in Phase Transitions and Critical Phenomena, C. Domb and M.S. Green, eds., Academic Press (1972), Vol. 1, pp. 7-109; C. Gruber, J. Stat. Phys. 14, 81 (1976).

3. D. Ruelle, Statistical Mechanics, Benjamin (1969): Thermodynamic Formalism, Addison-Wesley (to appear).

4. R.L. Dobrushin, Functional Anal. Appl. 2, 292 and 302 (1968).

5. O.E. Lanford and D. Ruelle, Commun. Math. Phys. 13, 194 (1969); O.E. Lanford, in Statistical Mechanics and Math. Problems, A. Lenard, ed., Springer (1973).

6. R.B. Israel, Comm. Math. Phys. 43, 59 (1975): S.A. Pirogov and Ya. G. Sinai, Funkts. Analiz. 25 (1974): D. Ruelle, On Manifolds of Phase Coexistence, Preprint (1975).

7. A. Messager and S. Miracle-Sole, Comm. Math. Phys. 40, 187 (1975).

8. G. Gallavotti and S. Miracle-Sole, Phys. Rev. B5, 2555 (1972): J. Slawny, Comm. Math. Phys. 34, 271 (1973): 46, 75 (1976): C. Gruber, A. Hinterman and D. Merlini, Comm. Math. Phys. 40, 83 (1975): C. Gruber and A. Hinterman, Physics 83A, 233 (1975).

9. J. Ginibre, Comm. Math. Phys. 16, 310 (1970).

10. R.B. Griffiths, J. Math. Phys. 8, 478 (1967); D.G. Kelley and S. Sherman, J. Math. Phys. 9, 466 (1969).

11. J.L. Lebowitz and A. Martin-Löf, Comm. Math. Phys. 25, 276 (1972).

12. H. van Beyeren, Comm. Math. Phys. 40, 1 (1975).

13. C.M. Fortuin, P.W. Kasteleyn and J. Ginibre, Comm. Math. Phys. 22, 89 (1971).

14. J.L. Lebowitz, J. Stat. Phys., 16, 3 (1977).

15. c.f. J.L. Lebowitz, in Mathematical Problems in Theoretical Physics, H. Araki, ed., Springer (1975).

16. J. Frohlich, B. Simon and T. Spencer, Comm. Math. Phys. 50, 79 (1976).

17. L. Onsager, Phys. Rev. 65, 117 (1944).

18. J.L. Lebowitz, Comm. Math. Phys. 28, 313 (1972).

19. A. Martin-Löf, Comm. Math. Phys. 25, 87 (1972).

20. J.L. Lebowitz and E. Presutti, Comm. Math. Phys. 50, 195 (1976).

21. A. Messager, S. Miracle-Sole and C. Pfister, to be published.

A SURVEY OF LOCAL COHOMOLOGY

John E. ROBERTS

Centre de Physique Théorique, CNRS, Marseille and
UER Expérimentale et Pluridisciplinaire de Luminy

1. Introduction

Local cohomology may be thought of as the quantum analogue of the cohomology of differential forms on Minkowski space. I shall first formulate an analogue in the framework of Wightman field theory and describe the more interesting algebraic analogue in the next section.

1.1 Definition. A Wightman field W which is a totally antisymmetric covariant tensor field of rank p will be called a local p-form. It will be called a closed local p-form if its exterior derivative vanishes, $dW = 0$, and an exact local p-form if there is a local $(p-1)$-form V such that $W = dV$.

The key observation is that interesting phenomena are linked with the constraints imposed on the cohomology of differential forms by the locality requirement of quantum field theory. This was first pointed out to me many years ago by R. Haag in connection with quantum electrodynamics. Maxwell's equations may be written $dF = 0$, $J = dF^*$, where J is the dual tensor of the electric current, $J = j^*$. On the physical Hilbert space, where Maxwell's equations are valid as operator equations, F is a closed local 2-form but not an exact local 2-form since there is no local electromagnetic potential. The fact that J is an exact local 3-form implies that the electric charge is a superselection quantum number, a point of view which has been stressed by Strocchi and Wightman [1]. It also implies that the charge-carrying operators, e.g. the electron field, are not local relative to the observables on the physical Hilbert space. Thus the most striking qualitative features of quantum electrodynamics are intimately linked with the ideas of local cohomology.

Conversely if j is some other conserved current for which there are charge-carrying local operators then $J = j^*$ is a closed local 3-form but not an exact local 3-form. The simplest example here is the current of the free charged scalar field.

As far as the 1-cohomology goes, Pohlmeyer [2] showed that, in a theory with mass gap, every closed local 1-form is an exact local 1-form. He also pointed out that this result is not valid if $s = 1$, where s denotes the number of space dimensions. This can be anticipated from the above discussion because, if $s = 1$, a closed local 1-form is the dual tensor of a conserved current.

Little is known in general about higher cohomologies. The practical method for showing that a closed local p-form is not an exact local p-form is to show that its commutation properties are too weak. This has been illustrated above for the case $p = s$. A more sophisticated variant is

1.2. Proposition. Let F and G be closed 2-forms which are relatively local and f, g smooth functions on Minkowski space with $s=3$ with support in a double cone $\mathcal{O}$ centred on the origin. Let c, c' be differentiable 2-simplexes and define

$$F_f(c) = \int F^{\mu\nu}(x+y)\, f(y)\, dc_{\mu\nu}(x), \qquad G_g(c') = \int G^{\mu\nu}(x+y)\, g(y)\, dc'_{\mu\nu}.$$

Suppose $\partial c - \partial c' \subset 2\mathcal{O}'$ then $[F_f(c), G_g(c')]$ depends only on the homology class of ∂c in $(\partial c' + 2\mathcal{O})'$ and the homology class of $\partial c'$ in $(\partial c + 2\mathcal{O})'$. If $F = dA$ and A is relatively local to G then $[F_f(c), G_g(c')] = 0$ whenever $\partial c - \partial c' \subset 2\mathcal{O}'$.

As a corollary we have a result first pointed out to me by D. Buchholz.

1.3. Corollary. Under the hypotheses of Proposition 1.2, if F and G are local with respect to an irreducible set of fields then $[F_f(c), G_g(c')]$ is a multiple of the identity whenever $\partial c - \partial c' \subset 2\mathcal{O}'$.

Proposition 1.2 yields a new proof of the well known fact [3] that the free electromagnetic field is a closed local 2-form but not an exact local 2-form. Taking for F the electromagnetic field, for G the dual tensor F^*, for ∂c the circle $x_0 = 0$, $x_2 = 0$, $(x_1 - 1)^2 + x_3^2 = 1$ and for $\partial c'$ the circle $x_0 = 0$, $x_1^2 + x_2^2 = 1$, $x_3 = 0$, a simple direct computation shows that

$$i[F_f(c), F_g^*(c')] = k \int f(x)\, d^4x \int g(x)\, d^4x \tag{1.1}$$

where $|k| \neq 0$ depends on the normalization of F and the sign of k on the relative orientation of ∂c and $\partial c'$. Hence neither F nor F^* is an exact local 2-form. The explicit proof for this case is instructive. Suppose $F = dA$; one cuts ∂c along $x_3 = 0$ getting two 1-simplexes b and b' so that $F_f(c) = A_f(\partial c) = A_f(b) + A_f(b')$. Now let c_1 and c_2 denote two hemispheres obtained by cutting $x_0 = 0$, $x_1^2 + x_2^2 + x_3^2 = 1$ along $x_3 = 0$ oriented so that $\partial c_1 = \partial c_2 = \partial c'$. Suppose it is c_2 which intersects b then if A is local

$$[A_f(\partial c), F_g^*(c')] = [A_f(b), F_g^*(c_1)] + [A_f(b'), F_g^*(c_2)] = 0$$

contradicting (1.1).

The physical interpretation of this result is that a local electromagnetic potential A does not exist because the coherent state ω_b created from the vacuum by $e^{-iA_f(b)}$ would have charges q and $-q$ at the endpoints of the path b. To see this one notes that $F_g^*(c_1) - F_g^*(c_2)$ is proportional to the total charge in $x_0 = 0$, $x_1^2 + x_2^2 + x_3^2 \leq 1$ which is zero since $dF_g^* = 0$. But if one computes each term separately in the fictitious state ω_b one deduces that

$$\omega_b(F_g^*(c_1)) - \omega_b(F_g^*(c_2)) = k\int f(x)\,d^4x \int g(x)\,d^4x .$$

I would emphasise (cf. [3]) that the absence of a local electromagnetic potential is a consequence of the operator equation $dF^* = 0$ and has little to do with the metric of the underlying Hilbert space. In view of the direction taken by lattice gauge theories in the Hamiltonian approach [4] , it is worth pointing out that if one is prepared to make do with unitary operators $W(h)$ corresponding to $e^{i\int A^\mu(x) h_\mu(x) d^4x}$ then one can avoid the indefinite metric by taking the Weyl system for the electromagnetic potential over $\mathcal{D}(\mathbb{R}^4) \otimes \mathbb{R}^4$

$$W(h)\,W(h') = e^{-\frac{i}{2}\sigma(h,h')}\,W(h+h')$$

$$\sigma(h,h') = \int [h^\mu(0,\vec{x})\dot{h}'_\mu(0,\vec{x}) - \dot{h}^\mu(0,\vec{x})h'_\mu(0,\vec{x})]\,d^3\vec{x} \tag{1.2}$$

and defining the following vacuum functional.

$$\omega_0(W(h)) = 0 \quad \text{if} \quad \partial^\mu h_\mu \neq 0. \tag{1.3}$$

If $\partial_\mu h^\mu = 0$ then $(s>2)$, we may write $h_\mu = 2\partial^\nu g_{\mu\nu}$ for an antisymmetric tensor $g_{\mu\nu} \in \mathcal{D}(\mathbb{R}^4) \otimes \mathbb{R}^6$ and set

$$\omega_0(W(h)) = \omega_0(e^{i\int F^{\mu\nu}(x) g_{\mu\nu}(x) d^4x}) , \qquad h_\mu = 2\partial^\nu g_{\mu\nu} \tag{1.4}$$

Of course $\lambda \to \omega_0(W(\lambda h))$ is not continuous at $\lambda = 0$ unless $\partial^\mu h_\mu = 0$ so there is no linear electromagnetic potential A just a unitary potential W that may be thought of as a field taking its values in the Abelian compact group dual to $\mathbb{R}^4$ endowed with the discrete topology.

This parallels the treatment in [5] of the free massless "scalar" field with $s = 1$. This is really a local 1-form A satisfying the self-dual equations $dA = 0$, $dA^* = 0$ and the missing primitive for A can be introduced either by using an indefinite metric or as in [5] by using Weyl operators leading to new superselection sectors. Of course, the new sectors of the free electromagnetic field are unphysical in the sense that $dF^* \neq 0$ there.

In interacting quantum electrodynamics, the dearth of rigorous results forces me to rely on a speculative chain of reasoning. Everyone believes that here too there is no local electromagnetic potential A . As there are electrically charged particles now, the reason is not just that $e^{iA_f(b)}$ tries to create charges at the vertices of b , but probably that such charges would, as in the free theory, not be quantized. However, there is a formal gauge-invariant expression with just the right properties for being a unitary potential for a quantized charge :

$$\psi(\partial_0 b)^* e^{ie\int A^\mu(x)\, db_\mu(x)} \psi(\partial_1 b) \tag{1.5}$$

at least if one replaces the electron field in (1.5) by its phase.
Thus one might envisage a true unitary potential $u(b)$, say, that might be thought of as a path-dependent field taking values in the circle. This is consistent with the demands of the non-Abelian 2-cohomology described in section 3. The moral for constructivists is that perhaps the Gupta-Bleuler formalism, despite its practical utility, will prove to have been a time-consuming blind alley and that serious attention should be focussed on quantum fields taking values in non-linear manifolds.

2. Algebraic Field Theory

The local cohomology considered in the last section has one fatal drawback : the cohomology classes are real vector spaces and it is only in exceptional circumstances, such as free quantum electrodynamics, that one meets such invariants in elementary particle physics. In practice, invariants usually arise as the invariants of continuous unitary representations of locally compact groups.

A W^*-category is the abstract way of looking at representation theory in Hilbert space. Its objects may be thought of as representations and its arrows (morphisms) as intertwining operators between the representations.

2.1. Definition. A W^*-category M is a category where the set (π,π') of arrows between any two objects π and π' of M is a weakly closed linear subspace of the set of bounded operators from a Hilbert space $H(\pi)$ to a Hilbert space $H(\pi')$. Further we require that $1_{H(\pi)} \in (\pi,\pi')$, $(\pi,\pi')^* = (\pi',\pi)$ and that composition in M coincides with the composition of bounded linear operators.

Notice that (π,π) is a von Neumann algebra for each object π of M . Most of the elementary results for von Neumann algebras allow a simple generalization to W^*-categories. Of course, one can characterize a W^*-category abstractly, suppressing all mention of the Hilbert spaces $H(\pi)$.

The usual concepts of representation theory such as unitary equivalence, quasiequivalence, disjointness etc. can be defined in terms of intertwining operators (cf. [6, §5]) and hence apply to a W^*-category. In particular, π and π' are unitarily equivalent $\pi \simeq \pi'$ if (π,π') contains an isometry, π is irreducible if $(\pi,\pi) \cong \mathbb{C}$ and π is a direct sum of π_i , $i \in I$ if there are isometries $W_i \in (\pi_i,\pi)$ such that $\sum_{i\in I} W_i W_i^* = 1_\pi$.

Algebraic field theory tries to describe the structural properties of elementary particle physics in terms of the "algebra of local observables" $\mathfrak{A}$. Let $\mathcal{K}$ denote the set of closed double cones in Minkowski space. If $\mathcal{O} \in \mathcal{K}$, then $\mathfrak{A}(\mathcal{O})$ is thought of as the von Neumann algebra generated by the observables one can measure within $\mathcal{O}$. If $\mathcal{O}_1 \subset \mathcal{O}_2$ then $\mathfrak{A}(\mathcal{O}_1) \subset \mathfrak{A}(\mathcal{O}_2)$ and

$$A_1 A_2 = A_2 A_1 \, , \quad A_1 \in \mathfrak{A}(\mathcal{O}_1) \, , \quad A_2 \in \mathfrak{A}(\mathcal{O}_2) \, , \quad \mathcal{O}_1 \subset \mathcal{O}_2' \tag{2.1}$$

reflecting Einstein's causality principle. This is summed up by calling $\mathfrak{A}$ a <u>local net</u> of von Neumann algebras over $\mathcal{K}$.

We now describe the technical details involved in formulating the local 1-cohomology in the algebraic framework. One must first define $\mathfrak{A}$ on a larger class of sets in Minkowski space, for example the set $\mathcal{C}$ of all compact sets. This can always be done by defining, for $F \in \mathcal{C}$

$$\mathfrak{A}(F) = \vee \{ \mathfrak{A}(\mathcal{O}) : \mathcal{O} \in \mathcal{K} , \mathcal{O} \subset F \} \tag{2.2}$$

and $\mathfrak{A}$ is then a local net over $\mathcal{C}$. The local 1-cohomology turns out to be independent of the way in which the extension is made.

Let Σ_n denote the set of n-simplexes in Minkowski space, i.e. the set of continuous maps from $\Delta^n = \{ (t^0, t^1, \dots, t^n) \in \mathbb{R}^{n+1} : t^i \geq 0 \, , \sum_{i=0}^{n} t^i = 1 \}$ into Minkowski space, and let $\partial_i : \Sigma_n \to \Sigma_{n-1}$, $i = 0, 1, 2, \dots, n$ be the face maps defined by

$$(\partial_i c)(t^0, t^1, \dots, t^{n-1}) = c(t^0, t^1, \dots, t^{i-1}, 0, t^i, \dots, t^n) \, , \quad c \in \Sigma_n \, . \tag{2.3}$$

Let $\mathcal{K}_0$ denote the set of double cones centred on the origin.

<u>2.3. Definition.</u> A <u>local 1-cocycle</u> z with values in $\mathfrak{A}$ is a mapping $z : \Sigma_1 \to \mathcal{U}(\mathfrak{A})$, the unitary group of $\mathfrak{A}$, satisfying the cocycle identity

$$z(\partial_0 c) z(\partial_2 c) = z(\partial_1 c) \, , \quad c \in \Sigma_2 \tag{2.4}$$

and the locality condition, there exist $\mathcal{O} \in \mathcal{K}_0$ with

$$z(b) \in \mathfrak{A}(\mathcal{O} + b) \, , \quad b \in \Sigma_1 \tag{2.5}$$

where $\mathcal{O} + b = \mathcal{O} + b(\Delta^1)$. We say z is localized in $\mathcal{O} \in \mathcal{K}_0$ if (2.5) holds for all $\mathcal{O}_1 \in \mathcal{K}_0$ with $\mathrm{int}\, \mathcal{O}_1 \supset \mathcal{O}$.

This definition improves that of [7] where the inner automorphisms of $\mathfrak{A}$ were used as coefficients for the local cohomology.

We consider the local 1-cocycles as the objects of a net of W^*-categories $Z^1_\ell(\mathfrak{A})$. An arrow in $Z^1_\ell(\mathfrak{A})$ from z' to z is a triple $(z|w|z')$ where $w : \Sigma_0 \to \mathfrak{A}$ satisfies

$$z(b)\,w(\partial_1 b) = w(\partial_0 b)\,z'(b), \qquad b \in \Sigma_1 \tag{2.6}$$

and there exists $\mathcal{O} \in \mathcal{K}_0$ with

$$w(a) \in \mathfrak{A}(\mathcal{O}+a), \qquad a \in \Sigma_0 \tag{2.7}$$

$(z|w|z')$ is localized in $\mathcal{O} \in \mathcal{K}_0$ if z and z' are localized in $\mathcal{O} \in \mathcal{K}_0$ and if (2.7) holds for all $\mathcal{O}_1 \in \mathcal{K}_0$ with $\operatorname{int}\mathcal{O}_1 \supset \mathcal{O}$. The composition law in $Z^1_\ell(\mathfrak{A})$ is defined by

$$(z|w|z') \circ (z'|w'|z'') = (z|ww'|z'') \tag{2.8}$$

where $(ww')(a) = w(a)w'(a)$. The involution of $Z^1_\ell(\mathfrak{A})$ is defined by

$$(z|w|z')^* = (z'|w^*|z) \tag{2.9}$$

where $w^*(a) = w(a)^*$. The complex linear structure is also defined similarly by pointwise operations. The norm is defined by $\|(z|w|z')\| = \|w(a)\|$ and this is independent of $a \in \Sigma_0$ by (2.6). If $Z^1_{\mathcal{O}}(\mathfrak{A})$ denotes the subcategory of defined by the arrows localized in $\mathcal{O} \in \mathcal{K}_0$, then $Z^1_{\mathcal{O}}(\mathfrak{A})$ is a W^*-category and $Z^1_\ell(\mathfrak{A})$ is a net of W^*-categories over $\mathcal{K}_0$.

<u>2.4. Definition</u>. Two local 1-cocycles are <u>cohomologous</u> if they are unitary equivalent as elements of $Z^1_\ell(\mathfrak{A})$. z is a <u>local 1-coboundary</u> if it is cohomologous to the trivial local 1-cocycle 1 defined by $1(b) = I$, $b \in \Sigma_1$.

The main results on the local 1-cohomology are summarized in

<u>2.4. Theorem</u>. If $s > 1$ and $\mathfrak{A}$ is the observable net then $Z^1_\ell(\mathfrak{A})$ is closed under direct sums and subobjects. It admits a strictly associative "tensor" product $\times$ with the trivial 1-cocycle as unit. $z \times z'$ and $z' \times z$ are related by a coherent natural unitary equivalence. There is a dimension function d taking values in $\mathbb{Z}_+ \cup \{\infty\}$ such that

$$d(z \oplus z') = d(z) + d(z'), \qquad d(z \times z') = d(z)\,d(z')$$

If $d(z) < \infty$, z is a finite direct sum of irreducible cocycles and there is a conjugate cocycle $\bar{z}$ defined up to unitary equivalence such that 1 is a subobject of $\bar{z} \times z$ and $\bar{z} \oplus \bar{z}' \simeq \overline{z \oplus z'}$, $\bar{z} \times \bar{z}' \simeq \overline{z \times z'}$.

This theorem is proved using the techniques of [8,9] together with arguments having the flavour of algebraic topology. As will be explained later, the irreducible cohomology classes should be thought of as being in 1-1 correspondence with the equivalence classes of irreducible continuous unitary representations of the gauge group of the first kind $\mathcal{G}$.

$\mathcal{G}$ may be spontaneously broken and is not necessarily a compact group. Theorem 2.4 leaves something to be desired because there are examples of cocycles z with $d(z)=\infty$ which cannot be decomposed into a direct sum of irreducibles and Theorem 2.4 then gives no indications about conjugate cocycles.

The object of study in [8] was the superselection structure of representations π of the observable net $\mathfrak{A}$ satisfying the following selectrion criterion : there exists $\mathcal{O} \in \mathcal{K}_0$ such that

$$\pi_0 \restriction \mathfrak{A}(\mathcal{O}'+a) \simeq \pi \restriction \mathfrak{A}(\mathcal{O}'+a), \quad a \in \Sigma_0 \tag{S}$$

Here $\mathfrak{A}(\mathcal{O}')$ denotes the C^*-subalgebra of $\mathfrak{A}$ generated by the $\mathfrak{A}(\mathcal{O}_1)$ with $\mathcal{O}_1 \in \mathcal{K}$ and $\mathcal{O}_1 \subset \mathcal{O}'$ and π_0 denotes the vacuum representation of $\mathfrak{A}$ on $\mathcal{H}_0$ which we regard as the defining representation so that the symbol π_0 may be omitted. Let

$$\mathfrak{A}^d(\mathcal{O}) = \mathfrak{A}(\mathcal{O}')', \quad \mathcal{O} \in \mathcal{K}.$$

$\mathfrak{A}^d$ is called the <u>dual net</u> of $\mathfrak{A}$. In [8] $\mathfrak{A}$ was assumed to satisfy <u>duality</u>, i.e. $\mathfrak{A} = \mathfrak{A}^d$. This is too restrictive as it rules out spontaneously broken gauge symmetries. Instead we assume that $\mathfrak{A}$ satisfies <u>essential duality</u>, i.e. $\mathfrak{A}^d = \mathfrak{A}^{dd}$. Essential duality, unlike duality, is a consequence of the arguments of Bisognano and Wichmann [10].

Using the techniques of local cohomology one can show

<u>2.5. Theorem</u>. If $s > 1$ and $\mathfrak{A}$ satisfies essential duality then every representation π satisfying (S) has a canonical extension to a representation $\tilde{\pi}$ of $\mathfrak{A}^d$ on the same Hilbert space satisfying (S) with $\tilde{\pi}_0$ replacing π_0.

If $s = 1$, $\mathcal{O}'$ is no longer path-connected but decomposes into two components denoted by $\mathcal{O}^r$ and $\mathcal{O}^\ell$, the right and left spacelike complements of $\mathcal{O}$. Correspondingly there are, in general, two different extensions $\tilde{\pi}^\ell$ and $\tilde{\pi}^r$ (Compare [7 ; Thm. 3.3]). If $\tilde{\pi}^\ell \neq \tilde{\pi}^r$, then π should be interpreted as a "soliton sector" of relevance to elementary particle physics as it represents a local finite-energy perturbation of the vacuum sector. The soliton aspect only becomes apparent when π is extended to a larger algebra $\mathfrak{A}^d$ containing some "field quantities". However this extension is done it exhibits non-local features usually expressed in terms of a homotopy invariant. This is analogous to the interpretation of the chemical potential in [11] where this too is a latent parameter which appears when a KMS state of the observable algebra is extended to a KMS state of the field algebra. Examples of soliton behaviour which fit into this pattern may be found in [12, 13].

Conversely if $s > 1$ any soliton sector must violate the selection criterion (S) ; this is what happens in gauge theories so Theorem 2.5 leads to the same qualitative conclusions as the energy argument of [14].

2.6. Corollary. The analysis of [7,8] applies equally well to observable algebras satisfying essential duality.

In fact this analysis may be understood in terms of the local 1-cohomology by applying Theorem 2.4 to $\mathfrak{A}^d$ rather than to $\mathfrak{A}$ because

2.7. Proposition. If $s > 1$, the equivalence classes of representations of $\mathfrak{A}$ satisfying (S) are in 1-1 correspondence with the cohomology classes of local 1-cocycles of $\mathfrak{A}^d$.

3. Towards a local 2-cohomology

The local 1-cohomology discussed above should be regarded as the footprints in the observable algebra of an algebra of fields acted on by a gauge group of the first kind. From a mathematical point of view, the two structures are probably equivalent for $s > 1$ provided one insists on a maximal field algebra with normal commutation relations. This has, however, only been established [15] in the case of an Abelian gauge group that is not spontaneously broken.

Hence to exhibit observable nets with a non-trivial local 1-cohomology one naturally turns to a field net $\mathfrak{F}$ acted on by a gauge group $\mathcal{G}$ of the first kind and defines $\mathfrak{A}$ by the principle of gauge invariance

$$\mathfrak{A}(\mathcal{O}) = \{ A \in \mathfrak{F}(\mathcal{O}) : \beta_g(A) = A , g \in \mathcal{G} \} . \tag{3.1}$$

To explain how local 1-cocycles are linked to the representation theory of $\mathcal{G}$ we need

3.1. Definition [16,17] . A Hilbert space H in $\mathfrak{F}(\mathcal{O})$ is a norm-closed linear subspace of $\mathfrak{F}(\mathcal{O})$ such that $\psi'^*\psi \in \mathbb{C}I$ if $\psi, \psi' \in H$ and $F\psi = 0$, for all $\psi \in H$ implies $F = 0$.

If $\beta_g(H) = H$, $g \in \mathcal{G}$ then β induces a continuous unitary representation of $\mathcal{G}$ on H and we may define a local 1-cocycle by

$$z(b) = \sum_{j \in J} \alpha_{\partial_0 b}(\psi_j)\, \alpha_{\partial_1 b}(\psi_j)^* , \qquad b \in \Sigma_1 \tag{3.2}$$

where $\{\psi_j\}_{j\in J}$ is an orthonormal basis of H and α_a denotes translation through $a \in \Sigma_0$. z is a direct sum of trivial 1-cocycles if and only if H is pointwise invariant under $\mathcal{G}$.

The evidence presented up till now indicates that the local 2-cohomology will be related to field algebras acted on by gauge groups of the second kind. To see this we return to quantum electrodynamics. We expect (1.5) to generate local quantities in the vacuum Hilbert space which are not elements of the observable net $\mathfrak{A}$ generated by the electromagnetic field F. Thus $\mathfrak{A}$ will not satisfy duality although it should still satisfy essential duality in the light of [10]. We have been interpreting $\mathfrak{A} \neq \mathfrak{A}^d$ to mean that there are spontaneously broken gauge symmetries present. This is eminently reasonable here too if we take the unitary potential $u(b)$ seriously. I would expect that if say $g \in C^\infty(\mathbb{R}^4)$ there are automorphisms β_g of $\mathfrak{A}^d$ leaving $\mathfrak{A}$ pointwise invariant such that

$$\beta_g(u(b)) = e^{i(g(\partial_0 b) - g(\partial_1 b))}\, u(b), \qquad b \in \Sigma_1. \tag{3.3}$$

Now set

$$y(b)(A) = u(b)\, A\, u(b)^*, \qquad A \in \mathfrak{A}, \qquad b \in \Sigma_1 \tag{3.4}$$

$$z(c) = u(\partial_0 c)\, u(\partial_2 c)\, u(\partial_1 c)^*, \qquad c \in \Sigma_2 \tag{3.5}$$

then $y(b)$ is an automorphism of $\mathfrak{A}$ and $z(c)$ a unitary operator of $\mathfrak{A}$. This leads to

3.2. Definition. A local 2-cocycle of $\mathfrak{A}$ is a pair (y,z) where for each $b \in \Sigma_1$ $y(b)$ is a morphism of $\mathfrak{A}$ and for each $c \in \Sigma_2$, $z(c)$ is a unitary of satisfying the 2-cocycle identities

$$z(c)\, y(\partial_1 c)(A) = y(\partial_0 c)\, y(\partial_2 c)(A)\, z(c), \quad A \in \mathfrak{A}, \quad c \in \Sigma_2 \tag{3.6}$$

$$z(\partial_0 d)\, z(\partial_2 d) = y(\partial_0\partial_1 d)(z(\partial_3 d))\, z(\partial_1 d), \qquad d \in \Sigma_3 \tag{3.7}$$

and the locality requirement : there exists $\mathcal{O} \in \mathcal{K}_0$ such that $y(b)(A) = A$ if $A \in \mathfrak{A}((\mathcal{O}+b)')$, $b \in \Sigma_1$ and $z(c) \in \mathfrak{A}(\mathcal{O}+c)$, $c \in \Sigma_2$.

In fact the coefficient object here is really a net of monoidal W^*-categories End $\mathfrak{A}$ (cf. [17, §3]) and the local 2-cocycles should be considered as the objects of a 2-category $Z^2_\ell(\mathfrak{A})$. However I prefer to omit these technicalities in favour of a simplified picture of how fields, gauge groups and observables seem to be related to local cohomology. In the case of gauge groups of the second kind this has yet to be tested on a model. The field net $\mathfrak{F}$ is constructed from the observable net $\mathfrak{A}$ by glueing on a net $\mathcal{H}$ of Hilbert spaces carrying continuous unitary representations of the gauge group $\mathcal{G}$. This may be thought of as an exact sequence

$$1 \to \mathfrak{A} \to \mathfrak{F} \to \tilde{\mathcal{H}} \to 1 \tag{3.8}$$

where the tilde indicates that I identify Hilbert spaces carrying equivalent representations of $\mathcal{G}$. $\tilde{\mathcal{H}}$ has an obvious quotient net structure induced by $\mathcal{H}$ and, in general, this net structure will be non-trivial. This can be seen by looking at (3.3) where the character

$$\chi_g(b) = e^{i(g(\partial_0 b) - g(\partial_1 b))} \tag{3.9}$$

can be expected to be an element of $\tilde{\mathcal{H}}(\mathcal{O})$ only if $\partial_0 b, \partial_1 b \in \mathcal{O}$. Now the exact sequence (3.8) induces an exact sequence in local cohomology

$$1 \to H^0_\ell(\tilde{\mathcal{H}}) \to H^1_\ell(\mathfrak{A}) \to H^1_\ell(\mathcal{H}) \to H^1_\ell(\tilde{\mathcal{H}}) \to H^2_\ell(\mathfrak{A}) \to H^2_\ell(\mathcal{H}) \tag{3.10}$$

As they stand, these exact sequences should be taken with a pinch of salt, but they do indicate the spirit of what is going on. The local 0-cocycles of $\tilde{\mathcal{H}}$ correspond to representations of $\mathcal{G}$ that can be realized anywhere within $\mathfrak{F}$. They induce local 1-cocycles of $\mathfrak{A}$; this is the content of (3.2). On the other hand (3.9) is a local 1-cocycle of $\tilde{\mathcal{H}}$ and induces a local 2-cocycle of $\mathfrak{A}$ as in (3.4) and (3.5).

These ideas should apply not only to quantum electrodynamics but also to general non-Abelian gauge theories where the analogue of (1.5) is well known. If $\mathcal{G} = C^\infty(\mathbb{R}^4, \mathcal{G}_\infty)$ and ρ is a finite-dimensional continuous unitary representation of $\mathcal{G}_\infty$, one can envisage, in place of (3.3), a Hilbert space $H(b)$ in $\mathfrak{A}^d(\mathcal{O}+b)$ carrying a representation of $\mathcal{G}$ with character

$$\chi_g(b) = \mathrm{Tr}\,(\rho(g(\partial_0 b)\, g(\partial_1 b)^{-1}))\,. \tag{3.11}$$

(3.3) and (3.11) suggest that $H^2_\ell(\mathfrak{A})$ will turn out to describe the representation theory of $\mathcal{G}_\infty$.

Pure Yang-Mills fields, like free quantum electrodynamics, will not generate non-trivial local 2-cocycles in the algebraic version of local cohomology because the morphisms $y(b)$ of (3.4) are an integral part of this cohomology and reflect the presence of local potentials. This may perhaps be considered as illustrating the dictum that the principle of gauge invariance of the second kind forces an interaction on the theory.

Naturally one of the most interesting questions is whether local 2-cohomology will prove adequate for describing sectors such as the electrically charged sectors of quantum electrodynamics which do not satisfy the selection criterion (S) because the charge-carrying fields are not strictly local with respect to the observables. The basic idea of creating charge by transferring it in along a path from spacelike infinity fits in well with the ideas of local 2-cohomology. However any theory of superselection structure would necessarily be complicated because one would have to battle with the Higgs phenomenon, confinement and the occurrence of soliton sectors.

REFERENCES

[1] F. STROCCHI, A.S. WIGHTMAN,
Proof of the Charge Superselection Rule in Local Relativistic Quantum Field Theory,
J. Math. Phys. 15, 2198-2224 (1974)

[2] K. POHLMEYER,
The Equation Curl $W_\mu(x) = 0$ in Quantum Field Theory,
Commun.math.Phys., 25, 73-86 (1972)

[3] F. STROCCHI,
Gauge Problem in Quantum Field Theory,
Phys. Rev. 162, 1429-1438 (1967)

[4] J. KOGUT, L. SUSSKIND,
Hamiltonian Formulation of Wilson's Lattice Gauge Theories,
Phys. Rev. D11, 395-408 (1975)

[5] R.F. STREATER, J.F. WILDE,
Fermion States of a Boson Field,
Nucl. Phys. B24, 561-575 (1970)

[6] J. DIXMIER,
Les C^*-algèbres et leurs représentations,
Gauthier-Villars, Paris 1964

[7] J.E. ROBERTS,
Local Cohomology and Superselection Structure,
Commun.math.Phys., 51, 107-119 (1976)

[8] S. DOPLICHER, R. HAAG, J.E. ROBERTS,
Local Observables and Particle Statistics I,
Commun.math.Phys. 23, 199-230 (1971).

[9] S. DOPLICHER, R. HAAG, J.E. ROBERTS,
Local Observables and Particle Statistics II,
Commun.math.Phys., 35, 49-85 (1974)

[10] J.J. BISOGNANO, E.H. WICHMANN,
On the Duality Condition for Quantum Fields,
J. Math.Phys., 17, 303-321 (1976)

[11] H. ARAKI, R. HAAG, D. KASTLER, M. TAKESAKI,
Extensions of KMS States and Chemical Potential,
Commun.math.Phys., 53, 97-134 (1977)

[12] J. FRÖHLICH,
New Superselection Sectors ("Soliton-States") in Two-Dimensional Bose Quantum Field Theory Models,
Commun.math.Phys., 47, 269-310 (1976)

[13] J.L. BONNARD, R.F. STREATER,
Local Gauge Models predicting their own Superselection Rules,
Helv. Phys. Acta, 49, 259-267 (1976)

[14] G.H. DERRICK,
Comments on Nonlinear Wave Equations as Models for Elementary Particles,
J. Math. Phys., 5, 1252-1254 (1964)

[15] S. DOPLICHER, R. HAAG, J.E. ROBERTS,
Fields, Observables and Gauge Transformations II,
Commun. math. Phys. 15, 173-200 (1969)

[16] S. DOPLICHER, J.E. ROBERTS,
Fields, Statistics and Non-Abelian Gauge Groups,
Commun.math.Phys., 28, 331-348 (1972)

[17] J.E. ROBERTS,
Cross Products of von Neumann Algebras by Group Duals,
Symposia Mathematica 22, 335-363, Academic Press, London, New York 1976.

Operator Algebras and Statistical Mechanics

Huzihiro ARAKI

Research Institute for Mathematical Sciences
Kyoto University, Kyoto 606, JAPAN

Abstract Some topics in quantum statistical mechanics related to theory of operator algebras are reviewed.

1. C*-dynamical system

A C*-dynamical system is a pair consisting of a C*-algebra $\mathfrak{A}$ and a pointwise strongly continuous one-parameter group of automorphisms α_t, $t \in R$, of $\mathfrak{A}$. Before discussing quantum statistical mechanics under this mathematical framework in subsequent sections, we shall spend a few words to explain mathematical and physical backgrounds for this formulation.

If $\mathfrak{A}$ is abelian, a C*-dynamical system $(\mathfrak{A}, \alpha_t)$ can be viewed as a (topological) dynamical system, consisting of a locally compact space Ω (the spectrum of $\mathfrak{A}$) with $\mathfrak{A} = C_0(\Omega)$ (the algebra of continuous functions vanishing at infinity) and one-parameter group of homeomorphisms $\hat{\alpha}_t$ of Ω with $(\alpha_t a)(\xi) = a(\hat{\alpha}_{-t}\xi)$ for $a \in \mathfrak{A}$ and $\xi \in \Omega$.

The set $S(\mathfrak{A})^{\alpha}$ of all states invariant under (the adjoint action of) α_t for all $t \in R$ is a non-empty compact convex set if $\mathfrak{A}$ is unital. Extremal elements of $S(\mathfrak{A})^{\alpha}$ is called ergodic. For abelian $\mathfrak{A}$, $\phi \in S(\mathfrak{A})^{\alpha}$ corresponds to an $\hat{\alpha}$-invariant probability measure μ on Ω with the relation $\phi(a) = \int a(\xi)d\mu(\xi)$ for $a \in \mathfrak{A}$ and ϕ is ergodic if and only if (Ω, μ, α_t) is ergodic. Non-commutative ergodic theory deals with properties of ergodic states and decomposition of invariant states into ergodic states. (See, for example, [1], [2], [3] and references quoted therein.)

Dynamical systems based on differential equations are the origin of the subject. The corresponding class of C*-dynamical system can be described in terms of the following local structure. Let L_0 be a Boolean lattice and $\mathfrak{A}(\Lambda)$ for $\Lambda \in L_0$ be a $*$-subalgebra of $\mathfrak{A}$ such that $\Lambda_1 \subset \Lambda_2$ implies $\mathfrak{A}(\Lambda_1) \subset \mathfrak{A}(\Lambda_2)$, $\mathfrak{A}(\Lambda_1 \vee \Lambda_2)$ is generated by $\mathfrak{A}(\Lambda_1) \cup \mathfrak{A}(\Lambda_2)$, $\mathfrak{A}(\Lambda_1 \cap \Lambda_2) = \mathfrak{A}(\Lambda_1) \cap \mathfrak{A}(\Lambda_2)$ and the $*$-subalgebra $\mathfrak{A}_0$ of $\mathfrak{A}$ generated by $\mathfrak{A}(\Lambda)$, $\Lambda \in L_0$ is dense in $\mathfrak{A}$. Let δ_0 be a linear map from $\mathfrak{A}_0$ to $\mathfrak{A}$ satisfying $\delta_0(AB) = \delta_0(A)B + A\delta_0(B)$, and $\delta_0(A)^* = \delta_0(A^*)$ such that there exists an $H_\Lambda = H_\Lambda^* \in \mathfrak{A}$ for each $\Lambda \in L_0$ satisfying $\delta_0 A = i[H_\Lambda, A]$ for all $A \in \mathfrak{A}(\Lambda)$. (Such δ_0 is called a normal $*$-derivation.) Then the infinitesimal generator δ of α_t defined by $\delta(\alpha_t A) = (d/dt)\alpha_t A$ is required to contain $\mathfrak{A}_0$ in its definition domain and coincides with δ_0 there. Thre is obviously a consistency condition for the system $\{H_\Lambda\}$; namely $H_{\Lambda'} - H_\Lambda$ commutes with all $A \in \mathfrak{A}(\Lambda)$ if $\Lambda' \supset \Lambda$. If $\mathfrak{A}$ is a UHF-algebra, then for any given α_t there exists such a local structure with atomic L_0 and finite dimensional $\mathfrak{A}(\Lambda)$ ([4]).

In a model of quantum statistical mechanics, Λ is typically a bounded region in space, and H_Λ is the sum of energy operator in the region Λ and interaction energy between Λ and outside. The first mathematical problem in the formulation of quantum statistical mechanics as a C*-dynamical system is the construction of α_t for a given consistent family of H_Λ. Relevant questions are whether δ_0 is closable, whether δ_0 has an extension to a generator, whether the extension of δ_0 to a generator is unique and whether the closure of δ_0 is a generator.

For a large class of interactions for spin-lattice system, these extension problems have been solved in the affirmative either by an explicit construction of α_t by a power series of δ_0 ([5], [6], [7]) or by a clever use of general methods ([8]). From mathematical point of view, these extension problems for a general class of interactions for spin-lattice system is of interest and is in some sense connected with another interesting open question ([9]): whether every α_t for a UHF algebra is approximately inner in the sense that there exists $H_n \in \mathfrak{A}$ such that $\lim \mathrm{Ad}\, e^{itH_n} = \alpha_t$. Despite of recent intensive study on the extension problems, we do not seem to have a simple way of checking whether the closure of δ_0 for explicitly given H_Λ is a generator. (This corresponds to the existence and uniqueness problem for solutions of differential equations in classical dynamical system.)

The formalism described above is adapted to quantum statistical mechanics of a lattice system. An extension to continuum system involves a number of hard problems. For Bose system, even a free time evolution nor its generator can not be defined on the C*-algebras generated by strictly local observables. For Fermi system, the free time evolution falls into the above formalism. However, the introduction of the interactions seem to ruin a nice situation of the free evolution. For example the generator applied on a typical local observable does not belong to the algebra. If the interaction is of finite range and a von Neumann algebra is adopted as local algebra, then the generator applied on a typical local observable becomes an unbounded operator affiliated with the local algebra. Hence for Fermi system with a finite range interaction, locally normal states of a family of von Neumann algebras of local observables and the time translation unitary group on each representation group such that its infinitesimal action on local observables coincides with the given δ_0 could be a possible scheme.

If we leave aside the problem of construction of α_t, theory about equilibrium states (in the sense of limit Gibbs states) have been given by many authors; for example by Miracle-Sole and Robinson for potentials with hard cores ([10], [11]), by Ginibre with a method of functional integration ([12]) and by Suhov for one-dimensional system ([13] and references quoted therein).

For classical statistical mechanics, the above formalism is not directly applicable, as $[H_\Lambda, A]$ would vanish due to abelian $\mathfrak{A}$. For classical lattice system, one method is to enlarge $\mathfrak{A}$ to a non-commutative algebra and consider α_t for the given $H_\Lambda \in \mathfrak{A}$ as automorphisms of the enlarged algebra. This will be explained in the next section. For classical continuum system, the time translation is defined by an infinite system of differential equations. There, the existence and uniqueness problem (including the well-definedness of the equation) is solved in some case for almost all solutions relative to some measure ([14]). In the language of C*-algebras, this amounts to constructing α_t for the weak closure in the cyclic representation associated with a given state.

2. Semi-Quantum Lattice Systems

For the sake of definiteness, we consider a semi-quantum lattice system which has been studied from C*-algebra point of view by Kishimoto ([15]). Using the notation of the last section, the Boolean lattice L_0 consists of all finite subset of a square lattice Z^ν and its atom is a point of Z^ν. For each $\Lambda \in L_0$, $\mathfrak{A}(\Lambda)$ is finite dimensional and $\mathfrak{A}(\Lambda_1 \cup \Lambda_2) = \mathfrak{A}(\Lambda_1) \otimes \mathfrak{A}(\Lambda_2)$ if Λ_1 and Λ_2 are disjoint, where $\mathfrak{A}(\Lambda_1)$ is identified with the subalgebra $\mathfrak{A}(\Lambda_1)\otimes 1$ of $\mathfrak{A}(\Lambda_1 \cup \Lambda_2)$ and $\mathfrak{A}(\Lambda_2)$ with $1\otimes\mathfrak{A}(\Lambda_2)$. This model includes classical and quantum spin lattice systems as special cases where each $\mathfrak{A}(\Lambda)$ is abelian and a full matrix algebra, respectively.

In physics, a function $\Lambda \in L_0 \to \Phi(\Lambda) = \Phi(\Lambda)^* \in \mathfrak{A}(\Lambda)$ is given (called a potential) and H_Λ is defined as follows.

(2.1) $$H_\Lambda = \lim_{\Lambda'\uparrow} \sum\{\Phi(\Lambda_1);\ \Lambda' \supset \Lambda_1 \in L_0,\ \Lambda_1 \cap \Lambda \neq \emptyset\}$$

where $\Phi(\emptyset) = 0$, and the above limit is assumed to exists. It is then easy to check that

(2.2) $$\delta^\Phi A = i[H_\Lambda, A]\ ,\ A \in \mathfrak{A}(\Lambda)$$

defines consistently a $*$-derivation on $\mathfrak{A}$.

If each $\mathfrak{A}(\Lambda)$ has a trivial center (the case of quantum system) and $\mathfrak{A}_0$ is in the domain of the generator δ of an α_t, then δ coincides with δ^Φ for some Φ and the choice of Φ becomes unique if we require vanishing of the partial expectation of $\Phi(\Lambda)$ on Λ_1 with non-empty intersection with Λ relative to a prescribed product state on $\mathfrak{A}$ ([16]). The set of such Φ form a Fréchet space relative to seminorms $r_\Lambda(\Phi) \equiv \|H_\Lambda\|$. In this sense the requirement for the convergence of (2.1) is very natural. Conventional assumption for guaranteeing the convergence of (2.1) is to consider the Banach space of Φ with the norm

(2.3) $$\|\Phi\|_0 = \sup\{\textstyle\sum\{\|\Phi(\Lambda)\|; n\in\Lambda\};\ n = \text{atom}\}\ .$$

Then

$$H_\Lambda = \sum\{\Phi(\Lambda_1);\ \Lambda_1 \in L_0,\ \Lambda_1 \cap \Lambda \neq \emptyset\}$$

is absolutely convergent.

In non-quantum cases, H_Λ and Φ are not determined uniquely from δ^Φ even if we make the above mentioned requirement on Φ. This indicates that some information relevant to physics is lost if we consider only δ^Φ, an extreme case being the classical system where $\delta^\Phi = 0$. We can introduce a mathematical device to avoid this loss of information in the following way ([15]).

Let $\mathcal{O}(n)$ be the center of $\mathfrak{A}(n)$ for an atom n and let us imbed $\mathfrak{A}(n)$ in a full matrix algebra $\mathcal{L}(n)$ such that $\mathfrak{A}(n) = \mathcal{O}(n)' \cap \mathcal{L}(n)$.

Let $\mathscr{L}$ be the product C*-algebra of $\mathscr{L}(n)$, which contains $\mathfrak{A}$ as a C*-subalgebra. Let G_n be the compact abelian group of automorphisms of $\mathscr{L}$ given by Ad u with all unitary elements u of $\mathscr{B}(n)$ and G be the topological product of G_n. Let ε be the average over G : $\varepsilon(a) = \int g(a)dg$ where dg is the normalized Haar measure. Then ε is the projection from $\mathscr{L}$ onto $\mathfrak{A}$ with $\varepsilon(\mathscr{L}(\Lambda)) = \mathfrak{A}(\Lambda)$. For a state ϕ of $\mathfrak{A}$, there corresponds a unique G-invariant extension of ϕ to $\mathscr{L}$ given by $\hat{\phi}(a) = \phi(\varepsilon(a))$. Let the extension of δ^{Φ} to $\mathscr{L}_0$ (the union of $\mathscr{L}(\Lambda)$, $\Lambda \in L_0$) be denoted again by δ^{Φ}.

The closure of δ^{Φ} (either on $\mathfrak{A}$ or on $\mathscr{L}$) is known to be a generator if either

$$\|\Phi\|_s = \sup\{\sum\{\|\Phi(\Lambda)\| e^{s(|\Lambda|-1)};\ \Lambda \ni n\};\ n=\text{atom}\}$$

is finite for some $s > 0$ (where $|\Lambda|$ is the number of atoms in Λ) or if there is a monotone sequence $\Lambda_n \in L_0$ tending to Z^{ν} such that the distance of H_{Λ_n} from $\mathfrak{A}(\Lambda_n)$ is bounded uniformly in n. It then follows that α_t exists, is unique and is approximated by $\mathrm{Ad}(\exp iH_{\Lambda_n})$.

Another method introduced by Kishimoto to study the semi-quantum lattice is to decompose $\mathfrak{A}$ according to the spectrum of the center $\mathscr{B}$ of $\mathfrak{A}$ which is generated by $\mathscr{B}(n)$ for all atoms n. The spectrum Ξ_n of $\mathscr{B}(n)$ can be represented by minimal projections ξ_n of $\mathscr{B}(n)$. The spectrum Ξ of $\mathscr{B}$ is the topological product of Ξ_n: each $\xi \in \Xi$ is specified by its component $\xi_n \in \Xi_n$. For each $\xi \in \Xi$, we consider a finite type I factor $M(n,\xi_n) = \xi_n \mathfrak{A}(n)$ for each atom n and the C*-algebra product $M_\xi = \bigotimes_n M(n,\xi_n)$, which is a UHF algebra. Then there exists a homomorphism ε_ξ from $\mathfrak{A}$ onto M_ξ satisfying $\varepsilon_\xi(h) = \xi(h)\cdot \mathbf{1}$ for all $h \in \mathscr{B}$. Denoting by Θ the set of functions $\xi \in \Xi \to \varepsilon_\xi(a) \in M_\xi$ for all $a \in \mathfrak{A}$, we obtain a continuous field of C*-algebras $(\{M_\xi\},\Theta)$ over Ξ.

As for the time translation, $\alpha_t^{\xi}(\varepsilon_\xi(a)) = \varepsilon_\xi(\alpha_t(a))$ defines α_t^{ξ} on M_ξ, its infinitesimal generator δ_ξ has the domain $D(\delta_\xi) = \varepsilon_\xi(D(\delta^{\Phi}))$, satisfies $\varepsilon_\xi \circ \delta^{\Phi} = \delta_\xi \circ \varepsilon_\xi$ there and is given by $\delta_\xi(a) = i[\varepsilon_\xi(H_\Lambda),a]$ for $a \in M_{\xi,\Lambda} \equiv \varepsilon_\xi(\mathfrak{A}(\Lambda))$. For each state ϕ, the restriction of ϕ to $\mathscr{B}$ is identified with a probability measure μ_ϕ on Ξ: $\phi(a) = \int \xi(a)d\mu(\xi)$ for $a \in \mathscr{B}$. Then there exists a measurable family of states ω_ξ of M_ξ such that $\omega(a) = \int \omega_\xi(\varepsilon_\xi(a))d\mu(\xi)$ for $a \in \mathfrak{A}$.

Thus we have three methods of investigating a semi-quantum lattice system: (1) a state ϕ on $\mathfrak{A}$ relative to α_t and $\{H_\Lambda\}$, (2) a G-invariant state $\hat{\phi}$ of $\mathfrak{A}$ relative to α_t and $\{H_\Lambda\}$, (3) a probability measure μ on the spectrum Ξ of the center $\mathscr{B}$ of $\mathfrak{A}$ together with the family of states ϕ_ξ of M_ξ relative to α_t^ξ and $\varepsilon_\xi(H_\Lambda)$.

Example of semi-quantum lattice system arises in the lattice model of alloys and liquid metals which has classical observables for the presence and absence of (possibly several kinds of) atoms at each lattice sites generating $\mathscr{D}$ and quantum observables for the description of electronic states of atoms. In such a model, $\xi \in \Xi$ describes the configuration of distribution of atoms on lattice sites, μ giving the probability distribution, and ϕ_ξ describes the state when such configuration ξ is fixed.

3. Equilibrium States

In this section, we discuss several different conditions, which are supposed to characterize equilibrium states of a semi-quantum lattice system, and mutual relations among these conditions. (See [16].)

Some of mutual relations depends on some assumptions on Φ which we shall first describe.

(A) Core assumption: The closure of δ^Φ is the generator of α_t.

(B) Translational invariance assumption: Corresponding to the lattice translation, we consider a representation of the group Z^ν by automorphisms τ_n, $n \in Z^\nu$, of $\mathscr{A}$ and $\mathscr{L}$ which map $\mathscr{A}(\Lambda)$ and $\mathscr{L}(\Lambda)$ to $\mathscr{A}(\Lambda+n)$ and $\mathscr{L}(\Lambda+n)$ for all $\Lambda \in L_0$, where Λ is identified with a finite subset of the lattice Z^ν and $\Lambda+n$ is the translate of Λ by a lattice vector n. The assumption is then $\tau_n H_\Lambda = H_{\Lambda+n}$ (which follows from $\tau_n \Phi(\Lambda) = \Phi(\Lambda+n)$) for all $\Lambda \in L_0$.

These assumptions will always be stated explicitly whenever necessary. In our discussion we do not need Φ itself. However Φ can be reconstructed from H_Λ in such a way that (2.1) is convergent. On the other hand, if Φ is given innitially, we always assume that (2.1) converges and hence H_Λ is defined. We always assume the vanishing of the partial expectation of H_Λ on Λ relative to a fixed product state (such as the trace state), which can be achieved by the subtraction of the partial expectation of H_Λ on Λ relative to the fixed product state (if it is non-zero for the given H_Λ) without changing the derivation defined by (2.2). If this is the case, $\Phi(\Lambda)$ constructed from H_Λ has vanishing partial expectation on Λ_1 whenever Λ_1 has non-empty intersection with Λ.

The above convention for H_Λ becomes relevant when we have assumption (B). Then the set of such H_Λ becomes a Banach space relative to the norm

$$(3.1)\qquad \|H_n\| \equiv \|H\| \quad (\equiv \|\Phi\|)$$

where n is any atom. An easy estimate leads to the uniform bound for energy density

$$(3.2)\qquad \|H_\Lambda\| \leqq (2|\Lambda|-1)\|H\|$$

and the vanishing of the surface energy per volume

$$d(H_\Lambda, \mathscr{A}(\Lambda))/|\Lambda|$$

as Λ increases to Z^ν with surface to volume ratio tending to zero (or more generally in the Van Hove limit) where $|\Lambda|$ denotes the number of lattice points (atoms) in Λ and $d(H_\Lambda, \mathscr{A}(\Lambda))$ the distance of H_Λ from $\mathscr{A}(\Lambda)$.

We now begin the discussion of equilibrium conditions.

A condition which does not refer to the local structure is the famous KMS condition, which we apply to the state ϕ of $\mathscr{L}$. Fannes and Verbeure [17] has recently shown the equivalence of the KMS condition to the stationarity $\phi \cdot \alpha_t = \phi$ plus a correlation inequality which goes back to Bogoliubov and Roepstorff. Taking hint from this work, Sewell [18] recently used the following condition on the generator δ of α_t:

$$(3.3)\qquad \phi(\delta a) = 0$$

$$(3.4)\qquad (i\beta/2)\{\phi((\delta a^*)a) - \phi(a^*\delta a)\} \geqq S(\phi(aa^*);\phi(a^*a))$$

where the relative entropy function $S(v;u)$ is defined to be $u(\log u - \text{lov } v)$ for $u > 0$, $v > 0$, to be 0 for $u = 0$ and to be $+\infty$ for $u > 0$, $v = 0$. This condition, if required for all a in the domain of δ, is equivalent to the KMS condition. If it is required only for $a \in \mathscr{L}_0$, it is equivalent to the KMS condition under the assumption (A). It is interesting that this condition can be stated for any *-derivation δ irrespective of whether there is an α_t.

Now we turn our attention to conditions involving local conditions but still sticking to the state ϕ of $\mathscr{L}$. The KMS condition is equivalent to the Gibbs condition which requires ϕ to be separating (faithful on weak closure of the associated representation) and $\phi^{\beta H_\Lambda}$

(the perturbation of $\hat{\phi}$ by βH_Λ) to be the product of the trace on $\mathcal{L}(\Lambda)$ and some functional on its commutant for all $\Lambda \in L_0$. Converse implication holds under (A). ([19]) The Gibbs condition implies the LTS condition which requires for every $\Lambda \in L_0$ that the free energy

$$(3.5) \qquad \tilde{F}_{\Lambda,\beta}(\psi) = \tilde{E}_\Lambda(\psi) - T\tilde{S}_\Lambda(\psi)$$

(in the volume Λ) to be minimal for $\psi = \hat{\phi}$ among all states ψ of $\mathcal{L}$ such that the restriction of ψ to $\mathcal{L}(\Lambda')$ coincides with that of $\hat{\phi}$ for all region Λ' outside of Λ, where $\tilde{E}_\Lambda(\psi) = \psi(H_\Lambda)$ (energy in Λ as an open system), $\beta = (kT)^{-1}$,

$$(3.6) \qquad \tilde{S}_\Lambda(\psi) = \lim_{\Lambda'\uparrow}\{S_{\Lambda'}(\psi) - S_{\Lambda'\setminus\Lambda}(\psi)\}$$

(entropy in Λ as an open system),

$$(3.7) \qquad S_\Lambda(\psi) = -k\psi(\log \rho_\Lambda^\psi)$$

(entropy in Λ as a closed system) and ρ_Λ^ψ is the density matrix of ψ for $\mathcal{L}(\Lambda)$. The LTS condition implies Sewell's condition for $a \in \mathcal{L}_0$ and hence is equivalent to KMS and Gibbs conditions under (A). It is an open question whether Gibbs, LTS and Sewell's conditions are equivalent without (A).

Under the assumption (B), the (Van Hove) limit

$$(3.8) \qquad P(H)\ (\equiv P(\Phi)) \equiv \lim |\Lambda|^{-1} \log \tau(e^{-\beta H_\Lambda}) ,$$

where τ is the unique trace state of $\mathcal{L}$, exists. A translationally invariant state $\hat{\phi}$ (i.e. $\phi\tau_n = \phi$ for all $n \in Z^\nu$) is a solution of variational principle if

$$(3.9) \qquad P(H) = s(\hat{\phi}) - \beta e(\hat{\phi})$$

where $ks(\psi)$ is the entropy density defined as the limit of $|\Lambda|^{-1}S_\Lambda(\psi)$ and $e(\psi)$ the energy density defined as the limit of $|\Lambda|^{-1}\tilde{E}_\Lambda(\psi)$ (both limits exist as Λ tends to Z^ν in Van Hove sense). The supremum of the right hand side of (3.9) over all translationally invariant $\hat{\phi}$ is $P(H)$; hence the name of variational principle. Under the assumption (B), KMS, Gibbs, LTS conditions and variational principle are all mutually equivalent for translationally invariant $\hat{\phi}$.

We now turn to $\mathcal{A}$. Here the Gibbs and LTS conditions for ϕ are equivalent to Gibbs and LTS conditions for $\hat{\phi}$, respectively. Hence, under assumption (A), they are mutually equivalent and equivalent to KMS, Gibbs, LTS and Sewell's condition for $\hat{\phi}$. The KMS condition for ϕ, however, is weaker than these as it does not provide any information on μ. The same holds for Sewell's condition. The Gibbs condition restricted to μ is the same as so-called DLR equations.

Under Assumption (B) and for translationally invariant ϕ, the variational principle for ϕ is equivalent to all conditions mentioned above for ϕ and $\hat{\phi}$ and to the variational principle for $\hat{\phi}$.

Finally we consider $\{\mathcal{A}_\xi\}$. Under the Assumption (B) and for translationally invariant ϕ, all above conditions are equivalent to the following variational principle for translationally invariant μ and a measureable family of states ϕ_ξ of $\mathcal{A}_\xi$ with translationally invariant $\phi = \int \phi_\xi d\mu(\xi)$:

$$\text{(3.10)} \qquad P(H) = s(\mu) + \int P(H^\xi) d\mu(\xi) ,$$

$$\text{(3.11)} \qquad P(H^\xi) = \bar{s}(\phi,\xi) - \beta e(\phi_\xi) .$$

Here H^ξ is defined by the family $\varepsilon_\xi(H_\Lambda) \equiv H_\Lambda^\xi$ on $\mathcal{A}_\xi$, the limit in the definition of $P(H^\xi)$ exists for μ-almost all ξ (although H_Λ^ξ is not translationally invariant due to ξ) and is translationally invariant, (3.10) is to be solved first and then (3.11) for μ-almost all ξ, $\bar{s}(\phi,\xi)$ is defined by the limit

$$\text{(3.12)} \qquad \bar{s}(\phi,\xi) = k^{-1} \lim_{\Lambda\uparrow} |\Lambda|^{-1} \lim_{\Lambda'\uparrow} |\Lambda'|^{-1} \sum \{S_\Lambda(\phi_{\xi\circ\tau_n}) \; ; \; n \in \Lambda'\} ,$$

which exists for μ-almost all ξ and is τ_n-invariant ($\xi\circ\tau_n$ is the character $a \in \mathcal{D} \rightarrow \xi(\tau_n(a))$), and $e(\phi_\xi)$ is the limit of $|\Lambda|^{-1}\phi(\varepsilon_\xi(H_\Lambda))$ as Λ tends to Z^ν (in the sense of Van Hove), which exists for μ-almost all ξ.

The Gibbs condition for $\mathcal{A}$ implies the Gibbs condition for each ω_ξ, $\xi \in \Xi$, relative to $\{\varepsilon_\xi(H_\Lambda)\}$ and the following DRL equation (which unfortunately depends on the solutions ω_ξ for fixed ξ problems): For the conditional probabilities $\mu_\Lambda(\xi|\xi')$ and $\mu_\Lambda^0(\xi|\xi')$ for a given configuration ξ' outside the region Λ (i.e. $\xi|\mathcal{D}(\Lambda^c) = \xi'$) obtained from μ and the restriction μ^0 of the trace state τ of $\mathcal{L}$ to $\mathcal{D}$,

$$\phi_\xi^{\varepsilon_\xi(H_\Lambda)}(1)\mu_\Lambda(\xi|\xi')/\mu_\Lambda^0(\xi|\xi')$$

is independent of ξ with the given ξ', where ϕ_ξ^h is the perturbation of ϕ_ξ by h. The converse has been proved only under some technical assumptions.

4. Stability question

Physically, equilibrium states are to be understood as states which are stable under disturbance of the system. This presupposes a framework where the time development is given (and contains all necessary physical information about the system). Thus in examples of previous two sections, we consider the case where $\mathcal{D}$ is trivial (i.e. $\mathcal{D} = \mathcal{A}$).

Haag, Kastler and trych-Pohlmeyer [20] have adopted the stability under a local perturbation of the dynamics and proved for any α_t-stationary state satisfying their stability criterion under somewhat technical assumption of the L_1-clustering up to order 4 (see [21]) that either the state must satisfy the KMS condition (relative to α_t) at some β or else the energy spectrum of the state (the spectrum of the generator of one-parameter group of unitarics canonically implementing α_t in the associated cyclic representation) is one-sided (i.e. positive or negative) —— that is either a ground or ceiling states ——, and vice versa, thus justifying the KMS condition as a condition characterizing equilibrium states. Here the local perturbation of dynamics is an $\alpha_t^{(h)}$ whose generator differs from that of α_t by an operator h in $\mathcal{A}$. The stability is formulated roughly as the existence of stationary states for $\alpha_t^{(\lambda h)}$ for each $\lambda > 0$ which smoothly converges to the state under question as $\lambda \to 0$, all within the normal functionals in the cyclic representation associated with given states. Physically, it is supposed to describe a stability when some dust or impurity is introduced locally into the system.

The grand canonical ensemble for a finite system contains two-parameters β (the inverse temperature) and μ (the chemical potential) and satisfies the KMS condition at β relative to a mixed transformation $\alpha_t\gamma_{\mu t}$ but not relative to the time translation α_t alone where γ_θ is the so-called gauge transformation of the first kind with a period 2π ($\gamma_{2\pi}=1$). To understand this situation, the following view point can be adopted:

We consider a system $(\mathcal{F}, \alpha_t, \gamma, G)$ of a C*-algebra $\mathcal{F}$, a time translation α_t for $\mathcal{F}$, a compact group G and its representation by automorphisms γ_g, $g \in G$, of $\mathcal{F}$ which commute with α_t. Physically, we may imagin a situation where several kinds of particles are present and G is generated by all gauge transformations of the first kind for individual kinds of particles. The commutativity of α_t and γ_g is the assumption of the conservation of the number of particles of each kind. In such a situation, it is reasonable to consider the stability under local gauge-invariant perturbations, which corresponds to dust which does not cause the change in number of particles of each kind. States stable under such perturbation would be labelled by chemical potentials for each kinds of particles besides the inverse temperature. On the other hand, we can also consider dusts containing catalyzer for reactions between particles. If the reaction still conserves certain linear combination of particle numbers, then equilibrium states should be characterized by a stability under local perturbations which is invariant under a subgroup G_1 of the gauge group G. (An extreme case would be the case of trivial G_1.)

The same scheme as Haag, Kastler and Trych-Pohlmeyer leads to the characterization of equilibrium states ϕ of $\mathcal{F}$ by the KMS-condition (apart from an alternative possibility of a ground or ceiling states) at some β for the restriction ω of the state ϕ to the subalgebra $\mathfrak{A}$ of $\mathcal{F}$ consisting of all gauge-invariant elements of $\mathcal{F}$ —— mathematically called the G-fixed point algebra, physically called the observable algebra in contrast to the field algebra $\mathcal{F}$. It is then an interesting mathematical problem to characterize all possible ϕ satisfying this condition, which is an extension problem of KMS states ω of $\mathfrak{A}$ to a (clustering) state ϕ of $\mathcal{F}$.

This problem has been solved in [22] and [23] under some assumption of asymptotic abelian property (for automorphisms commuting with α_t and γ_g, such as lattice or time translations). The result is that the state satisfies the KMS condition relative to some one-parameter subgroup in the product group of the time translation and the gauge group or possibly have a one-sided spectrum relative to some (or all) parts of the gauge group and the KMS situation holds relative to the other parts of the gauge group.

It is a natural question physically whether all stationary states have an interpretation either as stable or metastable states. If the time translation is asymptotically abelian, all states stable under local gauge-invariant perturbation discussed above are α_t-stationary. Since the only assumption on the gauge group G is that γ_g commute with α_t apart from the compactness, it would be of interest to find first the set $\{\alpha_t\}'$ of all automorphisms of $\mathcal{F}$ commuting with α_t —— the automorphism commutant —— and to ask whether all α_t-

stationary states are in the closed convex hull of KMS or ground or ceiling states of one-parameter subgroup of $\{\alpha_t\}'$. Kishimoto [24] has recently obtained an interesting results in this connection for quasi-free evolution α_{u_t} of CAR algebras, where u_t is a one-parameter group of unitaries on "the one-particle space". At the moment, results refers to the group H of all quasi-free automorphisms α_u with unitary u belonging to $\{u_t\}''$. Under the assumption that u_t tends weakly to zero as $t \to \infty$ (which is the case for free motion due to the absolute continuity of the one-particle spectral measure), H' is shown to consists of α_u with unitary u in $\{u_t\}'$ and all primary H-invariant states are quasi-free.

References

[1] D. Ruelle: Cargèse lectures in Physics (ed. D. Kastler, Gordon and Breach, New York, 1970) pp.169-194.
[2] E. Størmer: loc. cit. pp.195-213.
[3] S. Doplicher and D. Kastler: Comm. Math. Phys. 7(1968), 1-20.
[4] S. Sakai: Amer. J. Math. 98(1976), 427-440.
[5] R. F. Streater: Comm. Math. Phys. 6(1968), 233-247.
[6] D. W. Robinson: Comm. Math. Phys. 7(1968), 337-348.
[7] D. Ruelle: Statistical Mechanics (Benjamin, New York, 1969).
[8] A. Kishimoto: Comm. Math. Phys. 47(1976), 25-32.
[9] R. T. Powers and S. Sakai: Comm. Math. Phys. 39(1975), 273-288.
[10] D. W. Robinson: Comm. Math. Phys. 16(1970), 290-309.
[11] S. Miracle-Sole and D. W. Robinson: Comm. Math. Phys. 19(1970), 204-218.
[12] J. Ginibre: Statistical Mechanics and Field Theory (ed. C. DeWitt and R. Stora, Gordon and Breach, New York, 1971) pp.327-427.
[13] Y. M. Suhov: Limit Gibbs state for a class of one-dimensional systems of quantum statistical mechanics, preprint.
[14] O. E. Lanford: Proceedings of the International Congress of Mathematicians (ed. R. D. James, Canadian Math. Congress, 1975) vol.2, pp.377-381.
[15] A. Kishimoto: Equilibrium states of a semi-quantum lattice system, to appear in Rep. Math. Phys.
[16] H. Araki: Proceedings of the Second U.S.-Japan Seminar on C*-algebras and Applications to Physics, to be published.
[17] M. Fannes and A. Verbeure: Correlation inequalities and equilibrium states. To appear in Comm. Math. Phys.
[18] G. L. Sewell: KMS conditions and local thermodynamical stability of quantum lattice system II, to appear in Comm. Math. Phys.
[19] H. Araki: C*-Algebras and Their Applications to Statistical Mechanics and Quantum Field Theory. (ed. D. Kastler, North Holand, 1976) pp.64-100.
[20] R. Haag, D. Kastler and E. B. Trych-Pohlmeyer: Comm. Math. Phys. 38(1974), 173-193.
[21] D. Kastler and O. Bratteli: Comm. Math. Phys. 46(1976), 37-42.
[22] H. Araki and A. Kishimoto: Comm. Math. Phys. 52(1977), 211-232.
[23] H. Araki, D. Kastler, M. Takesaki and R. Haag: Comm. Math. Phys. 53 (1977), 97 - 134.
[24] A. Kishimoto: On invariant states and the commutant of a group of quasitfree automorphisms of the CAR algebra.

FOUNDATIONS OF EQUILIBRIUM QUANTUM STATISTICAL MECHANICS

by Daniel KASTLER

Equilibrium quantum statistical mechanics is traditionally based on the "Gibbs' Ansatz" prescribing as follows the (mean) value of the observable a in the thermodynamical equilibrium state $\phi_{\beta,\mu}$ with temperature $k^{-1}\beta^{-1}$ and chemical potential μ :

$$\phi_{\beta,\mu}(a) = \mathrm{Tr}\{e^{-\beta(H-\mu N)}a\} / \mathrm{Tr}\{e^{-\beta(H-\mu N)}\} \tag{1}$$

Here H and N denote respectively the hamiltonian and particle number operators of the system under consideration. For the expression (1) to make sense it is necessary that the system be "enclosed in a box". Box-quantization causes the spectrum of the hamiltonien H to be discrete - the first condition to be met for $e^{-\beta H}$ to be trace class - a fact which is then garanteed by the asymptotic distribution of energy eigenvalues due to the presence of the kinetic energy. Unfortunately however, the "system in a box" is inacceptable as a physical model (no thermodynamic behaviour, no return to equilibrium !) : one thus has to supplement the Ansat z (1) with the prescription of taking the limit of an "infinite box" (thermodynamical limit). It is only after this limiting procedure that the thermodynamical features appear. The limit state is unique, or non unique (with a dependance upon the boundary conditions adopted for box quantization) according to wether one has absence of presence of more than one phase for the corresponding values of β and μ.

Although the theory of the thermodynamical limit is now in growingly good mathematical shape (thanks to the efforts in "constructive statistical mechanics") one may wish to adopt a stand point different from the former one for establishing the foundations of equilibrium statistical mechanics. Indeed on, the one hand, one would like to remove (1) as an "Ansatz" and start instead from first principles. On the other hand, one would like to deduce the properties of the equilibrium states of large (~ infinite) systems without having to resort to the mutilating detour of enclosing in a box with mathematically ad hoc, unphysical, boundary conditions. We want to describe such a programm in what follows.

The first step in realizing this programm is the replacement of the complex Gibbs Ansatz + thermodynamic limit by a substitute relevant to infinite systems (the second step then consists in deducing the latter from first principles). Such a substitute was found in 1967 by Haag, Hugenholtz and Winnink [1] to be the KMS-(Kubo Martin Schwinger) condition, to which we now turn our attention [1].

The KMS-condition

We consider a C^*-algebra A with a continuous one-parameter group $t \to \tau_t$ of automorphisms (the reader is referred to the Appendix for the definition of these terms) ; and on the other hand a state ϕ of the algebra A. Given elements $a,b \in A$ we define as follows the functions F_{ab} and G_{ab} of a real variable

$$\begin{cases} F_{ab}(t) = \phi(b\ \tau_t(a)) \\ G_{ab}(t) = \phi(\tau_t(a)\ b) \end{cases}, \tag{2}$$

with β a real number.The state ϕ is then called β-KMS for $t \to \tau_t$ whenever there is a fonction U_{ab} of the complex variable, holomorphic in the open strip $0 < \mathrm{Im}z < \beta$, bounded continuous on its closure, such that

$$\begin{cases} F_{ab}(t) = U_{ab}(t+i) \\ G_{ab}(t) = U_{ab}(t) \end{cases}, \quad t \in R . \tag{3}$$

Instead of requiring the fonctions F_{ab} and G_{ab} to be the boundary values of a fonction holomorphic in a strip one can require that, for any $a,b \in A$ (with a analytic w.r.t.τ)

$$\phi(b\ \tau_{i\beta}\ (a)) = \phi(ab) \tag{4}$$

1) The Kubo-Martin-Schwinger condition had been proposed in the late fifties by these authors [2] [3] as a boundary condition for the determination of "Greens functions" as solutions of an infinite system of differential equations. Hugenholtz, Haag and Winnink realized the basic algebraic role of the KMS-condition which was then revealed by Takesaki [4] also to be basic in the theory of Von Neumann algebras.

((4) is immediately obtained by making $t = 0$ in (3) which in turn results from (4) with a replaced by $\tau_t(a)$). Still another equivalent formulation of the β-KMS condition is obtained by requiring that

$$\hat{F}_{ab}(E) = e^{\beta E}\,\hat{G}_{ab}(E) \qquad , \qquad a,b \in A \tag{5}$$

where $\wedge$ denotes Fourier transform and E denotes the variable conjugate to t [2].

After the statement of these equivalent versions of the β-KMS condition we explain how this condition replaces the Gibbs Ansatz (1). For this we note that the operators H and N are the respective infinitesimal generators of the group of unitaries $t \to e^{iHt}$, $\phi \to e^{iN\phi}$, which in turn define as follows the dynamical and gauge automorphism groups $t \to \alpha_t$ and $\gamma \to \gamma_\phi$

$$\begin{cases} \alpha_t(a) = e^{iHt}\; a\; e^{-iHt} \\ \gamma_\phi(a) = e^{-iN\phi}\; a\; e^{iN\phi} \end{cases} , \tag{6}$$

where a is an arbitrary element of the field algebra (the C^*-algebra of bounded fields) $\mathfrak{F}$. Since H and N commute, so do the automorphisms α_t and γ_ϕ ; and the operator $H-\mu N$ is the infinitesimal generator of the combined automorphism group

$$t \to \tau_t = \alpha_t\, \gamma_{\mu t} \quad , \tag{7}$$

with

$$\tau_t(a) = e^{i(H-\mu N)t}\; a\; e^{-i(H-\mu N)t} \quad , \qquad a \in \mathfrak{F} \; , \; t \in R . \tag{8}$$

For $a \in \mathfrak{F}$ analytic for $t \to \tau_t$ one then has

$$\tau_{i\beta}(a) = e^{-\beta(H-\mu N)}\; a\; e^{\beta(H-\mu N)t} \tag{9}$$

so that the state $\phi_{\beta,\mu}$ defined by (1) satisfies

2) Energy as conjugate to time !

$$\phi_{\beta,\mu}(b\ \tau_{i\beta}(a)) = \frac{\mathrm{Tr}\{e^{-\beta(H-\mu N)} b\ e^{-\beta(H-\mu N)} a\ e^{\beta(H-\mu N)}\}}{\mathrm{Tr}\{e^{-\beta(H-\mu N)}\}}$$

$$= \frac{\mathrm{Tr}\{e^{-\beta(H-\mu N)} a\ b\}}{\mathrm{Tr}\{e^{-\beta(H-\mu N)}\}} = \phi_{\beta\mu}(a\ b)$$

(where we used twice the permutability under the trace) : we see that $\phi_{\beta,\mu}$ as given by the Gibbs Ansatz (1) fulfills the β-KMS property w.r.t. the automorphism group (7). Since this property persists through the thermodynamic limit, we propose, following Haag, Hugenholtz and Winnink, to replace the complex Gibbs Ansatz + thermodynamic limit by the following "KMS principle".

Thermodynamical equilibrium states with temperature $k^{-1}\beta^{-1}$ and chemical potential μ are states of the C^*-algebra $\mathfrak{F}$ of fields which fulfill the β-KMS property w.r.t. the one-parameter automorphism group $t\to\tau_t = \alpha_t\ \gamma_{\mu t}$, where $t\to\alpha_t$ and $\phi\to\gamma_\phi$ are the respective dynamical and gauge automorphism groups.

This principle is now valid and makes sense for finite and infinite systems. On the other hand it actually carries the same information as the Gibbs principle to which it is actually mathematically equivalent for finite systems [3]. We will therefore adopt it as the basis of quantum equilibrium statistical mechanics, whose foundations will therefore be established by explaining this "KMS principle" from first principles. In order to do that, it will be convenient to first state the KMS-principle in the following parts (I) and (II) corresponding respectively to the temperature and the chemical potential. For this we introduce the notion of observable algebra $\mathfrak{A}$ as the gauge invariant part of the field algebra $\mathfrak{F}$ [4]

$$\mathfrak{A} = \{A \in \mathfrak{F}\ \ ;\ \gamma_\phi(A) \text{ for all } \phi \in T^1\} \tag{10}$$

3) i.e. in the case of semi-finite Von Neumann algebras.
4) We shall denote the elements of $\mathfrak{A}$ by capitals A,B etc. while general elements of $\mathfrak{F}$ are denoted by low case letters a,b etc.

Note that since α and γ commute $\mathfrak{A}$ is stable under the dynamical automorphism $t \to \alpha_t$; and on the other hand that τ_t reduces to α_t on $\mathfrak{A}$, since γ_ϕ acts there trivially. Thus we can split the above principle in

(I) The thermodynamical states with temperature $k^{-1}\beta^{-1}$ are states of the observable algebra $\mathfrak{A}$ which fulfill the β-KMS condition with respect to the dynamical one-parameter automorphism group $t \to \alpha_t$.

(II) A state of the previous kind extends to a state of the field algebra $\mathfrak{F}$ which fulfills the β-KMS condition with respect to a mixed time and gauge one parameter group of the type (7), where μ is the chemical potential.

In the next two paragraphs we now offer an explanation of these principles (I) and (II).

In fact we shall somewhat generalize principle (II), which was stated in the simplest case, that of a gauge group isomorphic to the one dimensional torus T^1. This case is that of a (non relativistic) system containing one species of particles. In the case of n species T^1 has to be replaced by T^n with the accompanying occurence of n chemical potentials. In fact we shall treat below the case of an arbitrary compact gauge group G, more canonical mathematically and of some physical interest in view of groups like SU_3, SU_4, etc.

The notion of temperature as deduced from dynamical stability [5][6][7][8].

The explanation we offer for the principle (I) is roughly the following: each sufficiently clustering invariant state ω [5] of an asymptotically abelian C^*-dynamical system which is stable for the local perturbations of the dynamics is either a β-KMS state for some real β, or a ground state (= a positive energy state), or a trace. The two latter alternatives correspond respectively to the limiting cases of 0 and infinite temperature.

In order to substantiate this statement and make it precise we need to specify the notions of asymptotically abelian system ; of clustering state (of various degrees of clustering) ; and of local perturbations of the dynamics and stability of a state against such perturbations.

A system of a C^*-algebra $\mathfrak{A}$ an a one-parameter group $t \to \alpha_t$ of automorphisms of $\mathfrak{A}$ is called asymptotically abelian whenever

$$[A, \alpha_t(B)] \underset{t=\infty}{\longrightarrow} 0 \tag{11}$$

for any two $A, B \in \mathfrak{A}$. This condition expresses the vanishing of quantum correlations between observables measured at a great time distances. The requirement (11) can be stated for different topologies (the weak topology, the norm topology, etc.) one has accordingly weak , norm-asymptotic abelianness, etc. One of the main results of the theory, already obtained under weak asymptotic abelianness [9][10][11], is the fact that extremal invariance of a state ϕ of $\mathfrak{A}$ (i.e. the fact that $\phi(\alpha_t(A)) = \phi(A)$ for all $A \in \mathfrak{A}$ and $t \in R$ and that ϕ cannot be written as a convex combination $\lambda\phi_1 + (1-\lambda)\phi_2$, $0 < \lambda < 1$, of different states ϕ_1, ϕ_2) is synonimous with weak clustering of ϕ i.e. the condition

$$\frac{1}{2T} \int_T^{+T} \phi(A\alpha_t(B))dt \underset{T=\infty}{\longrightarrow} \phi(A)\phi(B) \quad , \quad A, B \in \mathfrak{A} , \tag{12}$$

5) under asymptotic abelianness the clustering property is tantamount to extremal invariance (= pure phase). See below for specifications.

i.e; the fact that ϕ is a product state asymptotically and in the mean. The analogous pointwise (stronger) property

$$\phi(A\alpha_t(B) \xrightarrow[t=]{} \phi(A)\phi(B) \quad , \quad A,B \in \mathfrak{A} \quad , \tag{13}$$

is called clustering (or mixing) of ϕ [6]. In fact we shall need here more stringent versions of the concepts of both asymptotic abelianness and clustering, defined w.r.t. a norm-dense, α-invariant $*$-subalgebra $\mathfrak{A}_0$ of the C^*-algebra $\mathfrak{A}$ [7] : we want our dynamical system to be L^1-asymptotically abelian on $\mathfrak{A}_0$ in the sense that

$$\int_{-\infty}^{+\infty} \| [\alpha_t(A), B] \| \, dt < \infty \quad , \quad A,B \in \mathfrak{A}_0 \quad , \tag{14}$$

and require our state ω to be hyperclustering of order 4, i.e. to fulfill, for $n \leqslant 4$ the following inequalities

$$|\omega^T_{(n)}(\alpha_t(A_1), \dots \alpha_{t_n}(A_n))| \leqslant C\{1 + \operatorname{Sup}_{i<j} |t_i - t_j|\}^{-1-\delta} \quad , \tag{15}$$

where $A_1, \dots, A_n$ are arbitrary elements in $\mathfrak{A}_0$ and C, δ are positive constants independant of the latter. Condition (15), evidently stronger than (12), is therefore a kind of a strengthened form of extremal invariance (= the pure phase condition).

We now turn to the definition of stability of ω under local dynamical perturbations. The perturbation $t \to \alpha_t^{(h)}$ of the dynamical automorphism group $t \to \alpha_t$ by the self adjoint element $h \in \mathfrak{A}$ is by definition the (unique) automorphism group fulfilling the condition [13], [14], [15]

$$\left.\frac{d}{dt}\right|_{t=o} \alpha_t^{(h)}(A) = \left.\frac{d}{dt}\right|_{t=o} \alpha_t(A) + i[h,A], \tag{16}$$

(A an arbitrary α-differentiable element of $\mathfrak{A}$). Recalling the fact that the derivation (= infinitesimal automorphism) generated by a hamiltonian is obtained by bracketing by the latter, we see that (16) amounts

6) Together with asymptotic abelianness clustering entails that $\phi(A\alpha_t(B)C) \xrightarrow[t=\infty]{} \phi(AC)\phi(B), A,B,C \in \mathfrak{A}$. As was noted by Sergio Doplicher at an early stage of the theory [12]. This fact can be interpreted as a "static stability" (in contrast with the "dynamical stability" discussed below).
7) $\mathfrak{A}_0$ is e.g. the set of strictly local observables for $\mathfrak{A}$ an algebra of quasi-local observables.

to "adding the self adjoint $h \in \mathfrak{A}$ to the hamiltonian" 8) i.e; perturbing the dynamics locally, since the self adjoint elements of $\mathfrak{A}$ are interpreted as (norm limits of) local observables. Having defined our notion of local dynamical perturbations we now define that of stability of an α-invariant state ω : ω is called <u>stable for local perturbations of the dynamics</u> whenever the following holds : to each (sufficiently small) self adjoint $h \in \mathfrak{A}$ there is a (perturbed equilibrium state) $\omega^{(h)}$ such that

i) $\omega^{(h)}(\alpha_t^{(h)}(A)) = \omega^{(h)}(A)$, $A \in \mathfrak{A}$, $t \in R$

ii) $\omega^{(\lambda h)}(A) \xrightarrow[\lambda=0]{} \omega(A)$, $A \in \mathfrak{A}$,

iii) $\omega^{(h)}(\alpha_t(A)) \xrightarrow[t=\pm\infty]{} \omega(A)$, $A \in \mathfrak{A}$

These requirements are physically transparent : (i) expresses the stationarity of the perturbed equilibrium state $\omega^{(h)}$ in the perturbed dynamics, (ii) the proximity of $\omega^{(h)}$ to the unperturbed state ω, (iii) return to equilibrium of $\omega^{(h)}$ to ω (in future and past), once the perturbation is removed. We have now defined all the specifications of the following result, which is offered as a basic explanation of part (I) of the "KMS-principle" above

<u>Theorem I</u> - Let $\mathfrak{A}$ be a C^*-algebra, with $t \to \alpha_t(A)$ a continuous one-parameter group of automorphisms of $\mathfrak{A}$ such that the dynamical system $\{\mathfrak{A},\alpha\}$ is L^1-asymptotically abelian on $\mathfrak{A}_0$, $\mathfrak{A}_0$ a norm-dense α-invariant $*$-subalgebra of $\mathfrak{A}$. Let ω be an α-invariant state of $\mathfrak{A}$ and assume ω hyperclustering of order 4 and stable for local perturbations of the dynamics. Then

- either ω is β-KMS for some real β
- or ω is a state with one sided energy spectrum (i.e. the spectrum of the associated unitary implementation of α is one-sided)
- or ω is a trace state (i.e. $\omega(AB) = \omega(BA)$, $A,B \in \mathfrak{A}$).

8) $\alpha_t^{(h)}$ can accordingly be defined by $\alpha_t^{(h)}(A) = P_t^{(h)}\alpha_t(A)P_t^{(h)}$, where $P_t^{(h)}$ is the usual unitary of the "interaction representation" determined by $\frac{d\,P_t^{(h)}}{dt} = i\,P_t^{(h)}\alpha_t(h)$, $P_o^{(h)} = I$.

As noted above (and can be easily checked) the two latter alternatives correspond respectively to the limiting cases $\beta = \infty$ and $\beta = 0$ of zero and infinite temperature. We end up this section with a short (partly heuristic)sketch of the proof of Theorem I. Expressing condition (i) above in differential form using (16) with A replaced by the differentiable element $\int_S^T \alpha_t(A)dt$ of $\mathfrak{A}$ and h by λh $(S,T,\lambda \in R)$, we obtain the relation

$$\frac{i}{\lambda}\, \omega^{(\lambda h)}(\alpha_T(A) - \alpha_S(A) = \omega^{(\lambda h)}([h, \int_S^T \alpha_t(A)dt]) \;. \tag{17}$$

Taking the limits $T \to +\infty$, $S \to -\infty$, under which the l.k.s. vanishes by (iii) ; and then the limit $\lambda \to 0$, we obtain, using (ii), the relation (cf. the definitions (2) of the functions F_{Ah} and G_{Ah}):

$$\int_{-\infty}^{+\infty} [F_{Ah}(t) - G_{Ah}(t)]dt = 0 \;, \tag{18}$$

(already the special case $E = 0$ of the β-KMS condition (5)). In order to derive the full condition, we now use the following trick. Note that, we have, due to asymptotic abelianness and clustering

$$F_{A_1\alpha_u(A_2),h_1\alpha_u(h_2)} \xrightarrow[u=\infty]{} F_{A_1h_1}F_{A_2h_2} \;, \tag{19}$$

and analogously for the function G. The replacements $A = A_1\alpha_u(A_2)$ $h = h_1\alpha_u(h_2)$ in (18) followed by the limit $u \to \infty$ then imply using hyperclustering of order 4

$$\int_{-\infty}^{+\infty} [F_{A_1h_1}(t)\, F_{A_2h_2}(t) - G_{A_1h_1}(t)\, G_{A_2h_2}(t)]dt = 0 \;, \tag{20}$$

which, upon making $A_1 = A$, $h_1 = B$, $A_2 = B'$, $h_2 = \alpha_s(A')$ reads

$$F_{AB} * G_{A'B'} = G_{AB} * F_{A'B'} \;, \tag{21}$$

or else

$$\hat{F}_{AB}(E)\, \hat{G}_{A'B'}(E) = \hat{G}_{AB}(E)\, \hat{F}_{A'B'}(E) \;. \tag{22}$$

If division of both sides by $\hat{G}_{A'B'}(E)$ is allowed, this leads to the relation $\hat{F}_{AB}(E) = \Phi(E)\, \hat{G}_{AB}(E)$, with Φ a universal function (independant of $A,B \in \mathfrak{A}$) which is then easily seen to be of the form $e^{\beta E}$ e.g. by using the relation analogous to (20) with three-fold instead of two-fold products (derived from (20) as (20) was from (18)). The possibility, given $E \in R$, to find $A', B' \in \mathfrak{A}$ with $\widehat{G}_{A'B'}(E) \neq 0$ in the case of an energy spectrum which is not one-sided is afforded by the following result : the energy spectrum of an α-invariant clustering state ω (i.e. the spectrum of the unitary implementation of the automorphism group $t \to \alpha_t$ afforded by ω) is shown to be either one-sided or to cover the whole reals (cf. theorem 3 of [5]).

We conclude this section with a few remarks

1. Theorem I above evolves the notion of temperature but does not show its one-sidedness. In fact this theorem contains no statement about the range of β, which remains an open problem (the situation will be the same for the theory of the chemical potential afforded by theorem II below).

2. A variant of Theorem I less demanding in clustering (presumably needing only weak clustering i.e. extremal invariance) can be formulated in terms of a different stability requirement - that of stable coexistence with another macroscopic system [16].

3. The degree of stability (varying the types of perturbations and the type of "nearness" of $\omega^{(h)}$ to ω) can be investigated versus the situation w.r.t. phase transition : cf. [8] and a forthcoming work of Hugenholtz, Mebkhout, Robinson and the author.

The notion of chemical potential deduced by extending states from the observable to the field algebra [17], [18].

We now turn our attention towards part (II) of the "KMS principle" stated above. The underlying problem is in fact the classification of the equilibrium states of a given system pertaining to a specified temperature. Principle (II) states that these equilibrium states are distinguished (amongst possibly other things) by the value of a second parameter, the chemical potential [9]. As suggested by (II) the notion of chemical potential arises from looking at the extension of the state ω from the observable algebra to the field algebra (as we shall see this extension turns out to be unique). As mentioned earlier, it is interesting, both mathematically and physically, to envisage the case of a general compact gauge group [10], abelian or not. The relevant information about extensions of extremal β-KMS states from $\mathfrak{A}$ to $\mathfrak{F}$ is then contained in the following theorem, which we present as the foundation of the above "principle" (II).

Theorem II - Let $\mathfrak{F}$ be a C^*-algebra (the field algebra) on which there are continuous actions $t \in T \to \alpha_t$ and $g \in G \to \gamma_g$ of the respective groups R (the additive real line of time) and G (the gauge group). Assume that G is compact and that α and γ commute : $\alpha_t \circ \gamma_g = \gamma_g \circ \alpha_t$, $t \in R$, $g \in G$. Assume also that time acts on $\mathfrak{A}$ in a norm-asymptotically abelian way [11]

$$\| [a,\alpha_t(b)] \| \xrightarrow[t=\infty]{} \quad , \quad a,b \in \mathfrak{F}. \tag{23}$$

Denote by $\mathfrak{A}$ the C^*-algebra of gauge invariant elements of $\mathfrak{F}$ (the observable algebra) (cf.(9)). We then have that

9) one parameter in the case where the gauge group is the one-dimensional torus T^1, generally a finite number of parameters (n for $G = T^n$). Besides the chemical potential there may be other elements of classification tied up with phase transitions, which we do not discuss here.
10) gauge group of the first kind.
11) or, for that matter, norm-asymptotically grassmannian in the case of anticommuting fields, see [18].

(i) each extremal (= weakly clustering) α-invariant state ω of $\mathfrak{A}$ possesses an extremal α-invariant extension ϕ to $\mathfrak{F}$. Further, two such extensions ϕ_1 and ϕ_2 are such that $\phi_2(a) = \phi_1(\gamma_g(a))$ for some $g \in G$.
(ii) Let ω be an α-invariant state of $\mathfrak{A}$ which is extremal β-KMS for $t \to \alpha_t$ and assume that ω is faithful (i.e. $\omega(A^* A) = 0$, $A \in \mathfrak{A} \Rightarrow A = 0$). Let ϕ be an extremal α-invariant extension of ω to $\mathfrak{F}$ (of the type considered in (i)). Then ϕ is β-KMS for a one parameter group of automorphisms of $\mathfrak{F}$ of the type $t \to \alpha_t \gamma_{\xi_t}$, where $t \to \xi_t$ is a continuous one-parameter subgroup of the center of the stabilizer G_ϕ of ϕ ($G_\phi = \{g \in G;\ \phi(\gamma_g(a)) = \phi(a)$ for all $a \in \mathfrak{F}\}$).

We add a few comments to this somewhat involved statement :
Part (i) has nothing to do with the KMS-property, but simply states existence and uniqueness up to gauge of the extremal α-invariant extension ϕ to $\mathfrak{F}$ of an extremal α-invariant state ω of $\mathfrak{A}$. We recall that, in the asymptotically abelian context assumed in (23), extremal α-invariant is synonimous with weakly clustering for α-invariant states of $\mathfrak{A}$ as well as of $\mathfrak{F}$

The content of the conclusion in part (ii) of Theorem II becomes perhaps more transparent if we specialize G to the usual case of an n-dimensional torus T^n (n species of particles). In this case the possible one-parameter subgroups $t \to \xi_t$ are all of the form $t \to (\mu_1 t, \mu_2 t, .., \mu_n t)$: these subgroups are thus indexed by a set of n <u>chemical potentials</u>. The general case of a subgroup $t \to \xi_t$ is thus a non-commutative generalization of this situation, for which the chemical potentials are the coordinates of a vector in the Lie algebra of the gauge group G. On the other hand, the conclusion in (ii) to the effect that ξ_t belongs to the center of G_ϕ, is in fact a statement restricting the possible values of the chemical potentials : for instance in the simplest case $G = T^1$, this conclusion implies that $\mu = 0$ whenever ϕ is not gauge invariant. Concerning the assumptions in (ii), we recall the known fact that extremal β-KMS entails extremal α-invariant [12]. On the other hand we note

12) it does even entail primary, which in turn entails clustering.

that the requirement of faithfulness of ω is redundant in the case (probably realised in relativistic field theory) that $\mathfrak{A}$ is simple (= without non trivial ideals) : the β-KMS property entails namely that the left ideal of ϕ (consisting of the $A \in \mathfrak{A}$ with $\phi(A^*A) = 0$) is a bilateral ideal, which then reduces to the zero. In non relativistic field theory, it does occur that $\mathfrak{A}$ is non simple (this is in fact the case for the gauge invariant part of the CAR-algebra) : one then has to require faithfulness of ω in order to exclude the occurence of ground state-like properties w.r.t. gauge (see [17],[18]). Last but important comment : it is essential to know (as stated in (i)) that the extension ϕ of ω to $\mathfrak{F}$ is unique up to gauge : only then indeed is it garanteed that the concept of chemical potential evolved in (ii) pertains to the state ω rather than to an (otherwise possibly non unique) extension.

For a detailed proof of Theorem II, we refer to [17] and [18]. Let us simply mention that the existence of an extremal α-invariant extension ϕ is a more or less classical result based on the Krein-Milman theorem. For the uniqueness of the extension ϕ up to gauge and the mixed time and gauge β-KMS property of ϕ, the strategy of proof is as follows : one first expresses the assumptions (respectively the fact that $\phi_2(A) = \phi_1(A)$, $A \in \mathfrak{A}$; or the β-KMS property of the restriction of ϕ to $\mathfrak{A}$) for all observables $A \in \mathfrak{A}$, obtained by averaging in gauge arbitrary elements of $\mathfrak{F}$:

$$A = \int \gamma_g(a)dg \qquad , \qquad a \in \mathfrak{F} \quad . \tag{24}$$

One then exploits clustering of ϕ in a way analogous to the derivation of (20) from (18) encountered above : first replacing a by $a_1\alpha_u(a^2)$, $a_1, a_2 \in \mathfrak{F}$, $u \in R$, and then letting $u \to \infty$.

We conclude this section by describing an interesting interpretation of the chemical potential μ , in the case $G = T^1$, in terms of <u>cocycle Radon-Nicodyn derivatives</u> and <u>localized automorphisms</u> - concepts basic respectively in the modern developments in Von Neumann algebras [19] and in the algebraic theory of superselection sectors [20],[21]. In Theorem II above the concept of chemical potential evolves from looking at extensions of equilibrium states ω of $\mathfrak{A}$ to the fields in $\mathfrak{F}$. In the light of the algebraic theory of superselection sectors [20],[21] the fields (specifically the unitaries of a given charge in $\mathfrak{F}$) are obtained as implementing the localized automorphisms ρ of $\mathfrak{A}$ [13]. One therefore explects that the information gained from looking at the extension of ω to the fields can also be extracted from a study of the states $\omega\circ\rho$ which one obtains by composing the state ω with localized automorphisms ρ (say of charge n). A first result in that direction is the following : for ω β-KMS with respect to $t \to \alpha_t$, the state $\omega\circ\rho$ is quasi-equivalent to ω . This fact is easily understood physically : the automorphism "creates a charge n" , thus the passage from ω to $\omega\circ\rho$ will cause a considerable change if ω is the vacuum (in which case $\omega\circ\rho$ generates a "sector" of charge n and is thus disjoint from ω). But on the contrary in the present case of a temperature state ω with non zero mean density (and thus infinite global charge) this passage is quite mild so that $\omega\circ\rho$ is quasi-equivalent to ω [14]. Now this is known to entail mathematically the existence of a unitary cocycle (the so called Radon-Nicodym derivative of $\omega\circ\rho$ w.r.t.ω) denoted $(D(\omega\circ\rho):D(\rho))t$, which effects the passage of the modular automorphism group of ω to that of $\omega\circ\rho$ [19]. On the other hand, the localized automorphism ρ carries a unitary cocycle $u_t \in \mathfrak{A}$ relating ρ to its time translates :

$$\alpha_t \circ \rho \circ \alpha_t^{-1} = \rho \circ \mathrm{Ad}\, u_t \ . \tag{25}$$

It now turns [18] that the quotient of these two cocycles is a one-dimensional unitary group determining as follows the chemical potential: one has

$$(D(\omega\circ\rho) : D\omega)_t = e^{in\beta(\mu+c)t}\ \pi_\omega(U_{-\beta t}) \quad , \tag{26}$$

where π_ω is the representation of $\mathfrak{A}$ generated by ω and c is a real constant independant of ω.

13) the automorphism ρ of $\mathfrak{A}$ is called <u>localized in a region R</u> if it leaves the local algebra of R locally invariant whilst acting trivially on all the local algebras of the regions space-like to R.
14) Sergio Doplicher proposed a few years back to use this criterion of quasi-equivalence for studying density states [22].

Outlook. The rôle of ergodicity in quantum statistical mechanics.

The theorems I and II above offer an a priori justification for the "KMS principle", substitute of the "Gibbs Ansatz" as the fundament of equilibrium quantum statistical mechanics. However we do not consider this justification as optimal : although the requirement of local dynamical stability is indeed a natural physical condition to be put on equilibrium states, it would however be much more satisfactory if this somewhat ad hoc condition would automatically be fulfilled for some deeper physical reason. We conjecture that this is the case : stability of extremal invariant states should actually follow from a condition of "quantal ergodicity". Indeed if we consider, say, a fluid in a vessel, partitioning this fluid by thought into two macroscopic systems (occupying, say, the right and the left portions of the vessel), these two systems will be largely dynamically independant due to locality, so that the whole dynamical system is almost a tensor product of the right and the left system. Now "quantal ergodicity" should be the principle stating the pervading character of the interaction, e.g. the fact that one cannot split the system algebra $\mathfrak{A}$ into a tensor product $\mathfrak{A}_1 \otimes \mathfrak{A}_2$ of two algebras both globally invariant under time automorphisms[15]. As one sees, this "ergodicity" principle and the principle of locality (e.g. asymptotic abelianness in time) tend to nearly contradict one another : taken together these principles should accordingly imply a considerable amount of structure. We conjecture that in combination they will imply as a mathematical consequence that all extremal invariant states of the algebra $\mathfrak{A}$ be β-KMS for the time automorphisms and for some (possibly not finite) temperature β.

15) or, more likely, a wider automorphism group including the time translations.

Appendix

A C^*-algebra $\mathfrak{A}$ is a complex linear space endowed with a bilinear product ab, $a,b \in \mathfrak{A}$; an antilinear *operation $a \in \mathfrak{A} \to a^* \in \mathfrak{A}$ and a norm $\| \ \|$ (under which it is complete) such that one has the properties $\|ab\| < \|a\|.\|b\|$, $\|a^*\| = \|a\|$, $\|a^*a\| = \|a\|^2$, $a,b \in \mathfrak{A}$. Any C^*-algebra can be realized (in general in many non unitarily equivalent ways) as a norm closed $*$-algebra of bounded operators on some Hilbert space. A state ϕ of $\mathfrak{A}$ is a linear complex valued function on $\mathfrak{A}$ such that $\phi(a^*a) > 0$, $a \in \mathfrak{A}$. An automorphism α of $\mathfrak{A}$ is a linear multiplicative map $a \in \mathfrak{A} \to \alpha(a) \in \mathfrak{A}$ with $\alpha(a^*) = \alpha(a)^*$, $a \in \mathfrak{A}$. States and automorphisms are automatically continuous (= bounded) : in fact $\|\phi\|=1$ and $\|\alpha(a)\| = \|a\|$ for all $a \in \mathfrak{A}$. Given a locally compact group G a continuous action α of G on $\mathfrak{A}$ is a map $g \in G \to \alpha_g$ of G into the automorphisms of $\mathfrak{A}$ such that $\alpha_{g^{-1}s} = \alpha_{g^{-1}} \alpha_s$, $g,s \in G$, and $g \in G \to \alpha_g(a)$ is continuous for all $a \in \mathfrak{A}$. If G is the additive line, $t \to \alpha_t$ is also called a continuous one-parameter group of automorphisms of $\mathfrak{A}$. The state ϕ is called α-invariant whenever $\phi(\alpha_g(a)) = \phi(a)$ for all $g \in G$ and $a \in \mathfrak{A}$. Such a state generates a representation π of $\mathfrak{A}$ on a Hilbert space $\mathcal{H}$, together with a continuous representation U of G on $\mathcal{H}$, and a vector $\Phi \in \mathcal{H}$ cyclic for π (= such that the set $\pi(a)\Phi$, $a \in \mathfrak{A}$, is dense in $\mathcal{H}$) such that

$$\begin{cases} \phi(a) = (\phi|\pi(a)|\phi) \\ \pi(\alpha_g(a)) = U(g)\ \pi(a)\ U(g)^{-1} \qquad , \qquad a \in \mathfrak{A} \ , \ g \in G\ . \\ U_g\Phi = \Phi \end{cases}$$

These elements, essentially uniquely determined, are called the GNS (Gelfand Neumark Segal) construction of ϕ. A Von Neumann algebra $\mathfrak{M}$ is a C^*-algebra with a faithful representation π on a Hilbert space $\mathcal{H}$ such that $\pi(\mathfrak{M}) = \pi(\mathfrak{M})''$, the bicommutant of $\pi(\mathfrak{M})$. A state ϕ of $\mathfrak{M}$ is normal if it can be realized as $\phi(a) = \mathrm{Tr}\{T_\phi(a)\}$ with T a (positive) trace class operator on $\mathcal{H}$. A a normal state ϕ of $\mathfrak{M}$, faithful in the sense that $\phi(a^*a) = 0$, $a \in \mathfrak{M}$, entails $a = 0$, determines uniquely a continuous one-parameter group $t \to \sigma_t$ of automorphismes of $\mathfrak{M}$ for which ϕ is KMS (i.e. β-KMS with $\beta = -1$) : σ iscalled the modular automorphism group of ϕ[4]. Given two normal faithful states ψ, ϕ of $\mathfrak{M}$ with modular automorphism group σ^ψ, σ^ϕ one has

$$\sigma^\psi(a) = v_t\ \sigma^\phi(a)\ v_t^* \qquad , \qquad a \in \mathfrak{M},$$

where $t \to v_t = (D\psi : D\phi)_t$ is a continuous unitary cocycle in $\mathfrak{M}$ (i.e. $t \to v_t$ is continuous and $v_{t+s} = v_t\ \alpha_t(v_s)$ $t,s \in R$) called the Radon-

Nicodym derivative of ψ w.r.t. ϕ .

Aknowlegments. The author is indebted to numerous colleagues for useful discussions - particularly to his friends and coautors of the works [5], [6], [18] in the bibliography which furnished the matter of the present report.

BIBLIOGRAPHY

[1] R. HAAG, N. HUGENHOLTZ, M. WINNINK
On the Equilibrium States in Quantum Statistical Mechanics.
Commun. math. Phys. 5, 215 (1967).

[2] R. KUBO
J. Physic. Soc. Japan, 12, 570 (1957).

[3] P.C. MARTIN, J. SCHWINGER
Phys. Rev. 115, 1342 (1959).

[4] M. TAKESAKI
Tomita's Theory of modular Hilbert Algebras and its Applications.
Springer Lecture Notes in Math. n° 128 (1970).

[5] R. HAAG, D. KASTLER, E. TRYCH-POHLMEYER
Stability and Equilibrium States.
Commun. math. Phys. 38, 173 (1974).

[6] O. BRATTELI, D. KASTLER
Relaxing the Clustering Condition in the Derivation of the KMS Property.
Commun. math. Phys. 46, 37 (1976).

[7] D. KASTLER
Equilibrium States of Matter and Operator Algebras.
Symposia Mathematica XX, 49 (1976).

[8] R. HAAG, E. TRYCH-POHLMEYER
Hambourg Preprint.

[9] D. RUELLE
States of Physical Systems.
Commun. Math. Phys. 3, 1 (1966).

[10] S. DOPLICHER, D. KASTLER, D.W. ROBINSON
Covariance Algebras in Field Theory and Statistical Mechanics.
Commun. math. Phys. 3, 1 (1966).

[11] S. DOPLICHER, D. KASTLER, E. STÖRMER
Invariant States and Asymptotic Abelianness - and literature quoted therei
J. Funct. Anal. 3, 419 (1969).

[12] S. DOPLICHER - Private communication.

[13] H. ARAKI
Expansional in Banach Algebras.
Ann. Sci. Ecole Norm. Sup. 6, 1 (1973).

[14] H. ARAKI
Relative Hamiltonian for faithful Normal States of a von Neumann Algebra.
Pub. RIMS Kyoto University 9, 165 (1973).

[15] D.W. ROBINSON
Return to Equilibrium.
Commun. math. Phys. 31, 171 (1973).

[16] R. HAAG - Private communication.

[17] H. ARAKI, A. KISHIMOTO
Symmetry and Equilibrium States
Commun. math. Phys. 52, 211 (1977).

[18] H. ARAKI, R. HAAG, D. KASTLER, M. TAKESAKI
Extension of KMS States and Chemical Potential.
Commun. math. Phys. 53, 97 (1977).
[19] A. CONNES
Une classification des facteurs de type III.
Ann. Sci. Ecole Norm. Sup. 6, 133 (1973).
[20] S. DOPLICHER, R. HAAG, J.E. ROBERTS
Fields, Observables and Gauge Transformations I and II
Commun. math. Phys. 13, 1 (1969) and 15, 173 (1969).
[21] S. DOPLICHER, R. HAAG, J.E. ROBERTS
Local Observables and Particle Statistics I and II.
Commun. math. Phys. 23, 199 (1971) and 35, 49 (1974).
[22] S. DOPLICHER - Private communication.

UNBOUNDED DERIVATIONS OF C*-ALGEBRAS AND CORRESPONDING DYNAMICS

Ola Bratteli

Dept de Physique
Univ. d'Aix-Marseille II
Luminy, Marseille and
CPT, CNRS
31 Chemin J. Aiguier
13 Marseille, France

Richard H. Herman*

Department of Mathematics
The Pennsylvania State
University, University Park,
Pennsylvania 16802 U.S.A.

It is our purpose in this lecture to describe some aspects of the theory of unbounded derivations as developed over the past several years. One of the motivations for this subject is that in physical systems one is often given a Hamiltonian (read unbounded derivation) and then one has to find the corresponding time development for the system. The question of where this time development takes place is a serious one. However, we shall take as our basic setting a C*-algebra or a von Neumann algebra i.e. time development will be a one-paramenter *-automorphism group of the algebra in question. More precisely there is a homomorphic map from $\mathbb{R} \to \mathrm{Aut}\,(\mathfrak{A})$ (Aut $(\mathfrak{A})$ is the automorphism group of the C* or von Neumann algebra $\mathfrak{A}$), with certain continuity properties. For a C*-algebra we require that $t \to \alpha_t(x)$ is continuous for all $x \in \mathfrak{A}$ and in the von Neumann algebra case that $t \to (\alpha_t(x)\xi \mid \eta)$ is continuous for all $x \in \mathfrak{A}$, $\xi,\eta \in \mathfrak{H}$ (the Hilbert space where the von Neumann algebra is acting). Under these circumstances it follows from the Hille-Yosida theory that $\alpha_t = \exp\,(t\delta)$ where δ is a (possibly) unbounded *-derivation of $\mathfrak{A}$, closed in the appropriate topology.

By a *-derivation we mean a linear map δ densely defined (in the appropriate topology) on $\mathcal{D}(\delta) \subseteq \mathfrak{A}$ and such that

$$\delta(x^*) = \delta(x)^*,\ \delta(xy) = \delta(x)y + x\delta(y);\ x,y \in \mathcal{D}(\delta).$$

The Hille-Yosida theory tells us that if δ comes from α_t, then Range $(I+\alpha\delta) = \mathfrak{A}$ and $\|(I+\alpha\delta)(x)\| \geq \|x\|$, for all $\alpha \in \mathbb{R}|\{0\}, x \in \mathcal{D}(\delta)$. Further a closed *-derivation satisfying there last two conditions gives rise to a *-automorphism group. The Hille-Yosida theory was however designed to deal with general linear operators and not derivations on algebras. Thus we should not take the above statements as the final answer but see if we can make use of the algebraic structure to derive necessary and sufficient conditions for an automorphism group to arise from a given *-derivation.

The first question that faces us is one of closeability, as generators are closed operators. To this end we point out that this is far from automatic. Recall that an

*Partially supported by The U.S. National Science Foundation

operator is closeable if the closure of its graph in $\mathfrak{A}\times\mathfrak{A}$ does not contain any point of the form $(0,y)$, where $y \neq 0$. For the commutative case one can show that on $C(K)$, (K is the Cantor set) no non-zero derivation is closeable and on $C[0,1]$ there exists a non-closeable derivation extending ordinary differention. The former fact was used [4] to show that there is a uniformly hyperfinite C*-algebras ($\mathfrak{A}=\overline{\cup\,\mathfrak{A}_n}$, $\mathfrak{A}_n$ full matrix algebras $\mathfrak{A}_n \subseteq \mathfrak{A}_{n+1}$) and a non-zero derivation δ such that $\cup\,\mathfrak{A}_n \subseteq \mathcal{D}(\delta)$ and $\delta \mid \cup\,\mathfrak{A}_n = 0$. Thus δ is not closeable.

A positive result was obtained by Powers and Sakai.

1. Theorem [18]. If $\mathcal{D}(\delta)$ is closed under the square root operation on positive elements then δ is closeable.

Kishimoto [14] extended this to maps satisfying $\delta(a^*a) \geq \delta(a^*)a + a^*\delta(a)$ showing dissipativeness and $\|(I-\alpha\bar{\delta})(x)\| \geq \|x\|$, for $\alpha \in \mathbb{R}^+$, where $\bar{\delta}$ is the closure of δ. In contrast to the above theorem, Ota showed

2. Theorem. If δ is a closed *-derivation and $\mathcal{D}(\delta)$ is closed under the square root operation of positive elements then δ is bounded.

The proof of this last theorem relies heavily on a result of Cuntz [10].

In his thesis Chi proved the following

3. <u>Theorem</u>. If $\mathfrak{A}$ is a simple C*-algebra and there exists $\varphi \neq 0$ in $\mathcal{D}(\delta^*)$, then δ is closeable.

Proof: Recall that $\varphi \in \mathcal{D}(\delta^*)$ means that there exists a constant $L \geq 0$ such that $|\varphi(\delta(x))| \leq L\|x\|$. Moreover a densely defined operator is closeable if and only if the domain of its adjoint is total. Now one easily sees that the functionals $x \to \varphi(axb)$ for $a,b \in \mathcal{D}(\delta)$, belong to $\mathcal{D}(\delta^*)$. The simplicity of $\mathfrak{A}$ is then used to show that this set is total.

4. <u>Cor</u>: If $\mathfrak{A}$ is simple and δ has an invariant state (a state ω such that $\omega\circ\delta = 0$) then δ is closeable.

This corollary remains true if the condition that $\mathfrak{A}$ is simple is replaced by the condition that the cyclic [24] representation π_ω associated to ω is faithful. Sakai has conjectured that for $\mathfrak{A}$ a simple C*-algebra with unit, δ is closeable if and only if $R(\delta)^- \neq \mathfrak{A}$. It is easily seen that $R(\delta)^- \neq \mathfrak{A}$ if δ is a generator or an inner limit (see ahead) derivation, for in both of these cases invariant states exist.

It is often useful to have certain types of elements, e.g. projections, within the domain of a derivation. For this one needs a functional calculus. The basic theorem in this direction is contained in [3] and arises in the calculations of [17].

5. Theorem. Suppose $I \in \mathfrak{A}$, and δ is a closed *-derivation. Let $x = x^*$ be in $\mathcal{D}(\delta)$ and $f : \mathbb{R} \to \mathbb{C}$ such that $\int |p\hat{f}(p)| dp < \infty$, then $f(x) = \int dp\hat{f}(p) e^{ipx} \in \mathcal{D}(\delta)$ and $\delta(f(x)) = i \int dp\hat{f}(p) p \int_0^1 dt e^{itpx} \delta(x) e^{i(1-t)px}$

Proof. We write $e^{itx} = \lim_{n\to\infty} (I + \frac{itx}{n})^n$

and

$$\delta((I + \frac{itx}{n})^n) = \sum_{k=1}^{m} (I + \frac{itx}{n})^{k-1} \frac{it}{n} \delta(x) (I + \frac{itx}{n})^{n-k}$$

The last expression converges as $n \to \infty$ to

$$it \int_0^1 ds\ e^{istx} \delta(x) e^{i(1-s)tx} .$$

(The sum in question is a Riemann sum for the integral). Invoking the closedness of δ, this last integral is $\delta(e^{itx})$. For the general case one uses a Fourier expansion. As a corollary of this theorem one knows that twice continuously differentiable functions leave $\mathcal{D}(\delta)$ invariant. This is not true in general for once continuously differentiable functions as has been shown by McIntosh [26].

The algebraic structure which most closely reflects that of quantum lattice systems is that of a uniformly hyperfinite (UHF) C*-algebra (see the discussion on closeability). It is extremely useful to observe that if a matrix algebra $\mathfrak{A}_n \subseteq \mathcal{D}(\delta)$ then δ restricted to $\mathfrak{A}_n$ is given by an element of $\mathfrak{A}$. Indeed for $x \in \mathfrak{A}_n$, $\delta(x)$ $[ih_n, x]$ where $h_n = h_n^* = \frac{-i}{m(n)} \sum_{1}^{m(n)} \delta(e_{ij}) e_{ji}$ [19]

Here $\mathfrak{A}_n$ is an $m(n) \times m(n)$ matrix algebra.

6. Theorem. Let $\mathfrak{A}$ be a UHF C*-algebra, δ a closed derivation. There exists an increasing sequence of matrix subalgebras $\mathfrak{A}_n$, all containing the identity such that $\overline{\cup \mathfrak{A}_n} = \mathfrak{A}$ and $\cup \mathfrak{A}_n \subseteq \mathcal{D}(\delta)$.

The theorem as stated is proven in [4] using techniques similar to the proof of this theorem when δ is a generator as first shown by Sakai [21]. The theorem is established by means of the functional calculus on the domain of a closed derivation, Theorem 5.

A conjecture of Powers and Sakai is that all one parameter *automorphism groups are approximately inner in that there exists $h_n = h_n^* \in \mathfrak{A}$, such that $e^{ih_n t} x e^{-ih_n t} \to \alpha_t(x)$ as $n \to \infty$, for all $x \in \mathfrak{A}$. This would be true if one could construct the $\mathfrak{A}_n$ above so that $\cup \mathfrak{A}_n$ is a core for the generator δ.

In this direction Longo [15] has shown that there exists a unitary cocycle in $\mathfrak{A}$ $\{u_n(t)$ such that $u_n(t+s) = u_n(t)\alpha_t(u_n(s))\}$ such that $\alpha_t(x) = \lim_{n\to\infty} u_n(t)\ x\ u_n(t)^*$, for all $x \in \mathfrak{A}$.

At this point we need the following two facts about UHF C*-algebras [25]. First there exists a unique trace state, τ, on $\mathfrak{A}$. Secondly τ, may be used to defined a projection ε_n, of $\mathfrak{A}$ onto $\mathfrak{A}_n$, via the equality $\tau(ax) = \tau(a\varepsilon_n(x))$ holding for $a \in \mathfrak{A}_n$. We may now state a result of Kishimoto

7. Theorem. Suppose that δ is a* derivation of a UHF C*-algebra defined on an increasing sequence $\{\mathfrak{A}_n\}$ of matrix algebras which generate $\mathfrak{A}$, and $\delta|\mathfrak{A}_n = \mathrm{ad}(ih_n)|\mathfrak{A}_n$. Further suppose that there exists a constant M so that $\|h_n - \varepsilon_n(h_n)\| \leq M$ for all n. Then δ is closeable and its closure is a generator.

One can show [18], through the case of Phragmen-Lindelöf or Poisson integral theorems, that approximately inner dynamics have KMS states for all values of the inverse temperature, β, and that there are ground states. As a consequence of the Trotter-Kato theorem Kishimoto's result shows that the *-automorphism group $\exp(t\bar{\delta})$ is approximately inner (in fact $\exp(t\bar{\delta})(x) = \lim_{n\to\infty} e^{ith_n}\ x\ e^{-ith_n})$ and thus has KMS states. In fact for this case we have only one KMS state for each β - i.e. there is no phase transition.

8. Theorem. With the same assumptions as in Theorem 7, there exists one and only one KMS state for each $-\infty < \beta < \infty$, for the time evolution $x \to \exp(t\bar{\delta})(x)$.

This theorem was first shown by Sakai [23,24], in the special case where the $\{h_n\}$ commute. The general version was obtained later by Araki [1] and Kishimoto [14].

A related result of Jørgenson [12] is

9. Theorem. Suppose $\{\mathfrak{A},\alpha_t\}$ is an approximately inner dynamics and

$$\lim_{m,n\to\infty} \|e^{it(h_n-h_m)}\ x\ e^{-it(h_n-h_m)}\| = 0.$$

If α_t has a $\beta_0(\neq 0)$ KMS then $\mathfrak{A}$ has a trace and hence α_t has KMS states for all $\beta \neq 0$.

We turn briefly to general C*-algebra theory. An $x \in \mathfrak{A}$ is called analytic for δ if $\sum_n t^n \frac{\|\delta^n(x)\|}{n!}$ has a **non-zero** radius of convergence. For a closed *-derivation to be a generator it is <u>not</u> sufficient that it has a dense set of analytic vectors (take differentiation on $C[0,1]$) as opposed to the nice situation for symmetric operators.

However one does have

10. Theorem [4]. Let $\mathfrak{A}$ be a C*-algebra and δ a closed densely defined derivation such that $\|(I+\alpha\delta)(x)\| \geq \|(x)\|$ for $\alpha \in \mathbb{R} | \{0\}$ and $x \in \mathcal{D}(\delta)$. If $\mathcal{D}(\delta)$ contains a dense set of analytic vectors for δ then δ generates an automorphism group.

We spoke above of invariant states for a derivation. If ω is such a state, then the relationship $\omega \circ \delta = 0$ shows that the definition

$$iH\pi_\omega(x)\Omega = \pi_\omega(\delta(x)\Omega \qquad , x \in \mathcal{D}(\delta)$$

is good and defines a symmetric operator H in the representation space $\mathfrak{H}_\omega$ coming from the GNS construction using ω. Moreover

$$[iH,\pi_\omega(x)] = \pi_\omega(\delta(x)) \qquad x \in \mathcal{D}(\delta)$$

as is easily verified.

In the von Neumann algebra setting the corresponding result for analytic vectors is

11. Theorem [7]. Let δ be a derivation of a von Neumann algebra $\mathcal{M}$ given by $[iH,.]$ with H a symmetric operator. Suppose $\mathcal{M}$ has a cyclic vector Ω with $H\Omega = 0$. Assume that there is $\mathcal{D} \subseteq \mathcal{D}(\delta)$ such that $\mathcal{D}$ is a *subalgebra strongly dense in $\mathcal{M}$. If $\delta(\mathcal{D}) \subseteq \mathcal{D}$ and $\mathcal{D}\Omega$ consists of analytic vectors for H, then H is essentially self-adjoint and $\bar{H}$ satisfies

$$e^{it\bar{H}}\mathcal{M}e^{-it\bar{H}} = \mathcal{M}$$

In the presence of a β-KMS automorphism group of a C*-algebra, the resulting cyclic vector in the representation space is also known to be separating.

12. Theorem [6]. Let δ be a *-derivation of a von Neumann algebra $\mathcal{M}$ given by a self-adjoint operator H. Suppose $\mathcal{M}$ has a cyclic and separating vector Ω and $H\Omega = 0$. Further assume that $\mathcal{D}(\delta)\Omega$ is a core for H. Then $e^{iHt}\mathcal{M}e^{-iHt} = \mathcal{M}$ if and only if $[e^{iHt}, \Delta_\Omega^{it}] = 0$, when Δ_Ω^{it} is the modular operator corresponding to Ω.

A commutative version of this theorem was first proven by Gallavotti and Pulvirenti [11]. The case where Ω is a trace vector was handled by Bratteli and Robinson [5].

Perturbation Theory

We shall measure the difference of two automorphism groups with a view towards determining when one of their generators is a "perturbation" of the other. (This notion will shortly be made precise). This problem arises naturally in that we may not have the exact dynamics of a physical system and would thus like to know the consequences of two sets of dynamics being suitably close. This problem was first dealt with by Bucholz and Roberts. They proved the

13. Theorem [8]. Suppose m is either a von Neumann algebra or a simple C*-algebra. Then $\|\alpha_t - \beta_t\| \to 0$ as $t \to 0$ if and only if

$$\delta_\alpha(x) = v\delta_\beta(v^{-1}xv)v^{-1} + [ih,x], \qquad x \in \mathcal{D}(\delta_\alpha)$$

where $h = h^* \in m$, $v \in m$ is unitary and both have norm continuous orbits under α_t.

Motivated by work of Kadison and Ringrose [13] we want to consider what happens when $\|\alpha_t - \beta_t\|$ is small for t small and doesn't necessarily go to zero.

14. Theorem [7]. Let m be a von Neumann algebra with separable pre-dual. The following two statements are equivalent:

1. There are ε_1, $0 \le \varepsilon_1 < \sqrt{199/50} \simeq 0.28$ and $\delta_1 > 0$ such that $\|\alpha_t - \beta_t\| \le \varepsilon_1$, $0 \le t \le \delta_1$

2. There are ε_2 , $0 \le \varepsilon_2 < \sqrt{199/50}$ and $\delta_2 > 0$, an inner automorphism γ and an inner derivation δ of m such that $\delta_\alpha = \gamma\circ(\delta_\beta+\delta)\circ\gamma^{-1}$ and $\gamma\circ(\delta_\beta+\delta)\circ\gamma^{-1}$

$$\|\alpha_t\circ\gamma^{-1}\circ\alpha_t\circ\gamma - i\| \le \varepsilon_2 \qquad 0 \le t \le \delta_2$$

If this is the case then $W \in m$ can be chosen giving γ, so that

$$\|W - I\|^2 \le 200\ (1 - \sqrt{1 - \frac{\varepsilon_1^2}{4}}\ )$$

Note that the form of this result and that of Buchholz-Roberts is the same. However in our case the orbit of W need not be norm continuous.

To arrive at δ_α from δ_β we have provided a "twist" and a "bounded perturbation". These two things are explained by the following two results

15. Theorem [7]. If m is a von Neumann algebra, then $\|\alpha_t - \beta_t\| = 0(t)$ as $t \to 0$ iff $\delta_\alpha = \delta_\beta + \delta$, where δ is a bounded (hence inner) derivation of m.

16. Theorem [7]. Let m be a von Neumann algebra with separable predual. Suppose that $\|\alpha_t - \beta_t\| \le \varepsilon < \sqrt{199}/50$ for all t then $\alpha_t = \gamma \circ \beta_t \circ \gamma^{-1}$, where γ is a inner automorphism. A unitary $W \in m$ can be chosen giving γ so that $\|W - I\|^2 \le 200\ (1 - \frac{\sqrt{1-\varepsilon^2}}{4})$.

Space does not permit concluding the details but the proof of Theorem 14 goes by reduction to the $0(t)$ case. We indicate how this is done in the case of unitary groups on a Hilbert space.

17. Theorem [7]. Let $U_t = \exp\ (itH)$, $V_t = \exp\ (itK)$

1. There exists ε_1, $0 < \varepsilon_1 < r_2$, $\delta_1 > 0$ such that

$$\|U_t - V_t\| \le \sqrt{2} - \varepsilon_1, \quad 0 \le t \le \delta_1$$

($\Leftrightarrow$) 2. There exist ε_2, $0 < \varepsilon_2 < \sqrt{2}$, $\delta_2 > 0$, P bounded $P = P^*$ (bounded), W a unitary such that

$$H = W(K+P)W^*$$

$$\|U_t W^* U_{-t} W - I\| \le \sqrt{2} - \varepsilon_2 \qquad 0 \le t \le \delta$$

Proof:

This proof is also lengthy but the idea is to twist the group V_t by a unitary W by defining $\hat{V}_t = WV_tW^*$ and then show that $\|U_t - \hat{V}_t\| = 0(t)$ as $t \to 0$. It then follows [20] that the generators of U_t and $\hat{V}_t$ differ by a self-adjoint operator. The twisting operator W is obtained by taking the polar decomposition of
$\Omega = \frac{1}{\delta_1} \int_0^{\delta_1} dt U_t V_{-t}$.

For the case of automorphism groups are one needs to choose appropriate unitaries so that W gives an automorphism of m and P a derivation of m. One of the basic items is the construction of a unitary cocycle.

Theorem [7]. Let m be a von Neumann algebra with separable predual and α_t, β_t one parameter *-automorphism groups of m. Assume there exist $\delta > 0$, $\varepsilon > 0$, $\varepsilon < \sqrt{71}/18$ and

$$\|\alpha_t - \beta_t\| \le \varepsilon \quad \text{for} \quad 0 \le |t| \le \delta$$

Then there exist $t \to \Gamma_t$ a σ-weakly continuous map of $\mathbb{R}$ into $U(m)$ such that

$$\Gamma_{t+s} = \Gamma_t \alpha_t(\Gamma_s)$$

$$\beta_t(x) = \Gamma_t \alpha_t(x) \Gamma_t^*$$

$$\|\Gamma_t - I\| \leq 10\sqrt{2(1-\sqrt{1-\frac{\varepsilon^2}{4}}} \quad , \qquad |t| \leq \delta/4$$

A similar result is to be found in [8]. In the case above one first finds unitaries satisfying the last two conditions by Borel lifting theorems and then adjusts to obtain a cocycle. In the case of [8] a topological lifting is used.

Examples for which we refer the reader to [7] show that

1. There is a simple C*-algebra with automorphism groups α_t, β_t such that $\|\alpha_t - \beta_t\| = O(t)$ as $t \to 0$ by $\mathcal{D}(\delta_\alpha) \cap \mathcal{D}(\delta_\beta)$ is not even dense. (The analogue of Theorem does not hold for simple C*-algebras)

2. One can find a von Neumann algebra where

$\|\alpha_t - \beta_t\| = \varepsilon$ for all $t \in \mathbb{R} \mid \{0\}$ where ε is any fixed number between zero and two.

Finally we refer the reader to Sakai's excellent talk [24] for many aspects of the theory which we were not able to include here.

References

1. H. Araki, On the uniqueness of KMS States of One-dimensional quantum lattice systems. Commun. Math. Phys. 44 (1975) 1-7.

2. O. Bratteli, R. H. Herman and D. W. Robinson, Quasi-analytic vectors and derivatior of operator algebras, Math. Scand. 39 (1976) 371-381.

3. O. Bratteli and D. W. Robinson, Unbounded derivations of C*-algebras, Commun. Math. Phys. 42 (1975) 253-268.

4. ______________________ Unbounded derivations of C*-algebras II, Commun. Math. Phys. 46 (1976) 11-30.

5. ______________________ Unbounded derivations and invariant trace states. Commun. Math. Phys. 46 (1976) 31-35.

6. O. Bratteli and U. Haagerup, Unbounded derivations and invariant states - preprint.

7. O. Bratteli, R. H. Herman and D. W. Robinson, Perturbation of Flows on Banach Spaces and Operator Algebra - preprint.

8. D. Bucholz and J. Roberts, Bounded Perturbation of Dynamics, Commun. Math. Phys. 49 (1976) 161-177.

9. D. P. Chi, Derivations in C*-algebras, Thesis, Univ. of PA.

10. J. Cuntz, Locally C*-equivalent algebras, Journal of Functional Analysis 23 (1976).

11. G. Gallavotti and M. Pulvirenti, Classical KMS condition and Tomita-Takesaki Theory, Commun. Math. Phys. 46 (1976) 1-9.

12. P. Jørgensen, Trace states and KMS state for approximately inner dynamical one-paramenter groups of *-automorphisms, Commun. Math. Phys.

13. R. V. Kadison and J. R. Ringrose, Derivations and automorphisms of operator algebras. Comm. Math. Phys. 4 (1967) 32-63.

14. A. Kishimoto, Dissipations and derivations, Commun. Math. Phys. 47 (1976) 25-32.

15. R. Longo, On Perturbed derivation of C*-algebra, preprint.

16. S. Ota, Certain operator algebra induced by *-derivation in C*-algebras on an indefinite inner product space preprint.

17. R. T. Powers, A remark on the domain of an unbounded derivation, Journal of Functional Anal. 18 (1975) 85-95.

18. R. T. Powers and S. Sakai, Existence of ground states for approximately inner dynamics, Commun. Math. Phys. 39 (1975) 273-288.

19. ____________________, Unbounded derivations in operator algebras, Jour. Func. Anal. 19 (1975) 81-95.

20. D. W. Robinson, The approximation of flows. Jour. Func. Anal. 24 (1977) 280-290.

21. S. Sakai, On one-parameter subgroups of *-automorphisms on operator algebras and the corresponding unbounded derivations, Amer. J. Math 98 (1976) 427-440.

22. ____________________, On Commutative normal *-derivation II, Jour. of Func. Anal. 21 (1976) 203-208.

23. ____________________, On commutative normal * derivations III, Tohoku Math J. 28 (1976) 583-59.

24. ____________________, Recent development in the theory of unbounded derivation in C*-algebras. Talk delivered at the U. S. - Japan Seminar on C*-algebras and their application to theoretical physics, U. C. L. A. April, 1977.

25. C*-algebras and W*-algebras, Springer-Verlag, 1971, Band 60

26. A. McIntosh, Functions and Derivations of C*-algebras.

INTRODUCTION TO THE FLOW OF WEIGHTS ON FACTORS OF TYPE III

Masamichi TAKESAKI *) **)
Université d'Aix-Marseille II (Luminy)
and
Centre de Physique Théorique CNRS Marseille

1. Introduction and Preliminary

In this talk, I would like to give a brief survey on a joint work of A. Connes and myself, "The Flow of Weights on Factors of Type III", [3].

A von Neumann algebra means, by definition, a non-degenerate self-adjoint algebra $\mathcal{M}$ of operators on a Hilbert space $\mathfrak{H}$ which is closed under the weak operator topology, i.e. closed under the locally convex topology in $\mathcal{L}(\mathfrak{H})$ induced by the family of semi-norms : $x \in \mathcal{L}(\mathfrak{H}) \mapsto |(x\xi|\eta)|$, $\xi, \eta \in \mathfrak{H}$. The fundamental theorem of operator algebras due to J. von Neumann says that $\mathcal{M} = \mathcal{M}''$ where $\mathcal{S}' = \{x \in \mathcal{L}(\mathfrak{H}) : xy = yx$ for every $y \in \mathcal{S}$ for any $\mathcal{S} \subset \mathcal{L}(\mathfrak{H})$. Since every element of $\mathcal{M}$ is written as a linear continuation of two self-adjoint elements, the spectral decomposition theorem for self-adjoint operators asserts that $\mathcal{M}$ is generated by its projections. Namely, the projections of $\mathcal{M}$ form a fundamental building block. Indeed, F. Murray and J. von Neumann concentrated, in thier pioneering work, on the analysis of the projection lattice of a factor. A factor is, by definition, a von Neumann algebra with trivial center, i.e. $\mathcal{M} \cap \mathcal{M}' = \mathbb{C}$.

For projections $e, f \in \mathcal{M}$, we write $e \sim f$ if there exists a partial isometry $u \in \mathcal{M}$ with $u^*u = e$ and $uu^* = f$. If $e \sim e_1 \leq f$, then we write $e \precsim f$. A projection $e \in \mathcal{M}$ is said to be <u>finite</u> if $e \sim e_1 \leq e$ implies $e = e_1$. Otherwise it is called <u>infinite</u>. Based on the structure of the projection lattice, F. Murray and J. von Neumann classified all factors into the following four classes :

<u>Type I</u> : There exists a minimal projection in a factor $\mathcal{M}$. In this case, $\mathcal{M}$ is isomorphic to $\mathcal{L}(\mathfrak{K})$ for some Hilbert space $\mathfrak{K}$. Thus, the dimension of $\mathfrak{K}$ determines uniquely the structure of $\mathcal{M}$, so that $\mathcal{M}$ is said to be type I_n if dim $\mathcal{M} = n^2$.

Type II_1 : There exists no minimal projection and 1 is finite.

---

*) The author is supported in part by N.S.F.

**) <u>Permanent Address</u> : University of California, Los Angeles
California 90024, U.S.A.

Type II_∞ : There exists a finite projection but no minimal projection. In this case, $\mathfrak{M}$ is isomorphic to the tensor product of a factor of type II_1, and a factor of type I_∞. In principle, the study of a factor of type II_∞ is reduced to that of a factor of type II_1.

Type III : There is no finite projection. A factor $\mathfrak{M}$ is of type III if and only if $e \sim 1$ for every non-zero projection $e \in \mathfrak{M}$ provided that $\mathfrak{M}$ acts on a separable Hilbert space.

Example : Let $\{\Gamma, \mu\}$ be a standard measure space, i.e. a σ-finite measure space isomorphic to the real line R equipped with a Stieljes measure. Suppose a countable group G acts on Γ in such a way that i) each $g \in G$ is a bi-measurable transformation of $\{\Gamma, \mu\}$ and ii) the transformed measure $\mu \circ g$ is quasi-invariant to μ. The action of G is called free if every $g \neq e$ has no fixed points except null set, i.e. $N_g = \{\gamma \in \Gamma : g\gamma = \gamma\}$ is a null set. For a technical simplicity, we assume that G is freely acting. But this is merely to avoid a longer argument. Indeed, the so-called Krieger's construction takes care of the case that G is not free. Let

$$\mathfrak{h} = L^2(G \times \Gamma, \delta \otimes \mu)$$

with δ the counting measure of G. We define two kinds of operators $\pi(f)$, $f \in L^\infty(\Gamma, \mu)$, and $u(g)$, $g \in G$, on $\mathfrak{h}$ as follows :

$$\begin{cases} \pi(f)\xi(h,\gamma) = f(h\gamma)\xi(h,\gamma), \quad f \in L^\infty(\Gamma,\mu),\ \xi \in \mathfrak{h}; \\ u(g)\xi(h,\gamma) = \xi(g^{-1}h, \gamma), \quad g, h \in G. \end{cases}$$

We denote by $\mathfrak{M}(G, \Gamma, \mu)$ the von Neumann algebra only generated by $\pi(f)$ and $u(g)$. It follows that

(i) $$u(g)\pi(f)u(g)^{-1} = \pi(\alpha_g f), \qquad f \in L^\infty(\Gamma, \mu),\ g \in G,$$

where $\alpha_g(f)(\gamma) = f(g^{-1}\gamma)$; every $x \in \mathfrak{M}(G, \Gamma, \mu)$ has an expression :

(ii) $$x = \sum_{g \in G} \pi(f_g^x)\, u(g) \ ;$$

and

(iii) $$f_g^{xy} = \sum_{h \in G} f_h^x\, \alpha_h(f_{h^{-1}g}^y) \ ;$$

(iii) $$f_g^{x^*} = \alpha_g (f_{g^{-1}}^x)^* .$$

We then have the following :

$$\mathcal{M}(G, \Gamma, \mu) \quad \text{is a factor}$$

$$\Longleftrightarrow \quad G \text{ is ergodic.}$$

Assuming the ergodicity,

$\mathcal{M}(G, \Gamma, \mu)$ is of type I $\Leftrightarrow$ G is transitive, i.e. $\Gamma = G\gamma$;

$\mathcal{M}(G, \Gamma, \mu)$ is of type II_1 $\Leftrightarrow$ There exists a finite non-atomic invariant measure equivalent to μ ;

$\mathcal{M}(G, \Gamma, \mu)$ is of type II_∞ $\Leftrightarrow$ There exists a σ-finite infinite invariant measure equivalent to μ ;

$\mathcal{M}(G, \Gamma, \mu)$ is of type III $\Leftrightarrow$ There exists no σ-finite invariant measure equivalent to μ .

If one takes $\Gamma = G$ and μ the counting measure, then $\mathcal{M}(G, \Gamma, \mu)$ is of type I.

If one takes $\Gamma = \mathbb{T}$, the one-dimensional torus, with Lebesgue measure μ , and $G = \mathbb{Z}$ acting as an irrational angle rotation, then $\mathcal{M}(G, \Gamma, \mu)$ is of type II_1. More generally, let Γ be a compact group with Haar measure μ and G be a dense countable group of Γ acting as translation on Γ from the left. Then $\mathcal{M}(G, \Gamma, \mu)$ is of type II_1 .

Let $\{\Gamma, \mu\} = \{\mathbb{R}, m\}$ with m Lebesgue measure. If $G = \mathbb{Q}$, the additive group of rational numbers, acting on $\mathbb{R}$ by translations, then $\mathcal{M}(G, \Gamma, \mu)$ is of type II_∞ .

Let $\{\Gamma, \mu\} = \{\mathbb{R}, m\}$ and G be the affine rational transformation group, i.e. $G = \{(a, b) : a \neq 0, \ a, b \in \mathbb{Q}\}$ acting on $\mathbb{R}$ by $\gamma \in \mathbb{R} \mapsto a\gamma + b \in \mathbb{R}$. Then $\mathcal{M}(G, \Gamma, \mu)$ is of type III.

2. Relation between $\mathcal{M}$ and $\mathcal{M}'$.

Due to the fundamental fact that $\mathcal{M} = \mathcal{M}''$, the study of the relation between $\mathcal{M}$ and its commutant $\mathcal{M}'$ has vital importance in the theory of operator algebras. As a matter of fact, at the earlier stage of the development, the major effort of many mathematician was concentrated on the problem of understanding the relation between $\mathcal{M}$ and $\mathcal{M}'$. With Tomita's theory, one can now see much clearer a picture of $\mathcal{M}$ and $\mathcal{M}'$.

For each $\xi \in \mathfrak{H}$, let e_ξ (resp. e'_ξ) denote the projection of $\mathfrak{H}$ onto $[\mathcal{M}'\xi]$ (resp. $[\mathcal{M}\xi]$), where $[\cdot]$ denotes the closed subspace spanned by . . It follows that $e_\xi \in \mathcal{M}$ and $e'_\xi \in \mathcal{M}'$. One of the deep results due to Murray and von Neumann states that $e_\xi \precsim e_\eta \Leftrightarrow e'_\xi \precsim e'_\eta$. From this one concludes that

$$\mathcal{M} \text{ is of type I} \iff \mathcal{M}' \text{ is of type I ;}$$

$$\mathcal{M} \text{ is of type II} \iff \mathcal{M}' \text{ is of type II ;}$$

$$\mathcal{M} \text{ is of type III} \iff \mathcal{M}' \text{ is of type III .}$$

However, one concludes from Tomita's theory that $\mathcal{M}_{e_\xi}$ and $\mathcal{M}'_{e'_\xi}$ are anti-isomorphic.

If $[\mathcal{M}\xi_0] = [\mathcal{M}'\xi_0] = \mathfrak{H}$ for a fixed vector $\xi_0 \in \mathfrak{H}$, then Tomita's theory yields that there exists a unitary involution J, called the modular conjugation, and a non-singular positive self-adjoint operator Δ, called the modular operator such that

$$x^*\xi_0 = J\Delta^{1/2}x\xi_0, \qquad x \in \mathcal{M};$$

$$y^*\xi_0 = J\Delta^{-1/2}y\xi_0, \qquad y \in \mathcal{M}';$$

$$J\Delta J = \Delta^{-1};$$

$$\Delta^{it}\mathcal{M}\Delta^{-it} = \mathcal{M}, \qquad \Delta^{it}\mathcal{M}'\Delta^{-it} = \mathcal{M}';$$

$$J\mathcal{M}J = \mathcal{M}', \qquad J\mathcal{M}'J = \mathcal{M}.$$

If we put

$$\omega(x) = (x\xi_0 \mid \xi_0), \qquad x \in \mathcal{M};$$

$$\sigma_t^\omega(x) = \Delta^{it}x\Delta^{-it}, \qquad t \in \mathbb{R},$$

then $\{\sigma_t^\omega, \omega\}$ satisfies the KMS-condition. It is further shown that the one-parameter automorphism group $\{\sigma_t^\omega\}$ is uniquely determined by ω subject to the

KMS-condition. If one fixes the group $\{\sigma_t\}$ first, then ω is almost uniquely determined, i.e. it is unique up to a multiple by a central element. A.Connes' important cocycle Radon-Nikodym theorem states that given faithful normal positive linear functionals φ and ψ on $\mathcal{M}$ there exists a continuous one parameter family $\{u_t\}$ of unitaries in $\mathcal{M}$ such that

$$u_{s+t} = u_s \sigma_s^{\varphi}(u_t) \quad \text{and} \quad \sigma_t^{\psi} = \mathrm{Ad}(u_t) \circ \sigma_t^{\varphi}$$

the functions : $s \in \mathbb{R} \longmapsto \varphi(u_s x) \in \mathbb{C}$ and $s \in \mathbb{R} \longmapsto \psi(x u_s) \in \mathbb{C}$, $x \in \mathcal{M}$, are the boundary values of a holomorphic bounded function on the horizontal strip, $0 \le \mathrm{Im}\, z \le 1$. If $\mathcal{M} = L^{\infty}(\Gamma, \mu)$ and $\varphi(x) = \int x(\gamma)\, d\mu(\gamma)$ and $\psi(x) = \int x(\gamma) h(\gamma)\, d\mu(\gamma)$, then $u_s = h^{is}$, $s \in \mathbb{R}$. Subject to above relative KMS-condition, $\{u_s\}$ is uniquely determined by φ and ψ. For this reason, we write $u_t = (D\psi : D\varphi)_t$. Conversely, if $\{u_s\}$ is a cocycle, i.e. $u_{s+t} = u_s \sigma_s^{\varphi}(u_t)$, then there exists a faithful semi-finite normal weight ψ on $\mathcal{M}$ such that $(D\psi : D\varphi)_t = u_t$. In the frame of normal states or positive linear functions, the above Radon-Nikodym theorem does not behave well. Thus one is eventually lead to the theory of weights.

3. Structure of a Factor of Type III

Let $\mathcal{M}$ be a fixed von Neumann algebra of type III. On $L^2(\mathbb{R})$, we define two one-parameter unitary groups $\{U_s\}$ and $\{V_t\}$ as follows :

$$\begin{cases} U_s \xi(t) = \xi(s+t), \\ V_s \xi(t) = e^{ist} \xi(t), \quad \xi \in L^2(\mathbb{R}),\ s, t \in \mathbb{R}. \end{cases}$$

Let $\omega = \mathrm{Tr}(h\,\cdot)$, where $h = \exp(\frac{d}{dt})$, and $\tilde{\varphi} = \varphi \otimes \omega$ on $\mathcal{M} \otimes \mathcal{L}(L^2(\mathbb{R})) = \bar{\mathcal{M}}$. It follows from the Radon-Nikodym theorem that all $\tilde{\varphi}$'s are conjugate under the inner automorphism group $\mathrm{Int}(\bar{\mathcal{M}})$ of $\bar{\mathcal{M}} \cong \mathcal{M}$. The fixed point algebra $\bar{\mathcal{M}}_{\tilde{\varphi}} = \mathcal{N}$ under the modular automorphism group $\{\sigma_t^{\tilde{\varphi}}\}$, the centralizer of $\tilde{\varphi}$, is a von Neumann algebra of type II_{∞} and

$$\bar{\mathcal{M}} \cong W^*(\mathcal{N}, \mathbb{R}, \theta) \text{ -- the crossed product,}$$

where

$$\theta_t(x) = V_t\, x\, V_t^*, \quad x \in \mathcal{N}.$$

The algebra $\mathcal{N}$ admits a faithful semi-finite normal trace τ such that

$\tau \circ \theta_t = e^{-t}\tau$. Conversely, if $\{\bar{\mathcal{N}}, \mathbb{R}, \bar{\theta}\}$ is a semi-finite von Neumann algebra equipped with a one-parameter automorphism group $\{\bar{\theta}_t\}$ and a trace $\bar{\tau}$ such that $\bar{\tau} \circ \bar{\theta}_t = e^{-t}\bar{\tau}$ and $\mathcal{M} \cong W^*(\bar{\mathcal{N}}, \mathbb{R}, \bar{\theta})$, then there exists an isomorphism π of $\mathcal{N}$ onto $\bar{\mathcal{N}}$ such that $\bar{\theta}_t \circ \pi = \pi \circ \theta_t$. Thus $\{\mathcal{N}, \mathbb{R}, \theta\}$ is uniquely determined by $\mathcal{M}$. Connes' invariants $S(\mathcal{M})$, the modular spectrum, and $T(\mathcal{M})$, the modular period group, of a factor $\mathcal{M}$ is then given by the following :

$$S(\mathcal{M}) = \{ e^t : \theta_t = \text{id. on the Center of } \mathcal{N} \} \quad ;$$

$$T(\mathcal{M}) = \text{The Point Spectrum of } \{\theta_t\} \text{ on the Center of } \mathcal{N} \; .$$

4. Comparison and Flow of Weights.

A weight on a von Neumann algebra $\mathcal{M}$ is a map $\varphi : \mathcal{M}_+ \mapsto [0,\infty]$ such that

$$\varphi(x+y) = \varphi(x) + \varphi(y), \quad x, y \in \mathcal{M}_+ \; ;$$

$$\varphi(\lambda x) = \lambda \varphi(x), \qquad \lambda \geq 0,$$

where $0 \cdot (+\infty) = 0$. It is called semi-finite if the left ideal $\mathfrak{n}_\varphi = \{x \in \mathcal{M} : \varphi(x^*x) < +\infty\}$ is σ-weakly dense in $\mathcal{M}$; normal if $\varphi(\sup x_i) = \sup \varphi(x_i)$ for every bounded increasing net $\{x_i\}$ in $\mathcal{M}_+$; faithful if $\varphi(x^*x) > 0$ for every $x \neq 0$. If φ is a semi-finite normal weight, then the left kernel $N_\varphi = \{x \in \mathcal{M} : \varphi(x^*x) = 0\}$ is of the form $\mathcal{M}e$ for some projection e . The support, $s(\varphi)$, of φ is by definition $1 - e$. Then φ is faithful on $\mathcal{M}_{s(\varphi)}$. Let $\mathcal{W}(\mathcal{M})$ denote the set of all semi-finite normal weights on $\mathcal{M}$. If φ is a weight, then φ is extended to a unique linear functional $\dot{\varphi}$ on $\mathfrak{m}_\varphi = \mathfrak{n}_\varphi^* \mathfrak{n}_\varphi$ the linear span of $\{x \in \mathcal{M}_+ : \varphi(x) < +\infty\}$. We shall not distinguish φ and the extended linear functional $\dot{\varphi}$ on $\mathfrak{m}_\varphi$. To each $\varphi \in \mathcal{W}(\mathcal{M})$, there corresponds the modular automorphism group $\{\sigma_t^\varphi\}$ of $\mathcal{M}_{s(\varphi)}$. The centralizer of φ means the fixed point algebra of $\{\sigma_t^\varphi\}$ in $\mathcal{M}_{s(\varphi)}$.

Given φ and ψ in $\mathcal{W}(\mathcal{M})$, we write

$$\varphi \precsim \psi$$

if there exists a partial isometry $v \in \mathcal{M}$ such that $vv^* \in \mathcal{M}_\psi$ and

$$\varphi(x) = \psi(vxv^*) = \psi_v(x), \qquad x \in \mathcal{M} .$$

If $vv^* = s(\psi)$, then we write $\varphi \sim \psi$ instead.

We now assume that $\mathcal{M}$ is a σ-finite von Neumann algebra. Then there exist uniquely an abelian von Neumann algebra $\mathcal{P}_{\mathcal{M}}$ and a mapping $p: \varphi \in \mathcal{W}(\mathcal{M}) \mapsto p(\varphi) \in \mathcal{P}_{\mathcal{M}}$ from $\mathcal{W}(\mathcal{M})$ onto the lattice of σ-finite projections in $\mathcal{P}_{\mathcal{M}}$ such that

$$\varphi \otimes \mathrm{Tr} \precsim \psi \otimes \mathrm{Tr} \iff p(\varphi) \leq p(\psi)$$

where Tr means the usual trace on $\mathcal{L}(\mathfrak{h}_0)$ with $\mathfrak{h}_0$ a fixed separable infinite dimensional Hilbert space. To each $\varphi \in \mathcal{M}_*^+$, there corresponds a unique $\mu_\varphi \in (\mathcal{P}_{\mathcal{M}})_*^+$ such that

$$\begin{cases} \mu_{\varphi_1+\varphi_2} = \mu_{\varphi_1} + \mu_{\varphi_2} & \text{if} \quad s(\varphi_1) \perp s(\varphi_2) ; \\ \mu_\varphi \leq \mu_\psi & \iff \varphi \precsim \psi . \end{cases}$$

The correspondance : $\varphi \longmapsto \mu_\varphi$ may be viewed as a dimension function. Namely, if $\mathcal{M}$ is a semi-finite von Neumann algebra with center $\mathcal{C}$ equipped with a faithful semi-finite normal trace τ, then we associate to each projection $e \in \mathcal{M}$ a positive linear functional : $x \in \mathcal{C} \longmapsto \mu_e(x) = \tau(ex)$. Then $\mu_e \leq \mu_f \iff e \precsim f$.

We now further assume that $\mathcal{M}$ is <u>a separable factor of type III</u>. Since $\{\mathcal{P}_{\mathcal{M}}, p\}$ is canonical to $\mathcal{M}$, and $\varphi \precsim \psi \iff \lambda\varphi \precsim \lambda\psi$ for any $\lambda \geq 0$, there exists a one-parameter automorphism group $\{\mathcal{F}_t\}$ of $\mathcal{P}_{\mathcal{M}}$ such that

$$\mathcal{F}_t\, p(\varphi) = p(e^{-t}\varphi), \quad \varphi \in \mathcal{W}(\mathcal{M}), \ t \in \mathbb{R}.$$

We call $\{\mathcal{P}_{\mathcal{M}}, \mathcal{F}_t\}$ <u>the global flow of weights</u> on $\mathcal{M}$. We remark here, however that $t \longmapsto \mathcal{F}_t$ is highly discontinuous. From the fact that $\mathcal{M}$ is a factor, it follows that there exists a unique σ-finite projection $e_0 \in \mathcal{P}_{\mathcal{M}}$ such that $\mathcal{F}_t e_0 = e_0$, $t \in \mathbb{R}$. Then it corresponds to a faithful semi-finite normal weight ϖ on $\mathcal{M}$ with properly infinite centralizer. The unitary equivalence class of ϖ is unique, and it is characterized by the fact that $\varpi \sim e^{-t}\varpi$ for every $t \in \mathbb{R}$. It turns out that $\varphi \otimes \omega$ on $\overline{\mathcal{M}}$ appeared in §3 is equivalent to ϖ on $\overline{\mathcal{M}}$. The weight ϖ is called <u>dominant</u>.

We then have a nice situation that $t \in \mathbb{R} \mapsto \mathcal{F}_t\, p(\varphi) \in \mathcal{P}_{\mathcal{M}}$ is σ-weakly continuous if and only if $\varphi \precsim \varpi$, and that $\{(\mathcal{P}_{\mathcal{M}})_{e_0}, \mathcal{F}_t\}$ is ergodic continuous flow. We call this ergodic flow the <u>smooth flow of weights on</u> $\mathcal{M}$ and denote it by $\{P_{\mathcal{M}}, F_t^{\mathcal{M}}\}$. The smooth part of the flow of weights is characterized by the following :

$$\varphi \precsim \varpi \iff \{x \in \mathcal{M} : \int_{-\infty}^{\infty} \sigma_t^{\varphi}(x^*x)dt < +\infty\}$$

is σ-weakly dense in $\mathcal{M}$. For this reason, we call φ integrable if $\varphi \precsim \varpi$. Furthermore, it turns out that

$$\{P_{\mathcal{M}}, F^{\mathcal{M}}\} \cong \{\mathcal{C}, \theta\}$$

where $\mathcal{C}$ is the center of $\mathcal{N}$ which appeared in §3.

5. Construction of the Flow of Weights

We now compute the flow of weights for $\mathcal{M} = \mathcal{M}(G, \Gamma, \mu)$. Let

$$\rho(g,\gamma) = \frac{d\mu\cdot g}{d\mu}(\gamma), \quad g \in G, \quad \gamma \in \Gamma.$$

$$\rho(gh,\gamma) = \rho(g, h\gamma)\,\rho(h,\gamma), \quad g, h \in G, \quad \gamma \in \Gamma.$$

Consider the cartesian product space $\Gamma \times \mathbb{R}_+$ equipped with $\mu \otimes m$, where m is the Lebesgue measure on $\mathbb{R}_+$, and define transformations as follows :

$$\begin{cases} T_g(\gamma, s) = (g\gamma, \rho(g,\gamma)^{-1}s), & g \in G; \\ S_t(\gamma, s) = (\gamma, e^{-t}s), & t \in \mathbb{R}. \end{cases}$$

Clearly $\{T_g, S_t : g \in G, t \in \mathbb{R}\}$ acts on $\Gamma \times \mathbb{R}^+$, and $\{T_g \cdot g \in G\}$ and $\{S_t\}$ commute. Hence the restriction of S_t^* on $\mathcal{C} = L^\infty(\Gamma \times \mathbb{R}_+, \mu \otimes m)^G$, the fixed point algebra under the action $\{T_g^*\}$, is an ergodic flow, where S_t^* (resp. T_g^*) means the automorphism of $L^\infty(\Gamma \times \mathbb{R}_+, \mu \otimes m)$ given by

$$\begin{cases} (S_t^* f)(\tilde{\gamma}) = f(S_{-t}\tilde{\gamma}), & \tilde{\gamma} \in \Gamma \times \mathbb{R}_+, \quad t \in \mathbb{R}, \\ (T_g^* f)(\tilde{\gamma}) = f(T_g^{-1}\tilde{\gamma}), & g \in G, \quad f \in L^\infty(\Gamma \times \mathbb{R}_+, \mu \otimes m). \end{cases}$$

It is then proved that

$$\{\mathcal{C}, S_t^*\} \cong \{P_{\mathcal{M}}, F_t^{\mathcal{M}}\}.$$

Therefore, one can construct the flow of weights explicitly for the factors given by the group measure space construction.

In the case that $G = \mathbb{Z}$ or more generally G is abelian, W. Krieger's results state that the flow of weights $\{P_{\mathcal{M}}, F^{\mathcal{M}}\}$ is a complete invariant for the algebraic type of $\mathcal{M}(G,\Gamma,\mu)$ -Thus a nice generalization of a result of H.A. Dye from the finite measure preserving case.

6. Automorphism Group of $\mathcal{M}$.

Let $\mathcal{M}$ be a separable factor of type III. We fix a continuous decomposition of $\mathcal{M}$:

$$\mathcal{M} = W^*(\mathcal{N}, \mathbb{R}, \theta), \quad \mathcal{N} = \mathcal{M}_\varpi$$

which appeared in §3. Then $\{\mathcal{C}, \theta\} \cong \{P_{\mathcal{M}}, F^{\mathcal{M}}\}$ where $\mathcal{C}$ is the center of $\mathcal{N}$. Identifying $P_{\mathcal{M}}$ and $\mathcal{C}$, let $Z^1(F^{\mathcal{M}})$ denote the set of all unitary $F^{\mathcal{M}}$ - one cocycle, i.e.

$$Z^1(F^{\mathcal{M}}) = \{c : \quad t \in \mathbb{R} \longmapsto c_t \in \mathcal{C} \text{ unitary and }$$
$$c_{s+t} = c_s F_s^{\mathcal{M}}(c_t)\}.$$

Let $B^1(F^{\mathcal{M}}) = \{c \in Z^1(F^{\mathcal{M}}) : c_t = b^* F_t^{\mathcal{M}}(b)$ for some $b \in \mathcal{C}$. Given $c \in Z^1(F^{\mathcal{M}})$, we define an automorphism σ_c^ϖ by

$$\begin{cases} \sigma_c^\varpi(x) = x, & x \in \mathcal{N}; \\ \sigma_c^\varpi(u(s)) = c_s u(s), & s \in \mathbb{R}, \end{cases}$$

where $\{u(s)\}$ is the one-parameter unitary group appearing in the decomposition $\mathcal{M} = W^*(\mathcal{N}, \mathbb{R}, \theta)$. It is an easy exercise to show that σ_c^ϖ is inner if and only if c is coboundary i.e. $c \in B^1(F^{\mathcal{M}})$. By construction σ_c^ϖ leaves $\mathcal{N}$ pointwise fixed. Conversely one can show that

$$\mathrm{Aut}(\mathcal{M}/\mathcal{N}) = \{\alpha \in \mathrm{Aut}(\mathcal{M}) ; \alpha(x) = x, \ x \in \mathcal{N}\}$$
$$= \{\sigma_c^\varpi : c \in Z^1(F^{\mathcal{M}})\}.$$

The basis of the proof is the relative commutant theorem : $\mathcal{N}' \cap \mathcal{M} = \mathcal{C} \subset \mathcal{N}$. Deviding out by $B^1(F^{\mathcal{M}})$ and the inner automorphism group, we get the following exact sequence :

$$\{1\} \longrightarrow H^1(F^{\mathcal{M}}) \longrightarrow \mathrm{Out}(\mathcal{M}) \longrightarrow \mathrm{Out}_{\tau,\theta}(\mathcal{N}) \longrightarrow \{1\}$$

where $\mathrm{Out}_{\tau,\theta}(\mathcal{N})$ is the canonical image of

$$\mathrm{Aut}_{\tau,\theta}(\mathfrak{N}) = \{\alpha \in \mathrm{Aut}(\mathfrak{N}) : \ \tau\cdot\alpha = \tau, \ \ \alpha\theta_s = \theta_s\alpha, \ s \in \mathbb{R}\}$$

in $\mathrm{Out}(\mathfrak{N})$.

7. Generalized Functional Calculus of the Modular Automorphism Groups.

We keep the notations from the previous section. If $c \in Z^1(F^{\mathfrak{M}})$ is twice continuously differentiable as a function of $t \in \mathbb{R}$, then one can define σ_c^φ for any $\varphi \in \mathfrak{W}(\mathfrak{M})$ and $u_c = (D\varphi : D\psi)_c \in \mathfrak{M}$ for faithful $\varphi, \psi \in \mathfrak{W}(\mathfrak{M})$, where $(D\varphi : D\psi)_t$, $t \in \mathbb{R}$, and σ_t^φ are considered as the one corresponding to $c \in Z^1(F^{\mathfrak{M}})$ with $c_t = e^{it} \in \mathbb{C}$. We then have $u_{cd} = u_c\sigma_c^\psi(u_d)$. Although we leave the detail to the original paper [3] , we shall present some evidence why $\{\sigma_c^\varphi\}$ may be viewed as a <u>functional calculus</u> of $\{\sigma_t^\varphi\}$. Let $\mathfrak{M}$ be a semi-finite factor. Then $\{P_{\mathfrak{M}}, F_t^{\mathfrak{M}}\} = \{L^\infty(\mathbb{R}_+)$, product-translation by $e^{-t}\}$. Hence for every $c \in Z^1(F^{\mathfrak{M}})$ there exists $f \in L^\infty(\mathbb{R}_+)$ with $c_s = f^*F_s^{\mathfrak{M}}(f)$. On the other hand, every $\varphi \in \mathfrak{W}(\mathfrak{M})$ is of the form $\varphi(x) = \tau(hx)$, $x \in \mathfrak{M}$, where τ is a fixed faithful semi-finite normal trace on $\mathfrak{M}$, and $\sigma_t^\varphi(x) = h^{it}xh^{-it}$. We then have

$$\sigma_c^\varphi(x) = f(h)x f(h)^*, \qquad x \in \mathfrak{M}.$$

REFERENCES

[1] A. CONNES
Une classification des facteurs de type III,
Ann. Sci. Ecole Norm. Sup., 4ème Sér., 6 (1973), 133-252.

[2] A. CONNES and M. TAKESAKI
Flots des poids sur les facteurs de type III,
C.R. Acad. Sci., Paris, Séri. A, 278 (1974), 945-948.

[3] A. CONNES and M. TAKESAKI
The Flow of Weights on Factors of type III,
To appear in Tôhoku Math. J.

[4] W. KRIEGER
On ergodic flows and the isomorphism of factors.

[5] M. TAKESAKI
Duality for crossed products and the structure of von Neumann algebras of type III,
Acta Math., 131 (1974), 249-310.

THE von NEUMANN ALGEBRA OF A FOLIATION

Alain CONNES

__Abstract__ :

Every smooth foliation $\mathcal{F}$ of a manifold V gives rise very naturally to a von Neumann algebra $M = L^\infty(V/\mathcal{F})$. The weights on M correspond exactly to operator valued forms on the "manifold" of leaves of $\mathcal{F}$. We compute their modular automorphism group, this yields the continuous decomposition of M in terms of another foliation $\mathcal{F}_o$ of V and a one parameter group of automorphisms of $\mathcal{F}_o$. We then illustrate this decomposition with a few examples.

Let V be a (smooth, finite dimensional real) manifold and $\mathcal{F}$ a smooth foliation of V. We assume for simplicity that the union of the leaves of $\mathcal{F}$ which have non-trivial holonomy is negligible (for the smooth measures on V). This is always the case if $\mathcal{F}$ is analytic.

__Definition 1__ - __A random operator__ $T = (T_f)$ __a measurable family__ $(T_f)_{f\in\mathcal{L}}$ __where for each leaf__ $f \in \mathcal{L}$ __of__ $\mathcal{F}$, T_f __is an operator in__ $L^2(f)$. (*)

As an example, let $Y \in C_o^\infty(V,T(\mathcal{F}))$ be a section (smooth with compact support) of the tangent bundle of $\mathcal{F}$. Then the flow $F_t = \exp tY$ induces on each leaf f a one parameter group of unitaries U_f^t of $L^2(f)$. For each t, $U^t = (U_f^t)_{f\in\mathcal{L}}$ is a random operator.

The random operators are added and composed as follows :

$$(T_1+T_2)_f = T_{1f}+T_{2f} \quad \forall f \in \mathcal{L} \;, \quad (T_1T_2)_f = T_{1f}T_{2f} \quad \forall f \in \mathcal{L} \;.$$

The natural norm is $\|T\|_\infty = \text{Ess Sup} \|T_f\|$, defined as the smallest $\lambda \geqq 0$ such that $\|T_f\| \leqq \lambda$ holds almost everywhere (i.e. the union of leaves where this fails is negligible).

(*) Given a manifold X, we let $L^2(X)$ be the hilbert space completion of the vector space $C_o^\infty(X,|\Lambda|^{1/2})$ of smooth half densities with compact support, with the canonical scalar product : $<\omega_1,\omega_2> = \int \omega_1\bar{\omega}_2$.

We did not state the measurability condition, up to equality almost everywhere it means simply that the family $(T_f)_{f \in \mathcal{L}}$ can be exhibited. For instance it is in general impossible to construct a random operator $T = (T_f)_{f \in \mathcal{L}}$ such that T_f is of rank one for almost all f, because this would give a measurable subset S of V which meets almost all leaves in one and only one point, which is impossible if for instance $\mathcal{F}$ is ergodic (*) and $\dim \mathcal{F} < \dim V$. As in ordinary measure theory we shall say "random operator" instead of "class of random operators modulo equality almost everywhere".

Proposition 2 - With the above algebraic operations and norm the random operators form a von Neumann algebra $M = L^\infty(V/\mathcal{F})$.

The notation $L^\infty(V/\mathcal{F})$ indicates that M depends only on the couple $(V/\mathcal{F})$ it also suggest that M generalises the L^∞ space of a manifold. If $\mathcal{F}$ has dimension 0 then it coincides with $L^\infty(V)$ and is commutative. In general it will be type I only if the above measurable cross section exists, i.e. very scarcely.

In non commutative integration theory, the basic object is a von Neumann algebra M (it reduces to L^∞(X, class of the measure) in the commutative situation) and the role of the measure (possibly infinite) is played by the weights φ on M.

If $M = L^\infty$(X, class of measure) then weights on M correspond exactly to positive measures μ in the class. If in particular X is a smooth manifold and the class is the smooth one, then μ is entirely described by a 1-density, positive and measurable (or equivalently an odd form, positive with measurable coefficients).

We now describe how this correspondance extends to our situation. Let us consider the space $\mathcal{L}$ of leaves as a "manifold" (see [1]) of dimension $q = \text{Codim } \mathcal{F}$. The tangent bundle to $\mathcal{L}$ is obtained as follows : for each $f \in \mathcal{L}$ the holonomy gives a natural trivialisation of the transverse bundle $x \longrightarrow T_x(V)/T_x(\mathcal{F})$ along f. Thus a tangent vector $\xi \in \tau_f(\mathcal{L})$ should be a constant section $x \longrightarrow \xi_x$ of the above bundle. We can then speak of $\Lambda^q \tau$ where $q = \text{Codim } \mathcal{F} = \dim \tau_f$ $\forall f \in \mathcal{L}$.

Definition 3 - An operator density $T = (T_v)_{v \in \Lambda^q \tau}$ is a measurable map $v \longrightarrow T_v$ which to each $f \in \mathcal{L}$ and $v \in \Lambda^q \tau_f$ associates a positive self adjoint (**) operator T_v in the hilbert space $L^2(f)$, with $T_{\lambda v} = |\lambda| T_v$, $\forall \lambda \in \mathbb{R}$.

(*) i.e. any measurable union of leaves A which is not negligible has a negligible complement A^c in V.

(**) In general unbounded.

Thus, for each $f \in \mathcal{L}$, one has a ray of positive operators, $T_v = \lambda T_{v_o}$, $v \in \Lambda^q \tau_f$, $\lambda > 0$. One cannot speak of T_f but only of T_f up to multiplication by a positive scalar. The simplest example of an operator density comes from a choice of a positive measurable transversal density ρ on V : for each $x \in V$, ρ_x maps $\Lambda^q(T_x(V)/T_x(\mathfrak{F}))$ to $\mathbb{R}_+$ and $\rho_x(\lambda v) = |\lambda| \rho_x(v)$, $\forall \lambda$. Given such a ρ let for each $f \in \mathcal{L}$ and $v \in \Lambda^q \tau_f$, the operator T_v be the multiplication in $L^2(f)$ by the function $x \longrightarrow \rho_x(v)$ on f . One checks that it defines an operator density.

We shall now describe the correspondence between operator densities and weights on $M = L^\infty(V/\mathfrak{F})$. In the ordinary case $(\dim \mathfrak{F} = 0)$ an operator density is just a density on V and one can integrate it over V to get a scalar. Our first aim is to show the existence of a canonical trace on operator densities, i.e. to give a meaning to $\int \mathrm{Trace}(T_v)$ in general. Let T be an operator density, in general T_v is <u>not</u> trace class in $L^2(f)$, however it will happen (*) that T_v is locally of trace class i.e. that $\chi T_v \chi$ is of trace class for every characteristic function of compact support χ on f .

<u>Lemma 4 - Let $\mathcal{U} = (U_\alpha)_{\alpha \in I}$ be a locally finite partition of V where each U_α is contained in the domain of a foliation chart and is a measurable union $U_\alpha = \bigcup_{x \in S_\alpha} P_x$ where S is a transversal to $\mathfrak{F}$ and P_x is relatively compact in the leaf through x , then : $\sum_{\alpha \in I} \int_{S_\alpha} \mathrm{Trace}(P_x^\alpha T_v P_x^\alpha)$ is independent of the choice of $\mathcal{U}$</u> .

We noted P_x^α the orthogonal projection in $L^2(f_x)$ associated to the subset P_x of the leaf f_x through f . Note that the integral makes sense because $\mathrm{Trace}\,(P_x^\alpha T_v P_x^\alpha)$ is a one density on S_α .

This lemma defines the quantity $\int \mathrm{Trace}\ T$ for every operator density T . It is unitarily invariant under $M = L^\infty(V/\mathfrak{F})$ that is $\int \mathrm{Trace}(UTU^*) = \int \mathrm{Trace}(T)$ for every random unitary operator $U \in L^\infty(V/\mathfrak{F})$.

<u>Theorem 5 - For every operator density T the equality $\varphi(A) = \int \mathrm{Trace}(TA)$ (**) defines a (semi finite, normal) weight φ on $L^\infty(V/\mathfrak{F})$; every weight on $L^\infty(V/\mathfrak{F})$ occurs exactly once in this way and the modular automorphism group σ^φ of φ is given by</u> :

$$(\sigma_t^\varphi(A))_f = T(v)^{it} A_f\, T(v)^{-it} \qquad , \qquad v \in \Lambda^q \tau_f\ ,\ v \neq 0 \qquad .$$

(*) This will be automatic for all $f \in \mathcal{L}$ if $\mathrm{Trace}\ T_v < \infty$.

(**) More precisely $\mathrm{Trace}(T^{1/2} A T^{1/2})$.

Note that the choice of $v \neq 0$, $v \in \Lambda^q \tau_f$ does not affect this equality, since $T(\lambda v) = |\lambda| T(v)$, $\forall \lambda \in \mathbb{R}$.

In particular T will define a trace on $L^\infty(V/\mathfrak{F})$ iff $T(v)$ commutes with all A_f and thus is a scalar, for almost all f , as is seen by the following lemma :

<u>Lemma 6 - Let</u> $\mathcal{A}$ <u>be any norm separable</u> C^* <u>algebra generating the von Neumann algebra</u> $L^\infty(V/\mathfrak{F})$, <u>then for almost all</u> $f \in \mathcal{L}$ <u>the natural representation of</u> $\mathcal{A}$ <u>in</u> $L^2(f)$ <u>is irreducible.</u>

So the (normal) traces on $L^\infty(V/\mathfrak{F})$ correspond exactly to the measurable <u>scalar</u> densities on the "manifold" $\mathcal{L}$. Those scalar densities which are locally integrable, are exactly the absolutely continuous holonomy invariant transverse measures on $\mathfrak{F}$, by [7] . The Reeb foliation of S^3 is an example where there is no (non-zero) absolutely continuous holonomy invariant transverse measure, while, since $L^\infty(V/\mathfrak{F})$ is of type I , there are lots of measurable scalar densities on $\mathcal{L}$.

In general there will be no (non-zero normal) trace on $L^\infty(V/\mathfrak{F})$ and the next step is to determine the spectrum of the modular automorphism groups σ^φ. Theorem 5 and lemma 6, have the following corollary :

<u>Corollary 7 - Let</u> T <u>be an operator density,</u> φ <u>the associated weight (*) , and</u> $E \in \mathbb{R}$, <u>then</u> $\mathrm{Exp}(E) \in \mathrm{Sp}\, \sigma^\varphi$ <u>iff for every</u> $\varepsilon > 0$ <u>there exists a non negligible set of leaves</u> $f \in \mathcal{L}$ <u>such that</u> :

$$\forall v \in \Lambda^q \tau_f \ , \ \exists\, E_1, E_2 \in \text{Spectrum} \log T_v \ , \ E_1 - E_2 \in [E-\varepsilon, E+\varepsilon] \ .$$

The computation of $S(L^\infty(V/\mathfrak{F})) = S(M)$ [3] where

$$S(M) = \bigcap_{\varphi \text{ weight}} \text{Spectrum } \sigma^\varphi \ ,$$

identifies it as the ratio set of W. Krieger (See [6] and [2]). R. Bowen has computed this invariant for Anosov foliations [2] and obtained that generally these will be of type III_1 so that $S(M) = [0,+\infty]$. We shall give below examples of type III_λ , $\lambda \in]0,1[$ i.e. $S(M) = \{0\} \cup Z$ and III_0 : $S(M) = \{0,1\}$ for analytic foliations.

(*) We assume here that φ is faithful i.e. that for $v \neq 0$, $T(v)$ is non singular.

So we see that the spectrum of $\log T_v$ (which is self adjoint and is shifted by an additive constant when v is multiplied by λ) is far from arbitrary. (One can check that the existence of an operator density T with $\log T_v$ bounded below for almost all v is equivalent to the semi finiteness of $L^{\infty}(V/\mathfrak{F})$) . In the type III situation (i.e. no non-zero scalar valued density on $\mathcal{L}$) one is interested in the continuous decomposition of the von Neumann algebra $L^{\infty}(V/\mathfrak{F})$. See [8] and [5] . Recall that a weight φ on M is called integrable when the following left ideal of M is generating :

$$\{x \in M, \exists c > 0, \left\| \int_{-K}^{K} \sigma_t^{\varphi}(x^* x) dt \right\| \leq c, \forall K \in \mathbb{R}_+\} .$$

This condition is equivalent to the smoothness of the map $\lambda \longrightarrow \lambda\varphi$ from $\mathbb{R}^+$ to the flow of weights on M ([5]) .

<u>Theorem 8</u> - <u>Let</u> ρ <u>be a</u> C^{∞} <u>transversal density and</u> φ <u>the corresponding weight on</u> $L^{\infty}(V/\mathfrak{F})$. <u>Then</u> φ <u>is integrable iff the set of critical points of</u> ρ <u>is negligible</u>.

We have to define what we mean by a critical point of ρ . For each $v \in \Lambda^q \tau_f$, $f \in \mathcal{L}$ the function $\log \rho_x(v)$ on f has a certain set C_f of critical points. It does not depend on the choice of $v \neq 0$ since changing v adds a constant to $\log \rho$, thus we can speak of $C = \cup C_f$.

This theorem allows to determine the continuous decomposition of $L^{\infty}(V/\mathfrak{F})$ by means of the codimension $q+1$ "foliation" (with critical points) $\mathfrak{F}_\rho$ whose leaves partition each leaf of $\mathfrak{F}$ in the (not necessarily connected) level manifolds of ρ .

To treat our examples we make the further hypothesis that $C = \emptyset$ and that V is compact. The foliation $\mathfrak{F}_\rho$ does not have singular points in this situation and letting ω be the 1-form on $\mathfrak{F}$ (*) which is the gradient of $\log \rho$, we can find a smooth vector field Y on V , $Y_x \in T_x(\mathfrak{F})$ $\forall x \in V$ such that $< Y, \omega >$ is the constant function 1 . Let U^t be the one parameter group of random operators associated with Y as above, then as $\exp tY$ multiplies ρ by e^t one gets that $U^t \varphi U^{t*} = e^t \varphi$. Hence in this situation φ is a dominant weight, the semi finite von Neumann algebra of the continuous decomposition of $L^{\infty}(V/\mathfrak{F})$ identifies with the centralizer of φ i.e. with $L^{\infty}(V/\mathfrak{F}_\rho)$ and the one parameter group of automorphisms θ_t of this algebra is defined by the action on $L^{\infty}(V/\mathfrak{F}_\rho)$ of the flow $\exp Y$ of automorphisms of the foliation $\mathfrak{F}_\rho$. In particular the flow of weights of $L^{\infty}(V/\mathfrak{F})$ is the action of $\exp Y$ on the ergodic decomposition of the foliation $\mathfrak{F}_\rho$. We now describe examples showing how to construct a foliation of a compact manifold with given flow of weights. As a tool we use the simplest example of an Anasov foliation namely we let V_o be the quotient $SL(2,\mathbb{R})/\Gamma$ of $SL(2,\mathbb{R})$ by the discrete cocompact subgroup Γ and $\mathfrak{F}_o$ be the foliation on V_o coming from the action (on the left) of the subgroup of lower triangular matrices. This foliation is of type III_1 (cf. [2]) and from the above discussion it is easy to check that its continuous decomposition (with respect to a left invariant transverse density ρ_o) yields as $\mathcal{F}_{\rho_o}$ the foliation associated to the horocycle flow (i.e. the matrices $\begin{bmatrix} o & o \\ b & o \end{bmatrix} = X_b$) and that $\exp Y_o$ is the geodesic flow (i.e. the matrices $\begin{bmatrix} e^t & o \\ o & e^{-t} \end{bmatrix} = Y_t$). Now let K be an auxiliary compact manifold and $F = (F_t)_{t \in \mathbb{R}}$ a smooth flow on K. We construct now a foliation $\mathcal{F}$ on $K \times V_o$ as follows. It comes from an action of the group of matrices $\begin{bmatrix} e^t & o \\ b & e^{-t} \end{bmatrix}$ where $t,b \in \mathbb{R}$, where $\begin{bmatrix} 1 & o \\ b & 1 \end{bmatrix}$ act by $\text{identity}_K \times \text{horocycle}_b$, while the matrix $\begin{bmatrix} e^t & o \\ o & e^{-t} \end{bmatrix}$ acts by $F_t \times \text{Geodesic}_t$. It is clear that this gives an action of the above group. We assume for simplicity that F has a smooth invariant measure (it is easy to modify the above construction so that it works in general), let α be the corresponding 1-density on K. Then $\alpha \times \rho_o$ defines a transverse density ρ on $\mathcal{F}$; and the corresponding foliation $\mathcal{F}_\rho$ is just given by the flow : identity $\times$ horocycle, so that its ergodic decomposition gives us K back. The flow $\exp Y$ is simply $F_t \times \exp tY_o$ and thus its action on the ergodic decomposition of $\mathcal{F}_\rho$. i.e. the flow of weights of $L^\infty(V/\mathcal{F})$ is the flow F_t on K.

If in particular we take K to be a circle of length L while F_t acts by rotations with speed 1 we get a foliation $\mathcal{F}$ of the compact manifold $V = S^1 \times V_o$, which is of type III_λ, $\lambda = \exp(-L)$. As soon as F acting on K is ergodic, with $\dim K > 1$, we get a factor of type III_o as $L^\infty(V/\mathcal{F})$.

This shows that all types of factors occur from simple examples. The problem "when is $L^\infty(V/\mathcal{F})$ approximately finite dimensional" is very interesting and examples will be discussed in [4], see also [2]. For instance an analytic (one dimensional) complex foliation on a 2 dimensional complex compact manifold can fail to be a.f.d., while all real flows are a.f.d.

Nuclearity and the C*-algebraic Flip

Edward G. Effros
(Supported in part by NSF)

1. Introduction

The commutative C*-algebras and the $n \times n$ matrices M_n not only serve as the simplest examples of C*-algebras, they are in many ways the prototypes for all of the others. Thus direct integral theory and spectral theory may be regarded as commutative techniques for studying the global structure of von Neumann algebras, and the local (more precisely, the singly generated) structure of C*-algebras, respectively. On the other hand, dimension theory is largely concerned with the global problem of displaying a von Neumann algebra as the $n \times n$ matrices over another algebra. M_n also appears locally when one considers approximations by finite dimensional subalgebras in a hyperfinite von Neumann algebra or an AF C*-algebra. The latter theory is of restricted interest, since the AF algebras form a rather small class of algebras.

Recently, M. Choi and the author [5]-[9] have discovered a new method of matrix approximation that is better suited to C*-algebra theory. In this paper I shall explain this "nuclear" approach, and review one of its important consequences. In an attempt to demonstrate that nuclearity often occurs quite naturally, I shall then use it to prove a new result concerning Sakai's "flip" map for C*-algebras (Theorem 3.1).

Bibliography.

[1] R. BARRE - De quelques aspects de la théorie des Q-variétés différentielles et analytiques. Annales Inst. Fourier, tome 23 (1973).

[2] R. BOWEN - Anosov foliations are hyperfinite (preprint).

[3] A. CONNES - Une classification des facteurs de type III, Annales Scientifiques E.N.S. , 4ème série tome 6, fasc. 2, (1973), p.133-252.

[4] A. CONNES and D. SULLIVAN - (To appear).

[5] A. CONNES and M. TAKESAKI - The flow of weights on type III factors. Tohoku Math. Journal. 29 (1977) p. 473-575

[6] W. KRIEGER - Ergodic flows and the isomorphism of factors, Math. Ann. 223 (1976), p. 19-70.

[7] D. RUELLE and D. SULLIVAN - Currents, flows and diffeomorphisms, Topology Vol. 14, p. 319-327.

[8] M. TAKESAKI - Duality in cross products and the structure of von Neumann algebras of typr III, Acta Math. 131 (1973), p. 249-310.

2. Nuclearity and liftings

If V is any vector space, we let $M_n(V)$ be the $n \times n$ matrices with entries in V. If A and B are C*-algebras, a linear map $\varphi\colon A \to B$ is said to be <u>completely positive</u> provided for each n the map

$$\varphi_n\colon\ M_n(A) \to M_n(B)\colon\ [a_{ij}] \to [\varphi(a_{ij})]$$

is positive. There is a fairly extensive literature concerning this notion (see, e.g., [12]). Of particular importance is the following characterization of the completely positive maps $\varphi\colon M_n \to A$ and $\psi\colon A \to M_n$. Let ε_{ij} be the usual set of matrix units in M_n. We identify $M_n(A^*)$ with $M_n(A)^*$ by using the pairing

$$[a_{ij}] \cdot [f_{ij}] = \sum_{i,j} f_{ij}(a_{ij}) \quad .$$

<u>Theorem 2.1 [4, Th. 2], [5, Lemma 4.3]</u>: A linear map $\varphi\colon M_n \to A$ completely positive if and only if $[\varphi(\varepsilon_{ij})]$ is a positive element of $M_n(A)$. A linear map $\psi\colon A \to M_n$ is completely positive if and only if $\psi(a) = [f_{ij}(a)]$ where $[f_{ij}]$ is a positive function in $M_n(A^*)$.

A C*-algebra A is said to be <u>nuclear</u> if the diagrams of unital completely positive linear contractions

(2.1) 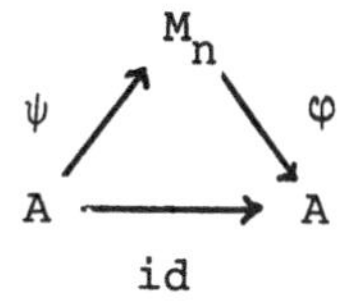 (n arbitrary)

approximately commute, i.e., given $a_1, \ldots, a_s \in A$ and $\varepsilon > 0$ we may find such a diagram with

$$\|\varphi \circ \psi(a_k) - a_k\| < \varepsilon \qquad k = 1,\dots,s$$

If A has a unit, it is not necessary to assume that φ and ψ are contractive (see the argument for [6,Th. 3.1] - given general φ and ψ, we have $\varphi \circ \psi = \varphi' \circ \psi'$ where $\psi'\colon A \to M_n$ is unital, and we may then replace φ' by $\varphi'/\|\varphi'\|$) . We have

Theorem 2.2 [6][7][8][11] (see also [18]): Suppose that A is a C*-algebra. Then the following are equivalent:

(1) A is nuclear

(2) For each representation π of A, $\overline{\pi(A)}$ is an injective von Neumann algebra

(2') A^{**} is injective

(3) If B is any other C*-algebra, the minimal and maximal tensor products of A and B coincide.

If A is separable, it is equivalent to assume

(4) For each factor representation π of A, $\overline{\pi(A)}$ is a hyperfinite von Neumann algebra.

Corollary 2.3 [11]: If G is a connected or a solvable second countable locally compact group, then its group C*-algebra $C^*(G)$ is nuclear.

The importance of the diagrams (2.1) is that they often enable one to generalize matrix approximation arguments for AF algebras to the much more general class of nuclear C*-algebras. Perhaps the most outstanding instance of this was found by M. Choi and the author [9] (for a more recent proof see [2]):

Theorem 2.4 (Completely Positive Lifting Theorem): Suppose that A is a separable nuclear C*-algebra and that J is a closed two-sided ideal in a C*-algebra B . Then any completely positive contraction $\varphi: A \to B/J$ has a completely positive contractive lifting $\psi: A \to B$.

This result has played a central role in the generalization of the Brown-Douglas-Fillmore theory to the separable nuclear C*-algebras (see [2],[13] - the earlier theory was primarily concerned with commutative algebras). The idea behind the proof is that any completely positive map $\varphi: M_n \to B/J$ has a lifting $\psi: M_n \to B$, since from Theorem 2.1, all one need do is lift the corresponding element in $M_n(B/J)^+$ to an element of $M_n(B)^+$. From this M.Choi and the author found it fairly easy to extend the result to AF C*-algebras A . However, by using the approximate diagrams (2.1) and careful matching arguments, we then obtained the full theorem. The result is false if one deletes the condition that A be nuclear [10],[19],[1] .

3. The flip

In a study of automorphisms, Sakai [16] proved that if R is a von Neumann algebra, and $R \mathbin{\bar{\otimes}} R$ is the usual spatial von Neumann tensor product, then the <u>flip</u>

$$\sigma: R \mathbin{\bar{\otimes}} R \to R \mathbin{\bar{\otimes}} R: \ r \otimes s \to s \otimes r$$

is an inner automorphism if and only if R is a type I factor. Connes then proved [11, Th. 5.1] the remarkable result that if R is a II_1 factor on a separable Hilbert space, then σ is weakly approximately inner, i.e., a point-weak limit of inner automorphisms if and only if R is hyperfinite. This result was literally "pivotal" to Connes' theory. It enabled him to "flip" elements of an injective II_1 factor into an algebra with matrix approximations.

Turning to C*-algebras Sakai went on to show that if A is a C*-algebra and $A \otimes_{\min} A$ is the spatial C*-algebraic tensor product, then the flip

$$\sigma: A \otimes_{\min} A \to A \otimes_{\min} A$$

is inner if and only if A is isomorphic to M_n for some n. Subsequently, Bunce [3] proved an analogous result for projective tensor products of Banach algebras. In the following we give an approximate, C*-algebraic version of his argument.

<u>Theorem 3.1</u>: Suppose that A is a unital C*-algebra for which the flip map

$$A \otimes_{\min} A \to A \otimes_{\min} A: \ a \otimes b \to b \otimes a$$

is approximately inner, i.e., a point-norm limit of inner automorphisms.

Then A is a nuclear C*-algebra.

Proof. Given elements $a_1,\dots,a_s \in A$ with $\|a_k\| \leq 1$ and $\varepsilon > 0$, we may by hypothesis assume that there exists a unitary $u \in A \otimes_{\min} A$ such that

$$\|u^*(a_k \otimes 1)u - 1 \otimes a_k\| < \varepsilon/3 \quad k = 1,\dots,s .$$

We may select $b_j, c_j \in A$, $j = 1,\dots,n$ with

$$\|u - \sum_j b_j \otimes c_j\| < \varepsilon/3$$

We then have

$$(3.1) \quad \|\sum_{i,j} b_i^* a_k b_j \otimes c_i^* c_j - 1 \otimes a_k\| < \varepsilon \quad k = 1,\dots,s$$

Given a state p on A, the map

$$p \otimes I : A \otimes A \to A: \ a \otimes b \to p(a)b$$

extends to a contraction $A \otimes_{\min} A \to A$ (this is just Tomiyama's slice map [17]) since for any $f \in A^*$ with $\|f\| \leq 1$, and $v \in A \otimes A$,

$$|f((p \otimes I)(v))| = |(p \otimes f)(v)| \leq \|v\|_{\min}$$

(this uses the polar decomposition of f - see [14,Prop. 8]). Applying $p \otimes I$ to the difference in (3.1),

$$(3.2) \quad \|\sum_{i,j} p(b_i^* a_k b_j) c_i^* c_j - a_k\| < \varepsilon \quad k = 1,\dots,s \ .$$

Defining

$$\psi: \ A \to M_n: \ a \to [\, p(b_i^* a b_j)\,]$$

and

$$\varphi : \ M_n \to A: \ \alpha \to \Sigma \alpha_{ij} c_i^* c_j \ ,$$

we may rewrite (3.2)

$$\|(\varphi \circ \psi)(a_k) - a_k\| < \varepsilon \qquad k = 1,\dots,s \ .$$

From Theorem 2.1, φ is completely positive, since

$$[\varphi(\varepsilon_{ij})] = [c_i^* c_j] = \begin{pmatrix} c_1,\dots,c_n \\ 0 \end{pmatrix}^* \begin{pmatrix} c_1,\dots,c_n \\ 0 \end{pmatrix} \geq 0$$

On the other hand any element of $M_n(A)^+$ is a sum of n matrices of the form $[c_i^* c_j]$ (see [15,Prop. 2.1]), and we have

$$\begin{aligned} [c_i^* c_j] \cdot [\, p(b_i^* \cdot b_j)\,] &= \Sigma\, p(b_i^* c_i^* c_j b_j) \\ &= p((\Sigma c_i b_i)^*(\Sigma c_j b_j)) \\ &\geq 0 \ , \end{aligned}$$

hence from Theorem 2.1, ψ is completely positive.

For $A = M_n$ the flip is an inner automorphism since a simple calculation shows that for any $c \in M_n \otimes M_n$

$$\sigma(c) = u^* c u \ ,$$

where

$$u = \sum_{i,j} \varepsilon_{ij} \otimes \varepsilon_{ji} \in A \otimes A$$

(this is the unitary that transforms $e_i \otimes e_j$ onto $e_j \otimes e_i$). On the other hand if $A = \overline{\cup A_n}$ (norm-closure) where the algebras A_n have approximately inner automorphisms, then it is obvious the same is true for A (unitaries in $A_n \otimes_{\min} A_n$ define inner automorphisms of $A \otimes_{\min} A$). It follows that any UHF algebra has an approximately inner flip (we are indebted to A. Connes for this observation).

In a subsequent paper [20] it will be shown that the flipping argument may be again used to find matrix approximations in certain C*-algebras.

References

1. J. Anderson, A C*-algebra A for which Ext A is not a group, to appear.

2. W. B. Arveson, Notes on extensions of C*-algebras, to appear.

3. J. Bunce, Automorphisms and tensor products of algebras, to appear.

4. M. D. Choi, Completely positive linear maps on complex matrices, Linear Algebra and Appl. 10 (1975), 285-290.

5. M. D. Choi and E. Effros, Injectivity and operator spaces, J. Fnal. Anal. 24 (1974), 156-209.

6. ______, Nuclear C*-algebras and the approximation property, Amer. J. Math., to appear.

7. ______, Separable nuclear C*-algebras and injectivity, Duke Math J. 43 (1976), 309-322.

8. ______, Nuclear C*-algebras and injectivity: the general case, Indiana Un. Math. J., to appear.

9. ______, The completely positive lifting property for C*-algebras, Ann. of Math, 104 (1976), 585-609.

10. ______, Lifting problems and the cohomology of C*-algebras, Can J. Math., to appear.

11. A. Connes, Classification of injective factors, Ann. of Math 104 (1976), 73-116.

12. E. Effros, Aspects of non-commutative order, to appear.

13. ______, Aspects of non-commutative geometry, to appear.

14. A. Guichardet, Tensor products of C*-algebras, Aarhus University Lecture Notes Series 12 (1969).

15. C. Lance, On nuclear C*-algebras, J. Func. Anal. 12 (1973), 157-176.

16. S. Sakai, Automorphisms and tensor products of operator algebras, Amer. J. Math. 97 (1975), 889-996.

17. J. Tomiyama, Applications of Fubini type theorem to the tensor products of C*-algebras, Tohoku Math J. 19 (1967), 213-226.

18. S. Wassermann, Injective W*-algebras, to appear.

19. ____________, Liftings in C*-algebras: a counterexample, to appear.

20. E. Effros and J. Rosenberg, to appear.

On the Connes spectrum of simple C*-dynamical systems

by

Dorte Olesen

If A is a C*-algebra and α a homomorphism of the topological group G into the group $\mathrm{Aut}(A)$ of *-automorphisms of A such that each function $t \to \alpha_t(x)$, $x \in A$ is continuous, we say that the triple (A,G,α) is a <u>C*-dynamical system</u>. If A is a W*-algebra, A_* its predual, and each function $t \to \varphi(\alpha_t(x))$, $x \in A$, $\varphi \in A_*$ is continuous, we call (A,G,α) a <u>W*-dynamical system</u>. A systematic introduction to C*- and W*-dynamical systems can be found in [17].

In this lecture we shall concentrate on C*-dynamical systems, using the W*-case only for background reference. Furthermore we shall always assume G to be locally compact abelian and denote by Γ its dual group. A few of the notions and results presented here have recently been generalized to non-abelian compact groups by several authors, but we will not go into this.

Two notions of spectrum for W* and C*-dynamical systems have proved particularly useful in recent years ([1], [5] see also [13], [14]).

(i) the <u>Arveson spectrum</u> <u>$\mathrm{Sp}(\alpha)$</u> which is the smallest closed subset of Γ such that

$$\alpha_f(x) = \int \alpha_t(x) f(t)\, dt \neq 0$$

for some x in A whenever f is an $L^1(G)$-function such that $(\text{support } \hat{f}) \cap (\mathrm{Sp}\, \alpha) \neq \emptyset$, and

(ii) <u>the Connes spectrum $\Gamma(\alpha)$</u> which is the intersection of the Arveson spectra of the restricted actions $\alpha|B$ where B ranges over the set $\mathcal{H}^{\alpha}(A)$ of all non-zero α-invariant hereditary C*-subalgebras B of A, thus

$$\Gamma(\alpha) = \bigcap_{B \in \mathcal{H}^{\alpha}(A)} Sp(\alpha|B).$$

(Recall that a C*-subalgebra B is called hereditary if $B = L \cap L^*$ for some closed left ideal L of A.)

To clarify the last definition note that if (A,G,α) is a W*-dynamical system, every B in $\mathcal{H}^{\alpha}(A)$ has a σ-weak closure of the form pAp with p in the set P of all non-zero α-invariant projections in A. In this case one has the simpler definition ([5])

$$\Gamma(\alpha) = \bigcap_{p \in P} Sp(\alpha|pAp).$$

As a further simplification it was proved in [5] that this coincides with the intersection over those p in P that belong to the centre of the fixed-point algebra.

Let us list some basic properties of these spectra

(a) α is trivial if and only if $Sp(\alpha) = \{0\}$.

(b) α is uniformly continuous, i.e.

$$\|\alpha_t - \iota\| \to 0 \quad \text{as} \quad t \to 0$$

if and only if $Sp(\alpha)$ is compact.

(c) Let for a fixed t in G $\sigma(\alpha_t)$ denote the spectrum of α_t in $B(A)$. Then

$$\sigma(\alpha_t) = \text{closure of } \{(t,\tau) \mid \tau \in Sp(\alpha)\}$$

(d) Let α be a single *-automorphism of A and $(A,\mathbb{Z},\alpha)$ the system generated by $n \to \alpha^n$. Then $\sigma(\alpha) = Sp(\alpha)$. Especially $\alpha = \iota$ if and only if $\sigma(\alpha) = \{1\}$.

(e) Let $G = \mathbb{R}$, and denote by δ the (skew-symmetric) derivation such that

$$\alpha_t = e^{it\delta}.$$

Let $\sigma(\delta)$ denote the spectrum of δ as a (possibly unbounded) operator on A, then by [9]

$$Sp(\alpha) = \sigma(\delta).$$

(f) If G is compact, Γ thus discrete, there is a decomposition of A into eigenspaces A_τ, $\tau \in Sp(\alpha)$,

$$A_\tau = \{x \in A \mid \alpha_t(x) = (t,\tau)x \quad \forall t \in G\},$$

and the mapping of A onto A_o is a projection of norm one.

(g) The Connes spectrum $\Gamma(\alpha)$ is a closed subgroup of Γ.

(h) $\Gamma(\alpha)$ is perturbation-invariant, i.e. if (A,G,α) and (A,G,β) are two C^*-dynamical systems such that for every t in G and x in A

$$\beta_t(x) = u_t\alpha_t(x)\ u_t^*$$

where u is a unitary 1-cocycle (i.e. $u_{t+s} = u_t\alpha_t(u_s)$) in the multiplier algebra $M(A)$ of A such that $t \to u_t x$ is continuous for x in A, then $\Gamma(\beta) = \Gamma(\alpha)$.

Let us remark at this point that one may obtain a characterization of $\Gamma(\alpha)$ in terms of the dual action on the C^*-crossed product $G \times_\alpha A$ which rather easily implies (g) and (h). We shall return to this in a while, but first we want to link the Connes spectrum with innerness of the automorphisms on a simple C^*-algebra, in order to get a feeling for what the notion is.

1. $\Gamma(\alpha)$ and innerness of automorphisms of simple C^*-algebras.

Assume in this section that (A,G,α) is a C^*-dynamical system, and that A has a unit (this assumption is not essential, but eases the notation). Let A_o throughout denote the fixed-point algebra.

1.1. Theorem [14] *Let* $\Gamma(\alpha)^o$ *denote the annihilator of* $\Gamma(\alpha)$ *in* G. *If* A *is a simple* C^**-algebra and* $Sp(\alpha)/\Gamma(\alpha)$ *is compact (in* $\Gamma/\Gamma(\alpha)$*) then*

$$\Gamma(\alpha)^o = \{t \in G | \alpha_t = \mathrm{Ad}\, u,\ u \in A_o^u\}.$$

1.2. Corollary. A $*$-automorphism α of a simple C^*-algebra is inner if and only if $\Gamma(\alpha) = \{1\}$.

1.3. Corollary. A $*$-automorphism α of a simple C^*-algebra such that $\Gamma(\alpha)$ is not the entire unit circle has a power which is inner.

1.4. Corollary. Let $(A,\mathbb{R},\alpha)$ be a uniformly continuous simple C^*-dynamical system, then $\Gamma(\alpha) = \{0\}$.

Take the minimal period of a periodic $*$-automorphism α to be the smallest natural number q such that $\alpha^q = \mathrm{Ad}\, u$, u a unitary in A_o. Then we obtain as a consequence of the above theorem also the following

<u>1.5. Proposition</u>. [15] If α is a periodic *-automorphism of a simple C*-algebra, then the following are equivalent

(i) The minimal period equals the period of α

(ii) $Sp(\alpha) = \Gamma(\alpha)$

(iii) If $\alpha^k = \text{Ad } w$ for w in A and $1 \leq k <$ period, then $\alpha(w) \neq w$.

If moreover the period is the product of distinct primes these are equivalent to

(iv) No α^k, $1 \leq k <$ period, is inner.

It is important to note that condition (iv) is not in general equivalent to (i)-(iii). E.g. there exists an automorphism α of the CAR-algebra $\mathcal{F}$ such that $\alpha^4 = \iota$, $\alpha^2 = \text{Ad } w$ for $w \in \mathcal{F}$ but $\alpha(w) = -w$.

Let us remark that all the above results are analogues of results for factors obtained in [5] and [7], and that similar results can be formulated for general W*-dynamical systems if one replaces the Connes spectrum by a somewhat larger subset (the Borchers spectrum) (see [2] and [17]).

<u>2. The ideal structure of crossed products</u>.

Let us consider an arbitrary C*-dynamical system (A,G,α). Recall that there is a natural representation $\hat{\alpha}$ of Γ as automorphisms of the C*-crossed product $G \times_\alpha A$ which yields the dual dynamical system $(G \times_\alpha A, \Gamma, \hat{\alpha})$. Briefly, if $A \subset B(\mathcal{H})$ for some Hilbert space $\mathcal{H}$ we regard the set $K(G,A)$ of continuous A-valued functions on G with compact support as operators on $L^2(G,\mathcal{H})$ by setting

$$(y\xi)(t) = \int_G \alpha_{-t}(y(s))\,\xi(t-s)\,ds$$

for $y \in K(G,A)$ and $\xi \in L^2(G,\mathcal{H})$. Then $G \underset{\alpha}{\times} A$ identifies with the C*-subalgebra of $B(L^2(G,\mathcal{H}))$ generated by $K(G,A)$. Now define

$$(\hat{u}_\sigma \xi)(t) = (t,\sigma)\xi(t)$$

and set $\hat{\alpha}_\sigma = \mathrm{Ad}\,\hat{u}_\sigma$. It is easy to check that each $\hat{\alpha}_\sigma$ defines an automorphism of $G \underset{\alpha}{\times} A$. Recent work ([12]) characterizes A as a certain C*-subalgebra of the fixed-point algebra under $\hat{\alpha}$ in the multiplier algebra $M(G \underset{\alpha}{\times} A)$.

Iterating we get the double dual system which (as shown in [18]) is covariantly isomorphic to $(A\otimes C(L^2(G)),G,\alpha\otimes \mathrm{Ad}\tilde{\lambda})$. In particular, $\Gamma(\hat{\hat{\alpha}}) = \Gamma(\alpha)$.

We are now able to formulate the above mentioned characterization of $\Gamma(\alpha)$ recently achieved in collaboration with G.K. Pedersen:

2.1. Theorem. Let (A,G,α) *be a C*-dynamical system. Then* $\gamma \in \Gamma(\alpha)$ *if and only if* $\hat{\alpha}_\gamma(I) \cap I \neq \{0\}$ *for every closed two-sided ideal* I *in* $G \underset{\alpha}{\times} A$.

Let us recall that for W*-dynamical systems $\Gamma(\alpha)$ is simply the kernel of the restriction of $\hat{\alpha}$ to the centre of the W*-crossed product ([6]). This characterization was also obtained for simple C*-dynamical systems (A,G,α) with G discrete in [10]. In [11], the above characterization was obtained for compact groups G.

In the following, a C*-dynamical system (A,G,α) is termed G-prime if any two non-zero α-invariant ideals have non-zero intersection, and G-simple if there are no non-trivial α-invariant ideals.

2.2 Proposition. A is G-prime (resp. G-simple) if and only if $G \times_{\alpha} A$ is Γ-prime (resp. Γ-simple).

2.3. Theorem. $G \times_{\alpha} A$ *is prime if and only if*

(a) A *is* G-*prime and (b)* $\Gamma(\alpha) = \Gamma$.

2.4. Theorem. (Dual version). A *is prime if and only if*

(a) $G \times_{\alpha} A$ *is* Γ-*prime and (b)* $\Gamma(\hat{\alpha}) = G$.

2.5. Theorem. Assume G *to be discrete. Then* $G \times_{\alpha} A$ *is simple if and only if*

(a) A *is* G-*simple and (b)* $\Gamma(\alpha) = \Gamma$.

The last results above have in preliminary version appeared in preprint form [16]. That theorem 2.5 does not extend to arbitrary locally compact abelian G was shown in [4] by the following

2.6. Counterexample. Let $\mathcal{F}$ denote the CAR-algebra, T the circle group and α the action of T on $\mathcal{F}$ as gauge-automorphisms. The crossed product $T \times_{\alpha} \mathcal{F}$ is prime but not simple.

3. Fixed-point algebras under compact actions.

Let us conclude by mentioning a few results concerning the fixed-point algebra A_o of (A,G,α) where G is compact abelian, as recently treated by several authors [10], [11], [15]. These are in several ways analogous to the crossed-product results.

<u>*3.1. Theorem.*</u> *[15] Assume* G *to be compact.* A_o *is prime if and only if*

(a) A *is* G*-prime and (b)* $\Gamma(\alpha) = Sp(\alpha)$.

<u>3.2. Theorem</u>. [11] Assume G to be cyclic of prime order or the circle group. Then the relative commutant A_o^c of A_o in A is commutative.

An earlier result than 3.2 was the following

<u>3.3. Proposition</u>. [15] Let G be as in 3.2, and assume A to be prime and $\Gamma(\alpha) = Sp(\alpha)$. Then A_o^c is trivial.

Relative commutants need not be so well-behaved, however: if $G = T^2$, A_o^c need not even be of type I ([11]).

Let us end by noting that the "simple" version of 3.1 does not hold for general compact groups: indeed, using the notation from 2.6 we have that $\mathcal{F}_o$ (the Current algebra) is prime but not simple ([3]). We do, however, have the following

<u>*3.4. Theorem.*</u> *[15] Assume* G *to be finite.* A_o *is simple if and only if*

(a) A *is* G*-simple and (b)* $\Gamma(\alpha) = Sp(\alpha)$.

Linking this with proposition 1.5 we obtain

<u>3.5. Corollary</u>. An involutory *-automorphism α of a simple C*-algebra is outer if and only if its fixed-point algebra is simple.

Thus we have reobtained the result originally proved in [8] that the even CAR-algebra is simple.

References.

1. W. Arveson: On groups of automorphisms of operator algebras. J. Functional Anal. 15 (1974), 217-243.

2. H.J. Borchers: Characterization of inner *-automorphisms of W*-algebras. Publ. RIMS, Kyoto Univ. 10 (1974), 11-49.

3. O. Bratteli: Inductive limits of finite dimensional C*-algebras. Trans. Amer. Math. Soc. 171 (1972), 195-234.

4. O. Bratteli: A non-simple crossed product of a simple C*-algebra by a properly outer automorphic action. Preprint.

5. A. Connes: Une classification des facteurs de type III. Ann. Sci. École Norm. Sup. 6 (1973), 133-252.

6. A. Connes and M. Takesaki: The flow of weights on factors of type III. Preprint.

7. A. Connes: Periodic automorphism of the hyperfinite factor of type II_1. Preprint.

8. S. Doplicher and R.T. Powers: On the simplicity of the even CAR-algebra and the free field model. Commun. Math. Phys. 7 (1968), 77-79.

9. D.E. Evans: On the spectrum of a one-parameter strongly continuous representation. Math. Scand. 39 (1976), 80-82.

10. A. Kishimoto and H. Takai: On an invariant $\Gamma(\alpha)$ in C*-dynamical systems. Preprint.

11. A. Kishimoto and H. Takai: Some topics on C*-dynamical systems based on a compact abelian group. Preprint.

12. M. Landstad: Duality theory of covariant systems. Thesis, Univeristy of Pennsylvania, 1974.

13. D. Olesen: On spectral subspaces and their applications to automorphism groups. Istituto Nazionale di Alta Matematica. Symposia Matematica, Volume XX (1976), 253-296.

14. D. Olesen: Inner *-automorphisms of simple C*-algebras. Commun. Math. Phys. 44 (1975), 175-190.

15. D. Olesen, G.K. Pedersen and E. Størmer: Compact abelian groups of automorphisms of simple C*-algebras. Inventiones math. 39 (1977), 55-64.

16. D. Olesen and G.K. Pedersen: Applications of the Connes spectrum to C*-dynamical systems. Preprint. To appear in J. Functional Anal.

17. G.K. Pedersen: An introduction to C*-algebra theory. Chapters VII and VIII. Lecture Notes.

18. H. Takai: On a duality for crossed products of C*-algebras. J. Functional Anal. 19 (1975), 25-39.

Quantum Field Theory of Massless Particles and Scattering Theory.

Detlev Buchholz

II. Institut für Theoretische Physik der Universität Hamburg,
D-2000 Hamburg 50, Federal Republic of Germany

0.) Introduction and Main Results

We present in this lecture a general method for the construction of asymptotic fields and collision states of massless particles [1,2]. Such a method might be of some general interest since the famous Haag-Ruelle collision theory [3,4] is only applicable to massive theories. But we hope that our results will also be useful for the solution of concrete problems in connection with the physics of massless particles. That this hope is not completely unfounded may be taken from the following remarks.

An old problem in quantum field theory which has occupied many theoretical physicists is the so called infrared problem: since it does not cost much energy to create a massless particle, infinitely many of them can be produced in collisions. Such states can not be described by vectors in Fock-space and one has to look for more adequate representations. There exists a vast literature on this subject and we desist from giving references. However, there seems to be no attempt to study this problem in the general setting of field theory. We believe that such a study could be fruitful because of the following reason: it turns out that asymptotic fields of massless particles can be defined on all states which are generated from the vacuum by local fields or (more generally) localized morphisms. The details of the model, such as the superselection structure and the massive part of the particle spectrum, are irrelevant for the construction. So the asymptotic fields are adequate quantities for a study of the infrared problem. Now there exist geometric relations between the local Heisenberg fields and the asymptotic fields, which can be interpreted as a field theoretic version of Huyghens' principle. These relations impose various restrictions on the physically admissable infrared representations. They could therefore be used as a starting point for a fresh look at the infrared problem in quantum field theory. One might object that quantum electrodynamics (which is the theory one thinks of first in connection with infrared troubles) is not a local field theory in a unitary gauge and therefore the above statements seem not to apply. However, this is not really true since the infrared problem in quantum electrodynamics already crops up in full (because of pair creation) in the sector which can be obtained from the vacuum by local, gauge invariant quantities.

A systematic analysis of observable consequences of spontaneously broken symmetries, like the Adler zeros of the S-matrix, could be another application of our results. (For a fairly complete review of the present status of the discussion see [5].) Spontaneous breakdown of a symmetry usually signals its appearence by a degeneracy of the vacuum. It is therefore gratifying that the presence of many vacua causes no difficulties as far as the construction of collision states is concerned. Even in models with a degenerate vacuum a collision theory for massless particles exists and seems to be an appropriate basis for further investigations.

In order to make precise to which models our arguments can be applied let us give now a brief list of the relevant assumptions:

i) The space of states is a Hilbert space $\mathcal{H}$, i.e. we have "positive metric".

ii) The basic Heisenberg fields φ are local:

$$[\varphi(x), \varphi(y)] = 0 \quad \text{for} \quad (x-y)^2 < 0 .$$

We do not treat the case where Fermi fields are also present because it would complicate the notation.

iii) The fields φ transform covariantly under translations,

$$\varphi(x) = \mathcal{U}(x)\, \varphi(0)\, \mathcal{U}(x)^{-1}.$$

Lorentz covariance is not needed for the argument, but we assume it here.

iv) The spectrum of the theory has the usual properties: the Hamiltonian is positive, $H \geq 0$, and there exists a unit vector Ω (the vacuum) which is invariant under translations. Uniqueness of Ω is not crucial, as was indicated above. However, we postulate it for simplicity. It is essential for the present investigations that there exists a non-trivial subspace $\mathcal{H}_1$, the space of massless one-particle states. The only property of these states which we shall use is that they are eigenvectors of the mass operator corresponding to the eigenvalue 0.

v) The smeared polynomials in the basic fields φ generate from the vacuum Ω a dense set of vectors in $\mathcal{H}$.

We also remark that we shall not work with the basic fields φ themselves but with bounded functions of them, like the Haag-fields. So we assume tacitly that such operators can be constructed without problems. With this structure the following results can be established.

Theorem I (Existence of collision states)

a) Let $\Phi_1, \dots \Phi_n \in \mathcal{H}_1$ be any collection of massless one-particle states. Then there exists a vector $\Phi_1 \overset{out}{\times} \dots \overset{out}{\times} \Phi_n \in \mathcal{H}$ describing the outgoing configuration of these particles at large positive times.

b) The vectors $\Phi_1 \overset{out}{\times} \dots \overset{out}{\times} \Phi_n$ are symmetric under permutations of the one-particle constituents Φ_i.

c) They transform covariantly under translations and Lorentz transformations,

$$\mathcal{U}(L) \cdot \Phi_1 \overset{out}{\times} \dots \overset{out}{\times} \Phi_n = (\mathcal{U}(L)\Phi_1) \overset{out}{\times} \dots \overset{out}{\times} (\mathcal{U}(L)\Phi_n).$$

d) They have scalar products

$$(\Phi_1 \overset{out}{\times} \dots \overset{out}{\times} \Phi_m, \Phi_1' \overset{out}{\times} \dots \overset{out}{\times} \Phi_n') = \delta_{mn} \cdot \sum_P (\Phi_1, \Phi'_{p(1)}) \cdots (\Phi_n, \Phi'_{p(n)}),$$

where the sum extends over all permutations of $(1, \dots n)$.

Thus the outgoing collision states of massless particles have the familiar Fock-structure. It needs no extra explanation that a similar statement holds for large negative times if "out" is replaced "in". Therefore the usual definition and physical interpretation of a scattering matrix for the massless particles is possible. The following result is perhaps more interesting:

Theorem II (Existence of asymptotic fields)

a) Let A be any local, bounded operator. Then there exists a densely defined operator A^{out} on $\mathcal{H}$ such that

$$A^{out}\Omega = P_1 A \Omega,$$

where P_1 is the projection onto $\mathcal{H}_1$.

b) $A^{out}(x) = \mathcal{U}(x) A^{out} \mathcal{U}(x)^{-1}$ is a solution of the wave-equation,

$$\Box_x A^{out}(x) = 0.$$

c) A^{out} transforms covariantly under the action of the automorphisms α_L inducing translations and Lorentz-transformations,

$$\alpha_L(A^{out}) = (\alpha_L(A))^{out}.$$

d) The operators A^{out} have c-number commutation relations,

$$[A_1^{out}, A_2^{out}] = \{(\Omega, A_1 P_1 A_2 \Omega) - (\Omega, A_2 P_1 A_1 \Omega)\} \cdot 1.$$

e) (Huyghens' principle). Let F be any local, bounded operator. If F is localized in a region, which has a positive timelike separation from the localisation region $\mathcal{O}$ of A, then

$$[A^{out}, F] = 0.$$

The operators A^{out} are the (smeared) asymptotic fields. If there exist several types of massless particles in the model (differing in helicity or charge quantum numbers) one can decompose the operators A^{out} into sums of elementary fields corresponding to each particle type. Since one does not gain any new insight from this procedure we shall dispense with it here.

It may be unexpected that the asymptotic fields A^{out} are always defined on a dense set of vectors in $\mathcal{H}$. This is in contrast to the massive case where the corresponding operators are only defined on collision states. There one needs the additional assumption of asymptotic completeness in order to be sure that the asymptotic fields act on all of $\mathcal{H}$. Even more unfamiliar is relation e), the field theoretic version of Huyghens' principle. This relation is a consequence of causality and the fact that massless particles always move with the maximal possible speed, the speed of light. As was mentioned above we believe that Huyghens' principle could be a useful tool for both, structural investigations and practical calculations. Two simple applications may be found in Ref. [6] and [7].

Now we come to the proofs of the above statements. The following remarks may be taken as a guide-line: we shall first establish the existence of certain operators A^{out} and of vectors $\Psi^{out} \in \mathcal{H}$ which will later turn out to be the asymptotic fields and collision states respectively. We shall then find connections between A^{out} and Ψ^{out} which will help us to calculate scalar products of the vectors Ψ^{out} and to prove Theorem I. Only then is it possible to verify that the operators A^{out} have all the properties given in Theorem II.

1) Construction of Asymptotic Fields

We turn now to the construction of operators which we shall later identify as the asymptotic fields corresponding to the massless particles. For this purpose we take any local, bounded operator A and define

$$A_t = -2t \cdot \int d\omega \, \dot{A}(t, t\underline{e}) \quad , \; t \in \mathbb{R}. \tag{1}$$

Here $d\omega = d\omega(\underline{e})$ is the normalized invariant measure on the unit sphere $S^2 = \{\underline{e}, |\underline{e}| = 1\}$; the dot denotes differentiation with respect to time. The expression (1) is designed in such a way that all contributions with a non-lightlike momentum transfer disappear in the limit of large t. To see what survives we apply A_t to the vacuum vector Ω. Bearing in mind that Ω is invariant under translations we get after a simple calculation

$$A_t \Omega = \{ e^{i(H - |\underline{P}|)t} - e^{i(H + |\underline{P}|)t} \} |\underline{P}|^{-1} H A \Omega, \tag{2}$$

where H is the Hamiltonian and $\underline{P}$ the momentum operator. Because of the rapid oscillations of the exponentials one expects that in the limit of large t only those components of $|\underline{P}|^{-1} H A \Omega$ remain for which energy and momentum are related by $H = |\underline{P}|$. This reasoning can be made rigorous using a Riemann-Lebesgue type of argument. It shows that

$$\underset{t \to \infty}{\text{w-lim}} \; A_t \Omega = P_1 |\underline{P}|^{-1} H A \Omega = P_1 A \Omega, \tag{3}$$

where P_1 is the projection onto $\mathcal{H}_1$. But weak convergence is not strong enough for our purposes. We take therefore a suitable time average of A_t, e.g.

$$\overline{A_t} = \varepsilon(t)^{-1} \cdot \int_t^{t+\varepsilon(t)} ds \, A_s \, , \tag{4}$$

where $\varepsilon(t)$ is any positive, slowly increasing function, e.g. $\ln(1+t^2)$. Then - by the mean ergodic theorem - we achieve strong convergence,

$$\underset{t \to \infty}{\text{s-lim}} \; \overline{A_t} \, \Omega = P_1 A \Omega. \tag{5}$$

In the next step we show that $\overline{A_t}, t \to \infty$ converges on a dense set of vectors in $\mathcal{H}$. To this end we study the localisation properties of $\overline{A_t}$. To be definite let us assume that the operator A (from which we started) is localized in a double cone $\mathcal{O}$. Then A_t is localized in the region $\{\mathcal{O} + (t, t \cdot \underline{e}), |\underline{e}| = 1\}$ and $\overline{A_t}$ in $\bigcup_{s \geq t} \{\mathcal{O} + (s, s \cdot \underline{e}), |\underline{e}| = 1\}$. It is crucial now, that the latter region has a spacelike complement for large t which increases as t increases. In the limit $t \to \infty$ it consists of the open future lightcone which has its apex on the top of $\mathcal{O}$. We denote this cone be $\mathcal{O}_+$ and call it the future tangent of $\mathcal{O}$. Using locality we see then that $\overline{A_t}, t \to \infty$ converges on the vectors $F\Omega$, where F is any bounded operator which is localized in $\mathcal{O}_+$:

$$s-\lim_{t\to\infty} \overline{A}_t F \Omega = s-\lim_{t\to\infty} F \overline{A}_t \Omega = F P_1 A \Omega . \tag{6}$$

Now the vectors $F\Omega$ form a dense set in $\mathcal{H}$ owing to the Reeh-Schlieder property of the vacuum. We may therefore define on this domain a linear operator A^{out} by

$$A^{out} = \lim_{t\to\infty} \overline{A}_t . \tag{7}$$

Of course we get a non-trivial operator A^{out}, only if $A\Omega$ has a non-vanishing component in $\mathcal{H}_1$. In the following we require furthermore that A is hermitian. Then A^{out} is hermitian too and therefore dosable; we denote its least closed extension also by A^{out} . These operators are the asymptotic fields we are interested in.

2) Construction of Collision States

Next we come to the construction of the collision states. It is not suprising that these states can be obtained as limits

$$\lim_{t\to\infty} \overline{A_{1t}} \cdots \overline{A_{nt}}\, \Omega , \tag{8}$$

where $A_1, \dots A_n$ are operators as just described. In order to verify this we need the following lemma.

Lemma 1:*) a) $\| \overline{A_{1t}} \cdots \overline{A_{nt}}\, \Omega \| \leq c$ uniformly in t .
b) $\lim_{t\to\infty} (\Omega, \overline{A_{1t}} \cdots \overline{A_{mt}}\, \Omega) = \sum (\Omega, A_{i_1} P_1 A_{i_2} \Omega) \cdots (\Omega, A_{i_{n-1}} P_1 A_{i_n} \Omega)$
if n is even. The sum extends over all partitions of $(1, \dots n)$ into ordered pairs. For odd n the limit vanishes.

The proof of the lemma is rather tedious and cannot be given here. So the following remarks may suffice:

---

*) Actually we cannot prove the lemma as it stands. However, if we restrict our attention to operators $A_1, \dots A_n$ which behave smoothly at the origin in momentum space, then the statement holds. We omit the details in order not to bury the basic ideas under too many technicalities.

i) Part a) of the Lemma is clearly a consequence of part b) since n is arbitrary. It suffices furthermore to study the vacuum expectation values $(\Omega, A_{1_t} \cdots A_{n_t}\Omega)$ without time-smearing because of the postulated slow increase of the function $\varepsilon(t)$ in
$\overline{A}_t = \varepsilon(t)^{-1} \cdot \int_t^{t+\varepsilon(t)} ds\, A_s$.

ii) Using the invariance of Ω under translations one gets

$$(\Omega, A_{1_t} \cdots A_{n_t}\Omega) = (-2t)^n \cdot \int d\omega_1 \cdots \int d\omega_n\, (\Omega, \dot{A}_1(t\cdot \underset{\sim}{e}_1) \cdots \dot{A}_n(t\cdot \underset{\sim}{e}_n)\Omega). \tag{9}$$

So in order to prove the lemma one has to establish certain clustering properties of spherical means of vacuum expectation values. We want to emphasize that the spherical averaging is crucial here because it suppresses all contributions coming from configurations $\underset{\sim}{e}_1, \dots \underset{\sim}{e}_n$ where some of the vectors are parallel. For such configurations the clustering properties of the vacuum expectation values are too weak and would not compensate the factor t^n.

iii) The leading contributions to (9) stem from two-point correlations like

$$\int d\omega_i \int d\omega_K\, (\Omega, \dot{A}_i(t\cdot \underset{\sim}{e}_i)\, \dot{A}_K(t\cdot \underset{\sim}{e}_K)\Omega). \tag{10}$$

These contributions are known to decrease like t^{-2}. Since there are at most $\frac{n}{2}$ such correlations in the expression (9) the integral in this relation decreases at least like $(t^{-2})^{n/2}$. This compensates the factor t^n in front of the integral.

After these qualitative remarks we shall prove now that the limit (8) exists. For the present we can establish only weak convergence. But it will turn out later that these sequences converge also strongly. We proceed by induction in the number n of operators $\overline{A_t}$ which we apply to the vacuum: for $n=1$ the statement was proved in the preceding chapter. So let us assume that

$$w\text{-}\lim_{t\to\infty} \overline{A_{1_t}} \cdots \overline{A_{n_t}}\, \Omega = \Psi^{out}(A_1, \dots A_n) \in \mathcal{H}. \tag{11}$$

Now if A_1 is localized in $\mathcal{O}$, say, then we take any bounded operator F which is localized in the future tangent $\mathcal{O}_+$ of $\mathcal{O}$. Using relation (6) and part a) of the lemma we get

$$\begin{aligned}\lim_{t\to\infty} (F\Omega, \overline{A_{1_t}} \cdots \overline{A_{n+1_t}}\Omega) &= \lim_{t\to\infty} (\overline{A_{1_t}} F\Omega, \overline{A_{2_t}} \cdots \overline{A_{n+1_t}}\Omega) \\ &= \lim_{t\to\infty} (A_1^{out} F\Omega, \overline{A_{2_t}} \cdots \overline{A_{n+1_t}}\Omega) = (A_1^{out} F\Omega, \Psi^{out}(A_2, \dots A_n)),\end{aligned} \tag{12}$$

and this shows that the scalar products between $\overline{A_{1_t}} \cdots \overline{A_{n+1_t}}\Omega$ and the dense set of vectors $\{F\Omega\}$ converge as $t \to \infty$. But the sequence $\overline{A_{1_t}} \cdots \overline{A_{n+1_t}}\Omega$ is uniformly bounded in t according to the above lemma and hence converges weakly. This completes the induction and establishes the existence of the collision states $\Psi^{out}(A_1, \dots A_m)$ defined in relation (11). We have labelled these states by the operators $A_1, \dots A_m$ from which they are constructed. But actually they merely depend on the one-particle states $P_1 A_1 \Omega, \dots P_1 A_m \Omega$ in a linear way.

3) Connection between Asymptotic Fiels and Collision States

So far we have established only weak convergence of the sequences (11). Therefore we cannot use part b) of Lemma 1 directly in order to calculated scalar products of the collision states. But it will help us that there exists an alternative way of constructing the collision states with the aid of the asymptotic fields A^{out}. This is the content of the subsequent lemma.

Lemma 2: a) $A^{out*}\Psi^{out}(A_1, \dots A_n) = \Psi^{out}(A, A_1, \dots A_m)$.
b) If $A_1, \dots A_n$ are localized in the future tangent $\mathcal{O}_+$ of the localisation region $\mathcal{O}$ of A, then also

$$A^{out}\,\Psi^{out}(A_1, \dots A_m) = \Psi^{out}(A, A_1, \dots A_m).$$

(Remember in this context that A^{out} is hermitian. Thus A^{out*} is an extension of A^{out}, $A^{out*} \supset A^{out}$.)

The first part of this lemma follows simply from equation (12) if one reads it from the right to the left and takes into account that the set of vectors $\{F\Omega\}$ is a core for the operator A^{out} by its very definition. The proof of the second part is not so easy. The crucial step is to show that the vector $\Psi^{out}(A_1, \dots A_n)$ is in the domain of A^{out}, i.e. we must verify that

$$|(A^{out*}\Phi, \Psi^{out}(A_1, \dots A_n))| \le c \cdot \|\Phi\| \tag{13}$$

for some constant c and every Φ in the domain of A^{out*}.
Because of the localisation properties of $A_1, \dots A_n$ we can estimate

$$\begin{aligned} |(A^{out*}\Phi, \Psi^{out}(A_1, \dots A_n))| &= |\lim_{t\to\infty}(A^{out*}\Phi, \overline{A_{1_t}} \cdots \overline{A_{n_t}}\Omega)| \\ &= |\lim_{t\to\infty}(\Phi, A^{out}\overline{A_{1_t}} \cdots \overline{A_{n_t}}\Omega)| \le \|\Phi\| \cdot \limsup_{t\to\infty} \|A^{out}\overline{A_{1_t}} \cdots \overline{A_{n_t}}\Omega\|. \end{aligned} \tag{14}$$

Using part a) of this lemma we get furthermore

$$\begin{aligned}
\| A^{out} \overline{A_{1_t}} \cdots \overline{A_{n_t}} \Omega \|^2 &= (A^{out} \overline{A_{1_t}} \cdots \overline{A_{n_t}} \Omega, A^{out} \overline{A_{1_t}} \cdots \overline{A_{n_t}} \Omega) \\
&= (A^{out} \overline{A_{n_t}} \cdots \overline{A_{1_t}}\, \overline{A_{1_t}} \cdots \overline{A_{n_t}} \Omega, A^{out} \Omega) \\
&= (\overline{A_{n_t}} \cdots \overline{A_{1_t}}\, \overline{A_{1_t}} \cdots \overline{A_{n_t}} \Omega, A^{out*} A^{out} \Omega) \\
&\leq \| \overline{A_{n_t}} \cdots \overline{A_{1_t}}\, \overline{A_{1_t}} \cdots \overline{A_{n_t}} \Omega \| \cdot \| A^{out*} A^{out} \Omega \| .
\end{aligned} \tag{15}$$

But we know from Lemma 1 that the last expression is uniformy bounded in t and this shows that inequality (13) holds. Thus the vectors $\Psi^{out}(A_1, \dots A_n)$ are in the domain of A^{out}, and part b) of the lemma follows from part a) and $A^{out*} \supset A^{out}$.

4) Scalar Products of Collision States

To show that the collision states have the correct scalar products is the most difficult part of the proof of Theorem I. We give a list of the facts which we have established so far and which we shall use in our argument in the same order:

i) If $A_1, \dots A_m$ are operators such that each A_i is localized in the future tangent of the localisation region of A_{i+1}, then

$$A_m^{out} \cdots A_1^{out} \Omega = \Psi^{out}(A_m, \dots A_1).$$

ii) For arbitrary $A_1, \dots A_m$ (with no restrictions on the localisation) we have

$$A_1^{out*} \cdots A_m^{out*} \Omega = \Psi^{out}(A_1, \dots A_m).$$

iii) The collision states were originally defined as

$$w\text{-}\lim_{t \to \infty} \overline{A_{1_t}} \cdots \overline{A_{n_t}} \Omega = \Psi^{out}(A_1, \dots A_n).$$

iv) Finally, the vacuum expectation values of the operators $\overline{A_t}$ converge for asymptotic t :

$$\lim_{t \to \infty} (\Omega, \overline{A_{1_t}} \cdots \overline{A_{n_t}} \Omega) = \sum (\Omega, A_{i_1} P_1 A_{i_2} \Omega) \cdots (\Omega, A_{i_{n-1}} P_1 A_{i_n} \Omega),$$

if n is even. For odd n the limit vanishes.

Now we take operators $A_1, \dots A_m$ as in i) and operators $A_{m+1}, \dots A_n$ with no further restrictions and calculate

$$\begin{aligned}
&(\Psi^{out}(A_m, \dots A_1), \Psi^{out}(A_{m+1}, \dots A_n)) \\
&= (A_m^{out} \cdots A_1^{out} \Omega, \Psi^{out}(A_{m+1}, \dots A_n)) \\
&= (\Omega, A_1^{out*} \dots A_m^{out*} \Psi^{out}(A_{m+1}, \dots A_n)) \\
&= (\Omega, \Psi^{out}(A_1, \dots A_n)) \\
&= \lim_{t \to \infty} (\Omega, \overline{A_{1_t}} \cdots \overline{A_{m_t}} \Omega) \\
&= \begin{cases} \sum (\Omega, A_{i_1} P_1 A_{i_2} \Omega) \cdots (\Omega, A_{i_{n-1}} P_1 A_{i_n} \Omega) & \text{for even n} \\ 0 & \text{for odd n.} \end{cases}
\end{aligned} \tag{16}$$

Thus the scalar products of the restricted class of collision states can be expressed in terms of one - particle scalar products. The extension of this result to arbitrary states is possible owing to continuity properties of $\Psi^{out}(A_1, \dots A_n)$ in the one-particle constituents $P_1 A_1 \Omega, \dots P_1 A_n \Omega$.

Proceeding from $\Psi^{out}(A_1, \dots A_n)$ to the n-particle collision states $P_1 A_1 \Omega \overset{out}{\times} \cdots \overset{out}{\times} P_1 A_n \Omega$ amounts to subtracting from $\Psi^{out}(A_1, \dots A_n)$ all contributions with a particle number less than n. This can be done by a procedure which is similar to normal ordering. It turns out then that the states $P_1 A_1 \Omega \overset{out}{\times} \cdots \overset{out}{\times} P_1 A_n \Omega$ have the scalar products given in Theorem I. Since the details are straight forward we omit them here. We also skip the proof of Lorentz-covariance because it is less interesting.

Knowing the scalar products of $\Psi^{out}(A_1, \dots A_m)$ it is now obvious that the sequences $\overline{A_{1_t}} \cdots \overline{A_{m_t}} \Omega$ converge strongly as $t \to \infty$, because they converge weakly and in addition

$$\lim_{t \to \infty} \| \overline{A_{1_t}} \cdots \overline{A_{m_t}} \Omega \| = \| \Psi^{out}(A_1, \dots A_m) \|. \tag{17}$$

5) Further Properties of the Asymptotic Fields

What remains to be done is to verify that the operators A^{out} have all properties expected from an asymptotic field. In particular we must show that they act on the collision states $\Psi^{out}(A_1,\dots A_n)$ like a free field. For special configurations of operators $A_1,\dots A_n$ we already know from Lemma 2 that

$$A^{out}\cdot\Psi^{out}(A_1,\dots A_n)=\Psi^{out}(A,A_1,\dots A_n), \tag{18}$$

as desired. So the only problem is to extend this result to arbitrary operators $A_1,\dots A_n$. This can be done using the above mentioned continuity properties of the vectors $\Psi^{out}(A_1,\dots A_n)$ in the one-particle constituents and the fact that A^{out} is closed.

As we have seen in the first step of our analysis the operators A^{out} are not only defined on collision states of massless particles but on a dense set of vectors in $\mathcal{H}$. This set contains also other states if the theory is asymptotically incomplete with respect to the massless particles. In order to analyse the properties of A^{out} on these other states it is useful to note that A^{out} is selfadjoint. This can be shown as follows: it is a simple consequence of the definition of A^{out} given in relations (6) and (7) that A^{out} commutes on its domain with any bounded operator F which is localized in $\mathcal{O}_+$, the future tangent of the localisation region $\mathcal{O}$ of A (Huyghens' principle). We may therefore estimate

$$\|(A^{out})^n F\Omega\|^2\leq\|F\|^2\cdot\|(A^{out})^n\Omega\|^2=\|F\|^2\cdot(2n)!\,2^{-n}(n!)^{-1}\cdot\|P_1A\Omega\|^{2n}, \tag{19}$$

and this shows that A^{out} has in its domain a dense set of analytic vectors: $\{F\Omega\}$. It follows then from a well known theorem due to Nelson that A^{out} is selfadjoint. (For Nelson's theorem see e.g. Ref. [8].) We may therefore proceed from the unbounded operators A^{out} to bounded functions of A^{out} like the resolvents $(i+A^{out})^{-1}$ or the Weyl-operators $e^{iA^{out}}$. These operators can be used to give a rigorous meaning to the statement that the asymptotic fields A^{out} have c-number commutation relations on $\mathcal{H}$,

$$[A_1^{out},A_2^{out}]=\{(\Omega,A_1P_1A_2\Omega)-(\Omega,A_2P_1A_1\Omega)\}\cdot 1. \tag{20}$$

That such a relation holds is plausible if one realizes that

$$[A_1^{out},A_2^{out}]\cdot F\Omega=F\cdot[A_1^{out},A_2^{out}]\Omega \tag{21}$$

for sufficiently many operators F, because of the timelike commutation relations between local and asymptotic fields just mentioned. The commutator $[A_1^{out}, A_2^{out}]$ acts on the vacuum Ω like a multiple of the identity, according to the results of the preceding chapter, and this leads to (20).

Equation (20) is also useful to establish locality properties of the operators A^{out}: it follows from a Jost-Lehmann-Dyson type of argument that the right hand side of (20) vanishes if A_1 and A_2 are localized in spacelike separated double cones. We omit the proofs of the remaining statements of Theorem II, because they are very technical.

References

1. D. Buchholz: Collision theory for massless Fermions. Commun. Math. Phys. 42 , 269 (1975)

2. D. Buchholz: Collision theory for massless Bosons. Commun. Math. Phys. 52, 147 (1977)

3. R. Haag: Quantum field theories with composite particles and asymptotic behaviour. Phys. Rev. 112, 669 (1958)

4. D. Ruelle: On the asymptotic condition in quantum field theory. Helv. Phys. Acta 35, 147 (1962)

5. H. Joos, E. Weimar: On the covariant description of spontaneously broken symmetry in general field theory. Nuovo Cimento 32A, 283 (1976)

6. D. Buchholz, K. Fredenhagen: Dilations and interaction. To be published in J. Math. Phys.

7. D. Buchholz, K. Fredenhagen: A note on the inverse scattering problem in quantum field theory. Submitted for publication in Commun. Math. Phys.

8. M. Reed, B. Simon: Methods of Modern Mathematical Physics II (Fourier Analysis, Self-Adjointness). New York: Academic Press 1975.

SCATTERING THEORY IN QUANTUM MECHANICS AND ASYMPTOTIC COMPLETENESS

J.M. COMBES
Département de Mathématiques, Centre Universitaire de Toulon
and
Centre de Physique Théorique, CNRS, Marseille

I - INTRODUCTION

It is a rather difficult task to describe the status of scattering theory in Quantum mechanics as it appears today. First the mathematics of such a theory are of interest for themselves in the sense that the fundamental question of this theory, namely unitary equivalence of two self-adjoint operators is a problem which existed even before its implications for the existence of the S-matrix were recognized. Once this was done the models furnished by the second order elliptic differential operators of Quantum mechanics have attracted a variety of mathematicians interested by the connection between this type of existence and uniqueness problem with Cauchy data at "infinity" and the above mentioned unitary equivalence problems. So it is not so surprising that the outcome of all those efforts for physics consists mostly in the conclusion, which everyone here will easily believe, that for a particle decreasing faster than $|x|^{-1-\varepsilon}$, $\varepsilon > 0$, the S-matrix exists and is unitary. The literature on this type of problem and those related to it (self-adjointness for more and more singular potentials, elimination of positive energy bound-states and singular continuous spectrum, etc...) is quite abundant and I need to apologize in advance for my very incomplete list of results and references.

The second reason which makes my task difficult is that you certainly want to know if simple proofs of asymptotic completeness have been found for N-particle systems and if such proofs give some insight on a general method applicable to quantum field theory models. In other words, is there a transparent and reasonably short extension to N-particle systems of the (beautiful !) work of L.D.Faddeev in 1963 [22]. Unfortunately the most recent and complete results I know are contained in a two hundred pages manuscript of I. Sigal [31] whose results will be described later.

One should not draw pessimistic conclusions from this fast and critical description. First because the powerful methods developped by mathematicians in the last decade provide intermediate results which certainly are as important as the well accepted unitarity of the S-matrix and contribute greatly to our understanding of N-particle scattering processes. Second the methods are still evolving and new techniques are in progress originating from other branches of P.D.E. mathematics. It can be expected also that present developments in analysis on infinite dimensional spaces, like those presented by Albeverio at this conference [51] will through the use of Feynman path integrals also clarify the undoubtedly very deep connection between quantum and classical scattering theory which I strongly believe (and will try to show at the end of this lecture) is the ingredient necessary to make decisive and fundamental progress in Scattering theory.

II - THE MATHEMATICAL PROBLEMS OF QUANTUM SCATTERING THEORY

The basic operator to be analysed is the "S-matrix" :

$$S = \Omega_+^*(J;H,H_0)\,\Omega_-(J;H,H_0)$$

where the "wave-operators" $\Omega_\pm(J;H,H_0)$ are defined as :

$$\Omega_\pm(J;H,H_0) = s\text{-}\lim_{t\to\pm\infty} e^{iHt} J e^{-iH_0t} P_{ac}(H_0) \tag{II.1}$$

Here H_0 and H are the hamiltonians for the free and interacting systems, J being an "identification operator" between the Hilbert spaces describing respectively the free and interacting states. The projection operator $P_{ac}(H_0)$ selects those states whose H_0 -spectral measure is Lebesgue - absolutely continuous.

Properties of wave-operators, in particular intertwining between the absolutely continuous parts of H_0 and H are reviewed in [1].

Obviously wave-operators exist only under certain very restrictive conditions on the pair $\{H_0, H\}$. One of them is the celebrated Kato-Birman theorem ([1]) asserting that if the difference $(H-z)^{-1}J - J(H_0-z)^{-1}$, $z \in \rho(H_0) \cap \rho(H)$, is trace-class, then not only the wave-operators exist but they are complete, i.e. their range is the absolutely continuous subspace of H (which we will denote by $\mathfrak{M}_{ac}(H)$). This implies unitarity of S (but it is not equivalent to it). The applicability of this criteria to one body-Schrödinger operators with local potentials (in which case $J = 1$) was at the origin of the work of Kuroda which culminated in [2] after some various kinds of improvements.

The non-direct applicability of Kato-Birman theorem to many-particle systems if one merely takes $J = 1$, was already implicit in the earliest formulation of multi-channel scattering theory [3]. Here in fact one expects that the continuous part of H is a direct sum of operators, each of them being unitarily equivalent to the Hamiltonian of some free system but in which some particles are bound by interparticle forces. The basic reason of this fact is that the total potential does not decay in every direction of the configuration space of the many-particle systems. So for each asymptotic partition α of the n-particle system into composite fragments there should exist the channel wave-operators :

$$\Omega_{\pm}^{\alpha} = s\text{-}\lim_{t \to \pm\infty} e^{iHt} e^{-iH_0^{\alpha} t} P^{\alpha} \tag{II.2}$$

where H_0^{α} is the kinetic energy operator for this system of non-interacting composite particles and P^{α} is the projection operator on the corresponding states. The S-matrix is then defined as :

$$S_{\alpha\beta} = (\Omega_{+}^{\beta})^{*} \, \Omega_{-}^{\alpha} \tag{II.3}$$

It's unitarity is a consequence of the asymptotic completeness relation :

$$\bigoplus_{\alpha} \text{Range}\,(\Omega_{\pm}^{\alpha}) = \mathcal{M}_{ac}(H) \tag{II.4}$$

which is one of the main problems to be discussed below. As we will see later multichannel scattering theory can be reformulated in the two Hilbert space formalism described at the beginning of this chapter with a suitable choice of H_0 and J .

III - STATIONARY METHODS

- The one-body problem

A quantum mechanical particle in a potential V is described by the elliptic second order differential operator on $L^2(\mathbb{R}^3)$:

$$H = -\Delta + V \tag{III.1}$$

There is an abundant literature on the definition of H as a self-adjoint operator when V is a real function. There is an almost exhaustive summary of results in [4] culminating with the condition that the positive part of V is $L^2_{loc}(\mathbb{R}^3)$ and the negative part is Δ -bounded ; then $-\Delta + V$ is essentially self-adjoint on $C_0^{\infty}(\mathbb{R}^3)$. For scattering by singular potentials or obstacles the Hamiltonian has to be defined in terms of forms ; this kind of problem is investigated in particular in [5] , [6] . The abstract stationary scattering theory tries to construct the wave-operators from their stationary form which reads in the one-body case :

$$\Omega_\pm (H, H_0) = s\text{-}\lim_{\varepsilon \to 0} i\varepsilon \int_{-\infty}^{+\infty} (H - \lambda \pm i\varepsilon)^{-1} \, dE_0(\lambda) \tag{III.2}$$

where $E_0(\cdot)$ is the spectral family of H_0 .

This form is derived from the Abel limit method and spectral theory (see e.g. [7]). It leads to the usual expression for scattering amplitudes

$$f(k, k') = \hat{V}(k-k') + \lim_{\varepsilon \to 0} (\phi_k, (H - k^2 - i\varepsilon)^{-1} \phi_{k'}) \tag{III.3}$$

where $\phi_k(x) = V(x) e^{ikx}$ and $k^2 = k'^2$

The point is that the limits as $\varepsilon \to 0$ of the integrands in (III.2) for example do not exist even as unbounded operators on $L^2(\mathbb{R}^3)$. However their weak limits do for certain states and this leads naturally to the introduction of auxiliary spaces such that these limits exist as bounded operators between those spaces. A natural choice of auxiliary spaces is suggested by the elliptic character of our operators ; in particular the weighted Sobolev spaces

$$\mathcal{H}_{m,m'} = \left\{ \varphi(x) \, ; \, (1 + |x|^2)^{m'/2} D^\alpha \varphi \in L^2 \, , \, 0 \le |\alpha| \le m \right\}$$

play a fundamental role in Agmon's analysis [8]. Notice that if one allows fractional derivatives, this class of spaces is invariant by Fourier transform. One has the following standard properties :

<u>Proposition 1</u> (trace theorem in Sobolev spaces [9])

If $\mathcal{O}$ is a bounded domain of $\mathbb{R}^n$ with C^∞ boundary Γ then for any $m > \frac{1}{2}$ and any $m' \in \mathbb{R}$ there exists a bounded linear map

$$I : \mathcal{H}_{m,m'}(\mathbb{R}^n) \to \mathcal{H}_{m-\frac{1}{2}, m'}(\Gamma)$$

such that

$$I\varphi = \varphi|_\Gamma$$

The other result is some kind of Sobolev inequality stated in [8]:

<u>Proposition 2</u>

For all $m' > \frac{1}{2}$ and all $\kappa > 0$ there exists a constant C depending only on m' and κ such that

$$\|\varphi\|_{\mathcal{H}_{2,-m'}} \le C \|(\Delta + z)\varphi\|_{\mathcal{H}_{0,m'}} \tag{III.4}$$

for all $\varphi \in \mathcal{H}_{2,0}(\mathbb{R}^n)$ and $K^{-1} < |z| < K$.
The first result has the obvious consequence that any φ in $\mathcal{D}(1+|x|^2)^{m/2}$ has for $m > 1/2$ a square integrable trace on the energy spheres $k^2 = E$, $E \neq 0$, which gives us a very convenient framework to work at fixed energy. The second gives a precise meaning to the boundary values of the free resolvant on the spectrum $\mathbb{R}^+$ of $-\Delta$. That such properties remain valid for some perturbations of the Laplacian is of course of prime importance as can be seen from (III.2) or from the subsequent expression (III.3) for scattering amplitudes. This problem is most conveniently analysed in the general context of "smooth operator" techniques, a concept introduced by T. Kato [10] which is now recognized as a standard tool in perturbation theory for the continuous spectrum :

Definition

Let H be a self-adjoint operator on the Hilbert space $\mathcal{H}$ and A a densely defined closed operator from $\mathcal{H}$ to another Hilbert space $\mathcal{H}'$ with $\mathcal{D}(A) \supset \mathcal{D}(H)$. Then A is said to be H-smooth if one of the following (common) values is finite :

a) $$\sup_{\substack{\varepsilon \neq 0 \\ \|\varphi\| = 1}} \int_{-\infty}^{+\infty} \|A(H-\lambda \pm i\varepsilon)^{-1}\varphi\|^2 \, d\lambda \tag{III.5}$$

b) $$\frac{1}{\pi} \sup_{\substack{\|\varphi\| = 1 \\ \varepsilon \neq 0,\ \lambda \in \mathbb{R}}} |\varepsilon| \, \|A(H-\lambda \pm i\varepsilon)^{-1}\varphi\|^2$$

c) $$\frac{1}{2\pi} \sup_{\|\varphi\| = 1} \int_{-\infty}^{+\infty} \|A e^{-iHt}\varphi\|^2 \, dt$$

d) $$\frac{1}{2\pi} \sup_{\substack{\|\varphi\| = 1 \\ \varepsilon \neq 0,\ \lambda \in \mathbb{R}}} |(A^*\varphi, [(H-\lambda+i\varepsilon)^{-1} - (H-\lambda-i\varepsilon)^{-1}]A^*\varphi)|$$

There is also a notion of "locally smooth" operator : A is H-smooth on the Borel set $I \subset \mathbb{R}$ if $A\,E(I)$ is H-smooth.

As an example let us consider Agmon's class of short-range potentials satisfying :

1) $V(x) = O(1+|x|^2)^{\frac{m}{2}}$, $m > \frac{1}{2}$, as $|x| \to \infty$

2) A local regularity assumption : $V \in L^2_{loc}(\mathbb{R}^3)$ ensuring in particular that H is self-adjoint.

Then Proposition 1 implies that for any compact interval $I \subset \mathbb{R}\backslash\{0\}$, the operator $A = |V|^{1/2}$ is Δ-smooth on I. In view of the next theorems one would like to know that A also is $-\Delta+V$ -smooth. Actually the key results of the theory of smooth-operators are

Theorem 1 [11]

If A is H -smooth on $I \subset \mathbb{R}$ then $E(I)\,\mathrm{Range}\,A^*$ is contained in $\mathcal{M}_{ac}(H)$.

This gives a criteria for absolute continuity for H on certain intervals. This result is complemented by a theorem giving sufficient conditions for unitary equivalence :

Theorem 2 (Kato [10])

Let H_0 and H be self-adjoint operator on Hilbert spaces $\mathcal{H}_0$ and $\mathcal{H}$ respectively. Let J be a bounded operator from $\mathcal{H}_0$ to $\mathcal{H}$ taking the domain of H_0 into the domain of $\mathcal{H}$. Then if

$$HJ - JH_0 = A^* A_0$$

where A_0 is H_0 -smooth and A is H -smooth, the wave-operators $\Omega_\pm(J; H, H_0)$ exist and are complete.

The same is true if $HJ-JH_0$ is a sum of such operators.

There is an obvious local version of this theorem if A_0 and A are simply locally smooth on some interval.

In order to apply these results to short-range potentials one can use the factorisation $V = A^* A_0$ with $A = |V|^{1/2}$ and $A_0 = \mathrm{sgn}V\,|V|^{1/2}$, and tries to solve :

$$A(H-z)^{-1}A^* = A(H_0-z)^{-1}A^* - A(H-z)^{-1}A^*A_0(H_0-z)^{-1}A^* \quad \text{(III.6)}$$

From (III.5d) what is needed in order to show that A is locally H -smooth is H_0 -smoothness of A_0 and bounded invertibility of $1+\varphi(\lambda \pm i0)$ with :

$$\varphi(\lambda \pm i0) = \lim_{\varepsilon \to 0} A_0 (H_0 - \lambda \pm i\varepsilon)^{-1} A \quad \text{(III.7)}$$

By Proposition 2 and Rellich compactness criteria Agmon shows that the limit (III.7) exists and defines a compact operator for almost all $\lambda \neq 0$. So Fredholm alternative can be used for such points to solve (III.6) . In addition he shows that the null set of singular points where $[1+\varphi(\lambda \pm i0)]\psi = 0$ has a non trivial solution coincides with the set of eigenvalues of H , thus excluding singular continuous spectrum. Actually under extra assumptions one can also exclude positive energy bound states so that H has absolutely continuous spectrum on $\mathbb{R}^+$ and is unitarily equivalent to $-\Delta$ when restricted to $\mathcal{M}_{ac}(H)$.

Other results using smooth operator techniques are those of T. Kato [10], [11] ,.(Kato gives uniform estimates for (III.7) which are stronger than those derived from (III.4) in the sense that they extend up to the threshold. This is a consequence of a stronger decay condition $V(x) = O(|x|^{-2-\varepsilon})$ on the potentials which allows to get a compact limit for (III.7) up to $\lambda = 0$. Notice that this implies finiteness of the discrete spectrum of H), and T. Kato and K. Yajima [12] in the weak coupling case, which apply also to non symmetric perturbations ; R. Lavine [13] uses commutator techniques to prove directly H -smoothness of a very general class of multiplicative operators A . The work of Kuroda [2] does not explicitly uses the smoothness concept although a decomposition $V = A^* A_0$ with $A = (1+|x|^2)^{m/2}$, $m > \frac{1}{2}$, is used which in view of Propositions 1 and 2 plays a basic role in solving (III.6). Finally I want to mention the important series of articles by M. Schecter (see e.g. [14]) generalizing most of the existing results to elliptic operators of arbitrary order and to a class of pseudo-differential operators.

Concerning scattering by long-range potentials $V(x) = O(1+|x|^2)^{-m}$, $m \leq 1/2$, the most recent approaches using adaptations of the above framework are those of Pinchuk [15] and Kitada [16]. It is well-known that for such potentials the usual wave-operators do not exist and a new definition is needed ([17] , [18]).

To get a crude idea of this fact in the stationary framework one can notice that the boundary value on $\mathbb{R}^+$ of the operator $\varphi(z)$ does not exist as a bounded operator since $\operatorname{Im} \varphi(\lambda \pm i0)$ is just the trace on $k^2 = \lambda$ of vectors in $\mathcal{H}_{m,0}(\mathbb{R}^n)$ with $m < 1/2$ and this trace is not continuous in this case (see [9] theorem 9.5, ch. I).

- N-body systems

The Hamiltonian for n-interacting particles is given by :

$$H = -\sum_{i=1}^{N} \Delta_i + \sum_{i<j=1}^{N} V_{ij} \qquad \text{(III.8)}$$

where the V_{ij} 's are the two-body potentials. From translational invariance, we can get rid of the degrees of freedom associated to the center of mass of the systems so that the Hilbert space for this system is $\mathcal{H} = L^2(\mathbb{R}^{3(N-1)})$. The lack of decay of $V = \sum V_{ij}$ in all directions of configuration space makes the one-body method not directly applicable ; in particular even for $V \geq 0$ the

operator $G(z)$ of (III.7) never has compact boundary values on $\mathbb{R}$.
One can get rid of this difficulty in many ways which will be described below. But for the moment we simply notice that if the couplings are strong enough, two-particles systems will have bound states and then the continuous part of is no more unitarily equivalent to the kinetic energy operator $H_0 = -\sum \Delta_i$. Rather one expects that it is unitarily equivalent to a direct sum $\bigoplus_\alpha H_0^\alpha$ where H_0^α is the kinetic energy operator for a "channel" α built with the n initial particles, but some of these particles being bound together. So let us introduce

$$\mathcal{H}_0 = \bigoplus_\alpha \mathcal{H}_\alpha \tag{III.9}$$

where $\mathcal{H}_\alpha$ is the Hilbert space of states for the "composite particles" in channel α ; it is identified as the subspace of vectors in $\mathcal{H}$ which are tensor products of wave-functions for those composite particles (solutions of the bound-state problem for subsystems).
One defines the identification operator $J : \mathcal{H}_0 \to \mathcal{H}$ as

$$J(\varphi_\alpha)_\alpha = \sum_\alpha \varphi_\alpha \tag{III.10}$$

and the "free" hamiltonian

$$\tilde{H}_0 = \bigoplus_\alpha H_0^\alpha \tag{III.11}$$

where H_0^α is the kinetic energy operator for particles in channel α (with a suitable substractive constant $-E_\alpha$, the sum of binding energies for composite particles). The wave-operator

$$\Omega_\pm(J; H, \tilde{H}_0) = s\text{-}\lim_{t \to \pm\infty} e^{iHt} J e^{-i\tilde{H}_0 t}$$

can be seen to reduce in each channel subspace $\mathcal{H}_\alpha$ to the wave-operators $\Omega_\pm^\alpha$ introduced in (II.2). Asymptotic completeness is equivalent to

$$\text{Range}\,(\Omega_\pm(J; H, \hat{H}_0)) = \mathcal{M}_{ac}(H)$$

A stationary form of this wave-operator can be written as in (III.2) for the one-body case. We will see that this reduction to a one-body like problem does not make the problems as simple. In particular, most authors have to make a stronger assumption on the decay of the potential than was necessary for the one body problem namely $V(X) = O(1+|X|^2)^{-m/2}$, $m > 2$; this is due to two reasons ; first this condition guarantees a finite number of channels. Second any energy sphere $\sum k_i^2 = E$ contains zero energy thresholds for two-body relative kinetic energies $(k_i - k_j)^2$ and to generalize Propositions 1 and 2 for such degenerate hypersurfaces one needs a stronger decay for V_{ij}. I emphasize that those decay assumptions for V_{ij} which are done in most works on many-body systems are technically convenient but by no way essential ; this is shown by Mourre [20] who has succeeded in handling those threshold singularities for potentials having the decay $O(1+|X|^2)^{-m/2}$, $m > 1$, and being repulsive at infinity. Mourre uses Lavine's techniques of [20] instead of Kato's uniform estimates [10] for boundary values of two-body resolvents.

Now we come to a fast description of the methods used to handle N-body problems. In the one-channel case which corresponds to $O(|X|^{-2-\varepsilon})$ potentials with sufficiently small coupling constant there is a direct generalisation of smooth-operator techniques as shown by Iorio and O'Carrol [21] following a remark of Kato [10]. The idea is to consider a new Hilbert space

$$\mathcal{H} = \bigoplus_a \mathcal{H}_a \quad , \quad \mathcal{H}_a = L^2(\mathbb{R}^{3(N-1)}) \ \forall a, \tag{III.12}$$

where the sum is over pairs of particles. The mapping

$$J_1^* \varphi = (\varphi, \varphi, \ldots \varphi)$$

and the matrix operator $[V] = [V_a \delta_{ab}]$ are then introduced and allow to rewrite the second resolvent equation as

$$(H-z)^{-1} = (H_0-z)^{-1} - (H_0-z)^{-1} J_1 [V] J_1^* (H-z)^{-1} \tag{III.13}$$

Writing $A = [|V_a|^{1/2}\delta_{ab}]J_1^*$ and $A_0 = [\mathrm{sgn}\, V_a |V_a|^{1/2}\delta_{ab}]J_1^*$, one obtains equation (III.6). Using suitable sets of Jacobi coordinates one shows that for weak coupling A_0 and A are respectively H_0 and H -smooth so that the wave-operators exist and are complete by theorem 2.

To treat real multi-channel systems one needs refinements of the above method since then J is no more the identity operator and according to (II 1) and Theorem 2 what one needs to show is $HJ - J\tilde{H}_0 = A^* A_0$ where A_0 is $\tilde{H}_0$-smooth and A is H -smooth. This unfortunately cannot be done since it would imply by Theorem 2 that $s.\lim_{t\to\pm\infty} e^{i\tilde{H}_0 t} J^* e^{-iHt} P_{ac}(H)$ exists ; this leads to a contradiction since one can show from prime principles that only the weak limit exists (or the strong Abel limit). One is then led to use different techniques. One of them is based on the celebrated Faddeev-Yakubovsky equations ([22] , [23]) which we describe here below in the case $N=3$. One of the advantages of Faddeev equations is to incorporate the solutions of the two-body problems into three-body equations. Then one can separate out in the kernel of Faddeev equations the different kind of singularities coming from two-body bound-states or two-body continuum.

Faddeev equations can be derived from the multiple collision expansion of Watson [27] which is a special way to perform partial summations of the Born series ; one gets for the transition operator, related to the resolvent by

$$T(z) = V + V(H-z)^{-1}V$$
$$(H-z)^{-1} = (H_0-z)^{-1} + (H_0-z)^{-1}T(z)(H_0-z)^{-1}$$

the expansion

$$T(z) = \sum_{p=1}^{\infty} \sum_{a_1 \neq a_2 \neq \cdots \neq a_p} T_{a_1}(z)(H_0-z)^{-1}T_{a_2}(z)\cdots(H_0-z)^{-1}T_{a_p}(z) \quad \text{(III.14)}$$

where $T_a(z)$ is the two-body transition operator for the pair a of particles.

Introducing $M_{ab}(z)$ as the sum of contributions in (III.14) with $a_1 = a$, $a_p = b$ we get finally

$$M_{ab}(z) = T_a(z)\delta_{ab} + \sum_{c \neq a} T_a(z)(H_0-z)^{-1}M_{cb}(z) \quad \text{(III.15)}$$

Those equations are most conveniently treated using the following factorization technique suggested by R. Newton [24] :

$$T_a(z) = |V_a|^{1/2}\, T_a'(z)$$

Then the equation for $(H-z)^{-1}$ takes the form

$$(H-z)^{-1} = (H_0-z)^{-1} + (H_0-z)^{-1}\sum_{a,b}|V_a|^{1/2}L_{ab}(z)(H_a-z)^{-1} \quad \text{(III.16)}$$

where the operators L_{ab} satisfy the equations :

$$L_{ab}(z) = \delta_{ab} + \sum_{c \neq a}|V_a|^{1/2}(H_a-z)^{-1}|V_c|^{1/2}L_{cb}(z) \quad \text{(III.17)}$$

whose kernel is obviously connected.

Another advantage of Eq.(III.16) is that it shows more explicitly the singularity structure of $(H-z)^{-1}$. Apart from the possible pole singularities coming from eventual non trivial solutions of the homogeneous equation associated to (III.17), $(H-z)^{-1}$ will have the singularities of $\bigoplus_{\alpha}(H_0^{\alpha}-z)^{-1}$ as expected from asymptotic completeness.

Equations (III.17) are solvable using two-Hilbert space and smoothness techniques as shown by Combescure and Ginibre [25]. It is unfortunate that for the reasons given above their proof of asymptotic completeness is indirect and requires, as Faddeev did [22] , the stationary form of wave-operators and regularity properties of the Green's function, instead of a direct use of Theorem 2.

- Other methods and results for the N-body problem

For the three-body problem there is an alternative approach by L. Thomas [26] using spectral integrals (J. Howland [40] and K. Yajima [28]). The four body case has been investigated by K. Hepp [29], G.A. Hagedorn [41] , along the lines of Faddeev [22]. He also proves asymptotic completeness for N-body problems with repulsive potentials ; this case is also investigated using commutator techniques by R. Lavine [13] and P. Ferrero, O. de Pazzis and D. Robinson [6] , this last paper containing a general discussion of singular potentials, in particular hard-cores. The most recent results on the N-body problem have been derived by I. Sigal [31] ; Sigal does not use explicitly N-body equations but constructs directly regularizers for the operators $H-z$, $\mathrm{Im}\, z \neq 0$, i.e. a family $F(z)$ of bounded linear operators such that

$$F(z)\,(H-z)^{-1} = A(z)$$

where $A(z)$ is a Fredholm operator. Of course there are many ways to do this and a suitable choice allows a well-behaved continuation of F and A up to the real axis. Sigal's analysis is quite long but presents in turn the advantage of being complete in the sense that no a-priori assumption is made on the spectra of Hamiltonians for subsystems. Under the usual $|x|^{-2-\varepsilon}$, $\varepsilon > 0$, decay condition for two-body potentials Sigal shows :

a) Except for non dense set of values of the coupling constants the discrete spectrum is finite. If furthermore the potentials are dilation analytic this is also valid for resonances. The exceptional values correspond to the sudden appearance of an infinitude of bound states for some values of the coupling constants ; this occurs if and only if at least two subsystems have simultaneously a quasi bound-state at their continuum threshold. It can be shown that in this case exchange forces (not direct forces) between quasi bound-states are long-range hence responsible for the Efimov effect.

b) In case there is no bound-state or quasi-bound states at thresholds asymptotic completeness holds.

- Spectral transformations

Among other results on N-body systems one can mention those linked to dilation analyticity. This method introduced originally by Bottino, Longoni and

Regge in 1962 to study analyticity of scattering amplitudes and Regge poles for non relativistic one-body systems turns out to be also very fruitful for the analysis of N-body systems. We summarize here shortly the description given in [32].

Consider the linear group :

$$\mathcal{L}_N = \left\{ (\tau, \lambda) \; ; \; \tau \in \mathbb{R}^{3(N-1)} \; , \; \lambda \in \mathbb{R}^+ \right\}$$

with the group law

$$(\tau, \lambda) * (\tau', \lambda') = (\lambda'^{-1} \tau + \tau' \; ; \; \lambda \lambda')$$

and its representation on $L^2(\mathbb{R}^{3(N-1)})$:

$$\mathcal{F}(\mathcal{U}(\tau, \lambda) \Phi)(P) = \lambda^{-3(N-1)/2} (\mathcal{F}\Phi)(\lambda^{-1} P + \bar{M}\tau)$$

where $\mathcal{F}$ denotes the Fourier transform and $\bar{M}$ is the mass matrix.

The main interest of this group for our purpose is that the family of operators

$$H_0^\alpha(\tau, \lambda) = \mathcal{U}(\tau, \lambda) \, H_0^\alpha \, \mathcal{U}^{-1}(\tau, \lambda)$$

has for any channel α an analytic continuation in the parameters τ, λ with spectrum

$$\sigma(H_0^\alpha(\tau, \lambda)) = E_\alpha + \lambda^{-2} \left\{ \zeta \in \mathbb{C} \; ; \; \operatorname{Re} \zeta \geq -\frac{\sigma \bar{M} \sigma}{2} \right. \quad \text{(III.18)}$$
$$\left. \text{and } |\operatorname{Im} \zeta|^2 \leq 2(\sigma \bar{M}_\alpha \sigma) \operatorname{Re} \zeta + (\sigma \bar{M}_\alpha \sigma)^2 \right\}$$

where $\sigma = \operatorname{Im}(\lambda \tau)$ and $\bar{M}_\alpha$ is the mass-matrix for composite particles in channel α. The set (III.18) is the interior of a parabola whose axis makes an angle $-2 \operatorname{Arg} \lambda$ with $\mathbb{R}^+$.

Now if one tries to analyse along the lines of Hunziker [33] the spectrum of the analytic continuation of $\mathcal{U}(\tau, \lambda) H \mathcal{U}^{-1}(\tau, \lambda)$ (which exists if two-body potentials are local and dilation analytic [34]) one finds that the essential spectrum of this continuation is exactly $\bigcup_\alpha \sigma(H_0^\alpha(\tau, \lambda))$. Notice that if this is to be expected from the eventual unitary equivalence of the absolutely continuous part of H and $\bigoplus_\alpha H_0^\alpha$, it is obtained here from prime principles and not from scattering theory : in fact the above result is true even if wave-operators do not exist ! In case they do, this result in turn suggests the existence of an analytic continuation for their kernels in momentum spaces which should be strongly related to analyticity properties of scattering amplitudes. So it is not surprising that the main outcome of this approach is for such properties ; in the case of two-body elastic or inelastic amplitudes this has been shown e.g. by J.M.Combes [32] and A. Tip [35] . Other results include spectral properties of H in particular absence of continuous singular spectrum [34] and of positive energy bound-states. Dilation analyticity techniques also appears as very useful for resonance energy calculation, resonances showing up in this approach as complex isolated eigenvalues of $H(0,\lambda)$ for $\mathrm{Im}\lambda \neq 0$.

IV - TIME-DEPENDENT METHODS

These methods try to solve directly the existence problem for wave-operators without recourse to resolvent methods. They are based on the integral representation

$$e^{iHt} J e^{-iH_0 t} = J - i\int_0^t e^{iHs}(HJ-JH_0)e^{-iH_0 s}ds \quad \text{(IV.1)}$$

which plays a basic role in the proof of Theorem 2.

For single channel scattering (J = 1) one of the most familiar approach to the existence problem for the S-matrix is Dyson's perturbation expansion obtained by iterating (IV.1). For repulsive interactions [29] or in the case of weak coupling ([21] , [36]) this expansion can be shown to converge to a unitary operator. This last property is obviously lost when none of the above conditions is satisfied and Dyson's expansion is not a good tool then to show completeness.

Another well-known and very ancient tool is the Cook's method ([37]) which is based on the observation

$$\left\| \int_\tau^t e^{iHs}(HJ-JH_0)e^{-iH_0 s}\varphi \right\| ds \leq \int_\tau^t \|(HJ-JH_0)e^{-iH_0 s}\varphi\| ds$$

So if $\int_T^\infty \|(HJ-JH_0)e^{-iH_0 s}\varphi\| ds < \infty$ for some T and for a dense set of vectors φ the wave-operators exist. This method has been adapted recently by Schecter [38] and Simon [39] to handle situations where

$HJ-JH_0 = A^* A_0$, where A is H -bounded

and $\int_T^\infty \|A_0 e^{-iH_0 s}\varphi\| ds < \infty$ for a dense set of φ's.

In this new form the theorem allows to prove existence of wave-operators when two-body potentials satisfy

$$(1+|x|^2)^{\frac{m}{2}} V(x) \in L^p$$

for $$0 < p \leq 2 \quad , \quad m > -\frac{1}{p}$$

One disadvantage of Cook's method is that since decay properties of $e^{-iH_0 t}\varphi$ play a fundamental role it does not work usually to show "completeness" since decay properties of $e^{-iHt}\varphi$ are usually unaccessible by known methods.

Cook's method has been adapted by many authors ([17] ,[18] , [42] , [43]) to treat long-range forces. There it is well-known that modified wave-operators have to be defined

$$\tilde{\Omega}_{\pm}(H,H_0) = \underset{t\to\pm\infty}{s\text{-}\lim}\, e^{itH} e^{-iW(t)} \tag{IV.2}$$

where $W(t)$ satisfies the Hamilton-Jacobi equation :

$$V\left(\frac{\partial W}{\partial p}\right) + p^2 - \frac{\partial W(t)}{\partial t} = 0 \tag{IV.3}$$

One can write $W(t) = p^2 t + \varphi(t)$; clearly $\varphi(t)$ is a divergent phase when V decays more slowly than the Coulomb potential at infinity.

This general formulation of time-dependent scattering theory is due to L. Hörmander [43] ; he uses stationary phase methods to prove convergence in (IV.2) His methods apply actually to more general elliptic systems than those obtained from the Laplacian in $\mathbb{R}^n$ and have been used in [44] by Berthier and Collet to study perturbations of pseudo-differential operators. Hörmander's method makes an optimal use of Cook's theorem since it covers all existence theorems previously known for both short-range and long-range potentials (modulo an adaptation to singular potentials as done by Simon [36]). A proof of asymptotic completeness along the lines of Hörmander would require informations about the symbol of the operator e^{iHt} ; let us hope they will be provided by further progress in pseudo-differential operator theory.

I would like to emphasize that the success of Hörmander's method can be traced to its semi-classical aspect in the sense that the stationary phase method is a special way to single-out from the quantum dynamics the contribution of classical orbits. A trivial example is the free evolution for $H_0 = -\Delta$

$$(e^{-iH_0 t}\varphi)(x) = (2\pi)^{-3} \int e^{i(xp - p^2 t)} (\mathcal{F}\varphi)(p)\, dp$$

which immediately gives for $x = pt$

$$(e^{-iH_0 t}\varphi)(pt) = c\,t^{-3/2}(\mathcal{F}\varphi)(p)\,e^{ip^2 t} + o(t^{-3/2}) \qquad \text{(IV.4)}$$

This is a precise indication, already known for acoustical scattering (see e.g. [45]), that energy density is carried away for large time along bicharacteristics of the partial differential operator, i.e. in our case along classical trajectories. The result (IV.4) is known in quantum scattering literature as the "cone theorem" of Dollard [46] (see also [47]). For any cone $\mathcal{C}$ in configuration space, the probability that particles are in $\mathcal{C}$ tends for large time to the probability that asymptotic momenta are in $\mathcal{C}$. This theorem was mostly used in studies about the connection between observed and theoretical cross-sections [47] and partly motivated the promising, but apparently abandoned, Algebraic scattering theory (see e.g. [48]) in which those asymptotic momenta were the basic objects to be studied instead of wave-operators.

Before discussing further those deep interelations between quantum and classical dynamics, let me recall that the main difficulties with the stationary methods originated from threshold singularities which just express the fact that some particles may have arbitrarily small velocities, so that the asymptotic regions can be reached in an arbitrarily large time. It is quite disappointing that so much work has to be done to get rid of these arbitrarily small sets in momentum space. This is typically a consequence of the undeterminacy principle and such difficulties do not show up in classical scattering ; here one has simple proofs of asymptotic completeness by W. Hunziker [49] and B. Simon [50]. As an example, one has [50] :

Theorem 3

Let the two-body potentials V_{ij} be such that

a) $\sup_x |\nabla V_{ij}(x)| < \infty$

b) For some $R_1 > 0$, $\alpha_1 > 2$ and $C_1 > 0$

$$|\nabla V_{ij}(x)| < C_1 |x|^{-\alpha_1} \quad \text{if } |x| > R_1$$

c) For some $R_2 > 0$, $\alpha_2 > 2$ and $C_2 > 0$

$$|\nabla V_{ij}(x) - \nabla V_{ij}(y)| < C_2 |x|^{-\alpha_2} |x-y| \quad \text{if } |x|, |y| > R_2$$

Then for each x_0 , $p \in \mathbb{R}^{3N}$ with $p_i \neq p_j$ for $i \neq j$, there exists one and only one solution $x(t)$, $t \in \mathbb{R}$, of classical equations of motion such that

$$|X(t) - X_0 - Pt| + |\dot{X}(t) - P| \underset{t \to \infty}{\longrightarrow} 0$$

Many channel systems are also investigated by W. Hunziker [49]. The question is whether one can use those results as an input in attempts to prove asymptotic completeness in the quantum theory. The first idea which comes to mind is to use Feynman path integrals techniques [51] with infinite trajectories, which remains to be studied along the lines of Albeverio-Hoegh-Krohn.

Let me describe here a strongly related method which has not been worked out completely yet but is suggestive of such possibilities. It is inspired from a work of Maslov [52] which deals with the $\hbar = 0$ limit of quantum dynamics. One can try to use scaling arguments to transform the $\hbar = 0$ limit into a $t = \infty$ limit ; I prefer instead to reformulate Maslov's idea in the context of one-body scattering theory.

Our aim is to analyse the connection between the solutions of Hamilton equations

$$\begin{cases} \dfrac{dX}{dt} = P(t) \\ \dfrac{dP}{dt} = -\nabla V(X(t)) \end{cases} \tag{I}$$

and Schrödinger equation

$$i\frac{\partial \psi}{\partial t} = (-\Delta + V)\psi \tag{II}$$

in the limit of large times. So it is natural to look at solutions of (I) with Cauchy data at $t = \infty$. From Theorem 3, we know that there exists a 2n parameter of such solutions (n is the space dimension here). We denote by $\mathcal{G}$ the subclass of trajectories such that :

$$|X(t) - Pt| + |\dot{X}(t) - P| \underset{t \to \infty}{\longrightarrow} 0 \tag{IV.5}$$

(The fact that we get rid of the n parameters associated to initial positions can be traced to the fact that only asymptotic momenta are measured in a scattering experiment). Let us make the following regularity assumption on trajectories in $\mathcal{G}$:

<u>Assumption</u>

For fixed $T > T_0$ and almost all $x \in \mathbb{R}^n$, $\exists$ a unique $X(x, t) \in \mathcal{G}$ such that $X(x, T; T) = x$.

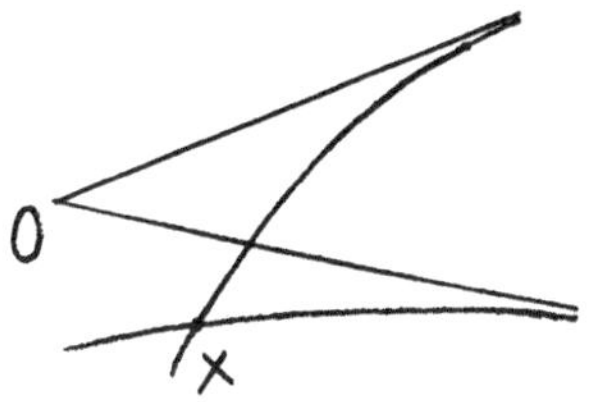

In other words through each point in $\mathbb{R}^n$, there is one and only one trajectory in $\mathcal{G}$ passing through X at time T. In other words the situation depicted on the figure is not allowed.

A similar statement has to be made (and is proved) in [52]. Since trajectories in $\mathcal{G}$ are parametrized by an asymptotic momentum we have then a one to one mapping

$$\Phi(T) \cdot X \in \mathbb{R}^n \rightarrow P \in \mathbb{R}^n$$

where P is the asymptotic direction of $X(x,T)$.

Example : $V=0$. Then for $T \neq 0$ one has $\Phi(T)(X) = \frac{X}{T}$ and $X(x,T;t) = xt/T$.

We now define the classical action associated to this family of trajectories as the solution of the Hamilton-Jacobi equation

$$\frac{|\nabla S(x,t)|^2}{2} + V(x) + \frac{\partial S}{\partial t}(x,t) = 0$$

which is pointwise asymptotic as $t \to \infty$ to the free action $S_0(x,t) = \frac{1}{2}\frac{x^2}{t}$. Finally let $\psi(\cdot,t)$ be a solution of (II) with $\psi \in \mathcal{M}_{ac}(H)$ (so that under reasonable conditions on V, $\psi(x,t) \to 0$ pointwise). Let us define

$$\tilde{\psi}(P,t) = \sqrt{Y(t)}^{-1} e^{iS(x,t)} \psi(x,t) \qquad \text{(IV.6)}$$

where $Y(t) = \mathrm{Det}\left|\frac{\partial \Phi(t)}{\partial x}\right|$, $P = \Phi(t)X$.

Notice that with classical dynamics instead of (II) $\tilde{\psi}(P,t)$ would be constant along classical trajectories, i.e.

$$\tilde{\psi}(\Phi(t)x,t) = \tilde{\psi}(P)$$

So if $\hbar = 1$ one has instead [52]

$$i\frac{\partial\tilde{\psi}}{\partial t}(P,t) = -\frac{1}{\sqrt{Y(t)}} \Delta_P \sqrt{Y(t)}\, \tilde{\psi}(P,t) \qquad \text{(IV.7)}$$

where Δ_P is the Laplacian in curvilinear coordinates P. Equation (IV.7) then incorporates all the diffusion effects due to quantum dynamics. It just describes what remains of the Schrödinger equation once you have subtracted the classical dynamics. The relevance of this equation for physical purposes lies in the following remark. First one expects that $\tilde{H}_0(t) = \frac{1}{\sqrt{Y(t)}} \Delta_P \sqrt{Y(t)}$ converges to zero in strong generalized sense as $t \to \infty$. As an example for $V = 0$, one has $\tilde{H}_0(t) = \Delta / t^2$. Accordingly as $t \to \infty$, $\tilde{\Psi}(P, t)$ should tend to limits $\tilde{\Psi}(P, \pm\infty)$. In view of the cone theorem one has for any cone $\mathcal{E}$

$$\lim_{t \to \pm\infty} \int_{\mathcal{E}} |\Psi(x,t)|^2 dx = \lim_{t \to \pm\infty} \int_{\Phi(t)\mathcal{E}} |\hat{\Psi}(p,t)|^2 dp$$

Since $\Phi(t)$ is expected to converge to the affine mapping $\Phi(t)x = x/t$ one obtains finally

$$\lim_{t \to \pm\infty} \int_{\mathcal{E}} |\Psi(x,t)|^2 dx = \int_{\mathcal{E}} |\hat{\Psi}(p, \pm\infty)|^2 dp$$

Since the probability to be in $\mathcal{E}$ at large times is directly related to the scattering cross sections [47] equation (IV.7) gives a complete description of the scattering experiment. Of course, all those arguments are very formal and mathematical details will be published later. But I hope they are enough convincing that both from a mathematical and physical point of view classical mechanics are more than just the $\hbar = 0$ limit of quantum mechanics.

- REFERENCES -

[1] T. KATO
Perturbation Theory for Linear Operators
Springer, New York (1966).

[2] S.T. KURODA
Scattering Theory for Differential Operators II.
J.Math.Soc.Japan 25, 643 (1973).

[3] H. EKSTEIN
Theory of Time-Dependent Scattering Multichannel Processes
Phys. Rev. 101, 880 (1956).

[4] B. SIMON, M. REED
Methods of Modern Mathematical Physics II. Fourier Analysis, Self-Adjointness
Acad. Press (1975).

[5] D.W. ROBINSON
Scattering Theory with Singular Potentials
Ann. I.H.P. 21 (1974).

[6] P. FERRERO, O. de PAZZIS, D.W. ROBINSON
Ann. I.H.P., 21, 217 (1974).

[7] J. HOWLAND
Banach Space Techniques in Perturbation Theory of Self-Adjoint Operators with Continuous Spectra
J. Math. Anal. and Appl., 20, 22 (1967).

[8] S. AGMON
Ann. Scuol. Norm. Sup. Pisa, Ser. IV, 2, 151.

[9] J.L. LIONS, E. MAGENES
Problèmes aux limites non homogènes et applications
Dunod, Paris (1968).

[10] T. KATO
Wave-Operators and Similarity for Some Non Self-Adjoint Operators
Math. Ann. 162, 258 (1966).

[11] T. KATO
Smooth Operators and Commutators
Studia Math. T. XXXI, 535 (1968).

[12] T. KAKO, K. YAJIMA
Spectral and Scattering Theory for a Class of Non Self-Adjoint Operators
Scient. Pap. of the College of Gen. Educ., Tokyo 26, 73 (1976).

[13] R. LAVINE
Commutators and Scattering Theory II
Ind. Univ. Math. 21, 643 (1973).

[14] M. SCHECTER
Scattering Theory for Elliptic Operators of Arbitrary Order
Com.Math. Helv. 49, 84 (1974).

[15] PINCHUK
Thesis, Univ. of California, Berkeley (1975).

[16] KITADA
To appear.

[17] J.D. DOLLARD
Asymptotic Convergence and the Coulomb Interaction
J. Math. Phys. 5, 729 (1964).

[18] V.S. BULSLAEV and V.B. MATVEEV
Wave-Operators for the Schrödinger Equation with Slowly Decreasing Potential
Teoret. Mat. Fiz. 1, 367 (1970).

[19] H. EKSTEIN
Scattering in Field Theory
Nuovo Cimento 4, 1017 (1956).

[20] E. MOURRE
Application de la méthode de Lavine au problème à trois corps
Ann. I.H.P. vol. XXVI, 3 (1977).

[21] IORIO, O' CARROL
Asymptotic Completeness for Multi-Particle Schrödinger Hamiltonians with Weak Potentials
Commun.math. Phys. 27, 137 (1972).

[22] L.D. FADDEEV
Mathematical Aspects of the Three Body Problem in the Quantum Theory of Scattering
Israel Scientific Translation (1965).

[23] O.A. YAKUBOVSKY
On the Integral Equations in the Theory of N-Particle Systems
J. Nucl. Phys. (USSR) 5, 1312 (1967).

[24] R. NEWTON
J. Math. Phys. 12, 1552 (1971).

[25] M. COMBESCURE, J. GINIBRE
Ann. Phys. 101 (1976).

[26] L.E. THOMAS
Asymptotic Completeness in 2 and 3 Particle Quantum Mechanical Systems
Ann. Phys. 90, 127 (1975).

[27] K.M. WATSON
Phys. Rev. 89, 575 (1953).

[28] K. YAJIMA
An Abstract Stationary Approach to 3-Body Scattering
To appear (presented at Oberwolfach Conference, July 1977).

[29] K. HEPP
On the Quantum Mechanical N-Body Problem
Helv. Phys. Acta 42, 425 (1969).

[30] W. THIRRING, P. URBAN
The Schrödinger Equation
Springer-Verlag (1977).

[31] I. SIGAL
Preprint E.T.H. (1977).

[32] J.M. COMBES
in Scattering Theory in Mathematical Physics
J.A. Lavita and J.P. Marchand Ed. , Reidel, Dortrecht (1974).

[33] W. HUNZIKER
Helv. Phys. Acta 39, 451 (1966).

[34] E. BALSLEV, J.M. COMBES
Spectral Properties of Schrödinger Hamiltonians with Dilation Analytic Potentials
Commun.math. Phys. 22, 280 (1971).

[35] A. TIP
A Note on the Analyticity of the Elastic Forward Electron-Atom Exchange Scattering Amplitude
Preprint FOM Institut voor Atom, Amsterdam (1977).

[36] R.T. PROSSER
Convergent Perturbation Expansions for Certain Wave-Operators
J.M.P. 5, 708 (1964).

[37] J.M. COOK
Convergence of the Möller Wave-Matrix
J.M.P. 36, 82 (1957).

[38] M. SCHECTER
A New Criterion for Scattering Theory
Yeshiva Preprint (1977).

[39] B. SIMON
Scattering Theory and Quadratic Forms : on a Theorem of Schecter
Yeshiva Preprint (1977).

[40] J. HOWLAND
Abstract Stationary Theory of Multichannel Scattering
J. Funct. Anal. 22, 250 (1976).

[41] G.A. HAGEDORN
Asymptotic Completeness for a Class of Four Particle Schrödinger Operators, Preprint Univ. Princeton.

[42] P. ALSHOM, T. KATO
Scattering by Long-Range Potentials
Proceed. Symp. in Pure Math. 23, AMS, 393 (1973).

[43] L. HORMANDER
The Existence of Wave-Operators in Scattering Theory
Math. Z. 146, 69 (1976).

[44] A. BERTHIER, P. COLLET
Wave-Operators for Momentum Dependent Long-Range Potentials
Preprint Univ. Paris VI.

[45] P. LAX, R. PHILLIPS
Scattering Theory.
Academic Press, New York, 1967.

[46] J.D. DOLLARD
Scattering into Cones
Commun.math.Phys. 12, 193 (1968) and
J. Math. Phys. 14, 708 (1973).

[47] J.M. COMBES, R. NEWTON, R. STOKHAMER
Scattering into Cones and Flux across Surfaces
Phys. Rev. D, 11, 366 (1975).

[48] H. EKSTEIN
Scattering without Scattering Operators
Annals of Physics 74, 303 (1972).

[49] W. HUNZIKER
in Scattering Theory in Mathematical Physics
Reidel, 1974.

[50] B. SIMON
Wave-Operators for Classical Particle Scattering
Commun.math.Phys. 23, 37 (1971).

[51] S. ALBEVERIO, R. HOEGH-KROHN
Proceedings of this Conference.

[52] V.P. MASLOV
The Quasi-Classical Asymptotic Solutions of some Problems
in Mathematical Physics
Zh. Vychisl. Mat. 1, 113 (1961).

STATIC SOLITONS IN MORE THAN ONE DIMENSION

L. O'Raifeartaigh

Dublin Institute for Advanced Studies

Dublin 4, Ireland

1. Introduction.

The most important development of the last decade in particle physics and field theory has undoubtedly been the advent of hidden-symmetric gauge theories[1]. The success of these theories in describing the weak and electromagnetic interactions has led to the belief that all the fundamental physical interactions are based on the hidden-symmetric gauge principle[2]. One of the more interesting by-products of this development has been the discovery that hidden-symmetric gauge theories admit static solutions to the field equations which are regular everywhere and for which the energy is finite[3]. Such solutions will be called solitons. This discovery provides a natural escape from a difficulty first pointed out by Derrick[4], namely, that for scalar fields alone, such solutions cannot exist in more than one space dimension. The hidden-symmetric gauge solutions exist for n space dimensions, where $1 \leq n \leq 4$, and the purpose of this lecture is to describe the static soliton solutions for $2 \leq n \leq 4$. To put the solutions in perspective, however, a section is first devoted to a description of the hidden-symmetric gauge theories and their role vis-à-vis the fundamental interaction.

2. Hidden-Symmetric Gauge Theories.

Let M(x) be an m-dimensional differentiable manifold with coordinates $x = x_k$, $k = 1 \ldots m$ and G be a compact connected Lie group. The ensemble consisting of a copy of G at each point x of M(x) is a principal G-bundle[5] over M(x), and the group G from which such a bundle is generated will be called a gauge-group. Let $A_\mu dx^\mu$ be a 1-form[5] on M(x) with values in the Lie algebra of G. The connection A_μ which maps the contravariant vectors dx^μ into the Lie algebra, will be called the gauge-potential, and it has the group transformation property

$$A_\mu \rightarrow G^{-1} A_\mu G + G^{-1} \nabla_\mu G \tag{2.1}$$

The curvature[5] of the 1-form, defined as

$$F_{\mu\nu} = \nabla_\mu A_\nu - \nabla_\nu A_\mu + [A_\mu, A_\nu] \tag{2.2}$$

will be called the gauge-field, and it has the covariant group transformation property

$$F_{\mu\nu} \rightarrow G^{-1} F_{\mu\nu} G \tag{2.3}$$

For any continuous unitary representation U(g) of G, the covariant derivative is defined as

$$D_{\mu} = \nabla_{\mu} + U(A_{\mu}) \tag{2.4}$$

and has the covariant group transformation property

$$D_{\mu} \rightarrow U(g)^{-1} D_{\mu} U(g), \qquad g \in G \tag{2.5}$$

This formalism is well-known to mathematicians and the question is why it should be of interest to physicists. The answer is that the formalism may be used to convert global symmetries to local (gauge) symmetries in a universal manner. More precisely, if $L(\nabla_{\mu}\phi, \phi, \nabla_{\mu}\psi, \psi)$ is a local Lagrangian density which is invariant with respect to a Lie group G, then the Lagrangian density

$$L(D_{\mu}\phi, \phi, D_{\mu}\psi, \psi) + \tfrac{1}{4}(F_{\mu\nu}, F_{\mu\nu}) \tag{2.6}$$

is invariant with respect to the local groups G(x) of the principal bundle. Here the bracket in the second term denotes inner product in the adjoint-representation space, and the second term is added to provide kinetic energy for the gauge-potential A_{μ} which appears without derivative in D_{μ}. On expansion, one finds from (2.2) that

$$(F_{\mu\nu}, F_{\mu\nu}) = 2(\nabla_{\mu}A_{\nu} - \nabla_{\nu}A_{\mu}, \nabla_{\mu}A_{\nu}) + 4([A_{\mu},A_{\nu}],\nabla_{\mu}A_{\nu}) + ([A_{\mu},A_{\nu}],[A_{\mu},A_{\nu}]) \tag{2.7}$$

which shows that for non-abelian groups G, the kinetic term induces a cubic and quartic self-interaction for the gauge field (with the cubic term linear in the first derivative). Lagrangians of the form (2.6) will be referred to as gauge-invariant Lagrangians and the question then is why physicists should be interested in them. The answer is that there is now some reason to believe that all four fundamental physical interactions - gravitational, weak, electromagnetic and strong - can be described by gauge-invariant Lagrangians. Let us consider these interactions in turn.

Gravitational. Einstein's gravitational theory has been known to be a gauge theory since its inception[6]. The principal bundle is identified as the group of local coordinate transformations at each point x of (3+1)-dimensional Riemannian space R(3, 1) so that G = GL(4,R), and the gauge-potential A_{μ} is identified as $\Gamma_{\mu\ \sigma}^{\ \lambda}\,\tau^{\sigma}_{\ \lambda}$

where $\Gamma^{\lambda}_{\mu\sigma}$ is the affine connexion[7] for R(3, 1) and τ^{σ}_{λ} are the generators of GL(4,R). However, Einstein's theory is more profound than the gauge theories defined above, because the group transformations are implemented on the manifold M(x) itself, and the connexion $\Gamma^{\lambda}_{\mu\sigma}$ is derived from the Riemannian metric[7], $\Gamma^{\lambda}_{\mu\sigma} = \{^{\lambda}_{\mu\sigma}\}$ where the bracketed quantities are the Christoffel symbols.

Electromagnetic. The electromagnetic interaction is well-known to be a gauge theory of the form (2.6) for the special abelian group G = U(1). In fact, for G = U(1) the Lagrangian (2.6) is just the standard Lagrangian for the interaction[8] of the electromagnetic vector potential A_μ with scalar and fermion fields.

Weak. The big discovery in recent years is that a satisfactory massive, renormalizable theory of the weak interactions (or more precisely of the weak and electromagnetic interactions) can be constructed by using gauge theory[9]. To account for presently known phenomena the group G = U(2) suffices, but the ultimate gauge group G may be larger. Of the four gauge-potentials belonging to U(2), one is identified with the electromagnetic vector potential, and two others with the charged intermediate vector bosons[9][10] of the weak interactions. The success of the weak interaction theory depends, however, not on the gauge symmetry alone but on the fact that the symmetry is hidden, a concept that will be discussed below.

Strong. There is no direct evidence for the hidden-symmetric gauge-character of the strong interactions, but there is a certain amount of indirect evidence[11] - the existence and confinement of colour, the broken nature of the strong interactions, the existence of solitons, and, of course, the belief that the strong interactions should fall into line with the other three. If the strong interactions do fall into line and all four fundamental interactions are based on the gauge principle, it would be tempting to go a step further and suppose that all four interactions are simply different manifestations of a single hidden-symmetric gauge theory. But for the moment this supposition is no more than speculation.

The concept of hidden symmetry[12] is vital for the success of gauge theory in describing the weak interactions. It is also vital for describing the phenomenon of superconductivity in electromagnetism and for the existence of soliton solutions of the field equations. It is formally defined as follows: Let $V(\phi)$ be the potential for the scalar fields ϕ in the Lagrangian (2.6), that is, let $-V(\phi)$ be the part of L which survives when $D_\mu\phi$, $D_\mu\psi$ and ψ are set equal to zero. Then $V(\phi)$ is invariant with respect to the gauge group G

$$V(\phi) = V(U(g)\phi) \qquad g \in G \tag{2.8}$$

where U(g) is the representation of G to which ϕ belongs. The symmetry is said to be hidden (spontaneously broken) if the minimum of $V(\phi)$ occurs for some value $\mathring{\phi}$ of ϕ which is not group invariant

$$U(g)\mathring{\phi} \neq \mathring{\phi} \tag{2.9}$$

for every g in G. In other words, a hidden symmetry is a symmetry of the Lagrangian which is not a symmetry of the potential minimum (of the vacuum in second quantized theory [12]). The subgroup H of G for which

$$U(h)\mathring{\phi} = \mathring{\phi} \qquad h \in H < G \tag{2.10}$$

that is, the stability or little group of $\mathring{\phi}$, is called the residual symmetry group of the Lagrangian.

For particle physics the great advantage of hidden symmetry is that it generates masses for the fields in the Lagrangian, particularly the gauge-fields, without destroying the formal gauge-invariance (and hence without destroying the renormalizability [13]). To see how this happens for the gauge fields, consider the conventional kinetic term $\frac{1}{2}(D_\mu\phi, D_\mu\phi)$ for the scalar fields ϕ. When the symmetry is hidden the 'true' field is $\theta = \phi - \mathring{\phi}$ and hence the kinetic terms become

$$\tfrac{1}{2}(D_\mu\theta, D_\mu\theta) + \tfrac{1}{2}(U(A_\mu)\mathring{\phi}, U(A_\mu)\mathring{\phi}) + (D_\mu\phi, U(A_\mu)\mathring{\phi}) \tag{2.11}$$

Since $\mathring{\phi}$ is a constant vector, the second term in (2.11) is a mass-term for the A-fields, and one sees that all the A-fields acquire masses except those for which $U(A_\mu)\mathring{\phi}$ is zero, that is, except those which belong to the residual symmetry group H. The third term in (2.11) is an induced interaction which plays an important role in the renormalization of the theory [13].

In electromagnetism hidden symmetry is responsible for the phenomenon of superconductivity [14] because ϕ contributes a part

$$j_k = e(\phi^*\nabla_k\,\phi - \phi\nabla_k\,\phi^*) + e^2A_k\,\phi^*\phi \tag{2.12}$$

to the electromagnetic current. Hence, if $\phi \cong \mathring{\phi}$ we have

$$j_k = K\,A_k \qquad \text{where} \quad K = e^2\mathring{\phi}^*\mathring{\phi} \neq 0 \tag{2.13}$$

and (2.11) is just the London [15] condition for superconductivity. In particular (2.13) admits a non-zero current for a zero electric field. In the superconducting case the field $\phi(x)$ is the density of Cooper pairs, and $V(\phi)$ is the Landau-Ginzburg potential [14].

3. Soliton Solutions of the Field Equations.

We turn now to the by-product of hidden-symmetric gauge theory discussed in the introduction, namely, that the field equations admit solutions for which the energy is finite. For simplicity the fermion fields and the electric (time) component A_0

of A_ν will be neglected, and the remaining fields will be assumed to be static. The Hamiltonian is then just the negative of the Lagrangian, and takes the form

$$H = H(F) + H(\phi) + H(v) = \int d^n x \{\tfrac{1}{4}(F_{k\ell}, F_{k\ell}) + \tfrac{1}{2}(D_k\phi, D_k\phi) + V(\phi)\} \quad k,\ell = 1\ldots n \tag{3.1}$$

where $n = m-1$ is the number of space-dimensions. The field equations are

$$D_k F_{\ell k} = (\phi, \tau D_\ell \phi) , \qquad D^2 \phi = \frac{\partial V}{\partial \phi} \tag{3.2}$$

where τ is the generator of G in the direction of F and the bracket denotes inner product in the space of the representation U(g) of G to which ϕ belongs. Soliton solutions of (3.2) will be defined as solutions for which the Hamiltonian (3.1) is finite $(0 < H < \infty)$.

A preliminary guide to the existence of soliton solutions can be obtained by considering the variation of H with respect to the scale transformations $\phi(x) \to \phi(x|\lambda)$, $A(x) \to \lambda A(x|\lambda)$. One finds by inspection that

$$\left(\frac{dH}{d\lambda}\right)_{\lambda=1} = (n - 4)\, H(F) + (n - 2) H(\phi) + n H(v) \tag{3.3}$$

Since the variation must vanish for any solution of the field equations, and H(F), $H(\phi)$ and H(v) are all positive one sees that the only possibilities for non-trivial solutions are the following:

$n = 1$ (in this case F is identically zero)

$\left.\begin{matrix} n = 2 \\ n = 3 \end{matrix}\right\}$ both F and ϕ must be non-zero

$n = 4$ the scalar field ϕ must be zero.

The case $n = 1$ is the usual case[16] of scalar-field solitons in one space-dimension. Derrick's observation[4] that these do not exist in more than one dimension may be verified directly from (3.3) by setting H(F) equal to zero. The case $n = 2, 3$ are the soliton solutions which are admitted by hidden gauge symmetry. The case $n = 4$ is a special case in which the 'Hamiltonian' (3.1) is actually the Euclidean action. The scalar fields are zero in this case so the gauge symmetry is not hidden.

Let us first consider the cases $n = 2, 3$. The reason that nontrivial solutions may exist in these cases is that, although the convergence of the integral in H(v) demands that $V(\phi) \to 0$ as $r \to \infty$, the hidden symmetry allows $\phi(x)$ to remain finite in this limit. In fact, since V(x) takes its minimum value zero for a non-zero value c^2 of the invariant (ϕ, ϕ), the condition $V(\phi) \to 0$ as $r \to \infty$ demands only that

$$\phi(r,\Omega) \to c\, \varepsilon(\Omega) \quad \text{where } c \neq 0 \text{ and } (\varepsilon(\Omega), \varepsilon(\Omega)) = 1 \quad \text{as } r \to \infty \tag{3.4}$$

Here Ω denotes the polar angles and $\varepsilon(\Omega)$ is a single-valued function of Ω with values in the Lie algebra. If $\varepsilon(\Omega)$ is not constant (or gauge-equivalent to a constant) a solution of the field equations with the boundary condition (3.4) will have a non-zero value of H. Thus the hidden symmetry allows the boundary condition (3.4), which prevents the energy from collapsing to zero.

The question is whether there exist any $\varepsilon(\Omega)$ which are not gauge-equivalent to constants, and whether there exist any solutions of the field equations with boundary values $\varepsilon(\Omega)$. We consider first the existence of non-trivial $\varepsilon(\Omega)$, or more precisely, the problem of classifying the $\varepsilon(\Omega)$ into gauge-inequivalent classes. This problem is solved[17] by noting that the convergence of the integral $H(\phi)$ in (3.1) requires that

$$[L + U(B(\Omega))]\,\varepsilon(\Omega) = 0 \quad \text{where} \quad B(\Omega) = \underset{r=\infty}{\mathrm{Lt}}\; r \times A \tag{3.5}$$

and L is the angular-momentum operator, and hence, by integration that

$$\varepsilon(\Omega) = U(g(\Omega))\,\varepsilon(0) \quad \text{where} \quad g(\Omega) \in G. \tag{3.6}$$

From eq. (3.6) it follows that $\varepsilon(\Omega)$ defines a continuous single-valued map of the unit sphere S_{n-1}, parametrized by Ω, into the space $G|H$ parametrized by $g(\Omega)|h$ where $h \in H$ and H is the little group of $\varepsilon(0)$. Hence the problem of classifying the in-equivalent $\varepsilon(\Omega)$ reduces to the problem of classifying the inequivalent maps $S_{n-1} \rightarrow G|H$. The latter problem is well-known to the mathematicians[18] as the problem of finding the (n-1)th homotopy group $\pi_{n-1}(G|H)$ of $G|H$, and has been solved for most groups of interest. For example, for $n = 2$ and $G|H = G = U(1)$ (electrodynamics in two space dimensions) the mapping $S_{n-1} \rightarrow G|H$ reduces to the mapping of the unit circle into itself, $S_1 \rightarrow S_1$, and the inequivalent maps correspond to the number of times the image circle is covered when the original circle is covered once. Thanks to the single-valuedness of the mapping, the number of times will be an integer $N(\varepsilon)$, and this integer serves to classify the inequivalent $\varepsilon(\Omega)$. Similarly for $n = 3$ and $G|H = SU(2)|U(1) = S_2$ the mapping $S_{n-1} \rightarrow G|H$ reduces to the mapping of the unit sphere in 3 dimensions into itself $S_2 \rightarrow S_2$, and the inequivalent mappings are characterized by an integer $N(\varepsilon)$ which counts the number of times that the image sphere is covered. In general, on account of the single-valuedness of $\varepsilon(\Omega)$ the inequivalent maps $S_{n-1} \rightarrow G|H$ are characterized by integers $N(\varepsilon)$ and these integers provide the required classification of inequivalent $\varepsilon(\Omega)$. In simple cases the integers $N(\varepsilon)$ can be expressed as explicit functionals of the $\varepsilon(\Omega)$. For example for the cases $n = 2$, $G|H = U(1)$ and $n = 3$, $G|H = SU(2)|U(1)$ above one has

$$N(\varepsilon) = \frac{1}{4\pi} \int_0^{2\pi} d\phi\{\varepsilon_1(\phi)\partial_\phi\, \varepsilon_2(\phi) - \varepsilon_2(\phi)\partial_\phi\, \varepsilon_1(\phi)\} \tag{3.7}$$

and

$$N(\varepsilon) = \frac{1}{4\pi} \int_{S_2} d\Omega \; (\varepsilon(\Omega), \vec{L\varepsilon}(\Omega)_\wedge \vec{L\varepsilon}(\Omega))\cdot x, \qquad \hat{x} = \vec{x}/r \tag{3.8}$$

respectively, where the bracket and wedge in (3.8) denote inner and cross-product in SU(2)-space. The integers $N(\varepsilon)$ are often called topological constants, and they are superselection operators[19].

Since the integers $N(\varepsilon)$ characterize the inequivalent $\varepsilon(\Omega)$, they characterize the inequivalent boundary conditions at $r \to \infty$ for the field equations (3.2). Hence the $N(\varepsilon)$ effectively characterize the soliton solutions, and it is not surprising that they have an important physical significance. For example, for the n = 2, $G|H = U(1)$ case above, if one uses the expression (2.12) for the electromagnetic current, together with the single-valuedness of $j(x)$ and Stokes theorem, one finds from (3.7) that

$$N(\varepsilon) = \frac{e}{2\pi} \int d^3x(\nabla \times A) \tag{3.9}$$

This equation shows that the magnetic flux is quantized in units of $2\pi|e$ and that $N(\varepsilon)$ is the number of flux quanta. Similarly in the case n = 3, $G|H = SU(2)|U(1)$ above, it can be shown[3] that

$$N(\varepsilon) = \frac{e}{4\pi} \int d^3x \left(\vec{\nabla} \cdot \vec{B}\right) \tag{3.10}$$

where $\vec{B}$ is the magnetic field. Eq. (3.10) shows that the magnetic charge is quantized in units of $4\pi|e$ and $N(\varepsilon)$ is the number of charge quanta. Because of the physical significance of $N(\varepsilon)$, the soliton solutions for n = 2 and n = 3 are called vortex and monopole solutions respectively.

For n = 4 the situation is somewhat different because there is no scalar field and hence no boundary condition $\varepsilon(\Omega)$. Furthermore the 4 dimensions n = 4 are considered not as the space-part of a 4+1-dimensional Minkowski space but as the Euclidean (imaginary time) continuation of the usual 3+1-dimensional Minkowski space. The 'Hamiltonian', which in this case is actually the Euclidean action, reduces to

$$H(F) = \int d^4x \; F_{k\ell} \, F_{k\ell} \qquad k,\ell = 1\ldots4 \tag{3.11}$$

and the boundary condition at $r \to \infty$ is simply $r^2F \to 0$. The reason that solitons can exist in this case is that the condition $r^2F \to 0$ demands for A only that

$$A_k(x) \to U^{-1}(g(\Omega)) \, \nabla_k \, U(g(\Omega)) \qquad g(\Omega) \in G, \tag{3.12}$$

and whereas in Euclidean space, or the unit sphere in 2 and 3 dimensions, (3.12) would imply that A(x) is gauge equivalent to zero, this is not true for the unit sphere S_3 in 4 dimensions. That is, there exist maps $g(\Omega)$ from S_3 to G which are not gauge-equivalent to the trivial map[5][20][21] $(\pi_3(G) \neq 0)$. These maps can be characterized by a topological constant N(F) which takes the form

$$N(F) = \int d^4x \, F_{k\ell} \tilde{F}_{k\ell} \quad \text{where} \quad \tilde{F}_{k\ell} = \tfrac{1}{2}\varepsilon_{k\ell st} F_{st} \tag{3.13}$$

and $\varepsilon_{k\ell st}$ is the Levi-Civitā symbol. Using the Schwarz inequality one has from (3.11) and (3.13)

$$H(F) \geq N(F) \tag{3.14}$$

which shows that if N(F) is non-zero the action cannot collapse to zero.

Having settled the question of boundary conditions at $r \to \infty$ for n = 2, 3, 4 one should consider the question as to whether there exist any solutions of the field equations with the required boundary conditions and with $H < \infty$. We shall consider only the case for which the gauge group is SO(n), where n is the space-dimension.

First let us consider the cases n = 2, 3 and make the Ansatz

$$A(x) = a(\Omega)\left(\frac{K(r)-1}{r}\right) \qquad \phi(x) = \varepsilon(\Omega)\left(\frac{S(r)}{r}\right) \tag{3.15}$$

where K(r) and H(r) are scalar functions and the denominator 1/r is inserted for convenience. Inserting this Ansatz in the field equations (3.2) one finds[22] that $a(\Omega)$ is determined uniquely to be

$$a_k^{\alpha\beta}(\Omega) = (\delta_{\alpha k}\, x_\beta - \delta_{\beta k}\, x_\alpha) / r \qquad \alpha,\ \beta,\ k = 1 \ldots n \tag{3.16}$$

If the function $\varepsilon(\Omega)$ is assumed to belong to the n-dimensional representation of SO(n) it also is completely determined, and it takes the form

$$\varepsilon_k(\Omega) = x_k/r. \tag{3.17}$$

Note that (3.17) corresponds to the unit map $N(\varepsilon) = 1$. Using the expressions (3.16) and (3.17) the field equations (3.2) reduce[3][22] to the coupled, non-linear, but <u>ordinary</u> differential equations

$$r^2 \frac{d^2K(r)}{dr^2} + (n-3)r\frac{dK(r)}{dr} = S^2K + (n-2)K(K^2-1) \tag{3.18}$$

$$r^2 \frac{d^2S(r)}{dr^2} + (n-3)r\frac{dS(r)}{dr} = (n-1)K^2S + r^4\frac{\partial V}{\partial S} \tag{3.19}$$

for the scalar functions K(r) and S(r). The boundary conditions for these equations are

$$K(0) = 1, \qquad K(r) \to 0, \quad S(r) \to cr \quad \text{as } r \to \infty \tag{3.20}$$

the boundary conditions at $r \to \infty$ following from (3.4) and (3.5). The equations (3.18), (3.19) can be derived from the Hamiltonian

$$H(K,S) = \int_0^\infty r^{n-3} dr \left\{ \frac{1}{(4-n)} \left[K'^2 + \frac{K^2 S^2}{r^2} \right] + (n-2)\frac{(K^2-1)^2}{2r^2} + \frac{(S'-S|r)^2}{2} + r\,V(S) \right\} \tag{3.21}$$

This Hamiltonian can be obtained by inserting the Ansatz (3.15) (3.16) (3.17) in the Hamiltonian (3.1), but the insertion is legitimate only because it has previously been verified that the Ansatz satisfies the field equations.

The problem then reduces to showing that eqns. (3.18) (3.19) have solutions which satisfy the boundary conditions (3.20) and for which the Hamiltonian (3.21) is finite. A rigorous proof that such solutions exist has been given by Fateev et al.[(23)]. The method of proof is to construct a Sobelev Hilbert-space Σ with the kinetic part of the Hamiltonian and show that any sequence of test-functions (K_r, S_r), $r = 1\ldots\infty$ for which $H(K_r, S_r) \to \inf H$, contains a subsequence which converges to a limit (K,S) in Σ, and that $H(K,S) = \inf H$. The proof relies heavily on the fact that each term in (3.21) is separately positive.

For n = 3, if the potential V is allowed to go to zero (after the boundary condition $S(r) \to cr$ has been extracted) there is an exact solution[(24)] of equations (3.18) (3.19), namely,

$$K(r) = \frac{cr}{\sinh cr} \qquad\qquad S(r) = \frac{cr}{\coth cr} - 1\,, \tag{3.22}$$

but no exact solutions are known for n = 2.

For n = 4 there is no ϕ-field and the field equation for the gauge field is

$$D_k\, F_{\ell k} = 0 \tag{3.23}$$

By making the Ansatz (3.15) and (3.16) for the A-field (except that α, β, k now run from 1 to 4) these equations reduce to the ordinary non-linear equation

$$r^2 \frac{dK(r)}{dr^2} + r\frac{dK(r)}{dr} = 2K(K^2-1) \tag{3.24}$$

with the boundary condition $K(r) \to -1, r \to \infty$, which follows from (3.12). Eq. (3.24) is the analogue of (3.18) in 4-dimensions, and it is easy to verify that it admits the exact solution

$$K(r) = \left(\frac{a^2 - r^2}{a^2 + r^2} \right) \tag{3.25}$$

where a is an arbitrary constant. This solution satisfies the boundary condition and is regular everywhere. Furthermore, one easily sees that $H(F) = N = 1$. The solution (3.25) was actually the first solution found for $n = 4$[21].

In general, however, the solutions for $n = 4$ are not found by making Ansätze such as (3.15) but by noting from (3.11) (3.13) and (3.14) that the minimum value of $H(K)$ is reached if, and only if,

$$F_{k\ell} = \pm\tilde{F}_{k\ell} \tag{3.26}$$

and hence that any solution of (3.26) will satisfy the field equations (3.23). Indeed, the only solutions of the field equations which are of interest are those which satisfy (3.26), since only these correspond to the absolute minimum of $H(K)$. Thus the problem reduces to solving the system (3.26). The advantage of (3.26) is two-fold. First, (3.26) are a set of first-order differential equations, and hence are much easier to solve than the field equations (3.23). In fact, various classes of solutions of (3.26) have already been found[25]. Second, if the operators D_k and $F_{k\ell} \pm \tilde{F}_{k\ell}$ are regarded as two consecutive operations of a Lie algebraic complex[20], then the number of solutions of (3.26) corresponding to each group G and each value of the topological constant $N(F)$ can be found by using the Atiyah-Singer index theorem[26].

The soliton solutions of (3.26) are called instantons or pseudo-particles, and although their physical meaning is not as clear as that of the vortices or monopoles, their contribution to the action integral can be interpreted as the amplitude for tunnelling between different gauge-vacua[27]. (Recall that in quantum mechanics tunnelling can be interpreted formally as motion in imaginary time.) The instantons are also thought to be connected with colour confinement, in particular with quark confinement[28].

References

(1) S. Weinberg, Scientific American, July (1974).
J. Bernstein, Rev. Mod. Phys. 46, 7 (1974).

(2) S. Weinberg, Rev. Mod. Phys. 46, 255 (1974).
L. Faddeev, JETP Lett. 21, 141 (1975).

(3) H. Nielsen, P. Oleson, Nucl. Phys. B61, 45 (1972).
A. Polyakov, JETP Letters 20, 194 (1974).
G. 't Hooft, Nucl. Phys. B79, 276 (1974).

(4) G. Derrick, J. Math. Phys. 5, 1252 (1964).

(5) W. Creub, S. Halperin, R. Vanstone, Connections, Curvature and Cohomology (Academic Press, 1972).

(6) A. Einstein, The Meaning of Relativity (Princeton Univ. Press 1953).
H. Weyl, Space-Time-Matter (Methuen London, 1922).

(7) E. Schrödinger, Space-Time Structure (Cambridge Univ. Press 1950).
J. L. Synge, Relativity - the General Theory (North-Holland, 1966).

(8) S. Schweber, Introduction to Relativistic Quantum Field Theory (Row-Peterson, 1961).

(9) J. C. Taylor, Gauge Theories of Weak Interactions (Cambridge Univ. Press 1976).
E. Abers, B. Lee, Phys. Reports. 9C, No. 1 (1973).

(10) R. Marshak, Riazzudin, C. Ryan, Theory of Weak Interactions (Wiley-Interscience, N.Y. 1969).

(11) J. Iliopoulos, Introduction to Gauge Theories, CERN Report 76-11 (1976).
S. Glashow, Ecole d'Eté de Physique de Particules, Gif-sur-Yvette, Inst. Nat. Phys. Nucléaire 1-N2-P3 (1975).

(12) S. Coleman, Secret Symmetry, Proc. Erice Summer School 1973.

(13) G. 't Hooft, Nucl. Phys. B33, 173, B35, 167 (1971).
G. 't Hooft, M. Veltman, Diagrammar, CERN Report 73/9 (1973).

(14) A. Fetter, J. Walecka, Quantum Theory of Many-Particle Systems (McGraw-Hill, 1971).

(15) F. London, H. London, Proc. Roy. Soc. A147, 71 (1935).

(16) A. Scott, F. Chu, D. McLaughlin, Proc. IEEE 61, 1443 (1973).

(17) M. Monastyrski, A. Perelmov, JETP Lett. 21, 94 (1975).

(18) J. Milnor, Topology (Univ. of Virginia Press, 1965).
S. T. Wu, Homotopy Theory (Academic Press, 1959).

(19) J. Arafune, P. Freund, C. Goebbel, J. Math. Phys. 16, 433 (1975).

(20) D. Husemoller, Fibre Bundles (Springer, 1966).

(21) A. Belavin et al., Phys. Lett. 59B, 85 (1975).

(22) L. Michel, L. O'Raifeartaigh, K. C. Wali, Phys. Letters 67B, 198 (1977).

(23) Y. Tyupkin, V. Fateev, A. Schwartz, Theor. Math. Phys. 26, 270 (1976).

(24) M. Prasad, C. Sommerfield, Phys. Rev. Lett. 35, 760 (1975).

(25) E. Witten, Phys. Rev. Lett. 38, 121 (1977).
G. 't Hooft (Harvard), F. Wilczek (Princeton), F. Corrigan, D. Fairlie (Durham), A. Karpf (Berlin), preprints.
R. Jackiw, C. Nohl, C. Rebbi, Phys. Rev. D15, 1642 (1977).

(26) P. Gilkey, The Index Theorem and the Heat Equation (Publish or Perish Inc., Boston, 1974).

(27) R. Jackiw, C. Rebbi, Phys. Rev. Lett. 37, 172 (1976).
G. 't Hooft, Phys. Rev. Lett. 37, 8 (1976).
C. Callan, R. Dashen, D. Gross, Phys. Lett. 63B, 334 (1976).

(28) A. Polyakov, Phys. Lett. 59B, 82 (1975).

Geometry of Yang-Mills fields

M.F. Atiyah

§1 Introduction

In this talk I shall explain how information about classical solutions of Yang-Mills equations can be obtained, rather surprisingly, from algebraic geometry. Although direct physical interest is restricted to the case of four dimensions I shall begin by discussing the two-dimensional case. Besides preparing the ground for the four-dimensional problem this has independent mathematical (and possibly physical) interest, and very complete results can be obtained.

I start by recalling the mathematical framework of the Yang-Mills equations in any number of dimensions (cf. also the lecture of Zakharov). We fix a compact connected Lie group G (for example SU(2)), a base manifold X of dimension n and a principal fibre bundle P over X with group G. We suppose moreover that X is oriented and has a Riemannian (or pseudo-Riemannian) metric, so that the Hodge $*$-operator is defined on differential forms: $*$ transforms a q-form into its "dual" (n-q)-form and $*^2 = \pm 1$. All this data being fixed we consider a connection A for the G-bundle P with curvature F. A is the "potential" of the Yang-Mills field F. In local coordinates A is given by a 1-form with values in the Lie algebra L(G) and F is a 2-form with values in L(G). By $*F$ we denote the corresponding (n-2)-form with values in L(G). Then as a Lagrangian density we take $F_\wedge *F$ where the exterior product of forms is combined with the Killing form in L(G) to produce an n-form on X. The Yang-Mills equations for A are then the associated Euler-Lagrange equations. They are second-order non-linear partial differential equations.

If X is actually compact Riemannian then

$$- \int_X F_\wedge *F = \|F\|^2$$

is the square of the natural L^2-norm of F, and the absolute minimum of $\|F\|^2$ (if attained) gives a solution of the Yang-Mills equations. Other solutions will arise from critical points of higher Morse index, and it is natural to examine the Yang-Mills equations from the point of view of Morse theory.

The standard Yang-Mills equations of physics arise when X is

Lorentz-space, but if we pass to imaginary time X becomes Euclidean 4-space R^4. Moreover, in dimension 4, the Yang-Mills equations are conformally invariant, hence R^4 can be replaced by the 4-sphere S^4 provided we impose suitable decay at infinity as a requirement for our solutions. This is the situation we shall pursue in §3.

§2 The two-dimensional case

The results described in this section have been obtained jointly with R. Bott and the full details are still being worked out.

In two-dimensions the Yang-Mills equations reduce to the assertion that $*F$, which is a section S of the adjoint vector bundle L(P) with fibre L(G), is covariant constant. This implies that the holonomy of the bundle (at a point $x_0 \in X$) is contained in the subgroup H of G_0 which centralizes $S(x_0)$ (here G_0 is the copy of G associated to x_0 so that $S(x_0) \in L(G_0)$). Moreover if T is the torus which is the closure of the one-parameter group $\exp(ts(x_0)) \subset H$, then the curvature lies in L(T) and satisfies the Yang-Mills equation for the <u>abelian</u> group T.

Thus solutions of the Yang-Mills equations for G are completely described by giving:

a) solutions of the (abelian) Yang-Mills equations for some torus $T \subset G$;

and b) a homomorphism $\pi_1(X) \to C(T)/T$, where C(T) is the centralizer of T in G and $\pi_1(X)$ is the fundamental group of X.

For example if X is the 2-sphere S^2 with its standard metric and G = SU(2) then (b) is vacuous and dim T = 0 or 1. The first case corresponds to the zero solution, the absolute minimum of $\|F\|^2$, while the second corresponds to a U(1)-solution which is characterized by an integer $k \geqslant 1$ (the Chern class). This description of the critical points of $\|F\|^2$ is precisely what the Morse theory (properly interpreted to take account of symmetries) predicts. Moreover it is essentially the same picture as occurs in the theory of geodesics on G. In a slightly different form it appears in a physical context in [5] where S^2 is the "sphere at infinity" in 3-space.

If X is a compact Riemann surface of genus $g \geqslant 2$ (and G = SU(2) for simplicity) then (b) becomes relevant when T is trivial and gives a homomorphism $\pi_1(X) \to SU(2)$. This connects up with the theory of stable holomorphic bundles on curves [6] and yields an interesting new approach.

I shall now explain a simple but significant relation between G-bundles P with connection A over a Riemann surface X and holomorphic bundles Q with group G^c (the complexification of G). Given P,A we can form the associated bundle P^c over X with group G^c and connection A^c. We can give P^c an almost complex structure J by using the complex structure of G^c along the fibres and the complex structure of X in the horizontal directions given by A. Then J satisfies the integrability condition, which by the Newlander-Nirenberg theorem [7] ensures the existence of an associated complex structure. Like J this complex structure is G^c-invariant and so P^c becomes a holomorphic G^c-bundle over X. Moreover the connection A is uniquely defined from the holomorphic structure of P^c and the reduction P. Thus G-bundles with connection are (up to isomorphism) equivalent to holomorphic G^c-bundles with a reduction to G. For example if G = U(n), G^c = GL(n,C) and U(n)-bundles with connection are equivalent to holomorphic n-dimensional vector bundles with hermitian metric.

The equivalence just described does not involve the Yang-Mills equation which comes in as an additional condition on the pair consisting (in case G = U(n)) of a holomorphic vector bundle and a hermitian metric. When $X = S^2$ one can view the Morse flow associated to $\|F\|^2$ as keeping the holomorphic structure fixed and decreasing the metric. The critical points of $\|F\|^2$ then appear as constrained minima corresponding to the classes of holomorphic bundles as given by Grothendieck's theorem [5]. When X has higher genus the situation is more complicated because holomorphic bundles have "moduli", but the essential picture persists and its details are being explored.

§3 The four-dimensional case

I turn now to consider the Yang-Mills equations for the 4-sphere S^4 with its standard metric. As mentioned earlier this is equivalent (by conformal invariance) to working in the Euclidean 4-space R^4 with suitable decay at infinity.

First we decompose the curvature F into its self-dual and anti-self-dual parts, i.e. we put $F = F^+ \oplus F^-$ where $F^{\pm}$ are the ($\pm$1)-eigenspaces of * (note that $*^2 = 1$ in R^4, whereas $*^2 = -1$ in Minkowski-space). Then we have

$$\|F\|^2 = \|F^+\|^2 + \|F^-\|^2$$

On the other hand the topological type of our SU(2)-bundle P over

S^4 is determined by an integer k and this can be computed, for any connection, by

$$8\pi^2 k = \|F^+\|^2 - \|F^-\|^2$$

It follows that $\|F\|^2 \geqslant 8\pi^2|k|$ and equality holds if and only if $F^+ = 0$ or $F^- = 0$ (depending on the sign of k). Thus for $k \geqslant 0$ the absolute minimum of $\|F\|^2$ is given by solutions of $F^- = 0$, while $F^+ = 0$ corresponds to $k \leqslant 0$. Explicit solutions have been constructed for all values of k [3][8] but these are not the most general solutions, as is shown by a parameter count [1][9][11].

I shall now explain briefly how the most general solution of the anti-self-dual Yang-Mills equation $F^+ = 0$ can be obtained from algebraic geometry. The basic idea comes from the Penrose twistor programme [10] and is due to R.S. Ward [12]. An account is given in [2]. One way to motivate our procedure is to recall that in 2-dimensions complex variable theory enters quite naturally (see §2). In R^4 we could introduce 2 complex variables, identify R^4 with $\mathbb{C}^2$ but there is no obvious natural choice, i.e. we could take $X_1 + iX_2$, $X_3 + iX_4$ or equally $X_1 + iX_3$, $X_2 + iX_4$ etc. If we ask for all ways of identifying R^4 with $\mathbb{C}^2$ linearly which are compatible with the Euclidean metric and orientation, this is equivalent to looking for an orthogonal 4×4 matrix J with $\det J = 1$ and $J^2 = -1$ (so that J gives multiplication by i). The set of all such J form the homogeneous space $SO(4)/U(2) \cong S^2$ and so can be parametrized by a complex variable u (including ∞). Thus we are led to introduce 3 complex variables: first u and then 2 further complex variables which are the complex coordinates of R^4 in the complex structure determined by u. More geometrically we have a complex structure on $S^2 \times R^4$. Note that this is <u>not</u> the complex product $S^2 \times \mathbb{C}^2$: it is in fact a non-trivial complex vector bundle over S^2. If we compactify R^4 to S^4 then $S^2 \times R^4$ gets naturally compactified to the complex projective 3-space $P_3(\mathbb{C})$. This is mapped onto S^4 with fibres which are complex projective lines.

Having introduced our complex coordinates in this fashion we now return to G-bundles P with connection A over S^4. We can lift (P,A) to give a G-bundle $\tilde{P}$ over $P_3(\mathbb{C})$ with connection $\tilde{A}$. Complexifying we get a G^c-bundle $\tilde{P}^c$ with connection $\tilde{A}^c$. Just as in §2 (since both base $P_3(\mathbb{C})$ and group G^c are complex manifolds) the connection $\tilde{A}^c$ defines an <u>almost complex structure</u> on $\tilde{P}^c$. Unlike the 2-dimensional case of §2 however the integrability of this almost complex structure is not automatic: the condition is that the curvature $\tilde{F}^c$ of $\tilde{A}^c$ should be a $L(G^c)$-valued 2-form of type (1,1) (i.e. involving locally only terms of the form $dZ^\alpha \wedge d\bar{Z}^\beta$ in the 3 complex coordinates Z^α of $P_3(\mathbb{C})$: this condition is clearly automatic when there is only one complex coordinate). Now $\tilde{F}^c$ is just the lift of the complexification of F on S^4, and one verifies that

$$\tilde{F}^c \text{ is of type } (1,1) \Leftrightarrow F^+ = 0.$$

Hence solutions of $F^+ = 0$ correspond to certain holomorphic G^c-bundles over $P_3(\mathbb{C})$. For $G = SU(2)$, $G^c = SL(2,\mathbb{C})$ and we end up with holomorphic 2-dimensional vector bundles over $P_3(\mathbb{C})$. By the basic theorems of Serre these are necessarily algebraic and this implies that F is given (in a suitable gauge) by rational functions. For further discussion of the algebraic geometry and its implications I refer to [2].

Note The difference between $F^+ = 0$ and $F^- = 0$ is only a matter of orientation.

Final comments

So far only solutions corresponding to absolute minima of $\|F\|^2$ have been constructed for $G = SU(2)$. On the other hand analogy with the 2-dimensional case described in §2 leads us to expect other critical points. If the Morse theory works also in 4-dimensions then we can predict something about the set of higher critical points. The existence of such new critical points is the outstanding open problem at the moment.

References

1. M.F. Atiyah, N.J. Hitchin and I.M. Singer, Deformations of Instantons, Proc.Nat.Acad.Sci.U.S.A. 1977.

2. M.F. Atiyah and R.S. Ward, Instantons and Algebraic Geometry, Comm.Math.Phys. 1977.

3. A. Belavin, A. Polyakov, A. Schwarz and Y. Tyupkin, Pseudoparticle Solutions of the Yang-Mills Equations, Phys.Lett. 59B (1975), 85-87.

4. P. Goddard, J. Nuyts and D. Olive, Gauge Theories and Magnetic Charge, CERN Preprint, 1976.

5. A. Grothendieck, Sur la classification des fibrés holomorphes sur la sphère de Riemann, Amer.J.Math. 79 (1957), 121-138.

6. M. Narasimhan and C. Seshadri, Stable and unitary vector bundles on a compact Riemann surface, Ann. of Math. 82 (1965), 540-567.

7. A. Newlander and L. Nirenberg, Integrability of almost-complex structures, Ann. of Math. 65 (1957), 391-404.

8. R. Jackiw, C. Nohl and C. Rebbi, Conformal Properties of Pseudoparticle Configurations, Phys.Rev.D 15 (1977), 1642-1646.

9. R. Jackiw and C. Rebbi, Degrees of Freedom in Pseudoparticle Systems, Phys.Lett. 67B (1977), 189-192.

10. R. Penrose, The Twistor Programme, Rep.Mathematical Phys. (to appear).

11. A. Schwarz, On Regular Solutions of Euclidean Yang-Mills Equations, Phys.Lett. 67B (1977), 172-174.

12. R.S. Ward, On Self-Dual Gauge Fields, Phys.Lett. 61A (1977), 81-82.

Mathematical Institute
Oxford University.

PERIODIC SOLITONS AND ALGEBRAIC GEOMETRY[(+)]

S.P. NOVIKOV

Landau Institute for Theoretical Physics

- Introduction -

This lecture surveys recently developed methods of constructing a broad class of periodic and almost periodic solutions of the KdV equations, characterized by the fact that the associated linear differential operator has a finite-zone structure.

The set of potentials with a given finite-zone spectrum form the Jacobian variety of a Riemann surface Γ , which is determined by the structure of the spectrum. Using the theory of Abelian functions the KdV flow on this variety is shown to represent a completely integrable Hamiltonian system.

1. - Periodic solutions of the KdV equation -

We shall be mainly concerned with the set of periodic or quasi-periodic solutions of the KdV equation

$$u_t = 6uu_x - u_{xxx}$$

$$u = u(x,t) \quad , \quad u_t \doteq \frac{\partial u}{\partial t} \quad , \quad u_x \doteq \frac{\partial u}{\partial x}$$

and with their topological and algebraic structure. A similar analysis can be carried through for other non-linear differential or finite-difference equations, in particular for the non-linear Schroedinger equation, for the Sine-Gordon equation, for the equation of the Toda lattice etc..

We recall briefly a few basic results of the Inverse Scattering Method, exibiting part of the structure of the KdV equation, quite independently from the function space in which the solution is sought.

a) The KdV equation is equivalent to the operator equation (Lax (2))

$$\frac{\partial L}{\partial t} = [A_1, L] \tag{1.1}$$

where $L = -\frac{d^2}{dx^2} + u(x,t) \quad , \quad A_1 = -4\frac{d^3}{dx^3} + 3\left(u\frac{d}{dx} + \frac{d}{dx}u\right)$

b) The KdV equation is equivalent (2), (3) to the equation

$$\frac{\partial u}{\partial t} = \frac{\partial}{\partial x}\frac{\delta I_1}{\delta u(x)} \tag{1.2}$$

where $I_1 = \int(\frac{1}{2}u_x^2 + u^3)\,dx$

In case the solution is sought in the space of periodic functions, the integral is extended over a period. Since $\frac{\partial}{\partial x}$ is an antisymmetric operator, equation (1.2) gives evidence, at least formally, that the KdV equation describes an Hamiltonian system. If I, J are functionals over a suitable function space, their Poisson Bracket is given by

$$\{I, J\} \doteq \int \frac{\delta I}{\delta u(x)}\frac{d}{dx}\left(\frac{\delta J}{\delta u(x)}\right)dx \tag{1.3}$$

(+) Expanded version by G.F.Dell'Antonio of notes written by Prof. S.P.Novikov at the Conference.

One can prove $\left((4),(3)\right)$ that there exists an infinite set $\{I_n\}$ n =-1, 0, ... of integrals of motion which are in involution, i.e. such that $\{I_n, I_m\} = 0$
One has

$$I_n = \int \left[(u^{(n)}(x))^2 + P(u^{(n-1)}, \cdots, u)(x) \right] dx \tag{1.4}$$

where $u^{(n)}(x) \doteq \frac{\partial^n u}{\partial x^n}$ and $P(\xi_{n-1}, \cdots \xi_0)$ is a polynomial in the variables indicated. In particular, one has

$$I_{-1} = \int u\,dx \ , \quad I_0 = \int \frac{u^2}{2} dx \ , \quad I_1 = \int \left(\frac{(u_x)^2}{2} + u^3 \right) dx \ ,$$

Clearly I_1 is the Hamiltonian for the KdV flow, and $\frac{dI_m}{dt} = 0$
c) One can consider the flow defined by the "higher-order" KdV equations

$$u_t = \frac{\partial}{\partial x} \left(\frac{\delta I_n}{\delta u(x)} + \sum_1^n c_j \frac{\delta I_{n-j}}{\delta u(x)} \right) \tag{1.5}$$

These flows commute pairwise.

d) The higher order KdV equations can be written $\left((2),(3)\right)$ in the equivalent form

$$\frac{\partial L}{\partial t} = \left[L, A_n + \sum_{j=1}^n c_j A_{n-j} \right] \tag{1.6}$$

where $A_K = \frac{d^{2K+1}}{dx^{2K+1}} + \sum_0^{2K} P_\ell (u, u^{(1)}, \cdots, u^{(K-1)}) \frac{d^\ell}{dx^\ell}$

If u is a stationary solution of the KdV equation, the operators A_K, k = 0,1,... and L commute pairwise, and define therefore a commutative algebra of operators.
Example - The solution of the equation $\frac{d}{dx}\left(\frac{\delta I_1}{\delta u(x)} + c \frac{\delta I_0}{\delta u} \right) = 0$ is the Weierstrass ellyptic function P(x). Correspondingly, a solution (the "soliton solution") of the KdV equation is $\frac{1}{2} u_0(x,t) = P(x-ct)$. We point out an important property of $u_0(x)$: the operator $L \doteq -\frac{d^2}{dx^2} + u_0(x)$ has only one forbidden zone in the Bloch spectrum.

2. - Bloch waves -

We recall some definitions and results $\left((5),(6),(1),(7)\right)$ from the theory of the Schroedinger equation with periodic potential.

Let $u(x)$ be periodic with period T, and define
$(\hat{T}\psi)(x) \doteq \psi(x+T)$
$\hat{T}$ is often called the "monodromy operator".
The "Bloch" eigenfunctions $\psi_\pm(x, x_0; E)$ are defined by

a) $L\psi_\pm = E\psi_\pm$
b) $\hat{T}\psi_\pm = \exp(i p(E) T)\psi_\pm$
c) $\psi_\pm(x_0, x_0; E) = 1$

The values of E for which p(E) is real define the "allowed zones". A connected component of an allowed zone is called an allowed band. Allowed bands are separated by "lacunae".

Eigenvalues corresponding to eigenfunctions with periods T or 2T either are double or are (the) end-points of the lacunae.

The following simple result $\left((5),(6),(1)\right)$ is important for what follows:
For any (real or complex) smooth periodic potential $u(x)$, the Bloch eigenfunctions $\psi_\pm(x, x_0; E)$ are meromorphic on a two-sheeted Riemann surface Γ, covering the entire E-plane, and having branch points at the end of the lacunae. In general the genus of Γ is infinite; however, if the number of lacunae is finite, then Γ is hyperellyptic and its genus is equal to the number of lacunae.

For potentials which are quasi-periodic and complex, one can generalize the definition of Bloch eigenfunctions substituting the requirement b) with

b') $\frac{\partial}{\partial x} \ln \psi_\pm(x, x_0; E)$ has the same periods as the potential $u(x)$.

We can now introduce an important definition:
An almost periodic potential u (real or complex) is called a "finite-band" potential if

1) the Bloch eigenfunctions defined through a), b'), c) above exist for all complex E;

2) $\Psi_\pm(x, x_0; E) \simeq \exp(i\sqrt{E}(x-x_0))$ when $E \to \infty$

3) $\Psi_\pm(x, x_0; E)$ is meromorphic, for fixed x, x_0, on a Riemann surface Γ doubly covering the E-plane;

4) the surface Γ has finite genus.

If the potential is periodic, this definition coincides with the usual one. We shall call Γ the "spectrum" of $L = \frac{d^2}{dx^2} + u(x)$, and n its genus. The branch-points of Γ are called "boundary of the zones". These definitions are suggested by the spectral analysis of L in the case in which $u(x)$ is real and periodic.

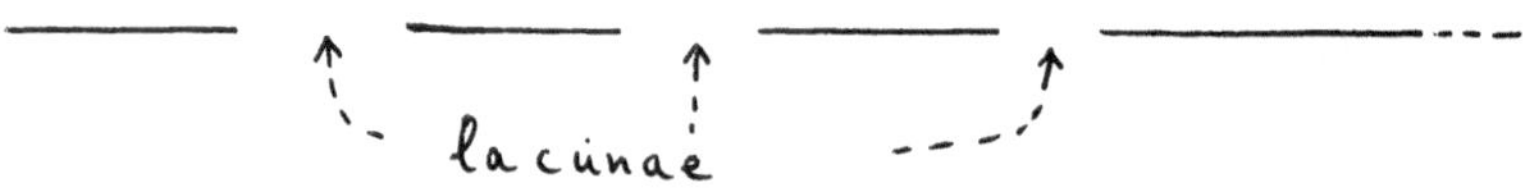

3. - Relation between the (higher-order) KdV flow and the flow of zeroes and poles of the Bloch eigenfunctions associated to finite-band potentials on a Riemann surface Γ -

The following results on the zeroes and poles of a Bloch eigenfunctions is essential for what follows:
Let $u(x)$ be a (real or complex) quasi-periodic potential, let Γ be its spectrum and n its genus. Then $\Psi_\pm(x, x_0; E)$ is meromorphic on $\Gamma \setminus \{\infty\}$, and has n poles and n zeroes on Γ. The location of the poles is given by n functions $\gamma_1, \ldots, \gamma_n$ in the following sense: the poles of $\Psi_\pm(x, x_0; E)$ are located at $E = \gamma_i(x)$, $i = 1 \ldots n$.

If $\gamma_i(x)$ are pairwise distinct for each i, either Ψ_+ or Ψ_-, but not both, will have a pole at $E = \gamma_i(x)$.

Also, the zeroes of $\Psi_\pm(x, x_0; E)$ are given by the same functions γ_i, in the sense that they are located at $E = \gamma_i(x_0)$. .

For a real periodic potential u, the poles and zeroes of $\Psi_\pm$ occur on the real axis, inside the lacunae.

Moreover, the knowledge of the functions $\gamma_i(x)$ completely determines the potential u. Symmetric functions of the γ_i's can be written in terms of u; in particular

$$u(x) = -2\sum_{i=1}^{n} \gamma_i(x) + \sum_{i=1}^{2n+1} E_i \tag{3.1}$$

$$\sum \gamma_i(x)\gamma_j(x) = \frac{1}{8}\left(3u^2(x) - \frac{d^2u}{dx^2}\right) + \frac{1}{2}\sum E_i E_j - \frac{3}{8}\left(\sum E_i\right)^2 \tag{3.2}$$

Eq. (3.1) is often referred to as "Hochstadt's trace-formula".

All these results follow from the analysis of the representation of $\Psi_\pm$ in terms of their Wronskian. In particular (3.1) and (3.2) follow from an asymptotic expansion in E^{-1}.

4. - Basic results of the Inverse Scattering Method -

We can now formulate the basic results [(7)(8)(9)] of the periodic Inverse Scattering Method. They will include an algebraic characterization of the manyfolds spanned by subsets of solutions of the KdV equation, as well as a choice of coordinates in which the KdV flow takes a particularly simple form. The new coordinates will be essentially action-angle variables for the (higher-order) KdV flows.

a) All stationary higher-order KdV equations represent completely integrable finite dimensional Hamiltonian systems (where the variable x is interpreted as time) ((6), (1)).

b) Any stationary solution of a higher-order KdV equation is ((9),(1)) a finite-band potential (possibly complex and quasi-periodic). Conversely (7) every finite-band potential is solution of some higher-order KdV equation.

c) The Block eigenfunction $\Psi_\pm$ (x,x ; E) is completely determined by its analytic properties (method of Achiezer (10). The function γ (x) which determines its zeroes and poles satisfies the equation

$$\frac{d\gamma_i}{dx} = 2i\sqrt{R(\gamma_i)}\prod_{j\neq i}(\gamma_j-\gamma_i)^{-1} \tag{4.1}$$

where $R(\lambda) \doteq \prod_{K=1}^{2n+1}(\lambda - E_K)$ and $E_1, \ldots, E_{2n+1}$ are the end points of the lacunae.

This result is obtained through a comparison of the meromorphic properties of $\Psi_\pm$ with those of their Wronskian.

d) Moreover, if u satisfies the KdV equation $u_t = 6\,u\,u_x - u_{xxx}$ the spectrum of L is an invariant and the t-dependence is described by the following equation for γ_j (1)

$$\frac{d\gamma_i}{dt} = \gamma_i\Big(\sum_{K\neq i}\gamma_K\Big)\,R(\gamma_i)\prod_{K\neq i}(\gamma_i-\gamma_K)^{-1} \tag{4.2}$$

We sketch a derivation of (4.2). This is done first for finite-zone periodic real potentials, and then formula (4.2) is extended to the collection of all finite band potentials. By b), every periodic potential with n zones is solution of an equation of the form

$$\sum_{\ell=-1}^{n} c_{n-\ell}\,\frac{\delta I_\ell}{\delta u(x)} = 0$$

It can be shown that the corresponding transition matrix $\hat{T}$ satisfies an equation of the form

$$0 = \Big[\sum_{-1}^{n} c_{n-\ell}\,\Lambda_\ell\,,\,\hat{T}\Big]$$

where the Λ_i's are matrix-valued functionals of u .

If u (x,t) satisfies the KdV equation, one will also have

$$\frac{\partial\hat{T}}{\partial t} = [\Lambda_1, \hat{T}] \tag{4.3}$$

The compatibility requirement

$$\frac{d\Lambda}{dt} = [\Lambda_1, \Lambda]\qquad,\quad \Lambda = -\sum_{-1}^{n} c_{n-\ell}\,\Lambda_\ell \tag{4.4}$$

determines therefore the dynamics of n -zone periodic potentials. The matrix Λ depends polynomially on E; the characteristic polynomial of the matrix Λ has the form det $(W\cdot I - \Lambda)$ ($= W^2+P(E)$) and the Riemann surface det $(W\cdot I - \Lambda) = 0$ coincides with the "spectrum" of the potential u (x) as defined above.

Writing χ (x,E) ($= i\frac{d}{dx}\ln\Psi$) in terms of the transition matrix $\hat{T}$ one obtains from (4.4) an equation for χ(x,E), and then for the γ_i's, using the representation of the Bloch eigenfunctions through their meromorphic properties in Γ .

This procedure leads to (4.2) in the case of periodic finite-zone real potentials. It is then shown that, if (4.2) holds for the zeroes and poles of finite-zone real or complex quasi-periodic potentials, the corresponding potentials satisfy the KdV equation.

5. - Complete integrability of the KdV periodic equation -

Using the Abel map and the Jacobi varieties one can caracterize the solutions of (4.1) in terms of the Riemann θ-function, and give a proof that (4.2) is a completely integrable system.

In fact (1), the Abel map provides action-angle variables for the KdV equations. One has in particular that, through the Abel map, Eqs.(4.1),(4.2) can be put in the form

$$\frac{d\eta}{dx} = \text{cost.} \qquad \frac{d\eta}{dt} = \text{cost.} \tag{5.1}$$

(η are Abel variables).

The procedure can be outlined as follows.

Let π be the canonical projection $\Gamma \rightarrow \mathbb{C}$ of Γ into the E-plane. Then, by (3.1) the potential u (x) is given by

$$u(x) = -2\sum_{j=1}^{n} \pi(P_j(x)) + \text{cost.} \tag{5.2}$$

and (4.1) have to be understood as equation in x_0 for the set $P_i(x_0)$ ($P_i(x_0)$ is the unique point of Γ which lies over $\gamma(x_0)$ and is a pole for $\psi_\pm(x, x_0; E)$).

When u is real and periodic, the point $P_j(x_0)$ lies over the cycle a_j in Γ obtained by fusing together the two "lacunae" ($[E_{2i}, E_{2i+1}]$,+) and ($[E_{2i}, E_{2i+1}]$,−) (the two signs correspond to the double covering of $\mathbb{C}$ by Γ).

The point $P_j(x_0)$ moves over a_j when x_0 varies. The "phase point" $\{P_1(x_0), \cdots, P_n(x_0)\}$ moves on torus $T^n = S_1^1 \times .. \times S_n^1$ (the product of the cycles a_j).

This picture can be extended to the general almost periodic finite-zone case and can be used to exibit the KdV flow as a rectlinear flow on a torus.

Let u be almost periodic and finite zone; let $\psi_\pm(x, x_0; E)$ be the Bloch eigenfunctions, and $P_i(x_0)$, i = 1....n, ($P_i(x_0) \in \Gamma$) be the position of their poles.

If $S^n(\Gamma)$ is the symmetric n^{th} power of Γ ; $S^n(\Gamma)$ is an algebraic variety and $\sigma(P_1, .. P_n) \doteq \sum_1^n \pi(P_j)$ is an algebraic function on it.

We recall briefly the construction of the Abel map.

One selects on Γ a base of cycles $a_1 .. a_n, b_1, .. b_n$ with the intersection matrix

$$a_i \circ a_j = b_i \circ b_j = 0 \quad , \quad a_i \circ b_j = \delta_{ij}$$

One considers then on Γ a basis of holomorphic differentials

$$\Omega_k = \sum_{l=1}^{n} C_{kl} \frac{E^{l-1}\, dE}{(P_{2n+1}(E))^{1/2}} \qquad k = 1 \cdots n$$

where the coefficients C_{kl} are chosen so that

$$\oint_{a_j} \Omega_k = 2\pi i\, \delta_{jk} \doteq A_{jk}$$

Define the "Jacobi matrix" as the matrix B with entries

$$B_{kl} = \oint_{b_l} \Omega_k$$

If $A^{(K)}$, $B^{(K)}$ are the vectors with m^{th} component A_{Km}, B_{Km} respectively, define $J(\Gamma)$ (the Jacobian of Γ) as the torus obtained by taking the quotient of $\mathbb{C}^n$ modulo the lattice generated by the vectors A^K, B^K.

Let $P_1^0 .... P_n^0$ be fixed points of Γ . The Abel map is then defined by

$$\underline{P} \mapsto \underline{\eta}(\underline{P}) \tag{5.3}$$

where $\underline{P} = \{P_1 \cdots P_n\}$, $\underline{\eta} = \{\eta_1 \cdots \eta_n\}$, and

$$\eta_k = \sum_{j=1}^{n} \int_{P_j^0}^{P_j} \Omega_k \tag{5.4}$$

One can prove that the set of zeroes of the Bloch eigenfunctions $\psi_\pm$ are represented by straight lines on the Jacobi variety. This result can be expressed saying that the Abel map "integrates" Eq. (4.1).

To obtain this result one makes use of the information quoted above on the poles, the zeroes and the asymptotic behaviour at E $=\infty$ of the Bloch functions $\psi_\pm(x, x_0 ; E)$. One considers in particular the pole structure and the asymptotic behaviour of the logarithmic differential

$$\omega \doteq \frac{d\ln\psi}{dE}\,dE$$

and one obtains (5.1) making use of (4.1) and of the following identity, well-known in algebraic geometry

$$\oint_{b_j} \Omega^{PQ} = \int_Q^P \Omega_j \qquad (5.5)$$

where Ω_j were defined previously and Ω^{PQ} is a differential on Γ whose only singularities are a pole with residue -1 at Q and a pole with residue + 1 at P, normalized by the condition $\oint_{a_j} \Omega^{PQ} = 0 \quad j = 1 \cdots n$.

As noticed previously, u (x) is a rational function on J (Γ); it can therefore be expressed in an algebraic form in terms of the Riemann Θ-function and its derivatives. The Θ-function is defined as

$$\theta(\eta_1 \cdots \eta_n) = \sum_{m_i \in Z} \exp\left(\frac{1}{2} \sum_{j,k} B_{jk}\, m_j m_k + \sum_k m_k \eta_k \right)$$

where B is the Jacobi matrix.

One obtains the following explicit form for $u(x)$:

$$u(x) = -2\frac{d^2}{dx^2} \ln \theta\left(\eta_1^0 + U_1\cdot(x-x_0) - K_1 , \cdots , \eta_n^0 + U_n\cdot(x-x_0) - K_n\right) + C$$

where $U_j \doteq \oint_{b_j} \Omega$, and Ω is the differential which has as only singularity a second-order pole at E $=\infty$ and is normalized by the conditions $\oint_{a_j} \Omega = 0 \;\forall j$. We conclude by stating the following result, which describes in a precise form the complete integrability of all the KdV equations:

Theorem:(1) The time evolution of finite-zone potentials according to the KdV equation or any of its higher order analogues, represents a motion on a torus along a rectilinear winding on the variety of all potentials with given spectrum Γ, which is isomorphic to the complex n-torus T^n ⫫

In particular, let ω_m be a differential on Γ whose only singularity is a pole of order 2m+2 at E $=\infty$. Choose (and then uniquely) ω_m so that $\oint_{a_j} \omega_m = 0 \;\forall j$. With this notation, the mth KdV equation (i.e. $\frac{\partial u}{\partial t} = \frac{\partial}{\partial x} \frac{\delta I_m}{\delta u(x)}$) reads

$$\dot{\eta}_k = W_k^{(m)} \qquad (5.6)$$

where

$$\omega_k^{(m)} \doteq \oint_{b_k} \omega_m$$

The proof follows essentially along the same lines as the proof of (5.1), which can be regarded as special case (m = 0). One studies the evolution, under the mth KdV equation, of the asymptotic form (at E $=\infty$) of the Bloch eigenfunctions. One proves that

$$\psi(x, x_0 ; E) \simeq \exp\{ iK\cdot(x-x_0) + iK^{2m+1}(t-t_0) \}$$

where $K^2 = E$.

Studying the asymptotic behaviour of the differential $d\ln\psi$ as done previously for theequation (5.1), and using the differential ω_m, one obtains (5.6).

From (5.6) and the well-known relation between η_k and $u(x)$, one obtains the following explicit solution of the KdV equation

$$u(x,t) = -2\frac{d^2}{dx^2}\Big(\ln \theta\big[(x-x_0)U_1 + (t-t_0)\omega_1^{(1)} + \eta_1^0 - K_1, \dots, (x-x_0)U_n + (t-t_0)\omega_n^{(1)} + \eta_n^0 - K_n\big]\Big) + C$$

- References -

(1) B.Dubrovin, V.Matveev, S.Novikov
Russian Math. Surveys 31 (1976) 59-146.

(2) P.D.Lax
Comm. Pure and Applied Math. 21 (1968) 467-490.

(3) C.S.Gardner
J.Math. Phys. 12 (1971) 1548-1551.

(4) V.Zakharov, L.Fadeev
Funkt. Anal. i Prilozen 5 (1971) 18-27.

(5) S.P.Novikov
Funkt. Anal. i Prilozen 8 (1974) 54-66.

(6) P.D.Lax
Lectures in Applied Math. 15 (1974) 85-96.

(7) B.V.Dubrovin
Funkt. Anal. i Prilozen 9 (1975).

(8) A.R.Its, V.B.Matveev
Trudy Math. Fiz. 23 (1975) 51-67.

(9) B.Dubrovin
Funkt. Anal. i Prilozen 9 (1975) 65-66.

(10) N.I.Akhiezer
Dokladi Akad. Nauk SSSR 141 (1961) 263-266.

YANG-MILLS EQUATIONS AS INVERSE SCATTERING PROBLEM

A.A.Belavin, V.E.Zakharov
The Academy of Sciences of the USSR L.D.Landau Institute for Theoretical Physics

Abstract:

Non-linear self-duality equations $F_{\mu\nu} = F^*_{\mu\nu}$ are shown to be conditions of compatibility of two linear equations. All the N-instanton fields are constructed explicity.

Introduction.

Here we propose a method of solving self-duality equations (1)

$$F_{\mu\nu} = \pm F^*_{\mu\nu} \tag{1}$$

for A_μ gauge fields in 4-dimensional Euclidean space. Classical trajectories described by these fields (instantons) correspond to the tunnel transitions between vacua with different topologies as it has been noticed by V.N.Gribov in the fall of 1975 (unpublished) and discussed thoroughly in subsequent papers (2, 3, 4). Instantons are conjectured to be responsible for the confinement of quarks (5, 6) and for the resolution of the $U(1)$ problem (2).

Self-dual fields (1) fall into classes defined by a topological charge

$$N = \frac{1}{8\pi^2} \int Sp\, F_{\mu\nu} F^*_{\mu\nu} \tag{2}$$

One-instanton solutions found in (1) are 8-parametric fields, parameters being four positions in space, scale and three Euler angles of global isotopic rotations. N-instanton fields should be 8N-parametric fields. This was proved in the spring of 1976 by A.S.Schwartz (unpublished) using Atiyah-Singer theorem on indices and was confirmed afterwards by direct counting the number of zero-energy excitations (7). Some multi-instanton solutions have been found previously (8), the most general of these being 5N-parametric solution of t'Hooft with fixed relative orientation of isospins.

The method developed allows finding all self-dual fields[+)]. Our basic observation is that non-linear equations (1) are just conditions of compatibility for some set of linear equations. For example, equations

$$D_\mu \psi = (\partial_\mu + A_\mu)\psi = 0 \tag{3}$$

are compatible, provided that operators $\mathcal{D}_\mu$ commute:

$$[\mathcal{D}_\mu, \mathcal{D}_\nu] \doteq F_{\mu\nu} = 0 \tag{4}$$

Once Eqs. (3) are solved, one could explicitly write down all the fileds with $F_{\mu\nu} = 0$: $(\partial_\mu + A_\mu)\psi = 0$ implies that

$$A_\mu = \psi \partial_\mu \psi^{-1} \tag{5}$$

i.e., we arrive at purely longitudinal fields, as has been expected.

+) The method to be presented is closely related to the solution of the inverse scattering problem for many-dimensional spaces developed in Ref. (9).

Self-duality equation as the compatibility condition

Let us start with Dirac equations for massless two-component fermions:

$$(D_4 + i\vec{D}\cdot\vec{\sigma})\,u = 0 \tag{6}$$

$$(D_4 - i\vec{D}\cdot\vec{\sigma})\,v = 0 \tag{7}$$

Let us search for the solutions of Eq. (7) of the following form

$$v = \begin{pmatrix}\lambda\\1\end{pmatrix}\psi(\lambda,x) \tag{8}$$

for all the values of λ . ($\psi(\lambda,x)$ is a 2x2 matrix in the isotopic space). Then Eq. (7) takes the form:

$$L_1\psi \doteq [\lambda(D_2 - iD_1) + (D_4 + iD_3)]\psi = 0 \tag{9}$$

$$L_2\psi \doteq [\lambda(D_4 - iD_3) - (D_2 + iD_1)]\psi = 0 \tag{10}$$

These equations are compatible, provided that

$$[L_1, L_2] = 0 \tag{11}$$

This commutator vanishes irrespective of λ only provided that $F_{\mu\nu} = -F^*_{\mu\nu}$. Starting with Eq. (6) one arrives in the same way at equation $F_{\mu\nu} = F^*_{\mu\nu}$. Thus, non-linear self-duality equations are substituted by linear equations (9) and (10).

Convenient and proper variables are complex coordinates and complex fields introduced as follows:

$$z_1 = \frac{x_2 + ix_1}{2} \qquad z_2 = \frac{x_4 + ix_3}{2} \tag{12}$$

$$B_1 = A_2 - iA_1 \qquad B_2 = A_4 - iA_3 \tag{13}$$

$$\partial_i = \frac{\partial}{\partial z_i} \qquad \nabla_i = \partial_i + B_i \qquad \bar{\nabla}_i = \bar{\partial}_i - B_i^+ \tag{14}$$

In these variables equation $F_{\mu\nu} = -F^*_{\mu\nu}$ takes the form:

$$\partial_1 B_2 - \partial_2 B_1 + [B_1, B_2] = 0 \tag{15}$$

$$\partial_i B_i^+ + \bar{\partial}_i B_i + [B_i, B_i^+] = 0 \tag{16}$$

In the linerized form one has:

$$(\lambda\nabla_1 + \bar{\nabla}_2)\,\psi(\lambda,x) = 0 \tag{17}$$

$$(\lambda\nabla_2 - \bar{\nabla}_1)\,\psi(\lambda,x) = 0 \tag{18}$$

One-instanton solution

Once ψ -function is found, fields B_i are given by

$$\lambda B_1 - B_2^+ = \psi(\lambda,x)\,(\lambda\partial_1 + \bar{\partial}_2)\,\psi^{-1}(\lambda,x) \tag{19}$$

$$\lambda B_2 + B_1^+ = \psi(\lambda,x)\,(\lambda\partial_2 - \bar{\partial}_1)\,\psi^{-1}(\lambda,x) \tag{20}$$

These equations imply that $\tilde{\psi}(\lambda) \doteq \psi^{+-1}(-\frac{1}{\lambda})$ satisfies Eqs. (17), (18), if $\psi(\lambda)$ does. Thus, restriction

$$\psi^{+}\left(-\frac{1}{\lambda}\right)=\psi^{-1}(\lambda) \qquad (21)$$

will be imposed on ψ -functions.

Eqs. (19) and (20) yield self-dual fields, provided ψ -functions are found which satisfy Eq. (21) and such that r.h.s. in Eqs. (19),(20) have at most linear dependence on λ +). To proceed further one should specify the analiticity properties of $\psi(\lambda,x)$ as functions of λ. We suppose that the only singularities in λ -plane are poles whose locations do not depend on x.

Let us start with few simplest cases. Suppose $\psi(x,\lambda)$ be independent of λ. Eq. (21) implies that in this case $\psi^{+}(x)=\psi^{-1}(x)$, and from Eqs. (19),(20) it follows that $B_i=\psi D_i\psi^{-1}$. Thus, we obtain all longitudinal fields with stress $F_{\mu\nu}=0$.

The next step is a two-pole matrix $\psi(\lambda,x)$ ++); poles taken to be located at $\lambda=\lambda_0$ and $\lambda=-\bar{\lambda}_0^{-1}$:

$$\psi(\lambda,x)=V\left(u\cdot I+\frac{\lambda-\lambda_0}{1+\lambda\bar{\lambda}_0}fA+\frac{1+\lambda\bar{\lambda}_0}{\lambda-\lambda_0}\bar{f}A^{+}\right) \qquad (22)$$

$$\psi^{-1}(\lambda,x)=\left(u\cdot I-\frac{\lambda-\lambda_0}{1+\lambda\bar{\lambda}_0}fA-\frac{1+\lambda\bar{\lambda}_0}{\lambda-\lambda_0}\bar{f}A^{+}\right)V^{-1} \qquad (23)$$

Here I, V and A are 2x2 matrices; I is a diagonal unit matrix, $V^{+}=V^{-1}$; u and f are some functions, $\bar{u}=u$. From equation $\psi\psi^{-1}=1$ it follows that $A^2=0$, and that $u^2=1+f\bar{f}\{A,A^{+}\}$. We impose normalization $\{A,A^{+}\}_{+}=1$. After some exercises one could verify that fields A_μ are nonsingular only provided $\lambda_0=0$; thus, we put $\lambda_0=0$.

The arbitrary unitary matrix $V(x)$ in expansions (22),(23) corresponds to gauge transformations; to fix a gauge we put $V(x)=1$. It is convenient to parametrize matrix A as follows:

$$A=\left(|a|^2+|b|^2\right)^{-1}\begin{vmatrix} ab & a^2 \\ -b^2 & ab \end{vmatrix} \qquad (24)$$

where $a(x)$ and $b(x)$ are functions to be specified below.

Let us substitute Eqs. (22)-(24) into Eqs. (19),(20). The residues of singular λ^{-2}, λ^{-1}, λ^{2} and λ^{3} terms should vanish identically. The λ^{-2} and/or λ^{3} terms lead to the equations:

$$A\partial_i A=0 \qquad (25)$$

which imply that $a(x)$ and $b(x)$ are analytic functions of two complex variables $\bar{z}_i$: $a(x)=a(\bar{z}_1,\bar{z}_2)$, $b(x)=b(\bar{z}_1,\bar{z}_2)$. The vanishing of λ^{-1} and/or λ^{2} terms leads to some differential equation, this equation integrates to the algebraic equation:

$$f^{-1}\cdot\sqrt{1+|f|^2}=\left(|a|^2+|b|^2\right)^{-1}\left(z_1(a\bar{\partial}_2 b-b\bar{\partial}_2 a)-z_2(a\bar{\partial}_1 b-b\bar{\partial}_1 a)+c(\bar{z}_1,\bar{z}_2)\right) \qquad (26)$$

where $c(\bar{z}_1,\bar{z}_2)$ is an integration constant. Thus, we are left with three analytic two-variable functions $a(\bar{z}_i)$, $b(\bar{z}_i)$ and $c(\bar{z}_i)$ to be determined.

A careful study of Eq. (26) demonstrates that it has solutions only provided that its r.h.s. is bounded in modulus from below by unity. This condition turns out to be extremely restrictive: only linear functions $a(\bar{z}_i)$ and $b(\bar{z}_i)$ are allowed; and function $c(\bar{z}_i)$ should be some constant. One of the possible solutions is given by:

$$a=\bar{z}_1,\quad b=\bar{z}_2,\quad c=2 \qquad (27)$$

Eight parameters the most general one-instanton solution depends on are: scale c, shifts $a\to\bar{z}_1-\bar{z}_1^0$, $b\to\bar{z}_2-\bar{z}_2^0$ and three isotopic rotations $a\to V_{11}\bar{z}_1+V_{12}\bar{z}_2$, $b\to V_{21}\bar{z}_1+V_{22}\bar{z}_2$ given by the unitary matrix V, $\det V=1$. Fields B_i given by solu-

---

+) We stress that fields B_i do not depend on the auxiliary parameter λ.

++) No single-pole solutions exist due to Eq. (21).

tion (27):

$$B_1 = [2(1+z_i\bar{z}_i)]^{-1}\begin{vmatrix} \bar{z}_1 & 0 \\ -2\bar{z}_2 & \bar{z}_1 \end{vmatrix}, \quad B_2 = [2(1+z_i\bar{z}_i)]^{-1}\begin{vmatrix} \bar{z}_2 & 2\bar{z}_1 \\ 0 & -\bar{z}_2 \end{vmatrix} \tag{28}$$

are just one-instanton solution of Ref. (1).

Reproduction of instantons

Suppose N-instanton fields B_i^0 have been found:

$$\lambda B_1^0 - B_2^{0+} = \Psi_N(\lambda\partial_1 + \bar{\partial}_2)\Psi_N^{-1} \tag{29}$$

$$\lambda B_2^0 + B_1^{0+} = \Psi_N(\lambda\partial_2 - \bar{\partial}_1)\Psi_N^{-1} \tag{30}$$

We look for (N+1)-instanton Ψ -matrix of the form:

$$\Psi_{N+1} = \Psi_1 \cdot \Psi_N \tag{31}$$

The resultant equations for (N+1)-instanton fields B_i are as follows:

$$\lambda B_1 - B_2^+ = \Psi_1(\lambda\nabla_1^0 + \bar{\nabla}_2^0)\Psi_1^{-1} \tag{32}$$

$$\lambda B_2 + B_1^+ = \Psi_1(\lambda\nabla_2^0 - \bar{\nabla}_1^0)\Psi_1^{-1} \tag{33}$$

where $\nabla_i^0 = \partial_i + B_i^0$.

We use for Ψ_1 an expansion reminiscent of Eq. (22):

$$\Psi_1 = u + \lambda f\tilde{A} + \frac{1}{\lambda}\bar{f}\tilde{A}^+ \tag{34}$$

$$\Psi_1^{-1} = u - \lambda f\tilde{A} - \frac{1}{\lambda}\bar{f}\tilde{A}^+ \tag{35}$$

Here $\tilde{A}^2 = 0$, and $u^2 = 1 + |f|^2\{\tilde{A},\tilde{A}^+\}_+$
Eq. (25) is substituted in this case by the equation

$$\tilde{A}\nabla_i^0\tilde{A} = \tilde{A}(\partial_i + B_i^0)\tilde{A} = 0 \tag{36}$$

In order to solve it one should bear in mind that Eq. (9) implies that

$$B_i = g\partial_i g^{-1} \tag{37}$$

where g is the matrix with $\det g = 1$. After transformation $\tilde{A} = gAg^{-1}$ we get for matrix A equation (25) their solution being already known:

$$A = (|a|^2 + |b|^2)^{-1}\begin{vmatrix} ab & a^2 \\ -b^2 & -ab \end{vmatrix} \qquad a = a(\bar{z}_i), \quad b = b(\bar{z}_i) \tag{38}$$

Eq. (26) is substituted by the equation:

$$f^{-1}(1 + |f|^2\{A,A^+\}_+)^{1/2} = (|a|^2 + |b|^2)^{-1}(z_1(a\bar{\partial}_2 b - b\bar{\partial}_2 a) - z_2(a\bar{\partial}_1 b - b\bar{\partial}_1 a) + c + \phi) \tag{39}$$

where ϕ is constrained by equations

$$A(g^{-1}\bar{\nabla}_2^0 g)A = (|a|^2 + |b|^2)^{-1}\partial_1\phi \cdot A \tag{40}$$

$$A(g^{-1}\bar{\nabla}_1^0 g)A = -(|a|^2 + |b|^2)^{-1}\partial_2\phi \cdot A \tag{41}$$

which result from Eq. (16). As well as in the one-instanton solution, Eq. (39) imposes strict restrictions on functions $a(\bar{z}_i)$, $b(\bar{z}_i)$ and $c(\bar{z}_i)$: its r.h.s. squared should have modulus not lower than

$$\{\tilde{A},\tilde{A}^+\}_+ = Ah^{-1}A^+h + h^{-1}A^+hA \qquad \text{with } h = g^+g \tag{42}$$

We have not succeeded in finding an explicit form of functions which satisfy this constraint. Nevertheless, one may be convinced that the procedure outlined works indeed step by step. Now we demonstrate how it works for $N=2$.

In this case fields B_i^0 are those given by Eq. (28). The matrix g coupled with these fields according to Eq. (37) is given by

$$g = (1+z_i\bar{z}_i)^{-1/2} \begin{vmatrix} \bar{z}_1 & z_2 \\ -\bar{z}_2 & z_1 + \frac{1}{\bar{z}_1} \end{vmatrix} \tag{43}$$

Constraints (4o) result in:

$$\phi = -\bar{z}_1^{-2}(1+z_i\bar{z}_i)^{-1}(\bar{z}_1 a - z_2 b)^2 \tag{44}$$

And, finally, we get:

$$\{A, A^+\}_+ = (1+z_i\bar{z}_i)^{-2}(|a|^2+|b|^2)^{-2}\left(|a\bar{z}_1 - bz_2|^2 + |az_2 + b(z_1+\bar{z}_1^{-1})|^2\right)^2 \tag{45}$$

A general solution of Eq. (39) is given by 8-parametric family of functions a, b and c: one of these threes is

$$a = \bar{z}_1 - \bar{z}_1^{-1}\quad , \quad b = \bar{z}_2 , \quad c = 2 + \bar{z}_1^{-2} \tag{46}$$

which gives a two-instanton solution with $O(3)\times O(2)$ symmetry.

Thus, we have established a recursion procedure of reproduction of instantons.

Conclusions

Despite all its limitations (explicit formulas for all N-istanton fields are not available) our approach looks rather promising. We hope that being developed in a proper way this approach would allow integrating complete Yang-Mills equations, constructing proper action-angle variables and finding out quasiclassical (and hopefully exact) solution of quantum problem. New conservation laws and hidden simmetry of Yang-Mills equations would be discovered in this way; this hidden symmetry (is it a two-dimensional infinite-parametric conformal group?) being responsible for 8N-parametric family of instantons with topological charge N. Linear equations (15), (16) might turn very useful when fluctuations near instanton fields are analyzed.

A number of assumptions are introduced above which at first sight are far from being self-explanatory. They are justified posteriori as our solutions saturate exact bound for the number of arbitrary parameters. Nevertheless, there are many interesting questions to be answered: how the number of poles of $\psi(\lambda, x)$ is connected with the topological charge N ; why the locations of poles should be independent of x ; what is the geometrical meaning of the constraint given by Eq. (39); why there exist single-component fermions in the self-dual gauge fields and so on.

After this work had been completed and ready for publication, news came (S. I.Gelfand, private communication) about interesting comment by Atiyah on the geometrical meaning of the self-duality equations considered on CP^3 spaces. We present below one of the possible connections of the CP^3 treatment of the duality equations with our approach.

Let us start with 8-dimensional space defined by four complex variables $\xi_a = (\xi_1, \cdots, \xi_4)$. Three-dimensional projective space CP^3 is obtained after identifying with each other all the points ξ_a which differ by a common complex factor. This space CP^3 is projected onto the space of our variables z_1 and z_2 as follows:

$$z_1 = \frac{\xi_1\bar{\xi}_3 + \bar{\xi}_2\xi_4}{|\xi_3|^2 + |\xi_4|^2} \qquad z_2 = \frac{\xi_2\bar{\xi}_3 - \bar{\xi}_1\bar{\xi}_4}{|\xi_3|^2 + |\xi_4|^2} \tag{A.1}$$

The Yang-Mills fields are lifted onto CP^3 according to:

$$C_a = \frac{\partial x_\mu}{\partial \xi_a} A_\mu = \frac{\partial z_i}{\partial \xi_a} B_i - \frac{\partial \bar{z}_i}{\partial \xi_a} B_i^+ \tag{A.2}$$

The self-duality equation takes on CP^3 the form:

$$\frac{\partial C_b}{\partial \xi_a} - \frac{\partial C_a}{\partial \xi_b} + [C_a, C_b] = 0 \tag{A.3}$$

so that

$$C_a = \psi \cdot \frac{\partial \psi^{-1}}{\partial \xi_a} \tag{A.4}$$

After insertion of formula (A.4) into Eq. (A.2) one arrives at Eqs. (17), (18) with the parameter λ identified as

$$\lambda = \xi_3 \cdot \xi_4^{-1}$$

We express our deep gratitude for useful discussions to V.Dutyshev, S.Matsaev, N.Nikolaev, S.Novikov, A.Polyakov and L.Faddeev.

References

1. A.Belavin, A.Polyakov, A.Schwartz, Y.Tyupkin (1975), Phys. Lett., 59B, 85.

2. t'Hooft, G. (1976), Phys. Rev. Lett., 37, 8.

3. Callan, C., R.Dashen and D.Gross (1976), Phys. Lett., 63B, 334.

4. jackiw,R. and C.Rebbi (1976), Phys. Rev. Lett., 37, 172.

5. Polyakov, A. (1975), Phys. Lett., 59B, 82.

6. Polyakov, A. (1976) NORDITA preprint.

7. Jackiw, R., C.Nohl and C.Rebbi (1976) MIT preprint.

8. B.Burlakov, V.Dutyshev, Report at the Meeting of Nuclear Physics Division of the Academy of Sciences, October (1976); ZhETF, in press.
   E.Witten (1976) Harward University preprint.

9. V.E.Zakharov, A.B.Shabat, in press.

NONLINEAR EVOLUTION EQUATIONS SOLVABLE BY THE INVERSE SPECTRAL TRANSFORM

F.Calogero
Istituto di Fisica, Università di Roma, 00185 Roma, Italy
Istituto Nazionale di Fisica Nucleare, Sezione di Roma

Abstract

The main ideas and some recent results on (classical) solitons are tersely surveyed.

Invited lecture presented at the International Conference on the Mathematical Problems in Theoretical Physics, Rome University, June 6-15, 1977.

1. Introduction

The idea of the inverse spectral transform was introduced ten years ago by Gardner, Greene, Kruskal and Miura [1] as a means for solving the initial value problem for the Korteweg-de Vries (KdV) equation

$$(1.1) \qquad q_t(x,t)+q_{xxx}(x,t)+q_x(x,t)q(x,t)=0 \quad .$$

This nonlinear partial differential equation was introduced at the end of the last century [2] to account for a phenomenon - the propagation of solitary waves in a shallow channel - that had been observed, described and investigated in the first half of the nineteenth century by Scott-Russel [3] . This same equation plays an important rôle in several other physical phenomena, including some occurring in the realm of plasma physics, which may account for the fact that Gardner, Greene, Kruskal and Miura were all working at the Plasma Physics Lab of Princeton University when they wrote their ice-breaking paper.

The inverse scattering transform method - or rather, as we prefer to call it, the inverse spectral transform (IST) method - was later shown to be applicable also to other nonlinear partial differential equations, thereby demonstrating that the GGKM result for the KdV equation was not a fluke. The first step in this direction was made by Zakharov and Shabat [4] who generalized the IST procedure and applied it to the so called nonlinear Schroedinger equation

$$(1.2) \qquad iq_t(x,t)+q_{xx}(x,t)+c|q(x,t)|^2 q(x,t)=0, \qquad c=+1 \text{ or } -1,$$

a nonlinear partial differential equation that also occurs in a large number of physical phenomena.

The way was thereby opened to the search and discovery of many other nonlinear partial differential equations, or rather classes of nonlinear partial differential equations, solvable by the same technique. Of the many papers in this area, we mention as particularly significant those by Zakharov and Shabat [5] , by Ablowitz, Kaup, Newell and Segur [6] and by Calogero and Degasperis [7] . The scope of application of the IST technique was thereby greatly expanded.

Much research was also devoted to the investigation of the remarkable properties possessed by the solutions of these nonlinear partial differential equations: the existence of an infinite number of local conservation laws [8] ; the existence of Bäcklund transformations [9], their implications (including a "nonlinear superposition principle"; see below), their generalizations and relation to the IST [10] ; and various qualitative aspects of the solutions, in particular their asymptotic behavior [11] . Of major importance from a conceptual point of view was the demonstration, first given by Zakharov and Faddeev [12] for the KdV equation, of the possibility to interpret a nonlinear partial differential equation solvable by the IST as a completely integrable Hamiltonian system (of course, with an infinite number of degrees of freedom). Also of fundamental importance was an early paper by Lax [13], displaying the basic rôle of the formula

$$L_t=[L,M] , \qquad (1.3)$$

where L and M are operators; it characterizes a time evolution that leaves invariant the spectrum of L (isospectral flow), and it has been a cornerstone on which many of the subsequent developments were founded.

This terse outline can of course do no justice to the many investigations - analytical, applicative and numerical - that have contributed over the last decade to the extraordinary development of this sector of applied mathematics (the number of relevant papers probably exceeds by now one thousand!). Moreover we have in this outline completely ignored certain important developments, such as the treatment of periodic and discrete cases, the extension to more than one space dimension, the relationship with algebraic and geometric approaches, the discovery of several completely integrable hamiltonian systems having the appearance of many-body problems and their intriguing relations to partial differential equations [14] , and, last but not least, the properties of nonlinear equations that are both solvable by IST and relativistically invariant (the first example being the Sine-Gordon equation [15]) and the whole field of research opened by the conjecture that elementary particles (or maybe only some of them) have something to do with solitons (a conjecture that appears naturally suggested by the particle-like behavior of solitons; see below).

For all these developments we must refer the reader to the original literature; there exist also some review papers [16], and a number of books (mostly Conference Proceedings) entirely devoted to these topics [17].

The purpose and scope of this lecture is to introduce a reader, not already familiar with this field of research, to some of the novel ideas that have been injected, in this context, into mathematical physics or, more generally, into applied mathematics. Our presentation is focussed on the interpretation of the (Inverse) Spectral Transform as an extension of the Fourier Transform; an extension adequate to "solve" certain classes of nonlinear partial differential equations just as the Fourier Transform "solves" certain classes of linear partial differential equations. For simplicity, we focus on the case with only one space variable and mainly on equations describing in the most straightforward manner the time evolution, namely on evolution equations having the general structure

$$(1.4) \qquad q_t(x,t)=F\left[q(x,t),\ q_x(x,t),\ q_{xx}(x,t),\ \dots\right] \quad .$$

We will of course consider a subclass of such equations, to be identified below. Note that we are assuming the function F in the r.h.s. not to depend explicitly on x and t; actually an explicit t-dependence could be easily taken care of in the framework of the classes of equations described below; but we prefer to dispense with inessential complications, since it is the ideas that we are trying to convey rather than the full generality of the results. And we have also stated that we shall limit our consideration to equations with two independent variables only; it should be emphasized that, in contrast to the case of linear partial differential equations solved by the Fourier Transform, the extension to more than one space variable is far from trivial.

This paper is organized according to the following scheme. In Section 2 we review tersely the solution of linear partial differential equations by Fourier Transform. In Section 3 we introduce and explain the idea of the Inverse Spectral Transform, and we identify classes of nonlinear evolution equations that are solvable by this technique. In Section 4 we focus on the major novel phenomenon associated with the class of nonlinear evolution equations solvable by the IST, namely the soliton. In Section 5 we describe a number of remarkable properties that characterize the nonlinear evolution equations solvable by IST : Bäcklund transformations, nonlinear superposition principle, generalized resolvent formula, nonlinear operator identities, conservation laws. Finally in Section 6 we outline the generalization of the approach to systems of coupled nonlinear evolution equations, and the novel behavior that characterizes the solitons in this more general case.

2. Why is the Fourier transform so important in mathematical physics?

Consider the linear partial differential equation

$$q_t(x,t)=-i\omega(-i\frac{\partial}{\partial x})q(x,t), \tag{2.1}$$

where $\omega(z)$ is an analytic function (say, a polynomial; or a rational function, in which case the denominator should be considered to act on the l.h.s.). The study of many natural phenomena can be reduced to the determination (on the whole real line, $-\infty<x<\infty$) of the solution $q(x,t)$ of (2.1) characterized by the initial condition

$$q(x,0)=q_o(x), \tag{2.2}$$

where $q_o(x)$ vanishes asymptotically, say

$$\lim_{|x|\to\infty}\left[|x|^{1+\varepsilon}\, q_o(x)\right]=0, \qquad \varepsilon>0. \tag{2.3}$$

Assume moreover, for simplicity, $\omega(z)$ to be real and odd, $\omega(z)= -\omega(-z)$; then (2.1) is also real, and $q(x,t)$ is real for $t>0$ if $q_o(x)$, eq.(2.2), is real (as we assume hereafter).

The solution of (2.1-3) is accomplished in three steps. The first step introduces the Fourier transform

$$\hat{q}(k,t)=\int_{-\infty}^{\infty}dx\ \exp(-ikx)q(x,t); \tag{2.4}$$

for t=0 this formula, together with (2.2), implies of course

$$\hat{q}(k,0)=\int_{-\infty}^{\infty}dx\ \exp(-ikx)q_{\circ}(x)\equiv\hat{q}_{\circ}(k), \tag{2.5}$$

so that $\hat{q}(k,0)$ can be considered as given. The second step notes that (2.1) implies

$$\hat{q}_t(k,t)=-i\,\omega(k)q(k,t), \tag{2.6}$$

that can be immediately integrated to yield

$$\hat{q}(k,t)=\exp\left[-i\,\omega(k)t\right]\hat{q}(k,0). \tag{2.7}$$

The last step reproduces the solution q(x,t) via the Fourier transform formula

$$q(x,t)=\int_{-\infty}^{\infty}\frac{dk}{2\pi}\ \exp(ikx)\hat{q}(k,t). \tag{2.8}$$

This expression, together with (2.7) and (2.5), provides the explicit solution of the problem (2.1-2).

We submit that the main reason why the Fourier transform is so important in mathematical physics is because it provides, as we have just illustrated, the appropriate technique to solve the problem characterized by (2.1) and (2.2); indeed let us reemphasize that this mathematical problem, as well as its generalizations that are also analogously solvable by Fourier methods, constitute the prototypical schematization of many natural phenomena.

As implied by the technique of solution that we have just described, the behavior of the solutions q(x,t) of (2.1-2) is essentially determined by the <u>dispersion function</u> $\omega(k)$: a typical solution q(x,t), characterized by initial data $q_{\circ}(x)$ that are localized in space and have a Fourier transform $\hat{q}_{\circ}(k)$ also localized around a value $k_{\circ}$, behaves generally as a "wave packet" moving with the <u>group velocity</u>

$$(2.8) \qquad v_g = d\omega(k)/dk \Big|_{k=k_o}$$

and dispersing asymptotically (the peak magnitude of its envelope is asymptotically localized around $x=x_o+v_g t$ and decreases as $t^{-\frac{1}{2}}$ for $t \longrightarrow \infty$). Let us note that this behavior, although of course well understood, is not immediately evident from the structure of the differential equation (2.1). This should be contrasted with the extremely simple time evolution of the Fourier transform $\hat{q}(k,t)$, see (2.7). It is thus seen that "things are much simpler in Fourier space than in configuration space"; indeed the main lesson to be drawn from the solvability of the mathematical problem (2.1-2) by Fourier techniques is, that the appropriate way to understand the corresponding physical phenomena is by translating them into the language of Fourier space.

Finally let us report certain elementary properties of the linear partial differential equation (2.1), that, as we shall see, have a natural generalization in the case of the nonlinear partial differential equations considered below.

First we note that, if $q(x,t)$ satisfies (2.1) and $q'(x,t)$ is related to $q(x,t)$ by the formula

$$(2.9) \qquad f(-i\partial/\partial x)q'(x,t)+g(-i\partial/\partial x)q(x,t)=0,$$

then $q'(x,t)$ also satisfies (2.1). The corresponding equation in Fourier space reads

$$(2.10a) \qquad f(k)\hat{q}'(k,t)+g(k)\hat{q}(k,t)=0,$$

or, equivalently,

$$(2.10b) \qquad q'(k,t)=-\left[g(k) / f(k)\right] \hat{q}(k,t),$$

clearly implying that, if $\hat{q}(k,t)$ satisfies (2.6), so does $\hat{q}'(k,t)$. Note that in these equations f and g are essentially arbitrary functions.

Secondly we note that, if $q_1(x,t)$ and $q_2(x,t)$ satisfy (2.1), so does

$$q(x,t)=c_1 q_1(x,t)+c_2 q_2(x,t); \tag{2.11}$$

this is the "superposition principle", corresponding to the linear character of (2.1).

Finally we remark that the solution of (2.1-2) can be formally written through the resolvent formula

$$q(x,t) = \exp\left[-it\,\omega(-i\,\partial/\partial x)\right] q_\circ(x), \tag{2.12}$$

corresponding to the straightforward integration of (2.1) with (2.2). Note incidentally that in the special case $\omega(z)=-z$, equation (2.1) becomes

$$q_t(x,t)=q_x(x,t), \tag{2.13}$$

so that its solution is

$$q(x,t)=q(x+t,0)=q_\circ(x+t). \tag{2.14}$$

Thus in this case (2.12) yields the operator formula

$$f(x+a)=\exp\left[a\ d/dx\right]\ f(x), \tag{2.15}$$

where we have written a in place of t and $f(x)$ in place of $q_\circ(x)$, to underline the arbitrariness of these quantities.

3. The Inverse Spectral Transform, and the Nonlinear Evolution Equations Solvable by it.

Consider the (singular) Sturm-Liouville problem characterized, on the whole line $-\infty < x < \infty$, by the Schroedinger equation

$$-\psi_{xx}(x,k) + q(x)\,\psi(x,k)=k^2\,\psi(x,k), \tag{3.1}$$

with the (real) "potential" $q(x)$ vanishing at infinity, say

$$\lim_{x \to \pm\infty} \left[|x|^{1+\varepsilon} q(x) \right] = 0, \qquad \varepsilon > 0. \tag{3.2}$$

As it is well known, the continuous part of the spectrum, corresponding to k real so that $k^2 \geq 0$, can be characterized by the "reflection" and "transmission" coefficients $R(k)$ and $T(k)$ defined by the asymptotic behavior of an appropriate solution of (3.1):

$$\psi(x,k) \longrightarrow \exp(-ikx)+R(k)\exp(ikx), \qquad x \longrightarrow +\infty, \tag{3.3a}$$

$$\psi(x,k) \longrightarrow T(k)\exp(-ikx), \qquad x \longrightarrow -\infty. \tag{3.3b}$$

These equations define $R(k)$ and $T(k)$ for positive k; for negative k these quantities are defined by analytic continuation, or equivalently by the reflection formula

$$R(-k)=R^*(k), \qquad T(-k)=T^*(k), \qquad k>0, \tag{3.4}$$

that is implied by the assumed reality of $q(x)$.

The spectral problem (3.1-2) may also admit a finite number of discrete negative eigenvalues ("bound states")

$$k_n^2 = -p_n^2, \qquad k_n = ip_n, \qquad p_n > 0, \qquad n=1,2...N; \tag{3.5}$$

the corresponding real wave functions $\psi(x,ip_n) \equiv \psi_n(x)$ being normalizable. The (real) quantities c_n, $\bar{c}_n$ are associated to the discrete eigenvalue p_n by the formulas

$$\int_{-\infty}^{\infty} dx \, |\psi_n(x)|^2 = 1, \tag{3.6}$$

$$\psi_n(x) \longrightarrow c_n \exp(-p_n x), \qquad x \longrightarrow +\infty, \tag{3.7a}$$

$$\psi_n(x) \longrightarrow \bar{c}_n \exp(p_n x), \qquad x \longrightarrow -\infty, \tag{3.7b}$$

For reasons that will become immediately clear we define as <u>spectral data</u> S the quantities

$$(3.8) \qquad S \; : \; \left\{ R(k), \quad -\infty < k < \infty; \;\; p_n, \; c_n, \;\; n=1,2\ldots N \right\}.$$

The <u>direct spectral problem</u> consists in the determination of the spectral data S from a given potential $q(x)$; it clearly has a unique solution, obtainable as indicated above via the solution of the Schroedinger equation (3.1).

The <u>inverse spectral problem</u> consists in the determination of a potential $q(x)$ from given spectral data S ; it also has a unique solution (provided the given spectral data satisfy some obvious properties, such as (3.4) and the existence of the Fourier transform of $R(k)$, see below). The method of solution is via the Gel'fand-Levitan-Marchenko integral equation

$$(3.9) \qquad K(x,x')+M(x+x')+\int_x^\infty dx'' K(x,x'')M(x''+x')=0, \qquad x \leq x',$$

where the function M is defined in terms of the spectral data S by

$$(3.10) \qquad M(y)=\sum_{n=1}^{N} c_n^2 \exp(-p_n y)+(2\pi)^{-1}\int_{-\infty}^{+\infty} dk \exp(iky)R(k).$$

The Fredholm equation (3.9) determines uniquely the function $K(x,x')$ (the dependence on the first argument, x, is parametric; the integral equation refers to the second argument, x'); and this function determines the potential $q(x)$ through

$$(3.11) \qquad q(x) = -2\frac{d}{dx} K(x,x).$$

The spectral data can thus be assigned with large arbitrariness, there existing always a corresponding potential. There are however natural choices of the spectral data, that produce potentials with particularly smooth behavior and fast asymptotic vanishing. These choices are functions $R(k)$ that admit analytic continuation off the real axis and have poles, in the upper half of the complex k plane, only on the imaginary axis, the locations and residues of these poles being related to the parameters characterizing the discrete part of the spectrum by the formula

$$\lim_{k\to ip_n} \left[(k-ip_n)R(k) \right] = ic_n^2 \quad . \tag{3.12}$$

Note that the identification of the poles of R(k) with ip_n implied by this formula is consistent with a comparison of (3.3) with (3.7), suggesting of course that also T(k) has poles at the same locations.

But even spectral data that violate this rule can produce very smooth potentials that vanish exponentially at infinity. For instance to the spectral data

$$R(k)=0; \qquad N=1, \qquad p_1=p, \qquad c_1=c, \tag{3.13}$$

there corresponds the potential

$$q(x)=-2p^2/\cosh^2 \left[p(x-x_\circ) \right] \tag{3.14}$$

with

$$c^2=2p \exp(2px_\circ), \tag{3.15}$$

as is immediately implied by the GLM integral equation (3.9) (that is in this case immediately solvable, the kernel M(x"+x') being separable).

The direct and inverse spectral problems institute a biunivocal correspondence between potential q and spectral data S, with the possibility to obtain one from the other by solving a <u>linear</u> problem: the Schroedinger equation (3.1) to go from q to S, the GLM integral equation (3.9) to go from S to q.

Assume now that the potential q depend on a parameter t ("time"), q=q(x,t); the corresponding spectral data then also depend on t, S= =S(t). The idea is now to identify time evolutions of q(x,t) that are described by (generally nonlinear) differential equations of type (1.4), such however that the corresponding evolutions of the spectral data are extremely simple. The initial value problem for (1.4) can then be solved by first going from $q(x,0)$ to $S(0)$, letting next evolve S from $S(0)$ to $S(t)$, and finally recovering $q(x,t)$ from $S(t)$. This procedure is clearly analogous to that described in the preceeding Section, except for the replacement of the (inverse and direct) Fourier transform by the (direct and inverse) spectral transform. The analogy is actually very close: indeed it will be seen that, if all nonlinear effects are neglected ("weak field limit"), the novel technique reproduces essentially the Fourier method.

What is now needed is therefore a technique to relate changes in the potential to changes in the spectral data. A very convenient tool to obtain this kind of information is provided by the following relationships, that are implied by generalized versions of the Wronskian theorem [18]:

$$(3.16)\qquad 2ikf(-4k^2)\left[R'(k)-R(k)\right]=\int_{-\infty}^{\infty}dx\,\psi'(x,k)\,\psi(x,k)f(\underline{\Lambda})\cdot\left[q'(x)-q(x)\right],$$

$$(3.17)\qquad (2ik)^2g(-4k^2)\left[R'(k)+R(k)\right]=\int_{-\infty}^{\infty}dx\,\psi'(x,k)\,\psi(x,k)g(\underline{\Lambda})\cdot\underline{\Gamma}\cdot 1,$$

$$(3.18)\qquad \underline{\Lambda}\cdot F(x)=F_{xx}(x)-2\left[q'(x)+q(x)\right]F(x)+\underline{\Gamma}\cdot\int_x^{\infty}dx'F(x'),$$

$$(3.19)\qquad \underline{\Gamma}\cdot F(x)=\left[q'_x(x)+q_x(x)\right]F(x)+\left[q'(x)-q(x)\right]\int_x^{\infty}dx'\left[q'(x')-q(x')\right]F(x')$$

In (3.16) and (3.17), q and q' are two different potentials, R and R' the corresponding reflection coefficients and ψ, ψ' the corresponding wave functions (solutions of (3.1) and (3.3)). The functions f and g are essentially arbitrary; to visualize the formulas in the simpler case, imagine they are polynomials. Finally $\underline{\Lambda}$ and $\underline{\Gamma}$ are integro-differential operators, whose action on a generic function $F(x)$ (required to vanish at infinity in the case of $\underline{\Lambda}$) is displayed by (3.18) and (3.19). Note that $\underline{\Lambda}$ and $\underline{\Gamma}$ depend themselves on q and q'; thus, if for instance f and g are polynomials, the factors multiplying the terms $\psi'(k,x)\,\psi(k,x)$ in the integrands in the r.h.s. of (3.16) and (3.17) are polynomials in the potentials q and q' and in their derivatives and integrals.

It should be emphasized that these relations are merely consequences of the Schroedinger equation (3.1) and of the boundary conditions (3.2) and (3.3). They clearly imply that, if the two potentials $q'(x)$ and $q(x)$ are related by the (nonlinear integrodifferential) equation

$$f(\underline{\Lambda})\left[q'(x)-q(x)\right]+g(\underline{\Lambda})\cdot\underline{\Gamma}\cdot 1=0, \tag{3.20}$$

the corresponding reflection coefficients are related by the linear formula

$$f(-4k^2)\left[R'(k)-R(k)\right]+2ikg(-4k^2)\left[R'(k)+R(k)\right]=0 \tag{3.21a}$$

implying

$$R'(k)=\left\{\left[f(-4k^2)-2ikg(-4k^2)\right]/\left[f(-4k^2)+2ikg(-4k^2)\right]\right\}R(k). \tag{3.21b}$$

The (arbitrary) functions f and g could of course depend also on other parameters (see below).

To derive evolution equations for $q(x,t)$ it is now convenient to set

$$q'(x)=q(x,t+\Delta t),\qquad q(x)=q(x,t). \tag{3.22}$$

In the limit $\Delta t \rightarrow 0$ the equations (3.16-19) then yield

$$2ikf(-4k^2)R_t(k,t)=\int_{-\infty}^{\infty}dx\,\psi^2(x,k,t)f(\underline{L})q_t(x,t), \tag{3.23}$$

$$(2ik)^2\,g(-4k^2)R(k,t)=\int_{-\infty}^{\infty}dx\,\psi^2(x,k,t)\,g(\underline{L})q_x(x,t), \tag{3.24}$$

$$\underline{L}\,F(x)=F_{xx}(x)-4q(x,t)F(x)+2q_x(x,t)\int_x^{\infty}dx'F(x'). \tag{3.25}$$

Clearly these equations imply that, to the nonlinear evolution equation for $q(x,t)$,

$$q_t(x,t)=\alpha(\underline{L})q_x(x,t), \tag{3.26}$$

there corresponds for R the linear equation

(3.27) $$R_t(k,t)=2ik\,\alpha(-4k^2)R(k,t),$$

that can be immediately integrated to yield

(3.28) $$R(k,t)=\exp\left[2ik\,\alpha(-4k^2)t\right]R(k,0).$$

Because $\underline{L}$ is an integro-differential operator, it might appear that (3.26) need not be a pure differential equation. It can however be shown that repeated application of $\underline{L}$ to q_x produces only a (non-linear) combination of q and its x-derivatives (all the integrations contained in powers of $\underline{L}$ can be performed exactly). Thus any polynomial choice of α yields a nonlinear evolution equation for $q(x,t)$ of type (1.4), to which there corresponds the very simple evolution (3.28) for $R(k,t)$. To visualize the first equations of this class, we note that

(3.29) $$\underline{L}q_x(x,t)=q_{xxx}(x,t)-6q_x(x,t)q(x,t),$$

(3.30) $$\underline{L}^2q_x(x,t)=q_{xxxxx}(x,t)-10q_{xxx}(x,t)q(x,t)-20q_{xx}(x,t)q_x(x,t)-$$
$$-30q_x(x,t)q^2(x,t),$$

implying incidentally that, except for a trivial rescaling of the dependent variable by a factor of 6, the choice $\alpha(z)=-z$ reproduces just the KdV equation (1.1).

Rational choices of $\alpha(z)$ also yield equations that can be reduced to pure differential form, but only through a change of dependent variable. For instance insertion of

(3.31) $$\alpha(z)=a/(b+z)$$

in (3.26), yields the differential equation

(3.32) $$Q_{xxxt}+4Q_{xt}Q_x+2Q_{xx}Q_t+bQ_{xt}+aQ_{xx}=0$$

for

$$(3.32a)\qquad Q\equiv Q(x,t)=\int_x^\infty dx' q(x',t).$$

It is now clear how the class of nonlinear evolution equations (3.26) can be solved via the Inverse Spectral Transform (IST) technique. Given

$$(3.33)\qquad q(x,0)=q_\circ(x),$$

the corresponding reflection coefficient

$$(3.34)\qquad R_\circ(k)=R(k,0)$$

is computed from (3.1) and (3.3) (with $q(x)\equiv q_\circ(x)$); then $R(k,t)$ is obtained from (3.28); finally, from $R(k,t)$, $q(x,t)$ is constructed, solving the inverse spectral problem (3.9-11). If bound states are present, to perform the last step it is also necessary to know at time t the parameters of the discrete spectrum. The time evolution of these parameters can be investigated by techniques analogous to those described above for $R(k,t)$. We report directly the final results:

$$(3.35)\qquad p_n(t)=p_n(0)\equiv p_n,$$

$$(3.36)\qquad c_n(t)=\exp\left[-p\ \alpha(4p^2)t\right]c_n(0)$$

(note incidentally their consistency with (3.28) and (3.12)). We also report for completeness the (extremely simple) equation describing the time evolution of the transmission coefficient , that can also be obtained by analogous techniques:

$$(3.37)\qquad T(k,t)=T(k,0)\equiv T(k).$$

Although this formula is not needed for the solution by IST of (3.26), it clearly displays an important property of the flow (3.26) (see below). Attention should also be called to the time-invariance (3.35) of the discrete eigenvalues, displaying the isospectral character of the flow (3.26) (this indicates the connection with the Lax approach of eq.(1.3)).

4. Behavior of the Solutions of Nonlinear Evolution Equations solvable by the IST: Solitons

The IST method of solution that has just been described implies that the behavior of the solutions $q(x,t)$ of a nonlinear evolution equation of the class (3.26) should be analyzed in terms of two components, one associated to the discrete spectrum of the corresponding Schroedinger spectral problem, the other to the continuum (see the definition (3.10) of the kernel M of the GLM integral equation, where these two contributions are quite clearly displayed).

Let us first consider the special solution that corresponds to $R(k,t)=R(k,0)=0$ (see (3.28)) and to only one discrete eigenvalue. It reads

$$q(x,t)=-2p^2/\cosh^2\{p\,[x-x_\circ+\alpha(4p^2)t]\}, \tag{4.1}$$

as implied by (3.13-15) and (3.35-36). It consists therefore of a single bump of constant shape, moving with the constant speed $v=-\alpha(4p^2)$; for instance, for the KdV equation $q_t+q_{xxx}-6q_xq=0$, $v=4p^2$ (see (3.29) and (3.26)), so that in this case the bump moves to the right with a speed proportional to its height. The name soliton has been coined for such an entity [19].

If there are N discrete eigenvalues rather than only one, but still $R(k,t)=R(k,0)=0$, the GLM equation (3.9) can still be solved in closed form, and, after each parameter c_n is equipped with the factor characterizing its time dependence (see (3.36)), it yields the N-soliton solution of the corresponding equation of the class (3.26). While for an explicit display of this expression in the general case we refer to the literature (see for instance the first paper of reference [16]), we report here a two-soliton case, with $p_1=1$ and $p_2=2$, for the KdV equation $q_t+q_{xxx}-6q_xq$. It reads

(4.2) $$q(x,t) = \frac{-12[3+4\cosh(2x-8t)+\cosh(4x-64t)]}{[3\cosh(x-28t)+\cosh(3x-36)]^2} .$$

For large t, this formula approaches the sum of two soliton of the form

$$-2p_j^2/\cosh^2[p_j(x-4p_j^2t)+\tfrac{1}{2}\eta_j\log 3], \qquad j=1,2,$$

with $p_1=1$, $p_2=2$, $\eta_1=\mathrm{sign}(t)$, $\eta_2=-\mathrm{sign}(t)$.

The fact that the N-soliton solution becomes asymptotically (as $|t| \longrightarrow \infty$) just the sum of N separated solitons, each with its different amplitude and speed, is of course generally true. At intermediate times the solitons may instead come together, and their separate identities are then not visible (see (4.2)); but in the remote future each recovers the same shape and speed it had in the remote past. The interaction of one soliton with the others does however have some effect: it produces a shift of its coordinate relative to what it would have been if it had moved alone all the time (see, for an instance of this effect, the example just reported: for the soliton with $p_1=1$ the shift is $\Delta x=-\log 3$, for the soliton with $p_2=2$ it is $\Delta x=\frac{1}{2}\log 3$).

Let us now consider the other case, when no solitons are present, namely when in the associated spectral problem there are no discrete eigenvalues (no "bound states"). Indeed let us take the case of a very weak field, neglecting all nonlinear terms (if in the Schroedinger equation the "potential" is weak, it can hardly sustain bound states; although actually, if the potential is everywhere negative, it does always sustain at least one bound state; so that the following analysis is, strictly speaking, applicable only to solutions q that are both weak and, at least predominantly, positive; indeed it has also some other flaw, see below).

In this linear approximation, the evolution equation (3.26) becomes

(4.3) $$q_t(x,t)=\alpha(\partial^2/\partial x^2)q_x(x,t),$$

namely it coincides with (2.1), with the identification

$$\omega(z)=-z\,\alpha(-z^2) \ . \tag{4.4}$$

Moreover the GLM equation (3.9), after the neglect of the nonlinear term, yields

$$K(x,x')=-M(x+x') \tag{4.5}$$

and, through (3.11) and (3.10), it yields

$$\hat{q}(k,t)=\frac{i}{2}kR(k/2,t) \tag{4.6}$$

where $\hat{q}(k,t)$ is just the Fourier transform (2.4) of $q(x,t)$.

The consistency of the formula (4.6), together with (4.4), with the time evolution (2.7) of $\hat{q}(k,t)$ and (3.28) of $R(k,t)$, should be noted; as well as the precise sense in which the IST approach appears now as an extension of the Fourier transform method.

The analysis of Section 2 is now applicable to understand the behavior of a "weak field" solution of the type considered here. For instance for the linearized KdV equation (corresponding to $\alpha(z)=-z$ in (3.26), implying though (4.4) $\omega(k)=-k^3$) the group velocity (2.8) reads

$$v_g=-3k_o^2 \tag{4.7}$$

and is therefore never positive. Thus, in contrast to the solitons, who move to the right (and remain asymptotically localized), the "weak field" solution moves, if at all, towards the left (and of course it disperses asymptotically).

(A warning should be voiced here: the discussion we have just given is actually overly cavalier, so that, although its main message is essentially valid, a more careful treatment would be required to explain away some paradoxical conclusions suggested by (4.6), such as the fact that $\hat{q}(k,t)$ vanish at $k=0$, which of course need not happen. The difficulty has to do with the fact that, if q is small but also slowly varying, terms like qq_x need not be negligible relative to

q_{xxx}, as would for instance be required to justify the reduction of the nonlinear KdV equation $q_t+q_{xxx}-6q_xq=0$ to the linear partial differential equation $q_t+q_{xxx}=0$. But this is not the appropriate place for a detailed analysis of these fine points, that are ignored hereafter).

So far we have considered only the extreme cases: either a solution dominated by the nonlinear effects, indeed composed only of essentially nonlinear entities, the solitons; or a solution that is instead so weak to make all nonlinear effects negligible. The general case contains generally both behaviors; given a generic initial condition $q_0(x)$ at $t=0$ the corresponding solution $q(x,t)$ eventually develops a number of solitons and a "background part", each of these components behaving asymptotically as described above (note that in the KdV case this implies in the remote future a complete separation of the solitons, that escape separately to the right, from the background, that disperses and moves, if at all, towards the left).

Depending from the nature of the particular phenomenon modeled by the equation being considered there will be an appropriate interpretation of the soliton and background parts of the solution. The general message that the applied mathematician should learn from the solvability by IST of the nonlinear evolution equation he happens to be interested in, is the convenience, to interpret the associated natural phenomenon, to think in terms of "k-space quantities" (related to the solution through the spectral problem defined in the preceeding Section), whose time evolution is much more transparent than that of the configuration space solution itself; and in particular to identify the two separate (although, of course, nonlinearly intertwined) behaviors, namely the emergence of the solitons (if any) and the dispersive evolution of the background component (if any), this latter being essentially determined, in the usual Fourier fashion, by the dispersion function $\omega(k)$ associated to the linearized version of the evolution equation under scrutiny. In some cases, as in the KdV case outlined above, these two parts of the solution separate asymptotically; indeed one part generally "disappears" by dispersion, while the other factors into a series of separated solitons, each characterized by its own amplitude and speed.

The scope of applications of the KdV equation and of other nonlinear evolution equations solvable by IST is by now quite large, ranging from hydrodynamics to elementary particle physics, from plasma physics to the study of the propagation of signals through nervous fibres, and so on [16,17]. Suffice here to report the delightful description of the first scientific observation of a soliton [3] :

" I was observing the motion of a boat which was rapidly drawn along a narrow channel by a pair of horses, when the boat suddenly stopped - not so the mass of water in the channel which it had put in motion; it accumulated round the prow of the vessel in a state of violent agitation, then suddenly leaving it behind, rolled forward with great velocity, assuming the form of a large solitary elevation, a rounded, smooth and welldefined heap of water, which continued its course along the channel apparently without change of form or diminution of speed. I followed it on horseback, and overtook it still rolling on at a rate of some eight or nine miles an hour, preserving its original figure some thirty feet long and a foot to a foot and a half in height. Its height gradually diminished, and after a chase of one or two miles I lost it in the windings of the channel. Such, in the month of August 1834, was my first chance interview with that singular and beautiful phenomenon...."

This observation was (partially) explained in 1895 by Korteweg and de Vries, who derived the equation that now bears their name to describe (in an appropriate, moving, reference frame) the motion of long water waves in a shallow canal (neglecting dissipation, but including both dispersive and nonlinear effects), and displayed the single-soliton solution to it [2] . But actually the real explanation came only with the discovery of the IST in 1967 [1], since it was thereby shown that a soliton-like behavior does not obtain only from the very special initial conditions that correspond to the pure single-soliton solution, being instead a feature generally present in the solution of the KdV equation .

5. Bäcklund transformations, nonlinear superposition principle, generalized resolvent formula, nonlinear operator identities, conservation laws.

In this Section we review tersely some important properties of the solutions of the nonlinear evolution equations solvable by IST, emphasizing whenever appropriate the analogy with corresponding results for the linear partial differential equations solvable by the Fourier transform (see Section 2).

Assume $q(x,t)$ to be a solution of the nonlinear evolution equation (3.26); then the corresponding reflection coefficient $R(k,t)$ evolves in time according to (3.28). Assume $q'(x,t)$ to be related to $q(x,t)$ by the formula (3.20), with the integro-differential operators $\underline{\Lambda}$ and $\underline{\Gamma}$ defined by (3.18) and (3.19) with $q'(x) \equiv q'(x',t)$ and $q(x) \equiv q(x,t)$. Then (3.21b) implies that the reflection coefficient $R'(k,t)$ also evolves in time according to (3.28) (with $R(k,t)$ and $R(k,0)$ replaced by $R'(k,t)$ and $R'(k,0)$).

But this in turn implies that $q'(x,t)$ must satisfy the same nonlinear evolution equation (3.26) as $q(x,t)$. (Actually this conclusion would require an analysis of the behavior of the discrete parameters; space limitations force us to omit hereafter this part of the analysis).

The relation (3.20) is thus seen to be a "Bäcklund transformation", namely a relationship between the two fields $q(x,t)$ and $q'(x,t)$ such that, if one of them satisfies a nonlinear evolution equation of the class (3.26), $q'(x,t)$ also satisfies the same equation.

This "Bäcklund transformation" is very general, since the two functions f and g that characterize it are essentially arbitrary. It clearly is the analog of the formula (2.9) relating two solutions of the linear partial differential equation (2.1). Its significance is best understood looking at it in k-space rather than in x-space: indeed (3.21b) is much simpler than (3.20).

The simpler Bäcklund transformation corresponds to the choice $g=cost$, $f=cost$, $f/g=2p$, in which case (3.21) and (3.20) yield respectively

$$R'(k,t)=-\left[(k+ip)/(k-ip)\right] R(k,t), \tag{5.1}$$

$$Q'_x(x,t)+Q_x(x,t)=-\tfrac{1}{2}\left[Q'(x,t)-Q(x,t)\right]\left[4p+Q'(x,t)-Q(x,t)\right]. \tag{5.2}$$

In writing the last equation we have used the more convenient dependent variable

$$(5.3) \qquad Q(x,t)=\int_x^{\infty} dx' q(x',t), \qquad Q_x(x,t)=-q(x,t)$$

(clearly with an analogous formula for the primed quantities); to obtain it from (3.20) one integration has been performed.

To obtain Q', and therefore also q', from given q (or equivalently Q), (5.2) must be integrated; note that this is an ordinary differential equation for the x-dependence of $Q'(x,t)$, with the time entering only parametrically through the time dependence of $Q(x,t)$. It is nonlinear; indeed it is a Riccati equation. The constant of integration (that generally depends on t; it is constant in x, not in t!) must be chosen so that Q' satisfy the boundary conditions $Q'(+\infty,t)=Q'_x(\pm\infty,t)=0$, that are clearly implied by the definition (5.3) and by (3.2), not be singular, and $q'(x,t)$ be a solution of (3.26) (this latter requirement determines the time evolution of the constant of integration). As suggested by (5.1) and (3.12), for $p>0$ the $q'(x,t)$ so determined generally exists and has one more soliton (characterized by the parameter p) than $q(x,t)$.

A special example, that serves to illustrate one possible use of Bäcklund transformations, is to start from $q(x,t)=0$, that is of course a (trivial) solution of (3.26). Then (5.2) is easily integrated to yield

$$(5.4) \qquad q'(x,t)=-2p^2/\cosh^2\{p[x-\xi(t)]\},$$

and insertion of this expression in (3.26) yields

$$(5.5a) \qquad \xi_t(t)=-\alpha(4p^2) \quad ,$$

so that

$$(5.5b) \qquad \xi(t)=x_\circ-\alpha(4p^2)t \quad .$$

Insertion of this formula in (5.4) yields just the single-soliton formula (4.1), consistently with the indications given above.

Imagine now to perform two Bäcklund transformations (5.2), characterized by parameters p_1 and p_2, one after the other. With obvious notation:

(5.6) $$Q_{1x}+Q_x=-\tfrac{1}{2}(Q_1-Q)(4p_1+Q_1-Q),$$

(5.7) $$Q'_x+Q_{1x}=-\tfrac{1}{2}(Q'-Q_1)(4p_2+Q'-Q_1),$$

(5.8) $$R'(k,t)=\left\{(k+ip_1)(k+ip_2)/\left[(k-ip_1)(k-ip_2)\right]\right\}R(k,t),$$

(5.9) $$Q_{2x}+Q_x=-\tfrac{1}{2}(Q_2-Q)(4p_2+Q_2-Q),$$

(5.10) $$Q''+Q_{2x}=-\tfrac{1}{2}(Q''-Q_2)(4p_1+Q''-Q_2),$$

(5.11) $$R''(k,t)=\left\{(k+ip_2)(k+ip_1)/\left[(k-ip_2(k-ip_1)\right]\right\}R(k,t),$$

where the first 3 formulas assume that the transformation with parameter p_1 is performed first (associating Q_1 to Q) and the transformation with parameter p_2 last (associating Q' to Q_1), while the last 3 equations correspond to the transformation with parameter p_2 being performed first (associating Q_2 to Q) and that with parameter p_1 last (associating Q'' to Q_2). But clearly (5.8) and (5.11) imply $R'(k,t)=R''(k,t)$, and this in turn implies $Q'(x,t)=Q''(x,t)$. But then the difference of the differences of (5.6) and (5.7) and (5.9) and (5.10) yields, after a little trivial algebra,

(5.12) $$Q'(x,t)=Q(x,t)-(p_1+p_2)\left[Q_1(x,t)-Q_2(x,t)\right]/\\ /\left\{p_1-p_2+\tfrac{1}{2}\left[Q_1(x,t)-Q_2(x,t)\right]\right\}.$$

This formula may be considered a "nonlinear superposition principle"; it yield explicitly $Q'(x,t)$ in terms of $Q(x,t)$, $Q_1(x,t)$ and $Q_2(x,t)$. To recapitulate: $Q(x,t)$ is related though (5.3) to a solution $q(x,t)$ of (3.26); $Q_1(x,t)$ and $Q_2(x,t)$ are related to $Q(x,t)$ by Bäcklund transformations (5.2) with parameters p_1 and p_2. $q'(x,t)$, related through (5.3) to the $Q'(x,t)$ given by the explicit formula (5.12), is also a solution of (3.26) (as well as $q_1(x,t)$ and $q_2(x,t)$).

A simple instance of application of this formula obtains starting again from $q(x,t)=Q(x,t)=0$, in which case $q_1(x,t)$ and $q_2(x,t)$ are given by the single-soliton formula (4.1), implying

$$(5.13) \qquad Q_j(x,t)=2p\left\{ \operatorname{tgh}\left[p_j(x-x_j+\alpha(4p_j^2)t\right] -1\right\}, \qquad j=1,2.$$

Inserting these expressions in (5.12) (with an appropriate choice of the constants x_1 and x_2; one of them must contain an imaginary part $i\pi/(2p_j)$ to prevent $Q'(x,t)$ from having a singularity for real x) yields for $Q'(x,t)$, or rather for its x-derivative, the two-soliton solution of (3.26). And the process can clearly be continued to obtain multisoliton solutions ("soliton ladder"; see, for instance, the first paper of Reference [9]).

Let us now return to the formulae (3.20) and (3.21), but considering a more general case, with f and g depending explicitly on t. Then comparing (3.21b) with (3.28) it is easily seen that, setting $R(k)=R(k,0)$ and

$$(5.13a) \qquad f(z,t)=\cos\left[\tfrac{1}{2}(-z)^{\frac{1}{2}}\,\alpha(z)t\right],$$

$$(5.13b) \qquad g(z,t)=(-z)^{-\frac{1}{2}}\sin\left[\tfrac{1}{2}(-z)^{\frac{1}{2}}\alpha(z)t\right],$$

there obtains $R'(k)=R(k,t)$. This implies that also in (3.20) (as well, of course, as (3.18) and (3.19)) one can set $q(x)=q(x,0)$ and $q'(x)=q(x,t)$. The resulting formula provides in closed, if highly implicit, form, an equation relating $q(x,t)$ to $q(x,0)$; it is natural to term it <u>generalized resolvent formula</u>, for it constitutes the equivalent, for the nonlinear evolution equation (3.26), of the resolvent formula (2.12) (to which it does indeed reduce, using (4.4), in the weak field limit, i.e. when all nonlinear terms are neglected).

Since for $\alpha(z)=1$ the evolution equation (3.26) becomes simply $q_t(x,t)=q_x(x,t)$ and has therefore the solution $q(x,t)=f(x+t)$ with f an arbitrary function, the generalized resolvent formula in this case becomes the <u>nonlinear operator identity</u> [10]

$$(5.14) \qquad f(x+a)=f(x)+(-\underline{\Lambda})^{-\frac{1}{2}}\, tg\left[\tfrac{1}{2}(-\underline{\Lambda})^{\frac{1}{2}}\, a\right] \underline{\Gamma}\cdot 1 \qquad ,$$

where the integro-differential operators $\underline{\Lambda}$ and $\underline{\Gamma}$ are now defined by the following formulae that detail their action on a generic function F(x):

$$(5.15) \qquad \underline{\Lambda}\, F(x)=F_{xx}(x)-2\left[f(x+a)+f(x)\right] F(x)+\underline{\Gamma}\int_x^{\infty} dx' F(x'),$$

$$(5.16) \qquad \underline{\Gamma}\, F(x)=\left[f_x(x+a)+f_x(x)\right] F(x)+\left[f(x+a)-f(x)\right]\int_x^{\infty} dx' \left[f(x'+a)-\right.$$

$$\left.- f(x')\right] F(x').$$

Since $f(x)$ is arbitrary, one can replace it by $\lambda f(x)$ and expand in powers of λ the r.h.s. of (5.14), recovering the identity (2.15) (terms proportional to λ) and getting in addition an infinity of intriguing nonlinear operator identities.

Finally we mention tersely an important property of the nonlinear evolution equations (3.26), namely the fact that they possess an infinite number of independent conservation laws. There exist several ways to derive these conserved quantities [16,17] ; the procedure we outline here is similar to some of the techniques that can be found in the literature.

The starting point is the time invariance (3.37) of the transmission coefficient T(k) (for all values of k). This implies the constancy of the quantity

$$(5.17) \qquad C(k)=\int_{-\infty}^{\infty} dx\, q(x,t)\, \varphi(x,k,t) \qquad ,$$

where φ is related to the function ψ of Section 3 by the formula

$$(5.18) \qquad \varphi(x,k,t)=\exp(ikx)\, \psi(x,k,t)\, /\, T(k),$$

and is therefore characterized by the differential equation

$$(5.19) \qquad \varphi_{xx}(x,k,t)-2ik\, \varphi_x(x,k,t)-q(x,t)\, \varphi(x,k,t)=0$$

and by the boundary condition

$$\lim_{x \to \infty} \left[\varphi(x,k,t) \right] = 1; \qquad (5.20)$$

for indeed it can be easily shown [18] that

$$C(k) = 2ik \left[1 - 1/T(k) \right] \quad . \qquad (5.21)$$

Since C(k) is a constant of the motion for all values of k, also constant are the coefficients c_n of its asymptotic expansion in k:

$$C(k) = \sum_{n=0}^{N} c_n (2ik)^{-n} + 0\left[(2ik)^{-N-1} \right]. \qquad (5.22)$$

These coefficients are clearly related to the coefficients $\varphi^{(n)}(x,t)$ of the asymptotic expansion of

$$\varphi(x,k,t) = \sum_{n=0}^{N} \varphi^{(n)}(x,t)(2ik)^{-n} + 0\ (2ik)^{-N-1} \qquad (5.23)$$

through

$$c_n = \int_{-\infty}^{\infty} dx\ q(x,t)\ \varphi^{(n)}(x,t). \qquad (5.24)$$

But insertion of (5.23) in (5.19) and (5.20) yields the relations

$$\varphi^{(0)}(x,t) = 1, \qquad (5.25)$$

$$\varphi_x^{(n+1)}(x,t) = \varphi_{xx}^{(n)}(x,t) - q(x,t)\varphi^{(n)}(x,t), \qquad (5.26)$$

$$\varphi^{(n)}(-\infty,t) = \varphi_x^{(n)}(-\infty,t) = \varphi_{xx}^{(n)}(-\infty,t) = 0, \quad n=1,2,\ldots,N, \qquad (5.27)$$

which can be solved by recursion.

Inserting the expression for $\varphi^{(n)}(x,t)$ obtained in this manner in (5.24) one gets explicit expressions of the c_n's in the form of integrals of polynomials of q(x,t) and its derivatives. Each of these quantities is a constant of the motion for the whole class of equations (3.26). As it happens, however, only the even numbered c_n's yield novel constants of the motion; each odd-numbered c_n can instead be expressed as a polynomial in the constants of lower index. The first three conserved quantities that are obtainable in this manner (and also by other techniques) are most simply written as follows:

$$\int_{-\infty}^{\infty} dx\ q(x,t) \quad ,$$

$$\int_{-\infty}^{\infty} dx\ q^2(x,t) \quad ,$$

$$\int_{-\infty}^{\infty} dx \left[2q^3(x,t)+q_x^2(x,t) \right] .$$

6. Zoomerons; Interacting Solitons

Up to now we have discussed the class of nonlinear evolution equations (3.26), that are solvable via the IST associated to the Schroedinger spectral problem on the whole line. Other classes of nonlinear evolution equations can be identified, that are solvable with the help of analogous spectral problems; and a treatment similar to that reported in the preceeding Sections can be applied also in these cases.

Such a class of nonlinear evolution equations, that has played a particularly important rôle in the development of the theory, is associated to the spectral problem characterized by the system of two first-order equations

(6.1a) $$\psi_x(x,k)+ik\ \psi(x,k)=q(x)\ \varphi(x,k),$$

(6.1b) $$\varphi_x(x,k)-ik\ \varphi(x,k)=r(x)\ \psi(x,k),$$

with appropriate boundary conditions. This class of equations includes some that appear in many physical applications, in particular the "modified Korteweg-de Vries equation"

(6.2a) $$q_t(x,t)+q_{xxx}(x,t)+q_x(x,t)q^2(x,t)=0,$$

the "nonlinear Schroedinger equation"

(6.2b) $$iq_t(x,t)+q_{xx}(x,t)\pm |q(x,t)|^2 q(x,t)=0,$$

and the "Sine-Gordon" equation

(6.2c) $$q_{tt}(x,t)-q_{xx}(x,t)=\pm\sin q(x,t).$$

For details the interested reader is referred to the original literature (see in particular [4], [6] and the second paper of [7]).

A more general class of solvable nonlinear evolution equations is associated to the matrix version of the Schroedinger spectral (or "scattering") problem on the line. Again the whole roster of results described in the preceeding Sections can be exhibited, and we refer for them to the original literature (see in particular the last two papers listed under [7]). But since in this case a novel phenomenon appears, we illustrate it by focussing on just one special equation, reporting without proof some results that follow in a rather straightforward manner from those given in the last two papers of [7] and in [20].

The equation that we discuss has an intriguing appearance:

(6.3) $$\left[\partial^2/\partial t^2-\partial^2/\partial x^2\right]\left[Z_{xt}(x,t)/Z(x,t)\right]+2\left[Z^2(x,t)\right]_{xt}=0 \quad .$$

The fact that it is of higher than first order in t, as well as its peculiar nonlinearity (Z appears in the denominator), originate from the elimination of other fields that are coupled to Z(x,t) in a multicomponent, but less peculiar, version of this same equation, that corresponds in fact more directly to the underlying spectral problem (indeed the "potential" in the matrix Schroedinger equation is a multicomponent object).

The initial data to be provided are

(6.4) $$Z(x,0)=Z_o(x), \quad Z_t(x,0)=Z_1(x), \quad Z_{tt}(x,0)=Z_2(x),$$

and it can be shown that, in order that a solution be obtainable by the IST (and therefore exist for all time), it is sufficient that the two functions

(6.5a) $$f(x)=Z_{1x}(x)\ /\ Z_o(x),$$

(6.5b) $$g(x)=Z_{2x}(x)\ /\ Z_o(x)-Z_{1x}(x)Z_1(x)\ /\ Z_o^2(x),$$

vanish asymptotically (as $x \longrightarrow \pm\infty$) and that they, together with $Z_o(x)$, be regular for all (real) values of x.

Note that (6.3) is invariant under the exchange of Z with -Z. Hereafter we take advantage of this to assume

(6.6) $$Z_o(\infty) \equiv Z_\infty > 0.$$

The solvability of (6.3) by IST implies that the whole collection of results detailed in the previous Sections can be extended to the solutions of this equation. In particular, one can consider separately the "soliton part" of the solution, with every soliton evolving independently from the rest of the solution, and the "background part". Hereafter we focus attention on the solitons, discussing the behavior of the single-soliton solution, and displaying also the two-soliton solution: indeed the novelty alluded to above refers just to the behavior of the solitons, whose amplitude now varies with time and, most important, whose position does not now move with constant speed (namely as that of free particles), but rather with a speed that changes in time (namely, as that of particles acted upon by an external force; see below).

Indeed the single-soliton solution of (6.3) is

(6.7) $$Z(x,t)=Z_\infty+p\ \sin\theta(t)\left[1-\text{tgh}\left\{p\left[x-\xi(t)\right]\right\}\right],$$

implying

(6.8) $$Z_x(x,t)=-p^2\ \sin\theta(t)\ /\ \cosh^2\left\{p\left[x-\xi(t)\right]\right\};$$

a comparison of the latter formula with (4.1) makes precise the analogy with the cases previously considered. The time evolution of the soliton position $\xi(t)$ and of the amplitude $\sin\theta(t)$ are determined by the following equations:

$$\theta_t(t)=2Z_\infty+2p\sin\theta(t)\ , \tag{6.9}$$

$$\xi_t(t)=-\cos\theta(t)\ . \tag{6.10}$$

These equations are of course very easily solved. But rather than exhibiting explicitly the functions $\xi(t)$ and $\theta(t)$ we prefer to report the following result, implied by them: the soliton position $\xi(t)$ evolves in time as the coordinate of a unit mass nonrelativistic particle of total energy

$$E=\tfrac{1}{2}\left[1-(Z_\infty/p)^2\right] \tag{6.11}$$

moving in the external potential $V(x-\xi_0)$ where

$$V(x)=-\frac{Z_\infty}{p}\,\text{sign}(\gamma)\,\exp\left[2p(\bar{x}-x)\right]+\tfrac{1}{2}\exp\left[4p(\bar{x}-x)\right]\ , \tag{6.12}$$

$$\gamma=Z_\infty/p+\sin\theta_0\ , \tag{6.13}$$

$$\bar{x}=(2p)^{-1}\log|\gamma|\ , \tag{6.14}$$

and of course ξ_0 and θ_0 are the "initial" values of ξ and θ,

$$\xi_0=\xi(0),\qquad \theta_0=\theta(0). \tag{6.15}$$

The potential $V(x)$ diverges as $x\longrightarrow-\infty$ and vanishes as $x\longrightarrow+\infty$ (we are always assuming $p>0$); if γ is positive, it has at $x=\bar{x}-(2p)^{-1}\log(Z_\infty/p)\equiv x_{min}$ the negative minimum $V(x_{min})=-\tfrac{1}{2}(Z_\infty/p)^2=$ $=E-\tfrac{1}{2}$. Thus if E, defined by (6.11), is positive, the soliton comes in from the right in the remote past and boomerangs back to the right in the remote future; if E is negative (implying $\gamma>0$; see (6.11) and (6.13)) the soliton is trapped, and it oscillates around $x_{min}+\xi_0$ (with period $\pi(Z_\infty^2-p^2)^{-\frac{1}{2}}$). Note that there is hardly any hint of this behavior in the appearance of the nonlinear differential equation (6.3), that is translation-invariant.

The (pure) two-soliton solution reads:

$$(6.16)\quad Z(x,t)=Z_{\infty}+(p_1+p_2)\left[1-\tau_1\tau_2\cos^2\left\{\tfrac{1}{2}\left[\theta_1(t)-\theta_2(t)\right]\right\}\right]^{-1}\left[\tau_1(1-\tau_2)\sin\theta_1(t)+\right.$$
$$\left.+\tau_2(1-\tau_1)\sin\theta_2(t)\right]$$

with

$$(6.17)\quad \tau_j = p_j/(p_1+p_2)\left[1-\mathrm{tgh}\left\{p_j\left[x-\xi_j(t)\right]\right\}\right], \qquad j=1,2.$$

Here of course p_1 and p_2 are the (positive!) parameters characterizing the two solitons, and the quantities $\xi_j(t)$, $\theta_j(t)$, $j=1,2$, evolve (independently) in time as indicated above (of course with p replaced by p_j). Note that, since the sign of E, eq.(6.11), depends on the value of p, the two solitons might also have different behaviors, one oscillating and the other boomeranging. There exists now a film ("Zoomerons"), computer-produced by J.C. Eilbeck (Heriot-Watt University, Edinburgh), that displays a representative collection of such behaviors. Examples of sequences taken from this film are given in the Figure .

The solitons of the KdV equation (and of the other equations of the class (3.26)) behave in some sense as independent free particles (although to be sure the shift of their positions induced by a "collision" indicates that they do experience an interaction); the solitons (or rather zoomerons) of the equation (6.3) (and of other classes of equations [7,20]) behave as particles in external potentials. Recently a special situation has been exhibited, where two solitons behave as particles interacting among themselves [21]; it remains however to be seen whether this finding opens indeed a novel perspective, that would be most interesting, or it is just a fluke, as it might well be.

REFERENCES

1 C.S.Gardner, J.M.Greene, M.D.Kruskal and R.M.Miura, "Method for Solving the Korteweg-de Vries Equation", Phys.Rev.Lett. 19, 1095--1097 (1967).

2 D.J. Korteweg and G. de Vries, "On the Change of Form of Long Waves Advancing in a Rectangular Canal, and on a New Type of Long Stationary Waves", Phil.Mag. 39, 422-443 (1895).

3 J.Scott-Russell, "Report on Waves", Report of the Fourteenth Meeting of the British Association for the Advancement of Science, London,1845,pp.311-390.

4 V.E.Zakharov and A.B.Shabat, "Exact Theory of Two-Dimensional Self--Focusing and One-Dimensional Self-Modulation of Waves in Nonlinear Media", Soviet Phys. JETP 34, 62-69 (1972) [Russian original: Zh. Eksp.Teor.Fiz. 61, 118-134 (1971)] .

5 V.E.Zakharov and A.B.Shabat, "A Scheme for Integrating the Nonlinear Equations of Mathematical Physics by the Method of the Inverse Scattering Problem. I", Func.Anal.Appl. 8, 226-235 (1974) [Russian original: Funk.Anal.Pril. 8, 43-53 (1974)] .

6 M.J.Ablowitz, D.J.Kaup, A.C.Newell and H.Segur, "The Inverse Scattering Transform - Fourier Analysis for Nonlinear Problems", Stud. Appl.Math. 53, 249-315 (1974).

7 F.Calogero, "A Method to Generate Solvable Nonlinear Evolution Equation", Lett.Nuovo Cimento 14, 443-448 (1975); F.Calogero and A.Degasperis, "Nonlinear Evolution Equations Solvable by the Inverse Spectral Transform. I & II", Nuovo Cimento 32B, 201-242 (1976) & 39B, 1-54 (1977); "Nonlinear Evolution Equations Solvable by the Inverse Spectral Transform associated to the Matrix Schroedinger Equation" (to appear in the Springer monograph edited by R.K.Bullough).

8 See, for instance: P.D.Lax, "Integrals of Nonlinear Equations of Evolution and Solitary Waves", Comm.Pure Appl.Math. 21, 467-490 (1968); R.M.Miura, C.S.Gardner and M.D.Kruskal, "Korteweg-de Vries equation and generalizations. II. Existence of conservation laws and constants of motion", J.Math.Phys. 9, 1204-1209 (1968).

9 See, for instance: H.D.Wahlquist and F.B.Estabrook, "Bäcklund Transformation for Solutions of the Korteweg-de Vries Equation", Phys. Rev.Lett. 31, 1386-1390 (1973); G.L.Lamb jr., "Bäcklund Transformations for Certain Nonlinear Evolution Equations", J.Math.Phys. 15, 2157-2165 (1974); H.H.Chen, "General Derivation of Bäcklund Transformations from Inverse Scattering Problems", Phys.Rev.Lett. 33, 925-928 (1974).

10 F.Calogero, "Bäcklund Transformations and Functional Relation for

Solutions of Nonlinear Partial Differential Equations Solvable via the Inverse Scattering Method", Lett. Nuovo Cimento 14, 537-543 (1975); F.Calogero and A.Degasperis, "Transformations between Solutions of Different Nonlinear Evolution Equations Solvable via the same Inverse Spectral Transform, Generalized Resolvent Formulas and Nonlinear Operator Identities", Lett. Nuovo Cimento 16, 181--186 (1976).

11 See, for instance: S.V.Manakov, Sov.Phys. JETP 38, 693 (1974); H. Segur and M.J.Ablowitz, "Asymptotic Solutions and Conservation Laws for the Nonlinear Schroedinger Equation", J.Math.Phys. 17, 710 (1976); M.J.Ablowitz and H.Segur, "Asymptotic Solutions of the Korteweg-de Vries Equation", Stud.Appl.Math. 1977 (in press); V.E.Zakharov and S.V.Manakov, "Asymptotic Behavior of Nonlinear Wave Systems Solvable by the Method of the Inverse Scattering Transform", Zh.Eksp.Teor.Fiz. 71, 203 (1976).

12 V.E.Zakharov and L.D.Faddeev, "The Korteweg-de Vries Equation: a Completely Integrable Hamiltonian System", Func.Anal.Appls. 5, 28-287 (1971) [Russian original: Funk.Anal.Pril. 5, 18-27 (1971)].

13 P.D.Lax, "Integrals of Nonlinear Equations of Evolution and Solitary Waves", Comm.Pure Appl.Math. 21, 467-490 (1968).

14 Recently there have been very interesting developments on this last topic: H.Airault, H.P.McKean and J.Moser, "Rational and Elliptic Solutions of the Korteweg-de Vries Equation and a Related Many-Body Problem", (NYU preprint, to be published); D.V.Choodnovsky and G.V.Choodnovsky, "Pole Expansion of Nonlinear Partial Differential Equations", Nuovo Cimento B (in press); F.Calogero, "Motion of poles and zeros of special solutions of nonlinear and linear partial differential equations and related "solvable" many-body problems", Nuovo Cimento B (in press).

15 See, for instance: M.J.Ablowitz, D.J.Kaup, A.C.Newell and H.Segur, "Method for Solving the Sine-Gordon Equation", Phys.Rev.Lett. 30, 1262-1264 (1973); L.D.Faddeev and L.A. Takhtajan, "Essentially Nonlinear One-Dimensional Model of Classical Field Theory", Commun. JINR Dubna E2-7998 (1974).

16 A.C.Scott, F.Y.F.Chu and D.W.McLaughlin, "The Soliton: a New Concept in Applied Science", Proc.IEEE 61, 1443-1483 (1973); B.A. Dubrovin, V.B.Matveev and S.P.Novikov, "Nonlinear Equations of Korteweg-de Vries Type, Finite-Zone Linear Operators and Abelian Varieties", Uspekhy Mat.Nauk 31, 55-136 (1976); R.Rajaraman, "Some Non--Perturbative Semi-Classical Methods in Quantum Field Theory (A Pedagogical Review)", Physics Reports 21, 227-313 (1975).

17 Nonlinear Wave Motion (A.C.Newell, ed.), Lect.Appl.Math. 15, AMS, Providence, R.I., 1974; Dynamical Systems, Theory and Applications (J.Moser, ed.), Lect.Notes in Physics 38, Springer, 1975; Bäcklund Transformations (R.M.Miura, ed.), Lect.Notes in Math. 515, Sprin-

ger, 1976. Two other books now in preparation are going to be published soon, one by Springer under the editorship of R.K.Bullough, and one by Pitman under the editorship of F.Calogero.

18 F.Calogero, "Generalized Wronskian Relations, One-Dimensional Schroedinger Equation and Nonlinear Partial Differential Equations Solvable by the Inverse Scattering Method", Nuovo Cimento 31B, 229-249 (1976). See also the papers of Ref.[7].

19 N.J.Zabusky and M.D.Kruskal, "Interactions of "solitons" in a collisionless plasma and the recurrence of initial states", Phys.Rev. Lett. 15, 240-243 (1966).

20 F.Calogero and A.Degasperis, "Coupled Nonlinear Evolution Equations Solvable Via the Inverse Spectral Transform, and Solitons that Come Back: the Boomeron", Lett.Nuovo Cimento 16, 425-433 (1976); "Bäcklund Transformations, Nonlinear Superposition Principle, Multisoliton Solutions and Conserved Quantities for the "Boomeron" Nonlinear Evolution Equation", Lett.Nuovo Cimento 16, 434-438 (1976).

21 F.Calogero and A.Degasperis, "Special Solution of Coupled Nonlinear Evolution Equations with Bumps that Behave as Interacting Particles", Lett.Nuovo Cimento 19,525-533 (1977).

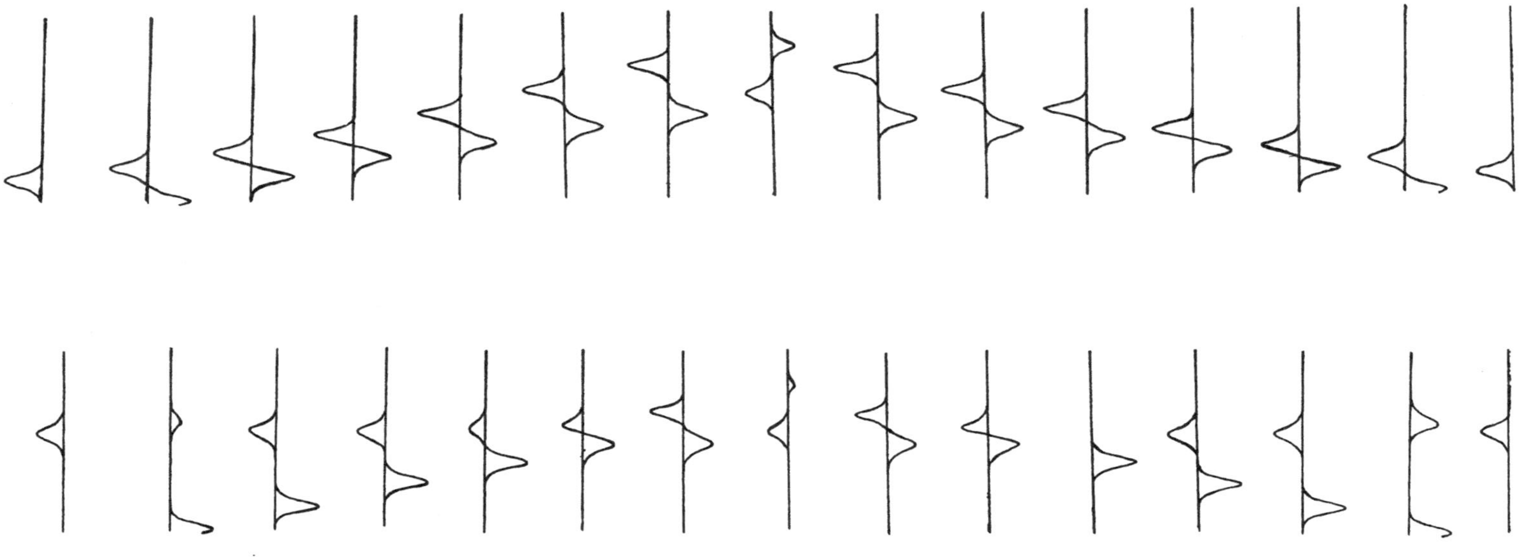

FIGURE CAPTION

Two series of shots (equally spaced in time) from the film Zoomerons by J.C.Eilbeck. The function displayed is the x-derivative of the two-soliton solution (6.16). $Z_{\infty}=1.$ $p_1=1.2.$ $p_2=1.05$ (first case: both solitons boomerang); $p_2=0.9$ (second case: one soliton boomerangs, the other is trapped).

QUANTIZATION OF PARTICLE-LIKE SOLUTIONS IN FIELD THEORY

L.D.Faddeev, P.P.Kulish
Leningrad Department of the V.A.Steclov Mathematical Institute

At present there are are many non-linear field equations in two space-time dimensions which can be solved by the inverse-scattering method. For these equations we know soliton and n-soliton solutions, their scattering characteristics, the classical S-matrix, action-angle variables, infinite series of locally conserved currents and some other features. Here is a list of such equations:

1) nonlinear Schrodinger equation

$$\text{a.}\quad i\psi_t = -\psi_{xx} - \varkappa|\psi|^2\psi, \qquad \psi(x,t)\in\mathbb{C}$$

$$\text{b.}\quad i\psi_t = -\psi_{xx} + \varkappa(|\psi|^2-\rho)\psi, \quad -\infty < x,t < +\infty$$

2) sine-Gordon equation

$$u_{tt} - u_{xx} + m^2 \sin u = 0, \qquad u(x,t)\in\mathbb{R};$$

3) "Lee model" or threewaves

$$\partial_t u_1 + v_1\partial_x u_1 = i u_2 u_3, \quad \partial_t u_3 + v_3 \partial_x u_3 = i u_1 u_2^*;$$

$$\partial_t u_2 + v_2 \partial_x u_2 = i u_1 u_3^*,$$

4) complex sine-Gordon equation $\psi(x,t)\in\mathbb{C}$,

$$\psi_{tt} - \psi_{xx} + m^2\psi(1-|\psi|^2) + \psi^*(\partial_\mu\psi)^2/(1-|\psi|^2) = 0;$$

5) matrix sine-Gordon equation

$$\partial_t V - \partial_x V = 2[A, U^+AU],$$

$$\partial_t U + \partial_x U = 2UV, \quad U^{-1} = U^T \in O(3), \quad V^+ = -V;$$

6) massive Thirring model for complex-valued function $\psi_\alpha \in \mathbb{C}$, $\alpha = 1,2$,
for function with value in the Grassmann algebra

$$\psi_\alpha(x,t)\in\mathcal{G}, \quad \alpha = 1,2$$

$$\psi_\alpha^+\psi_\beta + \psi_\beta\psi_\alpha^+ = \psi_\alpha\psi_\beta + \psi_\beta\psi_\alpha = 0; \quad \rho_\alpha = \psi_\alpha^+\psi_\alpha,$$

$$\begin{cases} i(\partial_t + \partial_x)\psi_1 = \psi_2 + \rho_2\psi_1 \\ i(\partial_t - \partial_x)\psi_2 = \psi_1 + \rho_1\psi_2 \end{cases}$$

7) equation of main chiral field $g(x,t) \in G$ (compact group)

$$L_\mu = g^{-1}\partial_\mu g \, , \quad \partial_\mu L_\mu = 0, \quad \partial_\mu L_\nu - \partial_\nu L_\mu = [L_\nu, L_\mu].$$

There are many other equations which can be solved by the inverse scattering thecnique, but the main problem is to find an appropriate reduction to a lagrangian form.

Some of these models have received a thorough quantum treatment. For instance the quantum problems corresponding to the N particles on a line with potential $-\varkappa\,\delta(x)$ (eq. (1a)) and to the Bose-gas of finite density with the interaction $\varkappa\,\delta(x)$ (repulsive, eq. (1b)) were completely solved. The semiclassical quantization performed in the angle-action variables leads to formulas for the energy of the N particle bound states and for the energy of excitations in the Bose-gas in full agreement with the quantum answer. In this case one can be even more precise. Namely, in the limit $\hbar \to 0$, $N \to \infty$, $\hbar N$ - constant

$$\psi(x,t \mid p, n) = \lim \hbar \int \langle p, N \mid \hat{\psi}(x,t) \mid p + \hbar\nu, N+1 \rangle d\nu ; \quad n = \hbar N,$$

where $\hat{\psi}(x,t) = \exp(-it\hat{H}/\hbar)\hat{\psi}(x)\exp(it\hat{H}/\hbar)$ is the Heisenberg field and $|p, N\rangle$ is the N particle bound state with momentum p we obtain the exact classical one-soliton solution of (1a).
In this way one also obtains n -solitons formulas.

Let us now consider the sine-Gordon and massive Thirring models. If we assume that there is no many-particle creation and that the S-matrix factorizes (this can be justified by the analysis of the classical systems) and suppose in addition natural analyticity properties, we can write down a (presumably) exact quantum S-matrix. This S-matrix gives rise to bound states which coincide with the semiclassical mass spectrum of the double soliton (breather)

$$M = \frac{16m}{\gamma} \sin\frac{n\gamma}{16} \; ; \quad n = 1, 2, \ldots, \left[\frac{8\pi}{\gamma}\right].$$

Thus there is a strong evidence that when a classical field theory is being quantized, solitons give rise to quantum particles, whose characteristics may be calculated semiclassically.

In our opinion the most universal way towards perturbation theory for the quantum case is to apply functional integral techniques. Our report will be mostly devoted to a method of quantization of soliton(*) developped by one of us (L.D.F.) together with V.E.Korepin at the Leningrad department of the V.A.Steklov Mathematical Institute.

Our main object is the transition amplitude

$$G(t_2, q_2 \mid t_1, q_1) = \int \exp\left(i\int_{t_1}^{t_2} \mathcal{L}\, dt\right) \prod_t dq(t) = \langle q_2 | e^{-iH(t_2 - t_1)} | q_1 \rangle,$$

(*) We use the term "soliton" as equivalent to the notion of particle-like solution. In a general process the number of solitons and their momenta are not conserved.

or rather its limit as $t_2 \to +\infty, t_1 \to -\infty$, $q_2 = \frac{p_2 t_2}{m} + q_{02}$, $q_1 = \frac{p_1 t_1}{m} + q_{01}$. Here q_2, q_1 are coordinates of one particle, or a set of coordinates if we deal with n particles, $\mathcal{L}$ is the lagrangian and H is the hamiltonian. The limit is related to transition amplitudes and the S-matrix by

$$\langle p_2 | S | p_1 \rangle = \lim_{t_2 \to +\infty, t_1 \to -\infty} \frac{G(t_2, \frac{p_2 t_2}{m} + q_{02} | t_1, \frac{p_1 t_1}{m} + q_{01})}{(2\pi) \cdot G_0(t_2, \frac{p_2 t_2}{m} | t_0, x_0) G_0(t_0, x_0 | t_1, \frac{p_1 t_1}{m})}$$

G_0 - propagator for free partcle. The answer does not depend on t_0, q_{0i}. These formulae are readily varified in the non-relativistic quantum mechanics; in field theory we propose them as a definition of the quantum S-matrix for solitons. We shall give explicit expressions for solitons in one-dimensional case

$$\mathcal{L} = \frac{1}{\gamma} \int [\tfrac{1}{2} (\partial_\mu u)^2 - v(u)]\, dx, \quad \min_{\{u = const\}} v(u) = v(u_j).$$

A structureless soliton is a solution of classical field equations completely characterized by its velocity v and the location of its energy density maximum (its center of location). For instance, if $v(u) = 1 - \cos u$ (sine-Gordon equation), then

$$u_s(x, t | v, x_0) = 4 \operatorname{arctg} \left(\exp [\varepsilon (x - vt - x_0)/\sqrt{1-v^2}] \right), \; \varepsilon = \pm 1.$$

To calculate the asymptotics of the transition amplitude of one soliton we use the stationary phase method, choosing a configutation which is approximately one-soliton at the moment t_1

$$u_2(x, t_2) = u_s(x, t_2 | v_2, q_2)$$

and which is also approximatly one-soliton at the moment t_2

$$u_1(x, t_1) = u_s(x, t_1 | v_1, q_1) = u_s\left(\frac{t_1 (x - x_1)}{\sqrt{t_1^2 - x_1^2}} \right), x_1 = v_1 t_1 + q_1; \; G(t_2, x_2 | t_1, x_1) = \int_{t_1, u_1}^{t_2, u_2} \exp(i \int_{t_1}^{t_2} \mathcal{L}\, dt) \prod_{x,t} du(x,t).$$

Then at large times the stationary point of the action will be given by the one-soliton solution. Changing the variables of integration $u(x,t) = u_s(x,t) + \sqrt{\gamma}\, \varphi(x,t)$ we obtain the known loop expansion

$$G(t_2, x_2 | t_1, x_1) = \exp\left(i \int_{t_1}^{t_2} \mathcal{L}(u_s)\, dt\right) \cdot \exp\left(i \sum_{n=0}^{\infty} \gamma^n W_n\right).$$

W_n is the sum of all connected n+1 – loop vacuum diagrams.

The corresponding diagrams are constructed with the help of the propagator

$$R(x|y) = \left(\gamma \frac{\delta \mathcal{L}}{\delta u(x) \delta u(y)} \Big|_{u = u_s} \right)^{-1} \equiv H^{-1}(u_s)$$

and the vertices $v^{(k)}(u_s)$, = 3,4,....In the relativistically invariant theories the first factor exp (iW_{-1}/γ) is of the form exp ($-iM_s(t_2-t_1)\sqrt{1-v^2}$). On the other hand we also know that the asymptotic of the transition amplitude for a quantized relativistic particle is of the form

$$G(t,x\,|\,0,0) \simeq \left(\frac{M}{2\pi i\sqrt{t^2-x^2}}\right)^{1/2} \exp\left(-iM\sqrt{t^2-x^2}\right)$$

$$t\to\infty\,,\quad x\simeq vt+x_0\,.$$

The factor in front of the exponential is important because it characterizes the spreading of the quantum particle wave packets. This factor as well as the quantum mass corrections can be obtained from the one-loop contribution W_0:

$$2\,W_0 = \ln\det{}^{-1/2} H(u_s)/H_0 = -\tfrac{1}{2}\,\mathrm{Tr}\ln H/H_0\,,\quad \tau=(t-vx)/\sqrt{1-v^2}$$

$$H(u_s) = \partial_\tau^2-\partial_\eta^2+v''(u_s)\equiv\partial_\tau^2+K\,,\quad H_0=\partial_\tau^2-\partial_\eta^2+v''(\text{"}0\text{"}).$$

The operator K has a zero eigenvalue ($-d_\eta^2+v''(u_s))u_s'=0,\ \eta=\frac{x-vt}{\sqrt{1-v^2}}$. That implies that K has no inverse operator; however for the operator H this eigenvalue lies on the continuous spectrum and in this case it is possible to construct H^{-1}. Let us wright down H in the form

$$H = d_\tau^2+K = d_\tau^2P+H(I-P)\,;\quad P=\frac{|u_s'\rangle\langle u_s'|}{\|u_s\|^2}\,.$$

Using the formulae

$$\det H/H_0 = \det d_\tau^2\,\det H(I-P)/H_0\,;\quad \det d_\tau^2=(t_2-t_1)/\sqrt{1-v^2}$$

we have

$$\exp(iW_0) = \sqrt{\frac{M}{2\pi i|t_2-t_1|}}\ \exp\left(-i\,\Delta M\,(t_2-t_1)\sqrt{1-v^2}\right).$$

The calculation of the 1 - loop correction to the soliton mass ΔM is done using the trace identities and gives

$$\Delta M = \frac{1}{2}\sum_n\sqrt{m^2-E_n}+\frac{ima^\infty}{2\pi}-\frac{im}{8\pi}\int_0^\infty d\beta\,\mathrm{ch}\beta\left[\frac{d}{d\beta}\ln\frac{a^2(\beta)}{a^2(-\beta)}+4\frac{a^\infty}{\mathrm{ch}\beta}\right],$$

where E_n are the eigenvalues (discret spectrum) of the operator K and a (β) is the Wronskian of the Jost solutions for the operator K .

The existence of eigenfunctions of the operator H with vanishing eigenvalues, such as du_s/dx, du_s/dt (zero-modes) leads to difficulties in constructing the perturbation theory in the non-covariant approach, where the time is separated and canonically conjugated coordinates are quantized. In such approach the relativistic invariance of the theory should be additionally verified and the center-of-mass coordinate and the momenta of the soliton should be considered separately from quantum fluctuation of the background field.

In the present approach none of these difficulties arise, because we use velocities not momenta to parametrize solitons.

If we consider the operator H at finite times t, acting on the space of functions $\varphi(x,t)$ vanishing at $t = t_2$, t_1, then the inverse operator R is correctly defined, because the zero-modes drop-out since du_s/dx does not vanish at finite times. In the limit $|t_2-t_1| \to \infty$ $R(x,t_2|y,t_1)$ gets an additional term, proportional to $|t_2-t_1|$, which nevertheless does not contribute to the sum of given order with respect to the Planck's constant. Therefore adding terms with zero-modes from the continuous spectrum we can simplify the propagator R. If we calculate the n-soliton scattering amplitude, then the stationary point for the action is the n-soliton solution and the number of the zero-modes from continuous spectrum is equal to the number of rising zero-modes and is equal to the number of free parameters in solution $\equiv$ the number of independent conservation laws. The addition to $R(x,t_2|y,t_1)$ a term with the rising zero-modes, e.g. $du_s/d\theta$, where θ is the soliton rapidity ($p = M_s \,\mathrm{sh}\,\theta$), changes the quantum mass correction ΔM and therefore shouldn't be done.

The one-loop correction to the scattering amplitudes for solitons with fixed velocities reduces to calculation of $\det^{-1/2}(H(u_{ss})/H_0)$, where u_{ss} is a classical solution decaying at $t \to \pm\infty$ into a sum of one-soliton solutions. This determinant can expressed through the asymptotics of the functions $\psi(x,t,\beta)$, satisfying the following homogeneous equation:

$$H\psi(x,t,\beta) = \left(\frac{d^2}{dt^2} - \frac{d^2}{dx^2} + v''(u_{ss})\right)\psi(x,t,\beta) = 0,$$

$$\psi(x,t,\beta) \underset{\substack{t\to-\infty \\ x\to-\infty}}{\sim} \exp(-i\tau(\beta,x,t)); \quad \psi(x,t,\beta) \underset{\substack{t\to-\infty \\ x\to+\infty}}{\sim} a(\beta)\exp(-i\tau) + \sum_{\substack{n \\ \beta_n \neq \beta}} b_n(\beta)\exp(-i\tau(\beta_n)),$$

$$\psi(x,t,\beta) \underset{\substack{t\to+\infty \\ x\to-\infty}}{\sim} \exp(-i\tau) + \int d\alpha\, C(\alpha,\beta)\exp(-i\tau(\alpha,x,t)),$$

$$\tau(\alpha,x,t) = m(t\,\mathrm{ch}\,\alpha - x\,\mathrm{sh}\,\alpha).$$

The quantum corrections are expressed through a (β), c (α,β). With the one-loop correction the soliton scattering amplitude has the form:

$$S(\varphi_1-\varphi_2) = \exp\Big\{-i\int_0^{\varphi_1-\varphi_2}\Delta(\alpha)\,d\alpha + \frac{1}{4\pi i}\int d\alpha \frac{d}{d\alpha}\ln a(\alpha-\varphi_1)\ln a(\alpha-\varphi_2) + \frac{1}{2}\int d\alpha\left[\ln(I+\hat{C})(\alpha,\alpha) - C^{\infty}\right]\Big\}.$$

Here the first term is the contribution of tree approximation, $\ln(I+\hat{C})(\alpha,\beta)$ is the kernel of the operator $\ln(I+\hat{C})$, and $C^{\infty} = \lim C(\alpha,\alpha)$, $\alpha \to \infty$.

As the example of the sine-Gordon equation shows, besides the simple (structureless) solitons, there may exist solitons with internal degrees of freedom, e. g. $u(x,t\,|\,v,q,T,\alpha)$ where the dependence of α is periodic with period 2π. One can define as in the case of structureless solitons, the transition amplitude $G(t_2,x_2,\alpha_2\,|\,t_1,x_1,\alpha_1)$, then the mass-spectrum of the periodic soliton is determined by the requirements:

$$\frac{d}{dT}\left(\frac{1}{n}\left[\ln G(nT,0,2\pi n\,|\,0,0,0)+M_n T\right]\right)=0,$$

$$\frac{1}{in}\left[\ln G(nT,0,2\pi n\,|\,0,0,0)+M_n T\right]=2\pi k$$

Considering the solution of the homogenious equation with the operator $\square + v''(u(x,\frac{t}{T},T))$ and their asymptotic characteristic it is possible to calculate the corresponding determinant and to obtain the one-loop corrections to the masses M_k and periods T_k of these solitons. The formulae are very complicated and we will not write them down here.

The renormalizability of the soliton quantum theory for scalar field is established in this approach in the same way as the renormalizability of the ground field. The propagator R at large momenta has the same behaviour as the usual propagator $(\square + m^2)^{-1}$, and the vertecies $\tilde{v}^{(n)}(p)$ behave like some constants. Therefore, using the functional integral representation for the generating functional of the S-matrix we see that the ultraviolet divergences of both the background field and the solitons are removed by the same countertems. But these counterterms have to be evaluated at different stationary points.

In regard to renormalizability we would like to point out some opportunities raised by models of field theory with an infinite number of conservation laws. If one assumes that the fact of existence of local conservation laws similar to those of a free field theory determines the type of interaction, then for the existence of these locally conserved currents in quantum theory it is necessary that quantum equations of motion coincide with the classical ones. Thus there are formally non-renormalizable theories where we can fix all the counterterms and renormalization constants in a unique way. For instance, the lagrangean of complex sine-Gordon model

$$\mathcal{L} = \int 2\left[\,|\partial_\mu \psi|^2/(1-|\psi|^2) - m^2|\psi|^2\right] dx$$

is non-renormalizable and there is an infinite sequence of locally conserved currents

$$\partial_\xi\left\{\partial_\eta \psi^* B/(1-|\psi|^2)^{1/2}\right\} = \frac{m}{2k}\,\partial_\eta\left\{1-2|\psi|^2-2i|\psi|(1-|\psi|^2)^{1/2}B\right\},$$

$$mkB = \partial_\eta\psi/(1-|\psi|^2)^{1/2} - (\operatorname{Im}\partial_\eta\psi)\frac{1-2|\psi|^2}{|\psi|(1-|\psi|^2)}B + \left[i\partial_\eta - \frac{\partial_\eta\psi^*}{(1-|\psi|^2)^{1/2}}\right]B^2,$$

$$B = \sum_{n=1}^{\infty} b_n(\xi,\eta)/(mk)^n;\quad t=\eta+\xi,\quad x=\eta-\xi.$$

In the case of more than two-dimensional spacetime we do not have exactly solvable field-theoretical models with solitons. The only example is the Kadomtzev-Petviashvily equation in 3-dimensional space-time, for which there exist a soliton solution falling off at the spatial infinity, and n-soliton solutions as well. But there are many examples of lagrangians, for which the 1-soliton solutions are explicity obtained or their existence is proved. Topological considerations play an important role here.

Let us consider a nonlinear chiral field g(x) with values in a compact group G. Then the scalar (with respect to this group) current

$$J_\mu(x) = \varepsilon_{\mu\nu\rho\sigma}\, tr\,[L_\nu, L_\rho] L_\sigma, \quad L_\nu = \partial_\nu g \cdot g^{-1}, \quad g(x) \in G$$

is conserved independently of the equations of motion (topological current), and the corresponding topological charge

$$Q = \int J_0(x)\, d^3x$$

conveniently normalized (it depends on G), takes on integer values. If we choose the lagrangian of the Skyrme model

$$\mathcal{L} = tr\left\{ \frac{1}{2\lambda^2} L_\mu^2 + \frac{\varepsilon^2}{2} \left([L_\mu, L_\nu] \right)^2 \right\}$$

then the energy functional for static configurations is estimated by its topological charge value

$$\frac{\varepsilon}{\lambda} |Q| = \frac{\varepsilon}{\lambda} \left| \int \varepsilon_{ik\ell}\, tr\left([L_i, L_k] L_\ell \right) d^3x \right| \leq$$

$$\leq \int tr\left(\frac{1}{2\lambda^2} L_i^2 + \frac{\varepsilon^2}{2} \left([L_i, L_k] \right)^2 \right) d^3x = H_{static}.$$

More interesting is the main chiral field model with gauge invariance $g(x) \to \Omega(x) g(x) \Omega(x)^{-1}$. There exists a lagrangian depending on the field g(x) and the Yang-Mills field, which is invariant with respect to the usual gauge transformation. Let us introduce the covariant generalizations of $\partial_\mu g \cdot g^{-1}$ and $g^{-1} \partial_\mu g$ (A_μ is the Yang-Mills field with values in the lie algebra of group G)

$$L_\mu(x) = \nabla_\mu g \cdot g^{-1} = (\partial_\mu g + A_\mu g - g A_\mu)\, g^{-1}$$

$$R_\mu(x) = g^{-1} \nabla_\mu g = g^{-1} \partial_\mu g + g^{-1} A_\mu g - A_\mu .$$

The conserved topological current has the form

$$J_\mu(x) = \varepsilon_{\mu\nu\rho\sigma}\, tr\left\{[L_\nu, L_\rho]L_\sigma - 3F_{\nu\rho}(L_\sigma + R_\sigma)\right\},$$

$$F_{\mu\nu}(x) = \partial_\mu A_\nu - \partial_\nu A_\mu + [A_\mu, A_\nu].$$

The corresponding topological charge in this case is equivalent to the previous one, and the convenient lagrangian is

$$\mathcal{L} = tr\left\{\frac{1}{2\lambda^2} L_\mu^2 + \frac{1}{4e^2} F_{\mu\nu}^2 + \frac{\varepsilon^2}{2}\left([L_\mu, L_\nu]\right)^2\right\}.$$

If we take into consideration the equality

$$tr\, L_\mu^2 = tr\, R_\mu^2,$$

and use the inequality $2ab \leq ca^2 + b^2/c$, $c > 0$, then for static configuration energy functional we obtain the estimate

$$\frac{\varepsilon}{\lambda(36 + e^2\varepsilon^2)^{1/2}} |Q| \leq H_{static}.$$

In a number of simpler cases such estimates or special substitutions allow to obtain explicitly or to prove the existence of particle-like solutions in multi dimensional space.

We do not give extensive bybliography here, but refer the reader to review papers [1-6] and some special articles.

References.

1) L.D.Faddeev. "Quantization of solitons" preprint IAS, 1975.
2) R.Dashen, B.Hasslacher, A.Neveu. Phys.Rev.D10, 4141, 4130, 1974.
3) R.Rajaraman. Phys.Reports, 21C, 227, 1975.
4) S.Coleman. Lecture Notes. Erice Summer School, 1975.
5) R.Jackiw. "Quantum meaning of classical field theory" submitted to Rev.Mod.Phys.
6) L.D.Faddeev, V.E.Korepin, "Quantum theory of solitons" submitted to Physics Reports.
7) L.D.Faddeev, L.A.Tachtadjan. Uspechi MN, 29, 249, 1974.
8) L.D.Faddeev, L.A.Tachtadjan. Theor.Math.Phys., 21, 160, 1974.
9) A.M.Polyakov. Pisma JETPH, 20, 430, 1974.
10) L.D.Faddeev, V.E.Korepin, P.P.Kulish. Pisma JETPH, 21, 302, 1975.

11) V.Fateev, J.Tupkin, A.S.Schwarz. Sov.J.Nucl.Phys.22, 321, 1975.
12) L.D.Faddeev, V.E.Korepin. Theor.Math.Phys.25,147,1975.
13) P.P.Kulish. Theor.Math.Phys.,26, 198, 1976.
14) P.P.Kulish, E.R.Nisimov. Theor.Math.Phys., 29, 161, 1976.

TOPICS IN INFINITE DIMENSIONAL ANALYSIS *, **

by

Sergio Albeverio and Raphael Høegh-Krohn
Matematisk Institutt, Universitetet i Oslo

and

Zentrum für interdisziplinäre Forschung, Universität
Bielefeld

ABSTRACT

We present some recent results in three different but connected domains of infinite dimensional analysis. In chapter I we report on work concerning the homogeneous generalized random fields of constructive quantum field theory. In particular a Dynkin entrance boundary construction is given and the global Markov property of the boundary measures is discussed. The connection of these

fields with the general theory of Dirichlet forms and diffusion processes on rigged Hilbert spaces is shortly described. In chapter II we report on the energy representation of Sobolev--Lie groups. These groups are defined as completions in the energy metric of the groups of C^1-mappings from Riemann manifolds into Lie groups of compact type. The energy representation extends the one given by Euclidean Markov fields to the case of fields with values in a Lie group. In chapter III we report on results concerning oscillatory integrals in infinitely many dimensions and their asymptotic expansions, with applications to the Feynman path integrals and the approach to the classical limit of quantum mechanics.

* Work supported in part by the Norwegian Research Council for Science and the Humanities.

** Talk given at the International Conference on the Mathematical Problems in Theoretical Physics, Rome, June 6-15, 1977.

I. Markov fields and diffusion processes

I.1 Local and global Markov fields

Markovian random fields as extensions of markovian processes to the case of more dimensional time parameter have received increasing attention in recent years, in different contexts like probability theory, information theory, statistical mechanics and quantum field theory. Although we shall limit ourselves here to particular examples of generalized random fields over R^2, many of the ideas and methods are valid more generally. We shall concentrate ourselves on the Euclidean Markov random fields as a particularly interesting class of homogeneous markovian (generalized) random fields over R^d .[1)] Let us here recall that the simplest Euclidean Markov field is the Gaussian one, called "free Euclidean (Markov) field". It is defined by a measure μ^o on the space of distributions $\mathcal{D}'(R^d)$ characterized by its Fourier transform

$$\int_{\mathcal{D}'(R^d)} e^{i\langle\xi,\varphi\rangle} d\mu^o(\xi) = e^{-\frac{1}{2}(\varphi,(-\Delta+m^2)^{-1}\varphi)},$$

where $\varphi \in \mathcal{D}(R^d)$, Δ is the Laplacian in $L^2\ (R^d)$, m is a positive constant, $(\ ,\ )$ is the scalar product in $L^2\ (R^d)$ and $\langle\ ,\ \rangle$ is the pairing between $\mathcal{D}$ and $\mathcal{D}'$. μ^o possesses a Markov property, in a sense we shall define below. From now on we take, in view of the applications, d = 2. Before we proceed to develop some concepts, let us stress that we have to be rather short in this lecture but the interested reader can find more details in [3] and in other publications we shall mention at the relevant points. Let μ be a probability measure on $\mathcal{D}'(R^2)$ and let $\xi(x)$, $x \in R^2$ be the associated generalized random field, such that $\varphi \to \langle\xi,\varphi\rangle$ is a linear random function. We shall consider measures μ for which this linear function extends to $\rho \to \langle\xi,\rho\rangle$ with $\langle\xi,\rho\rangle \in L^2(d\mu)$, for any bounded Borel measure ρ on R^2 with support in a compact and with the property that $\int d\rho(x)(-\Delta + m^2)^{-1}(x-y)\,d\rho(y) < \infty$, where $(-\Delta+m^2)^{-1}(x-y)$ is the kernel of $(-\Delta+m^2)^{-1}$. Let now Λ be an arbitrary Borel subset of R^2 and let $B(\Lambda)$ be the σ-algebra generated by the random functions $\langle\xi,\rho\rangle$ with $\operatorname{supp}\rho \subset \Lambda$. Let $E(\cdot\,|\Lambda)$ be the conditional expectation with respect to the σ-algebra $B(\Lambda)$. Let C be a piecewise smooth curve dividing R^2-C into two disjoint open regions $\Omega_+{}^C$ and $\Omega_-{}^C$. Let $f_\pm$ be $B(\Omega_\pm{}^C)$ - measurable functions, non negative or integrable. Then if

$$E\,(f_+ f_- \mid C) = E(f_+ \mid C)\; E\,(f_- \mid C) \tag{1}$$

for all $f_\pm$ and all closed C (i.e. all C which are contained in a finite region) then we say that μ has the local Markov property. If (1) holds also for all open C (hence in particular for straight lines) then we say that μ has the global Markov property. We remark that the local resp. global Markov condition (1) is equivalent e.g. with

$$E\,(f_- \mid \Omega_+{}^C \cup C) = E\,(f_- \mid C). \tag{2}$$

(1) and (2) are expressions of the "conditional independence of the fields with support in $\Omega_+{}^C$ and $\Omega_-{}^C$, given the fields on C".

Remark: A distinction between local and global Markov property has been pointed out by Newman [4].

The free Euclidean Markov field measure μ^o has the global Markov property, as pointed out by Nelson [5], [6].

I.2 Dynkin boundary for a local Markov field

We introduce a partial order $<$ into the family of all piecewise smooth closed curves C which divide R^2-C into two disjoint regions $\Omega_-{}^C$ and $\Omega_+{}^C$, $\Omega_-{}^C$ being the inner of the curve. We write $C_1 < C_2$ iff $\Omega_+{}^{C_1} \supset \Omega_+{}^{C_2}$. Let now μ be a fixed local Markov field. Then $C_1 < C_2$ iff $B(\Omega_+{}^{C_1}) \supset B(\Omega_+{}^{C_2})$. We then define the σ-algebra at infinity B_∞ by $B_\infty \equiv \bigcap_C B(\Omega_+{}^C)$. We have that $B(\Omega_+{}^C)$ converges monotonically downwards to B_∞ as C increases, hence $E(\cdot \mid \Omega_+{}^C)$ converges in the same way as orthogonal projections in $L^2(d\mu)$, hence strongly. One shows easily that the limit is the orthogonal projection $E(\cdot \mid \infty)$ i.e. the conditional expectation with respect to B_∞. By martingale theory one has easily that $E(\cdot \mid \Omega_+{}^C)$ also converges strongly in $L^1(d\mu)$ and pointwise μ-almost surely to $E(\cdot \mid \infty)$. Let now $L_\infty(\infty, d\mu)$ be the space of all $L_\infty(d\mu)$-functions which are measurable with respect to the σ-algebra at infinity. $L_\infty(d\mu)$ is a commutative C^*-algebra with unit, hence by Gelfand's theorem it is isomorphic with the Banach space $C(D(\mu))$ of continuous functions on a compact Hausdorff space $D(\mu)$. We call $D(\mu)$ the "Dynkin boundary" associated with μ[2]. $f \to E(f \mid \infty)$ is a bounded linear map from $L_\infty(d\mu)$ into $L_\infty(\infty, d\mu)$. To $E(f \mid \infty)$ there corresponds by Gelfand's isomorphism a function in $C(D(\mu))$, denoted by $E(f \mid \infty)(z)$, where z runs over $D(\mu)$. Hence $f \to$

$E(f|\infty)(z)$ is, for any z, a linear continuous functional on $L_\infty(d\mu)$, given by a measure $d\mu_z$ on $\mathcal{D}'(R^2)$ such that

$$E(f|\infty)(z) = \int_{\mathcal{D}'(R^2)} f(\xi)\, d\mu_z(\xi). \tag{3}$$

Consider now the restriction of μ to B_∞, realized through the isomorphism by a probability measure μ_∞ on $D(\mu)$. We have then, $\mathcal{D}'(R^2)$ being Suslin, the regular disintegration

$$\mu(\xi) = \int_{D(\mu)} \mu_z(\xi)\, d\mu_\infty(z)\,.. \tag{4}$$

We call the points z in $D(\mu)$ and the corresponding measures μ_z the <u>pure phases</u> of μ .

<u>Remark</u>: If μ is invariant with respect to the translations in R^2, then there exists also an ergodic decomposition of μ with respect to translations. The decomposition (4) is finer than this ergodic decomposition. They are equivalent when the translations act triviallyon $D(\mu)$, like e.g. for one-dimensional stochastic processes.

<u>Remark</u>: Concepts of σ-algebra at infinity have been introduced and discussed in the literature in particular by Dobrushin [8] and Lanford and Ruelle [9], especially in connection with statistical mechanics, see also e.g. [10] , [11]. In connection with field theory related concepts (like DLR-equations) were introduced and used especially by Newman [4], Guerra, Rosen and Simon [12], Fröhlich [13], [14], Glimm and Jaffe [15] and Accardi [16]. For Dynkin's boundary theory see e.g. [7].

<u>I.3 On the global Markov property of the pure phases</u>

Let μ be a locally Markov measure on $\mathcal{D}'(R^2)$. Let C be an open piecewise smooth curve dividing R^2-C into two disjoint regions, Ω_-^C and Ω_+^C. Let K_r be the inner of the circle with center at the origin and radius r, and let $\dot{K}_r$ be its boundary. Choose r so big that a piece of C is contained in K_r. Let Λ_+ open sets contained in $\Omega_+^C \cap K_r$. Let C_1 be a simple closed smooth curve coinciding with C in $C \cap \bar{K}_r$ and containing $\Omega_-^C \cap K_r$ in its interior. Let f_+ be $B(\Lambda_+)$-measurable and nonnegative or integrable. Then by the local Markov property we have

$$E\,(f_+\, f_- \mid C_1) = E\,(f_+ \mid C_1)\; E\,(f_- \mid C_1). \tag{5}$$

This implies

$$E\,(f_+\, f_- \mid C \cup \tilde{K}_r) = E\,(f_- \mid C \cup \tilde{K}_r)\; E\,(f_- \mid C \cup \tilde{K}_r) \tag{6}$$

where $\tilde{K}_r \equiv R^2 - K_r$. Then by martingale theory we get

$$E\,(f_+\, f_- \mid C \cup \infty) = E\,(f_+ \mid C \cup \infty)\; E\,(f_- \mid C \cup \infty), \tag{7}$$

where $E\;(\cdot \mid C \cup \infty)$ is the conditional expectation with respect to $B(C) \cup B_\infty$, in the case where the Borel σ-algebra at infinity is countably generated. By the definition of μ_z we have then, equivalently,

$$E_{\mu_z}\;(f_+\, f_- \mid C) = E_{\mu_z}\;(f_+ \mid C)\; E_{\mu_z}(f_- \mid C), \tag{8}$$

where, if ν is a measure, E_ν and $E_\nu\;(\cdot \mid C)$ denote expectation resp. conditional expectation with respect to ν. We have thus the following

<u>Theorem 1</u> Any pure phase μ_z is globally Markov, under the assumption that the Borel σ-algebra at infinity is countably generated.

I.4 Measures given by multiplicative functionals

Let μ be again a locally Markov measure. Let Λ be any Borel set of R^2. A multiplicative functional m is a map from the family of Borel sets into positive integrable $B(\Lambda)$ - measurable functions with the property that if $\Lambda_1 \cap \Lambda_2 = \emptyset$ then $m_{(\Lambda_1 \cup \Lambda_2)} = m_{\Lambda_1}\, m_{\Lambda_2}$. If m is a multiplicative functional, then it is easily seen that $d\mu_\Lambda \equiv m_\Lambda\, d\mu \;/ \int m_\Lambda\, d\mu$ is again locally Markov.

<u>Remark</u>: Multiplicative functionals are defined in analogy with those studied particularly by Dynkin for ordinary processes. They have been introduced in the theory of Euclidean generalized random fields by Nelson [5], [17].

Let now μ be the free Euclidean Markov field measure μ^o. Let $v\,(\cdot)$ be any linear combination with nonnegative coefficients of a polynomial bounded from below, an exponential function of the form $\int e^{\alpha s}\, d\nu(\alpha)$ and a trigonometric function of the form $\int \cos\alpha(s+\beta)\, d\sigma(\alpha)$, where ν, σ are arbitrary positive bounded measures with support in the interval $(-2\sqrt{\pi}\,,\; 2\sqrt{\pi}\,)$. It is known from constructive quantum field theory that there is a procedure (Wick ordering, Ito's multiple integrals, Segal's centering) by which to the formal object $v\,(\xi(x))$ there can be associated a generalized random variable $:v(\xi(x)):$ in such a way that $\int_\Lambda :v\,(\xi\,(x)):\, dx \in L^2(d\mu^o)$ and

$$m_\Lambda(\xi) \equiv \exp\left(-\int_\Lambda :v(\xi(x)):\, dx\right) \tag{9}$$

is a multiplicative functional. For the precise definition of m_Λ see the references below mentioned in connection with the limit $\Lambda \uparrow R^2$. As first remarked by Nelson, the measure $d\mu_\Lambda \equiv m_\Lambda\, d\mu^o / \int m_\Lambda\, d\mu^o$ is globally Markov. Control on the convergence of μ_Λ as $\Lambda \uparrow R^2$ has been first obtained for polynomials with small coefficients, even exponentials and trigonometric interactions by Glimm, Jaffe and Spencer [19], Albeverio and Høegh-Krohn [20] and Fröhlich and Seiler [21] respectively. In these cases (see also [4], [13]) weak convergence of μ_Λ as $\Lambda \uparrow R^2$ holds, as well as convergence of all moments. For these and other interactions the weak convergence of modified measures given by $d\mu_\Lambda \equiv m_\Lambda\, d\mu^o_X / \int m_\Lambda\, d\mu^o_X$ is also known, where $d\mu^o_X$ is the measure on $B(\Lambda)$ defined as $d\mu^o \upharpoonright B(\Lambda)$ but with the Laplacian replaced by the Laplacian with some self-adjoint boundary conditions X. See [17], [12], [22], [23], [24], [26], [27].

Choose now any of the interactions v for which the weak limit μ^v of μ_Λ as $\Lambda \uparrow R^2$ has been shown to exist and to be locally absolutely continuous with respect to the free Euclidean Markov field measure μ^o. This is the case for the weak polynomial interactions and a large class of other polynomial interactions, as well as for the exponential and trigometric interactions, see e.g. [19], [4], [13], [20], [27], [28]. Let Λ be any bounded open set, C be a simple closed smooth curve in the interior of Λ, D be the region enclosed by C. Let f be a B(D)-measurable function, non negative or integrable. Then by the local Markov property of μ_Λ we have

$$E_{\mu_\Lambda}(f \mid C) = E_{\mu^o}(f\, m_D \mid C) / E_{\mu^o}(m_D \mid C) = E_{\mu_D}(f \mid C) \tag{10}$$

and therefore

$$E_{\mu_\Lambda}(f) = \int E_{\mu_D}(f \mid C)\, d\mu_\Lambda^C, \tag{11}$$

where $d\mu_\Lambda^C \equiv d\mu_\Lambda \upharpoonright B(C)$. By taking the weak limit as $\Lambda \uparrow R^2$ and using the assumption that μ^v is locally absolutely continuous with respect to μ^o, we get

$$E_{\mu^v}(f) = \int E_{\mu_D}(f \mid C)\, d\mu_C^v \tag{12}$$

where $d\mu^{v}_{C} \equis d\mu^{v} \upharpoonright B(C)$. From this it follows

$$E_{\mu^{v}} (f \mid C) = E_{\mu_D} (f \mid C). \tag{13}$$

Let now C_o be a simple closed smooth curve in the interior of D and let Λ_+, Λ_- be two open sets in the interior of D, such that Λ_- is in the interior of C_o and Λ_+ in its exterior. Let $f_{\pm}$ be $B(\Lambda_{\pm})$ -measurable. Using the local Markov property of μ_D with respect to C_o we get then the local Markov property of μ^{v}, hence by Theorem 1 we have

Theorem 2 Let μ^{v} be the weak limit of a measure μ_{Λ} given by a multilicative functional of the free Euclidean Markov field, such that μ^{v} is locally absolutely continuous with respect to μ^{o}. Then μ^{v} is locally Markov. If the Borel σ -algebra at infinity is countably generated, then the corresponding pure phases μ_z are globally Markov.

Remark: By what we said above the local Markov property holds for the Euclidean measures obtained as weak limits for polynomial, exponential and trigonometric interactions. The first proof of the local Markov property was given for weak polynomial interactions by Newman [4] . The relation (12) looked upon as an equation for the measure μ^{v} is of the DLR-type, as discussed in field theory especially by Guerra, Rosen and Simon [12]. (Note that the distinction between local and global Markov and correspondents of above results hold also in the discrete case).

The question now arises: When is μ^{v} itself a pure phase? An answer to this question comes from a result of Dobrushin and Minlos [29] : For weak polynomial interactions there exists a space of measures such that the equation (12) has only one solution in this space, namely the one constructed by the Glimm, Jaffe and Spencer cluster expansion.

Remark: This result should also follow from the cluster expansion, along the lines of Fröhlich-Simon's multiplicative boundary conditions perturbations [27], see [3].

As a Corollary of this result we have the triviality of B_{∞} , since if B_{∞} were not trivial then for any solution μ of (12) in the given class we could find a non constant B_{∞} -measurable function h such that $h\mu$ would also solve (12) and be in the same class against the uniqueness result. Since the measure μ^{v} for weak polynomial interactions is locally absolutely continuous with respect to μ^{o} and locally Markov, we then have (Theorem 2), μ^{v} being a pure phase by the triviality of B_{∞}, that μ^{v} is globally Markov if the Borel σ -algebra at infinity is countably generated. The verification of the latter property is an open problem.

I.5 Some consequences of the global Markov property

Consequences of a postulated global Markov property of an Euclidean measure have been analysed by Nelson and others, see e.g. [5], [17], [32], [2], [13].[3] Let us briefly recall some of them. Let μ be a global Euclidean Markov measure. Consider the splitting $R^2 = R \oplus R$ and write a point $y \in R^2$ as $y = (t,x)$, $t \in R$, $x \in R$. Let $\xi(y) = \xi(t,x)$ be the Euclidean generalized random field given by μ. We can look at it as a stochastic process $t \to \xi_t(x)$ indexed by R, with state space $\mathcal{D}'(R)$. Then the global Markov property of μ implies that the Markov condition (1) holds when one takes C as the x-axis $\{t=0\}$. This and the translation invariance of μ give that $\xi_t(x)$ is a homogeneous Markov process, with invariant measure μ_o, the restriction of μ to the time zero fields i.e. $B(\{t=o\})$[4]. The invariance under time reflection gives that $\xi_t(x)$ is a symmetric process, associated with a symmetric Markov transition semigroup e^{-tH}, $t \geq 0$, with non negative, self-adjoint infinitesimal generator H in $L^2(d\mu_o)$. $L^2(d\mu_o)$ can then be naturally identified with the physical Hilbert space of the corresponding Wightman theory, a property called "cyclicity of the time zero fields". Then e^{itH}, $t \in R$ is identified with the unitary group of time translations of the corresponding Wightman models.

I.6 Connections with the theory of diffusion Dirichlet forms and diffusion processes

Let μ^v be the weak limit of measures μ_Λ of the type of Section I.4 and such that the restriction μ_o^v to the "time zero fields" $B(\{t=0\})$ exists, which is e.g. the case for polynomial [33], [13] and exponential interactions [20]. It has been shown by us in [35] that to μ_o^v there is associated a diffusion Dirichlet form which has the properties of the general theory of (diffusion) Dirichlet forms on rigged Hilbert spaces as given in [35], [36], extending Beurling--Deny-Fukushima's theory. We shall now summarize briefly some of the ensuing results for the case of the measures μ_o^v. for more details we refer to [35], [36], [3].

Let FC^k be the linear subspace of $L^2(d\mu_o^v)$ consisting of "C^k-cylinder functions" i.e. functions f such that $f = f \circ P$ for some finite dimensional projection P (depending on f) from $L^2(R)$ onto $PL^2(R)$, continuously extended to $\mathcal{D}'(R)$, and such that if $\tilde{f}$ is the restriction of f to the range of P, then $\tilde{f}$ is C^k on the finite dimensional space $PL^2(R)$. For any FC^1 the following gradient operator is naturally defined as a map from FC^1 into $L^2(R)\otimes L^2(d\mu_o^\nu)$:

$$(\nabla f)(x,\xi) \equiv \sum_{i=1}^{n} \varphi_i(x)\,(\varphi_i\cdot\nabla)f(\xi(x)) \equiv \frac{\delta f}{\delta\xi(x)},$$

where $f(\xi) = f(\langle\xi,\varphi_1\rangle,\dots,\langle\xi,\varphi_n\rangle)$, $\{\varphi_i\}$ being a base in $PL^2(R)$. $\varphi\cdot\nabla f$ is the directional derivative of f in the direction of the vector φ and $\frac{\delta}{\delta\xi(x)}$ is the variational derivative. It has been shown in [35] that ∇ is closable as an operator from $FC^1 \subset L^2(d\mu_o^\nu)$ into $L^2(R)\otimes L^2(d\mu_o^\nu)$.

This then gives that $\frac{1}{2}\nabla^*\overline{\nabla}$, where $\overline{\nabla}$ is the closure of ∇, is a self-adjoint, non negative operator in $L^2(d\mu_o^\nu)$. It is the self-adjoint operator uniquely associated with the closure of the form

$$\frac{1}{2}\iint\Big(\frac{\delta f}{\delta\xi(x)}\Big)^2 dx\, d\mu_o^\nu(\xi) \equiv \frac{1}{2}\int(\nabla f)^2 d\mu_o^\nu, \tag{14}$$

first defined on FC^1. The closed form is called the (diffusion) Dirichlet form given by μ_o^ν. It is also shown in [35] that the physical energy operator H of the Wightman models corresponding to μ^ν coincides with $\frac{1}{2}\nabla^*\overline{\nabla}$ on the dense domain FC^2 of $L^2(d\mu_o^\nu)$, so that we have the representation, for $f\in FC^2$,

$$(f, Hf) = \frac{1}{2}\iint\Big(\frac{\delta f}{\delta\xi(x)}\Big)^2 dx\, d\mu_o^\nu(\xi). \tag{15}$$

H and $\frac{1}{2}\nabla^*\overline{\nabla}$ are then "diffusion operators" in the sense that they are of the form of a "second order elliptic (infinite dimensional) partial differential operator", precisely, on FC^2,

$$Hf = \frac{1}{2}\nabla^*\overline{\nabla}f = -\frac{1}{2}\Delta f - \beta\cdot\nabla f \tag{16}$$

where

$$\Delta f(\xi) \equiv \int \frac{\delta^2 f}{\delta\xi(x)^2}\,dx \equiv \sum(\varphi_i\cdot\nabla)^2 f(\xi)$$

$$(\beta\cdot\nabla f)(\xi) \equiv \int \beta(x,\xi)\,\frac{\delta f}{\delta\xi(x)}\,dx \equiv \sum_{i=1}^{n}\beta(\varphi_i)(\xi)\,\varphi_i\,\nabla f(\xi)$$

where $\beta(\varphi)(\xi) \equiv -\frac{1}{2}(\varphi\cdot\nabla)^* 1(\xi)$, $1(\cdot)$ being the function identically one in $L^2(d\mu_o^\nu)$, and $\int\beta(x,\xi)\,\varphi(x)\,dx \equiv \beta(\varphi)(\xi)$.

Similarly one has the following representation for the Lorentz boost, see [3],

$$(f,\Lambda f) = \frac{1}{2} \iint x \left(\frac{\delta f}{\delta \xi(x)}\right)^2 dx \, d\mu_o^v(\xi) \tag{17}$$

on FC^1, and

$$(\Lambda f)(\xi) = -\frac{1}{2} \iint x \left(\frac{\delta^2 f}{\delta \xi(x)^2} + \beta(x,\xi) \frac{\delta f}{\delta \xi(x)} \right) dx \tag{18}$$

on FC^2. Moreover different regularity properties of μ_o^v can be proven, see [35], [36], [3]. In particular μ_o^v is $\mathcal{D}(R)$-quasi invariant, hence there exists a unitary representation in L^2 $(d\mu_o^v)$ of the abelian group $\mathcal{D}(R)$ by translation. Let $\pi(\varphi)$, $\varphi \in \mathcal{D}(R)$ be the infinitesimal generator of the strongly continuous one parameter unitary group representing the translations $\xi \to \xi + t\varphi$, $t \in R$, $\xi \in \mathcal{D}'(R)$. Then $\pi(\varphi)$ and $\langle \xi, \varphi \rangle$ satisfy the Weyl canonical commutation relations. For the case of the measure μ^v (with the assumed condition), it can be shown that $\pi(\varphi)$ would be essentially self-adjoint e,g. on the vectors $\{e^{i\langle \xi,\psi \rangle}, \psi \in \mathcal{D}(R)\}$ and that $\pi(\varphi)$ would coincide with the "physical field momentum operator" known from Glimm and Jaffe's work on the corresponding Wightman theory. We would then have the existence of the "Schrödinger representation" for weak polynomial interactions. The following equation of motion

$$i[\pi(\varphi), H] = :v':(\varphi) + \langle \cdot, (-\Delta + m^2)\varphi \rangle \tag{19}$$

would then hold, where v' is the derivative of the polynomial giving the interaction. The function one in $L^2(d\mu_o^v)$ is an analytic vector for $\pi(\varphi)$. Moreover $\mu_o^v(\xi + t\varphi)$

is analytic in t for all $|t| < C_\varphi$, C_φ being a constant and μ_o^v is "strictly positive" in the sense ([35]a)) that the conditional measure obtained by conditioning with respect to subspaces of finite codimension has strictly positive density relative to Lebesgue measure, on compacts.

<u>Remark:</u> The results in this section show in particular that parts of the canonical program of Coester and Haag [37] and Araki [38] by which "the physics is determined entirely by the vacuum measure" have been realized for weak polynomial interactions.

<u>Remark:</u> $\frac{1}{2} \nabla^* \bar{\nabla}$ generates a symmetric time homogeneous Markov process $t \to \eta_t$ which has been shown in [36] to be a strong Markov, standard Dynkin, in fact Hunt process with continuous paths, on a properly chosen compactification of $\mathcal{D}'(R)$, with

an associated analytic potential theory of Beurling-Deny, Fukushima and Silverstein type. We conjecture that the process ξ_t defined in Section I.5 and the diffusion process η_t are equivalent as processes. We recall that one already knows that their infinitesimal generators coincide on a dense subset. From [35] and [36] we have also that they both solve the stochastic diffusion equation

$$d\zeta_t = \beta(\zeta_t)\,dt + dw_t , \tag{20}$$

where w_t is the standard Brownian motion associated with $L^2(R)$.

II. The energy representation of Sobolev-Lie groups (fields with values in a Lie group)

We report here shortly on results contained essentially in [39] and we refer to this reference for more details. There are several motivations for this study we shall mention later. For the moment let us just point out that it is an attempt to at least partially extend the Euclidean Markov formalism to the case of fields $\xi(x)$ with values in a Lie group G and where the points x are taken in an orientable Riemannian manifold M. Let us first take the case where $G = R^n$, in which the extension we have in mind is immediate, yet useful for stating precisely the problem. Let $C_o^1(M,R^n)$ be the space of C^1-maps $x \in M \to \varphi(x) \in R^n$ which are zero outside compacts on M. For any $\varphi \in C_o^1(M, R^n)$ we denote the differential of this map at x by $d\varphi(x)$. Let $d\varphi(x)'$ denote the dual map and set $\langle d\varphi(x), d\varphi(x)\rangle \equiv$ (trace of $d\varphi(x)\, d\varphi(x)'$) as an operator from R^n into R^n. Define

$$\|\varphi\|^2 \equiv \frac{1}{2}\int_M \langle d\varphi(x), d\varphi(x)\rangle\, dx \tag{21}$$

and

$$(\varphi,\psi) \equiv \frac{1}{2}\left[\|\varphi\|^2 + \|\psi\|^2 - \|\varphi-\psi\|^2\right].$$

Completing $C_o^1(M, R^n)$ in the norm $\|\ \|$ we get a real Hilbert space $H_1(M, R^n)$ with norm $\|\ \|$ and scalar product $(\ ,\ )$. This reduces to the classical Dirichlet space if $M = R^d$, $n = 1$.

Let now μ^o be the standard Gauss measure associated with the dual of $H_1(M, R^n)$, formally given by $\exp(-\frac{1}{2}\|\varphi\|^2)\,d\varphi$. μ^o defines an associated random field $\xi(x)$, $x \in M$, with values in R^n. When $M = R^d$, $n=1$, $d \geq 3$ we have the free Markov field with $m = o$ on R^d. Analogously as in this well-known case, in the general case μ^o gives a unitary irreducible representation $\tilde{V}$ of $H_1(M, R^n)$ in $L^2(d\mu^o)$, by translations, namely $(\tilde{V}(\varphi)f)(\xi) = (d\mu^o(\xi+\varphi)/d\mu^o(\xi))^{1/2} f(\xi+\varphi), \varphi\in H_1(M, R^n)$. This representation is equivalent to the following one (corresponding to the "coherent vectors" formalism of free Euclidean fields). Let $e^{\varphi} \equiv \exp(\frac{1}{2}\|\varphi\|^2)\tilde{V}(\varphi)1$, where 1 is the function identically 1 in $L^2(d\mu^o)$. Let $\mathcal{F}_o(M, R^n)$ be the linear space of all complex linear combinations of vectors of this form. Define $(e^{\varphi}, e^{\psi})_{\mathcal{F}} \equiv \exp((\varphi, \psi))$, Extending $(\ , \ )_{\mathcal{F}}$ by linearity to $\mathcal{F}_o(M, R^n)$, dividing then out by elements for which the seminorm $(\ , \ )_{\mathcal{F}}$ vanishes and then completing we get a Hilbert space $\mathcal{F}(M, R^n)$, which is the representation space for the irreducible representation V of $H_1(M, R^n)$ by translations, defined by

$$V(\varphi)e^{\psi} = \exp\left(-\frac{1}{2}\|\varphi\|^2 - (\varphi, \psi)\right) e^{\varphi+\psi} . \quad (22)$$

V is the equivalent of $\tilde{V}$ in $\mathcal{F}(M, R^n)$. The representation being irreducible, all functions of the fields can be recovered from this representation and in this sense the representation V suffices for all information about the fields. We shall now see that this "energy representation" V has an extension to the case where R^n, the space in which the fields take their values, is replaced by a compact Lie group (or more generally a Lie group with compact Lie algebra, the direct product of a compact Lie group and an abelian one).

Consider $C_o^1(M, G)$, the space of once differentiable maps from M into G which are identically the unit in G outside compacts in M. $C_o^1(M, G)$ is a group of maps with multiplication defined pointwise. Let $\varphi\in C_o^1(M, G)$, then the tangent space to G at $\varphi(x)$ carries a natural scalar product given by transport by left multiplication by $\varphi(x)$ of the one given by the Killing form on the Lie algebra of G. Identifying the tangent spaces TM_x and $TG_{(x)}$ with their duals, we can look upon the dual $d\varphi(x)'$ of the differential maps $d\varphi(x)$ as a bounded map from $TG_{(x)}$ to TM_x, hence the Hilbert-Schmidt norm $\langle d\varphi(x), d\varphi(x)\rangle \equiv \mathrm{Tr}(d\varphi(x)\, d\varphi(x)')$ is defined in the space $B(TG_{\varphi(x)}, TG_{\varphi(x)})$ of bounded linear maps in $TG_{\varphi(x)}$ and we set

$$\|\varphi\|^2 \equiv \frac{1}{2}\int_M \langle d\varphi(x), d\varphi(x)\rangle\, dx . \quad (23)$$

Note that, denoting by $\varphi(x)^{-1}\nabla\varphi(x)$ the element in the Lie algebra obtained by transporting back by the action of $\varphi(x)^{-1}$ the vector $\nabla\varphi(x) = (\frac{\partial}{\partial x^1}\varphi(x), \ldots , \frac{\partial}{\partial x^d}\varphi(x))$ in $TG_{\varphi(x)}$, $\{x^i\}$ being normal coordinates in M, we have

$$\|\varphi\|^2 = \frac{1}{2}\int_M B\,(\varphi(x)^{-1}\nabla\varphi(x),\, \varphi(x)^{-1}\nabla\varphi(x))\,dx\,. \quad (24)$$

Moreover one proves, from the left and right invariance of the Riemann structure on G, that $\delta(\varphi,\psi) \equiv \|\varphi^{-1}\psi\|$ is a metric on C_o^1 (M,G). We call the metric completion H_1 (M,G) of C_o^1 (M, G) a <u>Sobolev-</u> <u>-Lie group</u> (this is clearly an extension of the abelian Sobolev space $H_1(M, R^n)$). Note that $\|\varphi\|^2$ is the "Euclidean action" connected with models considered in the physical literature (σ-model, chiral models, pure Yang-Mills fields). We shall now give the energy representation of $H_1(M, G)$, which, by above discussion for $G = R^n$, corresponds in a sense to a Euclidean measure formally given by $\exp(-\frac{1}{2}\|\varphi\|^2)d\varphi$. This energy representation of H_1 (M, G) is obtained exactly as in the case $G = R^n$, namely by considering the free $\mathbb{C}$-module $\mathcal{F}_o$ (M, G) with generators e^{φ}, $\varphi \in H_1$ (M, G) and scalar product $(e^{\varphi}, e^{\psi})_{\mathcal{F}} \equiv \exp((\varphi,\psi))$, with $(\varphi,\psi) \equiv \frac{1}{2}[\|\varphi\|^2 + \|\psi\|^2 - \|\varphi^{-1}\psi\|]$. Note that the symmetry of the scalar product comes the left and right invariance of the Riemann structure on G. Quotienting and completing with respect to $(\ ,\ )_{\mathcal{F}}$ yields then the Hilbert space $\mathcal{F}(M, G)$ of the unitary representation V of H_1 (M, G) by left translations:

$$V(\varphi)e^{\psi} \equiv \exp(-\tfrac{1}{2}\|\varphi\|^2 + (\psi,\varphi^{-1}))\,e^{\varphi\psi}\,. \quad (25)$$

This energy representation is by construction a non abelian extension of the representation given by Euclidean Markov fields in the case G = R. Moreover we have the following:

<u>Theorem 3</u>: If dim M = 1 then V coincides with the unitary representation of H_1 (M,G) given by left multiplication on the paths of the Brownian motion on the Lie group G.

For the proof of this Theorem we refer to [39].

<u>Remark</u>: For references concerning Brownian motion on Lie groups see e.g. [40] and references therein. Clearly above Theorem shows, at least in dim M=1, the Markovian character of the representation.

It seems to be unknown whether the representation is irreducible in this case. However for semisimple compact G the irreducibility has been proven for dim $M \geq 5$ by Ismagilov [41] and for dim $M \geq 2$ by Vershik, Gelfand and Graev [42].

The latter authors have also explored the connection with the representation of the Sugawara algebra of currents. In fact one of the motivations for the study of the energy representation is to provide representations for groups of mappings like the ones which arise in the theory of current algebras. The type of representation constructed here and independently in the mentioned papers of Ismagilov and Vershik, Gelfand and Graev is different from the largerly studied preceding "completely factorizable" ones, see e.g. [43], and references therein, in as much as "it corresponds to a measure exp $(-\frac{1}{2}\int\langle d\varphi(x), d\varphi(x)\rangle dx)$ rather than $\exp(-\frac{1}{2}\int\varphi(x)^2 dx)$". Such a new type of representations has also been discussed in [44] and [45].

III. Oscillatory integrals in infinitely many dimensions

III.1 Definition and properties of the oscillatory integrals

In ch. I we have been concerned with measures given in terms of an "Euclidean action" and in ch. II with an extension of this concept to the case of fields with values in compact Lie groups. From the physical point of view however more direct objects of study are the formal complex measures given by the physical action integral, according to the well known Feynman path formulation of quantum mechanics. Feynman path integrals are oscillatory integrals in infinitely many dimensions, e.g. the solution of the time dependent Schrödinger equation in R^d

$$i\hbar \frac{\partial}{\partial t}\psi(x,t) = -\frac{\hbar^2}{2m}\Delta\psi(x,t) + V(x)\psi(x,t) \qquad (26)$$

with initial condition $\psi(x,o) = \varphi(x)$ is given by the Feynman path integral

$$\psi(x,t) = \int_{\gamma(t)=x} e^{\frac{i}{\hbar}S_t(\gamma)} \varphi(\gamma(0))\, d\gamma \qquad (27)$$

over all paths $\gamma(\tau)$, $\tau\in[0,t]$ such that $\gamma(t)=x$, where $S_t(\gamma)$ is the action $\frac{m}{2}\int_0^t \left(\frac{d\gamma}{d\tau}\right)^2 d\tau - \int_0^t V(\gamma(\tau))\,d\tau$ along the path γ. We shall sketch here shortly a mathematical formulation of such integrals strong enough to permit the construction of asymptotic expansions in powers of $\hbar$ i.e. expansions around the classical limit, by a method of stationary phase in infinitely many dimensions, thereby justifying one of the most beautiful intuitions connected with the introduction (about 30 years ago) of the integrals. For details and applications of the mathematical theory sketched here, as well as for references to other work, see [46], [47].

Let $\mathcal{H}$ be a real separable Hilbert space, with scalar product $(\, , \,)$. Let $\mathcal{F}(\mathcal{H})$ be the space of all functions on $\mathcal{H}$ which are Fourier transforms of complex measures on $\mathcal{H}$ [5]. Such a space is a commutative Banach function algebra, with norm $\| \;\|$ given by the total variation $\| \;\|_{tot}$ of the corresponding measure i.e. for $f \in \mathcal{F}(\mathcal{H})$ such that f is the Fourier transform of a measure μ_f on $\mathcal{H}$ i.e. $f(\gamma) = \int_{\mathcal{H}} \exp(i(\gamma,\gamma'))\, d\mu_f(\gamma')$ we have $\|f\| \equiv \|\mu_f\|_{tot}$. For $f \in \mathcal{F}(\mathcal{H})$ the integral $\int_{\mathcal{H}} \exp(-\frac{i}{2}(\gamma,\gamma))d\mu_f(\gamma)$ is well defined. We use also the notation

$$\tilde{\int_{\mathcal{H}}} e^{\frac{i}{2}(\gamma,\gamma)} f(\gamma)\, d\gamma \tag{28}$$

for it i.e.

$$\tilde{\int_{\mathcal{H}}} e^{\frac{i}{2}(\gamma,\gamma)} f(\gamma)\, d\gamma \equiv \int_{\mathcal{H}} e^{-\frac{i}{2}(\gamma,\gamma)} d\mu_f(\gamma). \tag{29}$$

This, looked upon as a linear functional in f, was termed in [46] "the Fresnel integral of f" and $\mathcal{F}(\mathcal{H})$ was called the space of "Fresnel integrable functions". The reason for the notation (28) is the following: if $\mathcal{H} = R^n$ and $f \in \mathcal{F}(R^n) \cap L^1(R^n)$ then

$$\tilde{\int_{R^n}} e^{\frac{i}{2}(\gamma,\gamma)} f(\gamma)\, d\gamma = (2\pi i)^{-n/2} \int_{R^n} e^{\frac{i}{2}(\gamma,\gamma)} f(\gamma) d\gamma,$$

where the integral on the right hand side is the usual one. Thus the Fresnel integral reduces to $(2\pi i)^{-n/2}$-times the ordinary integral in the finite dimensional case, thus the sign $\sim$ above the integral indicates the presence of the normalizing factor $(2\pi i)^{-n/2}$ (in fact the Fresnel integral of $f = 1$ is one). Clearly on cylinder functions on $\mathcal{H}$ (i.e. functions $f \in \mathcal{F}(\mathcal{H})$ such that $f(\gamma) = f(P\gamma)$ for some finite dimensional projection P) we have again that the Fresnel integral reduces to an ordinary integral, namely, if the restriction $\tilde{f}$ of f to $P\mathcal{H}$ is in $\mathcal{F}(P\mathcal{H}) \cap L^1(\mathcal{H})$,

$$\tilde{\int_{\mathcal{H}}} e^{\frac{i}{2}(\gamma,\gamma)} f(\gamma)\, d\gamma = (2\pi i)^{-\frac{1}{2}\dim(P\mathcal{H})} \int_{P\mathcal{H}} e^{\frac{i}{2}(\gamma,\gamma)} \tilde{f}(\gamma)\, d\gamma . \tag{30}$$

In the general case, the Fresnel integral is a normalized linear bounded functional in f which has the contraction property that the integral of a product of functions is bounded in modulus by the product of the norms of the functions. The above notation and the observation that in finite dimension the Fresnel integral reduces to the integral of $e^{\frac{i}{2}(x,x)}f(x)$ against Lebesgue measure suggest invariance of the Fresnel integral with respect to rotations and reflections in $\mathcal{H}$ and the transformation, denoting (29) by $F(f)$, $F(f_\alpha) = e^{\frac{i}{2}(\alpha,\alpha)} F(e^{\frac{i}{2}(\cdot,\alpha)} f(\cdot))$ under translations $f \to f_\alpha$, with $f_\alpha(\gamma) = f(\gamma - \alpha)$. These and other transformation properties are proven, see [46], [48]. A final useful property is the "Fubini theorem" on iterated integrations:

$$\tilde{\int}_{\mathcal{H}_1 \oplus \mathcal{H}_2} e^{\frac{i}{2}(\gamma,\gamma)} f(\gamma)\, d\gamma = \tilde{\int}_{\mathcal{H}_2} e^{\frac{i}{2}(\gamma_2,\gamma_2)} \Big(\tilde{\int}_{\mathcal{H}_1} e^{\frac{i}{2}(\gamma_1,\gamma_1)} f(\gamma_1,\gamma_2)\, d\gamma_1 \Big) d\gamma_2 . \tag{31}$$

Let us finally remark that for the computation of Fresnel integrals approximations are available, e.g. through power expansions ([46],[47] Gaussian integrals ([48]) and Kato-Trotter expansions ([49], [50] , [51]).

Remark: The extension of the theory to the case where $\mathcal{H}$ is replaced by a real separable Banach space and (γ,γ) by a positive definite quadratic form on it is immediate [46].

The definition of Fresnel integrals recalled here from [46] is related especially to work of Ito [48] and C. Morette - De Witt [52]. For work related to the latter authors see also e.g. [53] - [55] and for developments related to our work see [50] and [51]. Recently Maslov and Chebotarev [56] have developed a formalism very closely related to ours.

Applications to the Feynman path integrals of quantum mechanics have been given in [46], [47]. They include in particular the justification of the formula (27), for potentials V which are Fourier transform of measures on R^d, in the form of the Fresnel integral

$$\psi(x,t) = \tilde{\int}_{\mathcal{H}} e^{\frac{i}{2\hbar}(\gamma,\gamma)}\, e^{-\frac{i}{\hbar}\int_0^t V(\gamma(\tau)+x)\,d\tau}\, \varphi(\gamma(0)+x)\, d\tau, \tag{32}$$

where $\mathcal{H}$ is the Hilbert space of absolutely continuous paths $\gamma(\tau)$, $\tau \in [0,t]$ with values in R^d, which end at the origin and have finite kinetic energy $\frac{1}{2}(\gamma,\gamma) = \frac{m}{2}\int_0^t \left(\frac{d\gamma}{d\tau}\right)^2 d\tau$.

Similarly, representations for the wave operators, the scattering amplitudes, the Green's functions are given in terms of Fresnel integrals [46].

III.2 Asymptotic expansions of the oscillatory integrals

Consider e.g. above formula (32) for the solution of Schrödinger's equation in terms of Fresnel integrals. We see that to study the approach to classical mechanics as $\hbar \to 0$ we have to study the behaviour of oscillatory integrals of the form

$$I(\hbar) \equiv \tilde{\int_{\mathcal{H}}} e^{\frac{i}{2\hbar}(\gamma,\gamma)} e^{-\frac{i}{\hbar}W(\gamma)} g(\gamma)\, d\gamma \qquad (33)$$

in the neighborhood of $\hbar = o$, where W and g are functions in $\mathcal{F}(\mathcal{H})$. This study in the finite dimensional case dim $\mathcal{H} < \infty$ is well known from the work of Maslov, Hörmander, Arnold, Duistermaat and others, see e.g. the references in [47]. We shall now briefly indicate how the infinite dimensional case is treated, referring for details to [47]. Suppose the measure μ_W of which W is the Fourier transform is such that $\int_{\mathcal{H}} |\alpha|^2 d|\mu_W|(\alpha) < \infty$. Then the Fréchet derivatives $dW(\gamma)$, $d^2 W(\gamma)$ exist, hence also $d\Phi(\gamma)$, $d^2\Phi(\gamma)$, where

$$\Phi(\gamma) \equiv \frac{1}{2}(\gamma,\gamma) - W(\gamma)$$

is the total phase function. Correspondingly as in the finite dimensional case we will now distinguish 3 cases:

A) $\Phi(\gamma)$ has only one critical (i.e. stationary) point γ_c (such that $d\Phi(\gamma) = o$ for $\gamma = \gamma_c$) and this critical point is regular (i.e. $|d^2\Phi(\gamma_c)| \neq 0$). Assuming, without restriction of generality, $\Phi(\gamma_c) = 0$ we then have the asymptotic expansion in powers of $\hbar$ given in the next theorem, which also contains a sufficient condition for the occurrence of the situation A):

Theorem 4 : If μ_W is such that $\int e^{\sqrt{2}\,\lambda|\alpha|} d|\mu_W|(\alpha) < \lambda^2$ for some $\lambda > o$, then $\Phi(\gamma)$ has one and only one critical point and this is regular. If moreover the measure μ_g of which g is the Fourier transform satisfies $\int e^{\sqrt{2}\,\lambda|\beta|} d|\mu_g|(\beta) < \infty$ then $I(\hbar)$ is analytic in $\operatorname{Im}\hbar < 0$ and C^∞ on the real axis and has the asymptotic expansion in powers of $\hbar$ (assuming $W(0) = dW(0) = 0$)

$$I(\hbar) = \sum_{m=0}^{N} \hbar^m (i/2)^m \sum_{n=0}^{\infty} (1/2)^n \frac{1}{n!(m+n)!} \left(\sum_{j=1}^{n} \nabla_{\gamma_j} + \nabla_\gamma \right)_2^{2(m+n)} \tag{34}$$

$W(\gamma_1) \ldots W(\gamma_n) g(\gamma) \big|_{\gamma=\gamma_c} + R_N$, with the bound on the remainder R_N, $|R_N| \le \hbar^{N+1} (N+1)!\, C_\lambda / \lambda^{2(N+1)}$, where C_λ is independent of $\hbar$ and N. The notation $(\quad)_2^{2(m+n)}$ means that one should eliminate, in the computation of the 2 (m+n)-power, all terms which do not contain each derivative ∇_{γ_j} to a power $\ge$ 2.

B) $\Phi(\gamma)$ has possibly infinitely many critical points, all of them regular. By an adaptation of the finite dimensional method of partition of the unit, the asymptotic expansion is the superposition of ones of type A), at least in the case where the critical points do not have limit points, which is the case e.g. when $\int d|\mu|(\alpha) < \lambda^2$ and $\int_{\mathcal{H}} e^{\sqrt{2}\lambda|\alpha|} d|\mu|(\alpha) < \infty$ for some $\lambda > 0$.

C) $\Phi(\gamma)$ has singular critical points i.e. critical points γ_c for which $|d^2\Phi(\gamma_c)| = 0$. The way to tackle this situation is, like in the finite dimensional case, to observe that critical singular points are unstable under small perturbations of the C^∞ topology and if W depends on k parameters and the degeneracy happened to be of codimension $k \ge 1$, then for a dense set of functions $y \to W(\gamma, y)$, we have that the critical points are regular. Hence if we study integrals of the form

$$(2\pi i)^{-k/2} \int_{R^k} dy\, e^{-\frac{i}{\hbar} f(y)} \chi(y) \tilde{\int_{\mathcal{H}}} e^{\frac{i}{2\hbar}(\gamma,\gamma)} e^{-\frac{i}{\hbar} W(\gamma, y)} g(\gamma) d\gamma \tag{35}$$

with $f \in C^\infty$ and $\chi \in C_0^\infty$ then for a suitable choice of f we will again be in the case of regular critical points and can then apply the methods of the situations A) and B), together with the finite dimensional theory.

In conclusion, in all situations A), B), C) one is able to give asymptotic expansions of the oscillatory integrals in powers of $\hbar$. We shall now mention briefly the application to the discussion of the asymptotics in $\hbar$ of quantum mechanics. The solution of Schrödinger's equation (26) can be written

$$\psi^{\hbar}(x,t) = \int_{R^d} dy\, \varphi(y)\, G_t^{\hbar}(x,y), \qquad (36)$$

where the Green's functions $G_t^{\hbar}(x,y)$ can be expressed in terms of a Fresnel integral by

$$G_t^{\hbar}(x,y) = (2\pi i \hbar t)^{-d/2}\, e^{\frac{i}{2\hbar}(y-x)^2/t} \widetilde{\int_{\mathcal{H}_o}} e^{\frac{i}{2\hbar}(\gamma,\gamma)}\, e^{-\frac{i}{\hbar}W(\gamma,y)}\, d\gamma \qquad (37)$$

with $W(\gamma,y) \equiv \int_0^t V(\gamma + (y-x)(1-\frac{\tau}{t}) + x)\, d\tau$ and where $\mathcal{H}_o$ is the Hilbert space of absolutely continuous paths $\gamma(\cdot)$ from $[o,t]$ into R^d, which start and end at the origin and have finite kinetic energy i.e. $(\gamma,\gamma) \equiv m \int_0^t (\frac{d\gamma}{d\tau})^2 d\tau < \infty$. We see that for initial conditions $\varphi(y) = \exp(-\frac{i}{\hbar} f(y))\chi(y)$ the integral over $R^d \oplus \mathcal{H}$ given by (36) and (37) has precisely the form (35), and one has then the following

Theorem 5: Assume the potential V on R^d is the Fourier transform of some complex measure μ_V such that $\int \exp(|\alpha|\varepsilon)\, d|\mu_V|(\alpha) < \infty$ for some $\varepsilon > o$. Then for the solution $\psi^{\hbar}(x,t)$ of the Schrödinger equation with generic initial condition $\psi^{\hbar}(x,o) = \exp(-\frac{i}{\hbar}f(x))\chi(x)$, $f \in C^\infty(R^d)$, $\chi \in C_o^\infty(R^d)$ we have an asymptotic expansion in powers of $\hbar$. To the first order in $\hbar$ we have

$$\psi^{\hbar}(x,t) = \sum_j D_j\, e^{-\frac{i}{2}\pi n(\gamma^{(j)})}\, e^{\frac{i}{\hbar} S_t(\gamma^{(j)})}\, e^{-\frac{i}{\hbar} f(\gamma^{(j)})} + o(\hbar),$$

where the sum is over all classical trajectories $\gamma^{(j)}$ ending at x and starting with momentum $-\nabla f(y_j)$ from some point $y^{(j)}$ (depending on x,t) (there are finitely many such trajectories). n is the Maslov index and $D_j \equiv |\det \partial\gamma_k^{(j)}/\partial y_l^{(j)}|^{-1/2}$, det being the determinant of the dxd-matrix $\partial\gamma^{(j)}/\partial y^{(j)}$, where the indices k, l refer to the components of $\gamma^{(j)}$ resp. $y^{(j)}$ in R^d.

Remark: This result is an example to the statement that expansions around the classical limit come out naturally from suitably defined Feynman path integrals. In this lecture we have only considered oscillatory integrals with phase functions of the type $B(\gamma,\gamma) - W(\gamma,y) - \Psi(y)$, where $B(\cdot,\cdot)$ is a positive definite quadratic form. The more general case where B is not necessarily positive definite has also been treated along similar lines in [46], where also some applications to the Feynman path integrals for an-harmonic oscillators and quantum fields were given.

Notes

1) Work on other types of markovian random fields is mentioned e.g. in [1]a), to which we refer, together with [2] also for background about Euclidean field theory and its development since the basic work of Symanzik, Nelson and Guerra. For relations with statistical mechanics see e.g. also [1]b).

2) Note that our name is inspired by Dynkin's entrance boundary theory, see e.g. [7] ..however $D(\mu)$ should not be confused with the (in general larger) boundary obtained by considering all measures given in terms of specifications by compatible systems of conditional probabilities like e.g. the ones given by Dobrushin-Lanford-Ruelle equations. It turns out that our definition is appropriate for the applications we have in mind.

3) For somewhat related work see also e.g.[34].

4) For previous discussions of time zero fields see e.g. [33], [20], [13], [2], [34].

5) It contains e.g. all finite linear combinations of positive-definite functions, which are continuous in the Minlos-Sazonov-Gross norm given by $(\gamma, B\gamma)$, for some $B > 0$ of trace class.

ACKNOWLEDGEMENTS

It is a pleasure to thank the Organizing Committee for the very friendly invitation. The first author would also like to express his gratitude to Professors R. Cairoli, S.D. Chatterji, Ph. Choquard and L. Streit for kind invitations to Lausanne and Bielefeld, where part of this work was done. Moreover he acknowledges gratefully the long standing hospitality of the Institute of Mathematics of Oslo University and the support by the Norwegian Research Council for Science and the Humanities.

REFERENCES

Chapter I

1 a) S. Albeverio, R. Høegh-Krohn, Probability and quantum fields, Lectures given at the "III cycle en Mathématiques et Physique", Lausanne, 1977 (in preparation).

b) S. Albeverio, R. Høegh-Krohn, Homogeneous random fields and statistical mechanics, J. Funct. Anal. 19, 242-272 (1975)

2 B. Simon, The $P(\varphi)_2$ Euclidean (Quantum) Field Theory, Princeton University Press, 1974.

3 a) S. Albeverio, R. Høegh-Krohn, Preprint in preparation.

b) S. Albeverio, R. Høegh-Krohn, Canonical relativistic quantum fields (to appear in Ann. Inst. H. Poincaré).

4 C. Newman, J. Funct. Anal. 14, 44 - 61 (1973).

5 E. Nelson, pp 413 - 42o in "Partial Differential Equations" D. Spencer Ed., Symp. in Pure Math. Vol, 23 AMS Publ., 1973.

6 E. Nelson, J. Funct. Anal. 12, 211-227 (1973).

7 E.B. Dynkin, pp. 5o7 - 512 in Actes Congrès Intern. Math. 197o, t. 2, (1971).

8 R.L. Dobrushin, Theor. Prob. Appl. 13, 197 - 224 (1968).

9 O.E. Lanford, D. Ruelle, Comm. Math. Phys. 13, 194-215 (1969).

1o C. Preston, Random fields, Springer Lecture Notes in Mathematics, 534 Berlin, 1976.

11 H. Föllmer, pp. 3o5 - 317 in Sémin. Prob. Strasbourg IX, Lecture Notes in Math. 465 Springer, Berlin, 1975.

12 F. Guerra, L. Rosen, B. Simon, Ann. Math. 1o1, 111 - 259 (1975).

13 J. Fröhlich, Helv. Phys. Acta 47, 265 - 3o6 (1974) Adv. Math. 23, 119 - 18o (1977)
J. Fröhlich, Ann. Inst. H. Poincaré 21, 271 - 317 (1974).

14 J. Fröhlich, Ann. Phys. 97, 1 - 54 (1976).

15 J. Glimm, A. Jaffe, Commun. Math. Phys. 44, 293 - 32o (1975)

16 L. Accardi, Local perturbations of conditional expectations, Marseille Preprint.

17 E. Nelson, pp. 94 - 124 in Ref. [18] .

18 G. Velo, A.S. Wightman, Edts., Constructive Quantum Field Theory, Springer, Berlin, 1973.

19 J. Glimm, A. Jaffe, T. Spencer, pp. 133 - 242 in Ref. [18].

2o S. Albeverio, R. Høegh-Krohn, J. Funct. Anal. 16, 39 - 82 (1974).

21 J. Fröhlich, E. Seiler, Helv. Phys. Acta 49, 889 - 924 (1976).

22 T. Spencer, Commun. Math. Phys. 39, 63 - 76 (1974).

23 J. Glimm, A. Jaffe, Ann. Inst. H. Poincaré A22, 1o9 - 122 (1975).

24 a) J. Glimm, A. Jaffe, T.Spencer, pp. 175 - 184 in Ref. 25 .

b) J. Glimm, A. Jaffe, T. Spencer, Commun. Math. Phys. 45, 2o3 - 216 (1975).

25 F. Guerra, D.W. Robinson, R. Stora, Edts., Les Méthodes mathématiques de la théorie quantique des champs, Collo Intern. du CNRS, N° 248, Marseille 1975, CNRS, 1976.

26 F. Guerra, L. Rosen, B. Simon, Ann. Inst. H. Poincaré A, 25, 231 - 334 (1976).

27 J. Fröhlich, B. Simon, Ann. of Math. 1o5, 493 - 526 (1977).

28 J. Fröhlich, Y.M. Park, Helv. Phys. Acta 5o, 315 - 329 (1977).

29 R.L. Dobrushin, R.A. Minlos, pp. 23 - 49 in Ref. [3o] .

3o B. Jancewicz, Ed., "Functional and probabilistic methods in quantum field theory, Vol. I, Acta Univ. Wratisl. N° 368, XII-th Winter School of Theo. Phys. Karpacz, 1975, Wroclaw, 1976.

31 E. Nelson, J. Funct. Anal. 12, 97 - 112 (1973).

32 B. Simon, Helv. Phys. Acta 46, 686 - 696 (1973).

33 J. Glimm, A. Jaffe, T. Spencer, Ann. of Math. 1oo, 585 - 632 (1974).

34 M.O' Carroll, P. Otterson, Comm. Math. Phys. 36, 37-58 (1974) W. Karwowski, Rep.Math. Phys. 7, 411-416 (1975)

K.Osterwalder,R.Schrader,Comm.Math.Phys.42,281-305 (1975)
G.C.Hegerfeldt,Comm.Math.Phys.35,155-171 (1974)

J.P.Eckmann, Relativistic quantum field theories in two space-time dimensions, Lecture Notes, Rome 1976 (to appear)

F.Constantinescu,W.Thalheimer,J.Funct.Anal.23, 33--38 (1976),

S.Nagamachi,N.Mugibayashi,Progr.Theor.Phys.53,1812-1826 (1975)

A.Klein,L.J.Landau,J.Funct.Anal.20, 44-82 (1975)

A.Klein, Bull. AMS 82, 762-764 (1976)

T.Hida,L.Streit,Nagoya Math.J. 68, (1977)

J.L.Challifour, J.Math.Phys.17, 1889- 1892 (1976)

Ph.Blanchard,Ch.Pfister, Processus gaussiens, équivalence d'ensembles et spécification locale, Bielefeld preprint(1977)

Ph.Courrège,P.Renouard; P.Priouret,M.Yor,Astérisque 22-23 , 1-245; 247-290 (1976)

G.Royer,Ann.ENS4^e Sér.8, 319-338 (1975)

I.Herbst, J.Math.Phys.17, 1210-1221 (1976)

35 a)S.Albeverio,R.Høegh-Krohn,Zeitschr.Wahrscheinlichkeitstheorie verw.Geb.40(1977). See also

b)S.Albeverio,R.Høegh-Krohn, pp. 11-59 in Ref.[25].

For the finite dimensional situation see also

c)S.Albeverio,R.Høegh-Krohn, J.Math.Phys.15, 1745- 1747 (1974)

d)S.Albeverio,R.Høegh-Krohn,L.Streit,J.Math.Phys.18,907-917(1977)

36 S.Albeverio,R.Høegh-Krohn, Hunt processes and analytic potential theory on rigged Hilbert spaces, ZiF Bielefeld preprint,Aug.1976 (to appear in Ann.Inst.H.Poincare B 13 (1977)).

37 F.Coester, R.Haag, Phys.Rev. 117, 1137-1145 (1960)

38 H.Araki, J.Math.Phys.1, 492-504 (1960).

Chapter II

39 S.Albeverio,R.Høegh-Krohn, Energy representation of Sobolev-Lie groups, ZiF Bielefeld preprint, May 1976 (to appear in Compositio Math.)

40 H.P.Mc Kean, Stochastic integrals, Academic Press, New-York (1969)

41 R.S.Ismagilov, Mat. Sb.100, N°1, 117-131 (1976) (russ.)

42 A.M. Vershik, I.M.Gelfand, M.I.Graev , Representation of

the group of smooth maps of a manifold into a compact Lie group, Moscow Preprint Inst. Prikl. Mat., Ak. Nauk N° 55 (1976) (russ.).

43 A.M. Vershik, I.M. Gelfand, M.I. Graev, Russ. Math. Surv. 28, N° 5, 83 - 128 (1973).

44 K.R. Parthasarathy, K. Schmidt, Commun. Math. Phys. 5o, 167 - 175 (1976).

45 R.F. Streater, Markovian representations of current algebras, J. Phys. A, 1o, 261 - 266 (1977).

Chapter III

46 S. Albeverio, R. Høegh-Krohn, Mathematical theory of Feynman path integrals, Lecture Notes in Mathematics Vol. 523, Springer, Berlin, 1976.
Some of the results also summarized in S. Albeverio, pp. 138 - 2o5 in Ref.[3o] .

47 S. Albeverio, R. Høegh-Krohn, Invent. Mathem. 4o, 59 - 1o6 (1977).

48 K. Ito, pp. 145 - 161, in Proc. Fifth Berkeley Symp. on Math. Stat. and Prob. Vol. II, part 1, Univ. Calif. Press, Berkeley 1967.

49 E. Nelson, J. Math. Phys. 5, 332 - 343 (1964).

5o a) A. Truman, J. Math. Phys. 17, 1852 - 1862 (1976),
b) A. Truman, J. Math. Phys. 18, 1499 - 15o9 (1977).

51 Ph. Combe, G. Rideau, R. Rodriguez, M. Sirugue-Collin, On some mathematical problems in the definition of Feynman path integral, Marseille Preprint (1976).

52 C. Morette - De Witt, Commun. Math. Phys. 28, 47 - 67 (1972); Commun. Math. Phys. 37, 63 - 81 (1974).

53 J. Tarski, pp. 169 - 18o in ""Functional Integration and its Applications", AM. Arthurs Edt., Oxford U.P., London, 1975.

54 P. Krée, pp. 163 - 192 in Séminaire P. Lelong, 1974/75, Lecture Notes in Mathematics, 524 Springer, Berlin 1976.

55 K. Brock, On the Feynman integral,Aarhus Univ. Publ. 1976.

56 V.P. Maslov, A.M. Chebotarev, Teor. i Matem. Fiz. 28, N°3, 291 - 3o7 (1976).

MATHEMATICAL FOUNDATIONS OF THE RENORMALIZATION GROUP METHOD IN STATISTICAL PHYSICS

Ya. G.Sinai

1. - Introduction -

The renormalization group method in statistical physics was developped by L. Kadanoff, M.Fisher, K.Wilson and others. There exists an enormous physical literature devoted to these questions. The surveys by Kogut and Wilson (1), Fisher (2), Ma (3), Brezin, Le Guillon, Zinn-Justin (4), Pokrovski and Patashinski (5) give a very good presentation of main ideas formal technique of perturbation theory which is used and of applications. The famous ε -expansion method by Wilson leads to results which are very satisfactory from the point of view of applications.

From the mathematical point of view the situation doesn't look so complete as one can think. There are at least three points in the theory under discussion which need more deep mathematical analysis.

1. The idea of ε-expansion assumes that the dimension of the space can take all real values. At the moment I don't see any rigorous approach to spaces of the fraction dimensions.

2. Usually one defines a renormalization group as a group acting in the space of hamiltonians. The investigation of this action near the fixed point is done using the linearized renormalization group. However if we try to apply this method to some concrete models we must have an expression for the hamiltonian which is valid in all domain of possible values of variables. It is easy to see that when these values are very large the form of the hamiltonian isn't defined by the fixed point.

3. Usual equations of renormalization group transformations are approximate. The rigorous theory must deal with exact equations, have nice estimates for remainder terms, etc.

The mathematical literature devoted to these problems is very poor. It should be mentioned the papers by Gallavotti Jona-Lasinio and their collegues (6)-(10), several papers on hierarchical models of Dyson by Bleher and Sinai (11),(12),(13) Collet and Eckmann (14) and some papers on scaling distributions by Sinai (15)(16), Dobrushin (17), (18). The first paper about the application of renormalization group to hierarchical models was done by G.Baker (19). Other references will be given in the text. Mathematical investigation of all these problems began only recently.

2. - Definition of Renormalization Group -

The renormalization group transformations were used very long ago in the classical branch of probability theory related to limit theorems for sums of independent or weekly dependent random variables. The method of renormalization group can be considered as a method for the investigation of limit theorems for strongly dependent random variables. Many problems of statistical mechanics and quantum field theory lead to probability distributions where there appear such strongly dependent random variables and one can hope to apply renormalization group method.

To begin with rigorous mathematical definitions we shall consider a probability space Ω consisting of points ω . Each ω is a configuration on the d-dimensional lattice $\mathbb{Z}^d$, $\omega = \{\omega(x), x \in \mathbb{Z}^d\}$, each variable $\omega(x)$ takes an arbitrary real value. By $\mathcal{F}$ we shall denote the usual σ-algebra of subsets of Ω.

Two sets of transformations act naturally in the space Ω. The first one is the group $\{T^x\}$ of space translations indexed by the points of the lattice. A single transformation T^y acts via the formula

$$(T^y\omega)(x) = \omega(x+y), \qquad x \in \mathbb{Z}^d.$$

Another set of transformations consists of scaling transformations. Let us fix a parameter α, $1 < \alpha < 2$, and consider a semigroup of transformations $\mathcal{O\!L} = \mathcal{O\!L}(\alpha) = \{\mathcal{O\!L}_k(\alpha)\} = \{\mathcal{O\!L}_k\}$, k $\geqslant$ 1 is an integer, where

$$(\mathcal{O\!L}_k\omega)(x) = \omega_k(x) = \frac{1}{k^{d\frac{\alpha}{2}}} \sum_{y=(y_1\dots y_d):\, kx_i \le y_i < k(x_i+1),\, i=1,\dots,d} \omega(y)$$

It is obvious that $\mathcal{O\!L}_{k_1}\cdot\mathcal{O\!L}_{k_2} = \mathcal{O\!L}_{k_1k_2}$. From the point of view of probability theory ω_k (x) is the normalized sum of random variables. The number α is the only parameter of the theory.

Definition 1. – The semigroup $\mathcal{O\!L}(\alpha)$ is called the renormalization group. Translations T^x and scaling transformations $\mathcal{O\!L}_k$ are connected by the commutation relation:

$$T^x\mathcal{O\!L}_k = \mathcal{O\!L}_k T^{kx}, \qquad kx = (kx_1,\dots,kx_d).$$

The semigroup $\mathcal{G}$ generated by all $\mathcal{O\!L}_k$ and T^x is isomorphic to the semigroup Γ of affine transformations of the $\mathbb{R}^d$ having the form: for $\gamma\in\Gamma$, $\gamma(t) = \frac{t+r}{k}$; $t=(t_1,\dots,t_d)\in\mathbb{R}^d$, $r=(r_1,\dots,r_d)\in\mathbb{Z}^d$. Under this isomorphism the transformation $\mathcal{O\!L}^k$ is isomorphic to the contraction $\gamma(t) = \frac{1}{k}t$ and the translation T^x is isomorphic to the shift $\gamma(t) = t+x$.

By the conjugate semigroup of transformations $\mathcal{G}^*$ we shall mean the semigroup acting in the space of probability distributions on Ω by the formula: for any probability distribution P and any g $\in$ G

$$(g^*P)(C) = P(g^{-1}C), \quad g\in G.$$

Definition 2. – A probability distribution P_o is called a lattice scaling distribution if it is invariant under the action of the group $\mathcal{G}^*$, or, in another words, if it is a fixed point of the group $\mathcal{G}^*$.

Remarks, analogies, ecc.

1. – The invariance of scaling distribution with respect to the group of translations means that $\{\Omega, \mathcal{F}, P_o, \{T^x\}\}$ is a strictly stationary random field in the sense of probability theory or a dynamical system with $\mathbb{Z}^d$ – time in the sense of ergodic theory.

2. – Random variables ω_k(x) are often called in statistical mechanics as block-spin variables. The invariance of P_o with respect to $\mathcal{O\!L}^*_k(\alpha)$ means that random variables ω_k(x) have the same common probability distribution as initial random variables ω(x).

3. – Assume that Q is an arbitrary strictly stationary random field and for some α, $1 < \alpha < 2$, the probability distributions $Q_k = \mathcal{O}\!\iota_k^{(\alpha)} Q_0$ converge weakly to a limit P_0. Then P_0 is the scaling distribution. Thus scaling distributions are limit probability distributions for sums $\omega_k(x)$ of random variables $\omega(y)$, where $\omega(y)$ are distributed according to some stationary probability distribution.

4. – Scaling probability distributions were introduced in the paper by Gallavotti and Jona-Lasinio (7) under the name of stable distributions and in my paper (15) under the name of automodel distributions. R.L.Dobrushin in his papers (17), (18) also uses the term "automodel distribution". Recently I received a letter from Professor B.Mandelbrot where he explained me that he and many of his colleagues use in similar problems in statistical hydromechanics the term "scaling distribution". It seems to me that the term "scaling distribution" is mostly appropriate and it will be used here and in subsequent pubblications.

5. – A very important question concerns the choice of the parameters α. If in the problem described above in 3. $E\,\omega(x) = 0$, $E\,\omega(x)\,\omega(y) \sim \frac{const}{\|x-y\|^{\delta d}}$; $\|x-y\| \to \infty$, then for $\alpha = 2 - \delta > 1$ expectation $E\,\omega_k^2(x) \sim$ const when $k \to \infty$. Thus the choice of α is defined by the decay of binary correlations.

As is well-known after the papers by Kadanoff, Fisher, Wilson very important properties of a scaling distribution follow from its stability properties. In the theory of limit theorems for sums of independent or weakly dependent random variables such questions doesn't arise at all because the scaling distributions which appear are always stable. Now we shall explain the notion of linearized renormalization group.

Let P_0 be a scaling distribution. We shall denote by $\mathcal{O}\!\iota^{**} = \{\mathcal{O}\!\iota_k^{**}\}$ the multiplicative semigroup of transformations acting in the space $L^1(\Omega, \mathcal{F}, P_0)$ via the formula:

$$(\mathcal{O}\!\iota_k^{**} f)(\bar{\omega}) = E(f \mid \omega_k(x) = \bar{\omega}).$$

Here $E(\cdot \mid \omega_k(x) = \bar{\omega}(x))$ means a conditional expectation when all $\omega_k(x) = \bar{\omega}(x)$ are fixed.

Assume that Q is a probability distribution which is absolutely continuous with respect to P_0 and $f(\omega) = \frac{dQ}{dP_0}(\omega)$. Then it is easy to see that $\mathcal{O}\!\iota_k^{*} Q$ is absolutely continuous with respect to P_0 and $d(\mathcal{O}\!\iota_k^{*} Q)/dP_0\,(\bar{\omega}) = (\mathcal{O}\!\iota_k^{**} f)(\bar{\omega})$.

Let f be a nice function of ω, for example a polynomial and $\mathcal{H}(f) = \sum_{x \in \mathbb{Z}^d} f(T^x \omega)$. $\mathcal{H}(f)$ is a formal expression. It can be considered as a hamiltonian generated by the potential f. Two different polynomials f_1, f_2 generate the same hamiltonian if and only if they are connected by a cohomology relation:

$$f_1(\omega) = f_2(\omega) + \sum_{s=1}^{d} \left(u_s(T^{e_s}\omega) - u_s(\omega) \right), \tag{2.1}$$

here $e_s = (\overbrace{0, \ldots\ldots 0}^{s-1}, 1, 0, \ldots\ldots 0)$, u_s are some polynomials.

The last assertion is a particular case of a theorem from (20). The space of hamiltonians $\mathcal{H}(f)$ is the factor-space which appear after factorization of the space of polynomials by equivalence relation (2.1). We shall put $\mathcal{O}\!\iota^{**}\mathcal{H}(f) = \sum_{x \in \mathbb{Z}^d} \mathcal{O}\!\iota_k^{**} f(T^x\omega)$. From the commutation relation it follows that $\mathcal{O}\!\iota_k^{**}(\mathcal{H}(T^y f)) = \mathcal{O}\!\iota_k^{**}(\mathcal{H}(f))$ for any $y \in \mathbb{Z}^d$.

Usually $\mathcal{O}\!\iota_k^{**}(\mathcal{H}(f))$ isn't a hamiltonian generated by a polynomial. Therefore it is necessary to consider a more wide class of potentials f generating hamiltonians $\mathcal{H}(f)$.

The choice of this class can depend on the scaling distribution under consideration. The natural demand to this class is its invariance under the action $\mathcal{O}^{**}$. Let us assume that the functional class $\mathcal{F}$ satisfying to this condition is chosen.

Definition 3. - The semigroup $\mathcal{O}^{**}$ acting in the space of hamiltonians $\mathcal{H}(f)$, $f \in \mathcal{F}$ is called the linearized renormalization group.

Definition 4. - The hamiltonian $\mathcal{H}(f)$ is called an eigen hamiltonian if $\mathcal{O}_k^{**}(\mathcal{H}(f)) = k^{\lambda}\mathcal{H}(f)$ for every k. The number λ is called the eigenvalue of the eigen hamiltonian $\mathcal{H}(f)$.

The eigen hamiltonians $\mathcal{H}(f)$ for which $\lambda > 0$ ($\lambda < 0$) are called unstable (stable). The eigen hamiltonians for which $\lambda = 0$ are called neutral or marginal. The terminology is taken from the usual theory of stability of dynamical systems. In physical literature people use sometimes the terms "irrelevant" or "relevant" hamiltonians instead of stable or unstable ones. In survey articles (1), (3) it is very well explained that universality hypothesis of critical indices means that for lattice scaling distribution which appear at β_{cr} the number of unstable eigen hamiltonians is equal precisely to one!

Now we shall introduce similar concepts for the case of random fields with continuous time. The most natural framework for the theory is the notion of random distributions in the sense of I.M.Gelfand (21) and K.Ito (22). To be more precise let us consider the Schwartz space S and the space S' of distributions, i.e. of continuous linear functionals on S. By the translation T^x, $x \in \mathbb{R}^d$ we mean the transformation which acts in the space S' via the formula:

$$(T^x\omega, f) = (\omega, T^x f), \quad \omega \in S', \ f \in S, \ (T^x f)(y) = f(y-x).$$

The group of all translations $\{T^x, x \in \mathbb{R}^d\}$ is isomorphic to $\mathbb{R}^d$. The scaling transformation $\mathcal{O}_\tau$ takes now the form:

$$(\mathcal{O}_\tau \omega, f) = (\omega, \mathcal{O}_\tau^* f), \quad (\mathcal{O}_\tau^* f)(x) = \tau^{-\frac{\alpha d}{2}} f(\tau x).$$

Here $0 < \tau < \infty$ and α plays the same role as before. The set of scaling transformations $\{\mathcal{O}_\tau\}$ is the one-parameter group. Let us denote by $\mathcal{G}$ the group generated by all $\{T^x\}$ and $\{\mathcal{O}_\tau\}$. It is easy to see that $\mathcal{G}$ is isomorphic to the group of affine transformation of $\mathbb{R}^d$ having the form: for every $g \in \mathcal{G}$, $g(\vec{x}) = \tau\vec{x} + \vec{z}$, $\tau > 0$ a number. By $\mathcal{G}^*$ we shall denote the conjugate group acting in the space of probability distributions on S'.

Definition 5. - Probability distribution P_0 is called a continuous scaling distribution if it is invariant under the action of $\mathcal{G}^*$, if it is a fixed point of the group $\mathcal{G}^*$.

See the papers (16), (17) about the history of the notion. In what follows the most intresting case is the case when $E(\omega, f)^2 < \infty$ for all $f \in S$. Using the expectation $E((\omega, f)\cdot(\omega, g))$ we can introduce the scalar product in the space S. By $\mathcal{H}_2$ we shall denote the completion of the space S with respect to this scalar product. Assume that each indicator

$$\chi(y) = \begin{cases} 1 & h x_i \leq y_i < h(x_i + 1), \ i = 1, \ldots, d \\ 0 & \text{in other cases,} \end{cases}$$

$x = (x_1, \dots, x_d) \in \mathbb{Z}^d$, $h > 0$ is a size of the lattice, belongs to $\mathcal{H}_2$. The following lemma is almost obvious.

Lemma 1. - Let P_0 be a continuous scaling distribution. For every $h > 0$ and $x \in \mathbb{Z}^d$ let us consider the random variable $\omega(x) = \lim_{n\to\infty} (\omega, f_n)$ where $f_n \in S$ converge in the Hilbert space $\mathcal{H}_2$ to the indicator χ. Then the induced probability distribution of random variables $\omega(x)$, $x \in \mathbb{Z}^d$ is a lattice scaling distribution.

Now acting in the spirit of probability theory we can formulate two main problems related to scaling distributions:

1) to describe the class of scaling distributions or at least to construct as much scaling distributions as possible;

2) for every scaling distribution to describe its domain of attraction, i.e. the set of initial stationary distributions which converge weakly to the given scaling distribution under the action of the renormalization group.

Both problems are closely connected with each other. Apparently one can consruct many scaling distributions. The importance of the scaling distribution is determined by its domain of attraction.

Having in mind the applications of the theory to phase transitions we can define more precisely the problem in this case. Let P_0 be a lattice scaling distribution. The problem is to describe the set $\mathcal{U}$ of hamiltonians of classical lattice spin systems such that for every $H \in \mathcal{U}$ there exist $\beta_{cr} = \beta_{cr}(H)$ and a limit Gibbs state Q_0 corresponding to the hamiltonian $\beta_{cr} H$ such that $\mathcal{O}_k^* Q_0$ converge weekly to P_0. Such formulation of the problem differs from the traditional formulation when one begins with a hamiltonian and tries to investigate the behaviour of the model at β_{cr}.

3. - Gaussian Scaling Distributions -

The description of gaussian continuous scaling distributions is indeed well-known in probability theory. First results were received by Kolmogorov in (23), more complete results were proven by M.S.Pinsker (24) and A.M.Jaglom (25). The simplest approach which is based upon the theory of random distributions is contained in the book by I.M.Gelfand and N.Ja.Vilenkin (26). In gaussian case the probability distribution is determined uniquely by its covariance $E((\omega,f)\cdot(\omega,g)) = \int \tilde{f}(\lambda)\tilde{g}(-\lambda)\, d\sigma(\lambda)$ (we assume that $E((\omega,f)) = 0$). Here $\tilde{f}$ means the Fourier transform of f, σ is a spectral measure. The invariance of probability distribution under scaling transformations means that σ is an homogeneous measure. The simplest case appears when $d\sigma(\lambda) = \rho(\lambda)\, d\lambda$ and $\rho(\lambda)$ is a positive homogeneous function of the power $(\alpha - 1) d$.

If we construct using Lemma 1 the lattice scaling distribution from the continuous scaling distribution with the spectral density ρ we shall have the gaussian lattice scaling distribution which will have the spectral density

$$(3.1) \qquad \rho_1(\lambda) = \prod_{s=1}^{d} \left| e^{2\pi i \lambda_s} - 1 \right|^2 \sum_{m \in \mathbb{Z}^d} \frac{\rho(\lambda + m)}{\prod_{s=1}^{d} (\lambda_s + m_s)^2}$$

It is precisely the formula from (15). The important property of ρ_1 is the presence of singularity at $\lambda = 0$ which leads to a slow decay of correlations.

The Hamiltonian corresponding to (3.1) can be written formally as

$$(3.2) \qquad \mathcal{H}(\omega) = \sum a(x-y)\,\omega(x)\,\omega(y),$$

$$a(x) = \int e^{-2\pi i (\lambda, x)} \rho_1^{-1}(\lambda)\, d\lambda; \quad a(x) \sim \frac{\text{const}}{\|x\|^{\alpha d}}, \ \|x\| \to \infty.$$

The gaussian lattice scaling distribution is the limit Gibbs state in the sense of Dobrushin-Lanford-Ruelle for the hamiltonian (3.2). The important case appears when the function $\rho(\lambda) = (\sum_{s=1}^{d} a_s \lambda_s^2)^{-1}$. Here $\alpha = 1 + \frac{2}{d}$. The interaction a(x) decays exponentially which means that (3.2) is the short-range hamiltonian.

Now we shall consider the action of linearized renormalization group for the case of lattice scaling distributions. Let us fix a gaussian lattice scaling distribution P_0 with the spectral density (3.1). The following lemma is well-known in the theory of gaussian probability distributions (see (28)).

Lemma 2. – Let $f(\omega)$ be an Hermite polynomial of n-th power for the Gaussian distribution P_0 . Then $\mathcal{O}\!\mathcal{L}_k^{**}(f(\omega))$ is again the Hermite polynomial of the same power.

Lemma 2 means that the space of Hermite polynomials of the given power is invariant under the action of the group $\mathcal{O}\!\mathcal{L}^{**}$.

Let $h^{(0)}(\lambda)$ be an homogeneous function of the power γd, $\gamma < -1$, $\lambda = (\lambda_1, \dots, \lambda_d)$ and

$$h^{(1)}(\lambda;\tau) = \sum_{m \in \mathbb{Z}^d} h(\lambda+m) e^{2\pi i(\tau, \lambda+m)}, \quad \tau \in \mathbb{R}^d.$$

We put

$$\rho_1^{(1)}(\lambda) = \rho_1(\lambda) \prod_{s=1}^{d} (1 - e^{-2\pi i \lambda_s})^{-1}$$

(see (3.1)),

$$h(\lambda;\tau) = h^{(1)}(\lambda;\tau) (\rho_1^{(1)}(\lambda))^{-1}$$

$$\Gamma_h(x;\tau) = \Gamma_h(x-\tau) = \int e^{-2\pi i(\lambda,\tau)} h(\lambda;\tau)\, d\lambda, \quad x \in \mathbb{Z}^d,$$

$$\zeta_h(\tau,\omega) = \sum_{x \in \mathbb{Z}^d} \Gamma_h(x-\tau)\, \omega(x).$$

Thus for every homogeneous function h we have a random field with continuous time $\zeta_h(\tau;\omega)$. Because $\zeta_h(\tau;\omega)$ depend linearly on ω the induced distribution of random variables $\zeta_h(\tau;\omega)$ is the gaussian distribution.

Lemma 3. – $\mathcal{O}\!\mathcal{L}_k^{**}(\zeta_h(\tau;\omega)) = k^{-(\gamma+\frac{\alpha}{2}+1)d} \zeta_h(\frac{\tau}{k}; \mathcal{O}\!\mathcal{L}_k \omega)$.

The assertion of the lemma is proven by direct calculations.

Lemma 4. – $\mathcal{O}\!\mathcal{L}_k^{**}(:\zeta_h^n(\tau;\omega):) = k^{-(\gamma+\frac{\alpha}{2}+1)d} :\zeta_h^n(\frac{\tau}{k}; \mathcal{O}\!\mathcal{L}_k\omega):$ for every n.

Lemma 4 follows easily from lemmas 2,3. Here: ζ_h^n : is n-th Hermite polynomial of the gaussian random variable ζ_h .

Lemma 5. – Hamiltonian

$$\mathcal{H}(\omega) = \int_{\mathbb{R}^d}\!\!\cdots\!\int :\zeta_h^n(\tau;\omega):\, d\tau$$

is an eigen hamiltonian with the eigenvalue $d[1 - n(\gamma + \frac{\alpha}{2} + 1)] = \varkappa_n$.

Proof. We have

$$\mathfrak{A}_k^{**} \int_{\mathbb{R}^d}\!\!\cdots\!\int :\xi_h^n(\tau;\omega):d\tau = \int_{\mathbb{R}^d}\!\!\cdots\!\int \mathfrak{A}_k^{**}:\xi_h^n(\tau;\omega):d\tau =$$

$$= k^{-n(\gamma+\frac{\alpha}{2}+1)d} \int_{\mathbb{R}^d}\!\!\cdots\!\int :\xi_h^n\left(\frac{\tau}{k};\mathfrak{A}_k\omega\right):d\tau =$$

$$= k^{d[1-n(\gamma+\frac{\alpha}{2}+1)]}\cdot\int_{\mathbb{R}^d}\!\!\cdots\!\int :\xi_h^n(\tau;\mathfrak{A}_k\omega):d\tau, \qquad \text{Q.E.D.}$$

Lemma 5 shows that strictly speaking the linearized renormalization group has a continuous spectrum because γ may be arbitrary. It is possible to show that only eigen hamiltonians with $\gamma=-\alpha$ are essential. For this γ we have $\varkappa_{2m} = d(1-2m(1-\frac{\alpha}{2}))$ which shows that the number of unstable eigen hamiltonians is finite. We have $\varkappa_2 = d(\alpha-1) > 0$. It means that the number of unstable eigen hamiltonians of lemma 5 is equal precisely to one if $\varkappa_4=(2\alpha-3)\leq 0$, or $\alpha\leq 3/2$. We consider only even n = 2m because we deal with hamiltonians invariant under the change of sign of all coordinates (± - simmetry). Thus for $\alpha = 1+\frac{2}{d}$ we have $\alpha\leq\frac{3}{2}$ if $d\geq 4$. These calculations are of the same nature as calculations in the physical literature where one shows that for $d\geq 4$ Landau theory of phase transitions is valid. From the point of view of the theory under consideration the last assertion is equivalent to the appearence at β_{cr} of a gaussian scaling distribution.

The investigation of the spectrum of linearized renormalization group in such spirit as above was done in (15). Our exposition here follows the paper by Dinaburg and Sinai (unpublished). Similar consideration and the deep investigation of the spectrum of linearized renormalization group are contained in the paper by P.M.Bleher (27). P.M.Bleher in (27) constructed many other eigen hamiltonians.

4. - ε - expansions near $\alpha = 3/2$.

The content of this paragraph is very close to the paper (29) by Fisher, Ma, Nickel and to the paper by Zak (30). For $\alpha=\frac{3}{2}$ the eigenvalue $\varkappa_4 = 0$, or the eigen hamiltonian $\int\!\!\int :\xi_h^4(t):dt$ is a marginal hamiltonian. General theory of bifurcations give a formal approach to the construction of a branch of non-gaussian scaling lattice distribution parametrized by the parameter $\varepsilon=\alpha-\frac{3}{2}$. Let us write the hamiltonian of such distribution in the form

$$\mathcal{H}(\omega) = \mathcal{H}_0(\omega) + \mathcal{H}_1(\omega)$$

where $\mathcal{H}_0(\omega)$ is the hamiltonian (3.2), $\mathcal{H}_1(\omega)$ is a correction. We shall use the invariance of $\mathcal{H}(\omega)$ under the renormalization group. For any $k > 1$ we can write formally putting $\omega_k = \mathfrak{A}_k\omega$

$$(4.1)\qquad e^{-\mathcal{H}(\omega_k)} = e^{-\mathcal{H}_0(\omega_k)-\mathcal{H}_1(\omega_k)} = \int_{\omega:\ \mathfrak{A}_k\omega=\omega_k} e^{-\mathcal{H}_0(\omega)-\mathcal{H}_1(\omega)}\,d\omega.$$

The last expression is a functional integral. In order to perform the integration we shall write

$$(4.2)\qquad \omega(x) = \sum_{y\in\mathbb{Z}^d} c_k(x,y)\,\omega_k(y) + \psi(x)$$

where the coefficients $c_k(x,y)$ are chosen in such a way that $\mathcal{H}_0(\omega) = \mathcal{H}_0(\omega_k) + \mathcal{H}_0(\psi)$. It isn't difficult to write down explicit expressions for $c_k(x,y)$. Lt us write (4.2) briefly as $\omega = c_k\omega_k + \psi$. Substituting (4.2) into (4.1) we have

$$e^{-\mathcal{H}_0(\omega_k)-\mathcal{H}_1(\omega_k)} = e^{-\mathcal{H}_0(\omega_k)} \int_\psi e^{-\mathcal{H}_0(\psi)-\mathcal{H}_1(C_k\omega_k+\psi)}\,d\psi$$

or

$$(4.3)\qquad e^{-\mathcal{H}_1(\omega_k)} = \int e^{-\mathcal{H}_0(\psi)-\mathcal{H}_1(C_k\omega_k+\psi)}\,d\psi .$$

The last expression can be considered as a non-linear integral equation for the hamiltonian $\mathcal{H}_1$. Similar equation appears in the theory of hierarchical models of Dyson (19),(12). In that case the corresponding integral is one-dimensional or finite-dimensional. Here we have an infinite-dimensional functional integral. For its solution one can use standard technique of perturbation theory.

Let us look for $\mathcal{H}_1$ in the form

$$(4.4)\qquad \mathcal{H}_1 = (a_1\varepsilon + a_2\varepsilon^2 + a_3\varepsilon^3 + \dots)\int :\xi^4(t,\omega):dt + \varepsilon^2 q_2 + \varepsilon^3 q_3 + \dots .$$

Perturbation theory shows how to find numbers a_1, a_2,...... and hamiltonians q_2, q_3........... We shall not discuss it here (see (15)).

There are two main problems which arise in connection with the formal expression (4.4).

1. To construct limit Gibbs state corresponding to (4.4). The series are certainly asymptotic. Let us use again an analogy with hierarchical models of Dyson. In that case one can write also an asymptotic series for non-gaussian solutions of corresponding non-linear integral equation.

The main result of our common paper with Bleher (12) is a theorem which shows how to deal with such series in hierarchical models. Roughly speaking we have shown that the finite sum of ε-expansion give a nice representation for the solution in the domain depending on ε. Outside this domain one has only an estimation from above for the solution. We hope that the same assertion is true for the hamiltonian (4.4). Namely (4.4) gives the good presentation for scaling hamiltonian when the values of $\omega(x)$ are not too large. If they are large the hamiltonian (4.4) must have another expression. In general, the problem is completely open for the solution.

2. A very important question concerns the domain of attraction of a limit Gibbs state corresponding to the hamiltonian (4.4).

It isn't a priori clear whether hamiltonians with interaction decaying as $\frac{1}{r^{\alpha d}}$ converge under the action of the renormalization group to the hamiltonian (4.4). It may happen that some renormalization of the power of interaction is needed.

There exist some other examples of non-gaussian scaling distributions. The first example was constructed by M. Rosenblatt (31). Other examples of similar type were proposed by Dobrushin (18). In paper by Karwowski and Streit (32) one can find another example of a non-gaussian scaling distribution.

But isn't clear whether these scaling distributions can appear in problems of the theory of critical point.

All problems presented in the text I discussed many times with many colleagues. I want to express my sincere gratitude for these very useful discussion to my colleagues in Moscow P.Bleher, E.Dinabaurg, R.Dobrushin, I.M.Lifshitz, R.Minlos, V.Malyshev, A.Poliakov and colleagues whom I met in the Institute des Hautes Etudes Scientifiques in Bures-sur-Ivette E.Brezin, H.Epstein, G.Gallavotti, G.Jona-Lasinio, J.Lebowitz, D.Ruelle.

I thank also the Organizing Committee of the МΛФ Conference for the invitation to present this lecture.

- References -

(1) J.Kogut and K.Wilson: Phys. Rept. 12C, 75 (1974).
(2) M.Fisher: Review of Modern Physics v.46, N.4 (1974).
(3) Ma S.-K.: Rev. Mod. Phys. 45, 589 (1973).
(4) E.Brezin, J.C.Le Guillou, J.Zinn-Justin: Phase Transitions and Critical Phenomena, edited by Domb and Green, Vol VI.
(5) Patashinski and Pokrovski, Russian Phys. Surveys (1977).
(6) G.Gallavotti, Knops: Comm Math. Phys. 36, 171 (1974).
(7) G.Gallavotti, G.Jona-Lasinio: Comm. Math. Phys. 41, 301 (1975).
(8) M.Cassandro, G.Jona-Lasinio: Preprint Bielefeld (1976).
(9) M.Cassandro, G.Gallavotti: Nuovo Cimento 25B, 691 (1975).
(10) G.Jona-Lasinio: Nuovo Cimento 26B, 99 (1975).
(11) P.Bleher, Ja.Sinai: Comm. Math. Phys. 33, 23 (1973).
(12) P.Bleher, Ja.Sinai: Comm. Math. Phys. 45, 247 (1975).
(13) P.Bleher: Proc. Moscow Math. Society Vol. 33 (1975).
(14) P.Collet, J.P.Eckmann: The ε -expansion for the Hierarchical Model. Preprint Université de Geneve (1977).
(15) Ja.G.Sinai: Theory of Probability and Applications, 21, 63 (1976).
(16) Ja.G.Sinai: "Statistical Physics" Proc. IUPAP Conference, Budapest (1975).
(17) R.L.Dobrushin: Proc. of IV International Symposium on Information Theory, Repino (1976).
(18) R.L.Dobrushin: Annals of Probability (in print).
(19) G.Baker: Phys. Rev. B5, 2622 (1972).
(20) V.V.Anshelevitsh, Ja.G.Sinai: Russian Math. Surveys Vol. XXXI, N.4 (1976).
(21) I.M.Gelfand; Doklady, V.100, 853-856 (1955).
(22) K.Ito: Mem. Col. Sci. Univ. Kyoto, Ser. A 28, 209-223 (1954).
(23) A.K.Kolmogorov: Doklady 26, 115 (1940).
(24) M.S.Pinsker: Izv. Acad. Sci. Sez. Math. 19, 319 (1955).
(25) A.M.Jaglom: Theory of Prob. and Appl. 2, 292 (1957).
(26) Gelfand, Vilenkin: Distributions Vol IV.
(27) P.Bleher: Proc. of VI-th Conference on Information Theory Repino (1976).
(28) Jn. A.Rozanov: Proc. Steklov Inst. V. CVIII, Moscow (1968).
(29) M.Fisher, S.Ma, B.Nickel: Phys. Rev. Lett. 29; 917 (1972).
(30) J.Zak: Phys. Rev. B8, 1 (1973).
(31) M.Rosenblatt: Proc. IV Berkeley Symp. on Prob. and Math. Stat. 431, Berkeley (1961).
(32) Karwowski, Streit; a Renormalization Group Model with a Non-Gaussian Fixed Point. Preprint. Universitat Bielefeld (1976).

ON THE RENORMALIZATION GROUP FOR THE HIERARCHICAL MODEL

P. Collet, J.P. Eckmann
Departement de Physique Theorique
Universitè de Genève
1211 GENEVE 4, Switzerland

A detailed lecture note on the subject is in preparation. We therefore give only a summary of results obtained so far (they duplicate, in part, with new proofs, results of Bleher and Sinai). The critical fixed point of the Hierarchical model has a single spin distribution satisfying the equation

$$\phi(z) = \pi^{1/2} \int du\, e^{-u^2} \phi(z c^{-1/2} + u)\, \phi(z c^{-1/2} - u) = (N\phi)(z).$$

For $c = 2^{\frac{1}{2}}(1-\varepsilon)$, $\varepsilon > 0$, this equation has a nontrivial solution

$$\phi_\varepsilon \approx 1 - \varepsilon \vartheta H_4(\gamma^{1/2} x) + O(\varepsilon^2) \quad \text{in} \quad L_2(\mathbb{R}, e^{-\gamma x^2} dx)$$

$$\gamma = 1 - 2^{-1/2}, \quad \vartheta \neq 0,$$

H_4 the fourth Hermite polynomial. The function ϕ_ε is entire in z and real analytic in $\varepsilon_0 > \varepsilon > 0$, C^∞ at the boundary.

The scaling limits determine the critical indices through the linearization of the map $N(\phi)(z)$ around ϕ_ε. The thermodynamic limits are shown to exist for all $\beta \neq \beta_{crit}$ by following the flow described by N in the large.

References

- P.M.Bleher, Ja.G.Sinai: "Investigation of the critical point in models of the type of Dyson's hierarchical model", Commun. Math. Phys. 33, 23 (1973).
- P.M.Bleher, Ja.G.Sinai: "Critical Indices for Dyson's Asymptotically Hierarchical Models", Commun. Math. Phys. 45, 347 (1975).
- P.Collet, J.P.Eckmann: "The ε -Expansion for the Hierarchical Model" Commun. Math. Phys. (to appear).

FLUCTUATIONES IN CURIE-WEISS EXEMPLIS[1]

Richard S. Ellis[2]
Dept. of Mathematics and Statistics
University of Massachusetts
Amherst, Massachusetts 01003

Charles M. Newman[3]
Dept. of Mathematics
Indiana University
Bloomington, Indiana 47401

1. Introduction

The primary topic of this paper is the statistics of mean field (or more accurately, Curie-Weiss) models. Although such models are often considered to possess trivial statistics (their fluctuations always being normally distributed), we shall see that when "properly" viewed, this is not the case. In fact, the probabilistic structure of these models is surprisingly rich and we present a detailed analysis of this structure with the hope that it will be helpful in analyzing analogous phenomena in less trivial models. Many of the results given here first appeared in [EN1] which can be consulted for detailed proofs. Sections 2-5 consist primarily of background material while sections 6-9 contain our main results.

2. General Ising Models

In this paper we will limit our attention to general Ising models (with pair interactions). For each $n = 1, 2, \dots$ we have a collection of (spin) random variables $\{X_i^n : i \in V_n\}$ where the V_n's are finite subsets of $\mathbb{Z}^d$ which tend to $\mathbb{Z}^d$ (in some appropriate sense) as $n \to \infty$; the joint distribution of $\{X_i^n\}$ is

$$(1) \qquad \frac{1}{Z_n} \exp\Big[\sum_{i,j \in V_n} J_{ij}(n) x_i x_j\Big] \prod_{i \in V_n} d\rho(x_i)$$

with ρ a finite measure on $\mathbb{R}^1$, and

$$(2) \qquad Z_n = \int \cdots \int \exp\Big(\sum J_{ij}(n) x_i x_j\Big) \prod d\rho(x_i)$$

---

[1]Presented at the Capitolium, June 11, 1977.

[2]Research supported in part by NSF Grant MPS 76-06644.

[3]Research supported in part by NSF Grant MPS 74-04870 A01.

assumed finite for all n. We will consider three examples:

<u>Example 1</u>. $J_{ij}(n) = 0 \quad \forall i,j,n$; $\{X_i^n\}$ is a set of independent random variables with common distribution $d\rho/\int d\rho$.

<u>Example 2</u>. $J_{ij}(n) = 1/2|V_n| \quad \forall i,j \in V_n$, where $|V|$ denotes the cardinality of V ; this is a Curie-Weiss (or mean field) model (see [K]).

<u>Example 3</u>.

$$J_{ij}(n) = \begin{cases} 1, & \|i-j\| = 1 \\ 0, & \|i-j\| \neq 1 \end{cases}$$

, where $\|\cdot\|$ denotes the Euclidean distance in $\mathbb{Z}^d$; this is a nearest neighbor model, which we list as a standard example of a short range, (and for $d \geq 2$) nontrivial model expected to have a probabilistic structure <u>qualitatively</u> similar to the mean field model structure presented in this paper.

In our formulation, all thermodynamic parameters are embedded in the ρ-dependence; thus a spin-½ model $\{\sigma_i^n\}$ at inverse temperature β in an external field h has $\sigma_i^n = \sqrt{\beta} X_i^n(\rho_{\beta,h})$ with

$$(3) \qquad d\rho_{\beta,h} = \exp(\sqrt{\beta}h)[\delta(x-\sqrt{\beta}) + \delta(x+\sqrt{\beta})]/2$$

and a "spin-1" model (as used in [BEG] to analyze the tricritical point of liquid helium) has

$$(4) \qquad d\rho = \exp(\sqrt{\beta}h)[a\delta(x) + (1-a)\{\delta(x-\sqrt{\beta}) + \delta(x+\sqrt{\beta})\}/2] \quad .$$

In taking the thermodynamic limit one lets $n \to \infty$ and considers the limits, both of thermodynamic quantities, such as the free energy,

$$(5) \qquad f = f(\rho) = \lim_{n\to\infty} - \frac{1}{|V_n|} \log Z_n \quad ,$$

and probabilistic quantities, such as the spins themselves,

$$(6) \qquad \{X_i^\infty : i \in \mathbb{Z}^d\} = \lim_{n\to\infty} \{X_i^n\} \quad ;$$

this latter limit is typically in the sense of (local) weak convergence of the corresponding probability distributions.

3. <u>Critical Points and Block Spins</u>

The (thermodynamic) phase diagram is the locus of singularities (i.e. points of non-smoothness) of f in the space of thermodynamic parameters, e.g. β,h - space for (3) or β,h,a-space for (4) . In a Landau model one has for example

$$(7)\qquad f = f(\gamma_1,\ldots,\gamma_{2k-1}) = \inf_{x\in\mathbb{R}} \left(x^{2k} + \sum_{j=1}^{2k-1} \gamma_j x^j\right)$$

so that the phase diagram is the closure of the set of $(\gamma_1,\ldots) \in \mathbb{R}^{2k-1}$ for which the polynomial has multiple global minima. Critical points are boundary points of the phase diagram and the "type" of a critical point is an integer describing the topological nature of the phase diagram in the neighborhood of that critical point; the point $(\gamma_1,\ldots,\gamma_{2k-1}) = (0,\ldots,0)$ is a critical point of type k (see [BS] for a catastrophe-theoretic discussion of this classification). The critical point in (pair interaction, ferromagnetic) spin-½ models is a type-2 (ordinary) critical point while mean field (and presumably short range) spin-1 models as described by (4) possess a unique type-3 (tri-) critical point (see [BEG] and section 6 below).

A critical point at $\rho = \rho_c$ is analyzed thermodynamically by describing the behavior of $f(\rho)$ as $\rho \to \rho_c$; critical exponents are defined by the leading order behavior. The critical exponent δ, for example, is defined by

$$(8)\qquad f(\beta_c,h) = (\text{linear in } h) - C|h-h_c|^{1+1/\delta} + (\text{higher order}),\qquad h \to h_c .$$

A critical point can be analyzed statistically in terms of block spins. This probabalistic viewpoint has been emphasized by Jona-Lasinio and others (e.g., see [J-L]) ; it underlies the renormalization group approach [Ka, WF] and is the basic framework of this paper.

Just as there are both "short" and "long" long range orders distinguished by whether the thermodynamic limit is taken prior to or together with the long range order limit [SML, sec. V] so there are two natural choices of block spin variable: the "short" block spin,

$$(9)\qquad S_n^s = \sum_{i\in V_n} X_i^\infty$$

and the "long" block spin,

$$(10)\qquad S_n^\ell = \sum_{i\in V_n} X_i^n \quad .$$

It has been pointed out, in the context of hierarchical models, that these two variables may exhibit differing asymptotic behavior [GK] ; we shall see that this is strikingly the case in Curie-Weiss models.

4. Classical Results for Independent Variables

We consider the case of Example 1 above and recall some standard results of probability theory; a good general reference is [F] . In this case, $S_n^s = S_n^\ell = S_n$ and for simplicity, we suppose that ρ is nondegenerate (not concentrated at a single point) and $\int d\rho = 1$. When the random variables Y_n converge weakly to the random variable X with distribution $d\nu(x)/\int d\nu(x)$, we write $Y_n \to X$ or $Y_n \to d\nu(x)$.

If $\int |x| d\rho < \infty$, then the Law of Large Numbers is valid:

$$\text{(11)} \quad \exists m \in \mathbb{R} \ni S_n/|V_n| \to m \quad ; \qquad \text{(LLN)}$$

in this case, $m = \int x d\rho$. If $\int x^2 d\rho < \infty$, then the Central Limit Theorem is valid:

$$\text{(12)} \quad \exists \sigma > 0 \ni \frac{S_n - |V_n| m}{|V_n|^{\frac{1}{2}}} \to \exp(-x^2/2\sigma^2)\, dx \quad ; \qquad \text{(CLT)}$$

in this case $\sigma^2 = \int (x-m)^2 d\rho$. If $\int x^2 d\rho = \infty$, $\int |x| d\rho < \infty$, and ρ satisfies certain regularity conditions [F, ch. 17] then a Noncentral Limit Theorem is valid:

$$\exists\, \alpha \in (1,2) \text{ and nondegenerate } \nu \ni$$

$$\text{(13)} \quad \frac{S_n - |V_n| m}{|V_n|^{\alpha/2}} \to d\nu \quad ; \qquad \text{(NLT)}$$

ν is universal (i.e. it depends on ρ only through α and one scaling parameter [and possibly one asymmetry parameter]) and in this case is a nongaussian (possibly asymmetric) stable distribution of exponent α (see [F]) .

5. Presumed Situation in Nontrivial Models

In nontrivial models such as Example 3 above it is believed that the validity or nonvalidity of (11) - (13) for $S_n = S_n^s$ and/or S_n^ℓ is related to the nature of the particular ρ (or point in thermodynamic parameter space or phase) in question as follows:

validity of LLN $\longleftrightarrow$ pureness of phase
validity of CLT $\longleftrightarrow$ noncriticality of phase
validity of NLT $\longleftrightarrow$ criticality of phase

One sort of universality which is expected here is analogous to that appearing in the classical case; namely, the limiting ν at a critical point should depend on ρ only through α and a (scaling) parameter. α in turn should be related to the critical exponent δ $(\alpha = 2\delta/(\delta+1))$ and these should depend on ρ only through the critical type (and the dimension d). This dependence implies that as ρ varies over all allowed measures for fixed d, there will be only a discrete set (one for each critical type) of allowable α's and (modulo scaling parameters) limiting ν's - a distinctly nonclassical phenomenon; of course, this is formally no longer the case if d is varied continuously. Another nonclassical phenomenon predicted on the basis of the renormalization group approach is that ν in (1.13) may be Gaussian even with $\alpha \neq 1$ - e.g. at an ordinary critical point of the classical Ising model with $d > 4$. For clarity, it should be pointed out that the limiting ν's for the nontrivial models discussed in this section as well as for the Curie-Weiss models treated below are definitely not the classical stable distributions which occur in section 4.

In analogy with the extension of a classical result such as (13) to cover the case of "triangular arrays" (see [F]), one may consider the asymptotics of $S_n(\rho_n)$ as ρ_n approaches a critical point. In the classical case, this extension leads to the class of infinitely divisible distributions as limiting ν's. In the case of statistical mechanics, we suppose $\rho = \rho(\gamma_1,\ldots,\gamma_{2k-1})$ with $\rho(\gamma_1^c,\ldots,\gamma_{2k-1}^c)$ type-k critical; then there should be constants $\epsilon_1,\ldots,\epsilon_{2k-1}$ (related to the various critical exponents) so that letting $\rho_n = \rho(\gamma_1^n,\ldots,\gamma_{2k-1}^n)$ with

$$\text{(14)} \qquad \gamma_i^n = \gamma_i^c + \lambda_i/|V_n|^{\epsilon_i} \qquad (i=1,\ldots,2k-1) \quad ,$$

one expects to obtain a 2k-1 parameter family of limiting measures which (in renormalization group language) lie along the "unstable manifold" in an infinitesimal neighborhood of the critical point:

$$\text{(15)} \qquad \frac{S_n(\rho_n) - |V_n| m}{|V_n|^{\alpha/2}} \to d\nu(\lambda_1,\ldots,\lambda_{2k-1}) \quad .$$

We shall see below that essentially all the qualitative structure discussed in this section actually occurs in Curie-Weiss models.

6. Curie-Weiss and Mean Field Models

Since the d-dependence in Curie-Weiss models is essentially nonexistent, we replace $\mathbb{Z}^d$ by $\{1,2,\ldots\}$ and V_n by $\{1,\ldots,n\}$. Thus the model consists of $\{X_i^n: i = 1,\ldots,n\}$ with joint distribution

$$(16)\quad \frac{1}{Z_n} \exp\Big[\Big(\sum_{i=1}^{n} x_i\Big)^2/2n\Big] \prod_{i=1}^{n} d\rho(x_i) \quad .$$

Essentially all our results extend naturally to rotator (vector valued spin) models; for the sake of brevity, we do not explicitly present these extensions. Our <u>only</u> assumptions on ρ are that it is nondegenerate, that $d\rho = 1$ (for simplicity) and that $\int \exp(x^2/2)d\rho < \infty$ (so that $Z_n < \infty \forall n$) ; ρ need not be even. The following is easily derived (see [K,EN1]) .

<u>Proposition</u>. Let

$$(17)\quad G_\rho(z) = z^2/2 - \log \int_{-\infty}^{\infty} \exp(zx)d\rho(x) \quad ; \quad z \in \mathbb{R} \quad .$$

G_ρ is real analytic and $G_\rho \to +\infty$ as $|z| \to +\infty$ so that it has only a finite number of global minima ;

$$(18)\quad f(\rho) \equiv \lim_{n\to\infty} -\frac{1}{n} \log Z_n = \inf_{z\in\mathbb{R}} G_\rho(z) \quad .$$

In the phase diagram defined by (18) we clearly have coexisting phases related to multiple global minima of G_ρ and critical phases related to nonquadratic global minima of G_ρ. We write

$$(19)\quad \rho \sim (m_1,k_1;\ldots;m_\ell,k_\ell) \qquad \text{(all the } m_i\text{'s distinct)}$$

when the set of global minima of G_ρ is $\{m_1,\ldots,m_\ell\}$ and

$$(20)\quad G_\rho(z) = G_\rho(m_i) + \lambda_i(z-m_i)^{2k_i}/(2k_i)! + o[(z-m_i)^{2k_i}] \; ; \; z \to m_i \, ,$$

with $\lambda(m_i) \equiv \lambda_i > 0 \; \forall i$. We call $k(m_i) \equiv k_i$ the <u>type</u> and $\lambda(m_i)$ the <u>strength</u> of the minimum m_i ; ρ is said to be <u>pure</u> if G_ρ has a unique global minimum and <u>semipure</u> if it has a unique global minimum of maximal type. A pure (or semipure) measure is said to be <u>centered</u> at m , the location of the global minimum (of maximal type). For example, the measure of (3) is pure of type 2 and centered at 0 when $\beta = 1$, $h = 0$ while the measure of (4) is pure of type 3 and centered at 0 when $\beta = 3$, $h = 0$, $a = \frac{2}{3}$.

The following theorem describes the (microscopic) thermodynamic limit of Curie-Weiss models; it follows immediately from an easily derived formula for the joint distribution of $\{X_1^n,\ldots,X_j^n\}$:

$$(21) \qquad \frac{(n/2\pi)^{\frac{1}{2}}}{Z_n} \int_{-\infty}^{\infty} \exp(-nG_\rho(m)) \Big[\prod_{i=1}^{j} \Big(e^{mx_i} d\rho(x_i) / \int e^{mx} d\rho(x) \Big) \Big] dm \quad .$$

The relation between Curie-Weiss and mean field models on the microscopic level implied by this theorem was apparently first derived in [EK]; it extends the macroscopic result of [K] . We define $d\tau_\rho$ to be the (weak) limit of $\exp(-nG_\rho(m))dm/\int \exp(-nG_\rho(m))dm$ which when $\rho \sim (m_1,k_1;\ldots;m_\ell,k_\ell)$ is easily seen to be

$$(22) \qquad d\tau_\rho(m) = \sum_{j=1}^{\ell} a_j \, \delta(m - m_j)$$

with $a_j = \bar{a}_j / \Sigma \bar{a}_j$ and

$$(23) \qquad \bar{a}_j = \begin{cases} 0 , & k_j \text{ not maximal} \\ [\lambda(m_j)]^{-1/2k_j} , & k_j \text{ maximal} \end{cases} \quad .$$

<u>Theorem 1</u>. $\{X_i^n : i = 1,\ldots,n\} \to \{X_i^\infty\}$, in the sense of weak convergence of finite dimensional joint distributions, where the joint distribution of $\{X_i^\infty\}$ on $\mathbb{R}^\infty$ is

$$(24) \qquad \int_{-\infty}^{\infty} \Big\{ \prod_{i=1}^{\infty} \Big[\exp(mx_i) \, d\rho(x_i) / \int \exp(mx) \, d\rho(x) \Big] \Big\} d\tau_\rho(m) \quad ;$$

in particular, if ρ is semipure and centered at m_0 , then $\{X_i^\infty\}$ is a set of independent random variables with common distribution $\exp(m_0 x) \, d\rho(x) / \int \exp(m_0 x) \, d\rho(x)$.

In the semipure case, m_0 is a solution (unique, if m_0 is the only local minimum of G_ρ) of $G_\rho'(m) = 0$ or equivalently

$$(25) \qquad m = \int x \exp(mx) \, d\rho(x) / \int \exp(mx) \, d\rho(x) \quad .$$

This is the "self-consistent field formula" used in the usual mean field approach; Theorem 1 thus explains the "equivalence" of Curie-Weiss and mean field models on the microscopic level - providing the thermodynamic limit is taken prior to the consideration of other asymptotics. As an immediate corollary of Theorem 1, the classical LLN and CLT, and (25), one obtains the following uninteresting asymptotics for <u>short</u> block spins in mean field models.

<u>Corollary</u>. Let $S_n^s = X_1^\infty + \ldots + X_n^\infty$; then

(26) $\quad S_n^s/n \to d\tau_\rho$

(so that (11) is valid if and only if ρ is semipure). If ρ is semipure and centered at m, then

(27) $$\frac{S_n^s - nm}{n^{\frac{1}{2}}} \to \exp(-x^2/2\bar{\sigma}^2)\,dx$$

with $\bar{\sigma}^2 = \int (x-m)^2 \exp(mx)\,d\rho(x) / \int \exp(mx)\,d\rho(x)$.

7. <u>The Asymptotics of Long Block Spins</u>

We let $S_n^\ell = X_1^n + \ldots + X_n^n$; the following results all follow directly from the easily established fact that if Z is a unit normal random variable independent of S_n^ℓ, then $S_n^\ell/n + Z/\sqrt{n}$ has distribution proportional to $\exp(-nG_\rho(x))dx$. The moral of the next theorem is that although the LLN remains the same for S_n^ℓ as for S_n^s, the CLT/NLT is strikingly different for S_n^ℓ and in particular exhibits the kind of critical behavior absent in (27) . Note that even at a noncritical point $(k = 1)$, the right hand sides of (27) and (28) differ.

<u>Theorem 2</u>. $S_n^\ell/n \to d\tau_\rho$. If ρ is semipure, centered at m and of type k, then

(28) $$\frac{S_n^\ell - nm}{n^{1-1/2k}} \to \begin{cases} \exp(-x^2(1-\bar{\sigma}^2)/2\bar{\sigma}^2)\,dx\,, & k = 1 \\ \\ \exp(-\lambda(m)x^{2k}/(2k)!)\,dx\,, & k \ge 2 \end{cases}.$$

<u>Remarks</u>. We discuss below a universal (k-dependent) upper bound on $\lambda(m)$. For $\rho = [\delta(x-1) + \delta(x+1)]/2$ (and $k=2$) , Theorem 2 is a somewhat more elegant version of the Simon-Griffiths result on

φ^4 field theories as classical (spin-½) Ising models [SG]; for ρ the tricritical measure of (4), Theorem 2 yields a result on φ^6 field theories as "spin-1" Ising models. The rotator version of Theorem 2 (for k = 2) is an improved version of the Dunlop-Newman result on $(\vec{\varphi}\cdot\vec{\varphi})^2$ field theories as classical rotators [DN].

The next theorem sketches the mean field version of (15); see [EN1] for a more complete presentation.

<u>Theorem 3</u>. If ρ is pure, centered at m, and of type k, then for any real $\lambda_1,\ldots,\lambda_{2k-1}$, $\rho_n \to \rho$ can be chosen so that

$$(29)\qquad \frac{S_n^{\ell}(\rho_n) - nm}{n^{1-1/2k}} \to \exp(-\lambda(m)x^{2k}/(2k)! - \sum_{j=1}^{2k-1} \lambda_j x^j/j!)\,dx \quad ;$$

these are the only possible limits (for arbitrary choices of $\rho_n \to \rho$).

Theorems 2 and 3 can be extended in various ways through the use of conditioning. In particular for arbitrary (not necessarily semi-pure) ρ, and m an arbitrary (not necessarily global or of maximal type) type k <u>local</u> minimum of G_ρ, one may choose ϵ small enough so that the conditional distribution of $(S_n^{\ell} - nm)/n^{1-1/2k}$ given that $|S_n^{\ell}/n - m| < \epsilon$ tends to the right hand side of (28). Such conditional limit theorems will be presented in [EN2]; analogous results for short range interaction models would be useful in the study of metastability.

8. <u>Renormalization Group Approach</u>

If we choose $\alpha \in [1,2)$ and suppose that $S_n^{\ell}/n^{\alpha/2}$ has an (unnormalized) probability density φ_n, then it is easily determined that

$$(30)\qquad \varphi_{2n}(z) = \int_{-\infty}^{\infty} \sqrt{b_n/\pi}\, \exp(-b_n u^2)\varphi_n(z/\sqrt{c}+u)\,\varphi_n(z/\sqrt{c}-u)\,du$$

with $c = 2^{2-\alpha}$ and $b_n = \text{const.}\ n^{\alpha-1}$. We note that if b_n is replaced by 1 in (30), we obtain the fixed point equation of the hierarchical model (see Eckmann's lecture in this conference). In our case, if $1 < \alpha < 2$ $(2 > c > 1)$ and we let $n \to \infty$, (30) formally yields the fixed point equation,

$$(31)\qquad \varphi_\infty(z) = [\varphi_\infty(z/\sqrt{c})]^2 \quad ;$$

whose solutions are $\exp(-\lambda|z|^p)$ for $p = 2/\log_2 c \in (2,\infty)$. It is interesting that only $p = 2k$ actually arises in the rigorous mean field limit theorems and that even then there is an upper bound on λ ; the moral is apparently to beware of extraneous solutions of renormalization group fixed point equations.

9. Global Analysis of the Phase Diagram

The results of this section are based on the theory of moments of Markov-Krein, Karlin-Studden, etc. (see [KS]) ; a more complete presentation is given in [EN1] . The following theorem gives global information concerning the mean field phase diagram as ρ varies over measures with finite support; it gives the upper bound on the $\lambda(m)$ appearing in (28) ; and it guarantees the existence of measures which are critical of type k for each k, such that (when combined with Theorem 2) one has the result that φ^{2k} field theories are (modified) "spin-(k-1)/2" Ising models (recall the discussion of section 6 for $k = 2,3$) .

Theorem 4. Given any $(m_1,k_1;\ldots;m_\ell,k_\ell)$, there is a unique probability measure ρ_p supported on exactly $k = k_1 +\ldots + k_\ell$ points such that $\rho_p \sim (m_1,k_1;\ldots;m_\ell,k_\ell)$; if $\rho \neq \rho_p$ and also $\rho \sim (m_1,k_1;\ldots;m_\ell,k_\ell)$ then ρ is not supported on less than $k + 1$ points and $\lambda_\rho(m_i) < \lambda_{\rho_p}(m_i)$ for each $i = 1,\ldots,\ell$. In particular, there is for each k a unique probability measure supported on k points which is pure, centered at 0 and of type k ; its support is the zero set of the kth Hermite polynomial.

We note that in the analogous theorem for rotator mean field models, the minimally supported (spherically symmetric) measure which is critical of type k at the origin (and is thus related to $(\vec{\varphi}\cdot\vec{\varphi})^k$ field theories) has support on exactly k/2 spherical shells - the origin counting as a half shell. We end this paper with two open questions.

Question 1: The universal bounds of Theorem 4 imply universal bounds on certain critical parameters (e.g. a lower bound on the C appearing in (8)) of type-k ρ's in terms of the type-k ρ of minimal support. Do analogous results hold true in nontrivial models?

Question 2: Theorem 5 suggests that the mean field phase diagram as ρ varies over the (2k-1)-parameter space

$$\left\{\sum_{i=1}^{k} p_i \delta(x - x_i) : \sum p_i = 1\right\}$$

is *globally* topologically equivalent to the Landau/catastrophe phase diagram of (7) ; is this also true for (some) nontrivial models? is it also true when $\{\Sigma p_i \delta(x-x_i)\}$ is replaced by

$$\{\exp\,(-x^{2k} - \sum_{j=1}^{2k-1} \gamma_j x^j)\,dx\} \quad ?$$

Acknowledgement. This paper has benefited from discussions with G. Emch, E. Lieb and others; the title was chosen with the assistance of G.W. Lawall of the University of Massachusetts at Amherst. One of the authors (C.M.N.) would like to thank the National Science Foundation, Indiana University, and the Indiana University Foundation for their financial assistance in attending this conference.

REFERENCES

[BEG] Blume, M., Emery, V.J., Griffiths, R.B.: Ising model for the λ transition and phase separation in He^3-He^4 mixtures. Phys. Rev. A 4, 1071-1077 (1971).

[BS] Benguigui, L., Schulman, L.S.: Topological classification of phase transitions. Phys. Lett. 45A, 315-316 (1973).

[DN] Dunlop, F., Newman, C.M.: Multicomponent field theories and classical rotators. Commun. Math. Phys. 44, 223-235 (1975).

[EK] Emch, G.G., Knops, H.J.F.: Pure thermodynamic phases as extremal KMS states. J. Math. Phys. 11, 3008-3018 (1970).

[EN1] Ellis, R.S., Newman, C.M.: Limit theorems for sums of dependent random variables occurring in statistical mechanics. U. Mass/ Indiana preprint (1977) submitted to Z. Wahrscheinlichkeitstheorie.

[EN2] Ellis, R.S., Newman, C.M.: in preparation.

[F] Feller, W.: An Introduction to Probability Theory and its Applications, Vol. II. New York: Wiley, 1966.

[GK] Gallavotti, G., Knops, H.: The heierarchical model and the renormalization group. Rivisita del Nuovo Cimento 5, 341-368 (1975).

[J-L] Jona-Lasinio, G.: Probabalistic approach to critical behavior. 1976 Cargèse summer school on "New Developments in Quantum Field Theory and Statistical Mechanics".

[K] Kac, M.: Mathematical mechanisms of phase transitions in: Statistical Physics, Phase Transitions and Superfluidity (Chretien, M., Gross, E.P., Deser, S., eds.). New York: Gordon and Breach, 1967.

[Ka] Kadanoff, L.P.: Scaling, universality, and operator algebras in: Phase Transitions and Critical Phenomena, Vol. 5A (Domb, C.,

Green, M.S., eds.). New York: Academic 1976.

[KS] Karlin, S., Studden, W.J.: Tchebycheff Systems: with Applications in Analysis and Statistics. New York: Interscience 1966.

[SG] Simon, B., Griffiths, R.B.: The $(\varphi^4)_2$ field theory as a classical Ising model. Commun. Math. Phys. 33, 145-164 (1973).

[SML] Schultz, T.D., Mattis, D.C., Lieb, E.H.: Two-dimensional Ising Model as a soluble problem of many fermions. Rev. Mod. Phys. 36, 856-871 (1964).

[WF] Wilson, K.G., Fischer, M.E.: Critical exponents in 3.99 dimensions, Phys. Rev. Lett. 28, 240-243 (1972).

ON THE PROBLEM OF THE MATHEMATICAL FOUNDATION OF THE GIBBS POSTULATE IN CLASSICAL STATISTICAL MECHANICS

R.L.Dobrushin and Y.M.Suhov

Institute for Problems of Information Transmission
USSR Academy of Sciences, Moscow 111024/USSR

I. Introduction

The rigorous mathematical study of problems of equilibrium Statistical Mechanics takes as a basic premise the postulate due to Gibbs which predicts that the large particle system must have, in an equilibrium state, "Gibbs probability distribution" determined by the interparticle interaction. Just this postulate in its modern form allows to treat the equilibrium Statistical Mechanics as a natural part of the theory of random fields.

A generally accepted statement in the physical literature is that the Gibbs postulate follows from the laws of the Newton classical mechanics but there is no interpretation of such an assertion which is satisfactory from the mathematical point of view. The problem of the rigorous mathematical deduction of the Gibbs postulate from the laws of Mechanics and natural general assumptions is now open and seems to be very interesting and difficult. At present, a new approach to this problem is developping which is based on the explicit consideration of infinite particle systems. The purpose of our note is to describe briefly some results obtained here in the last time and to give a correct formulation of a basic problem. Among a number of preceding papers which have influenced the content of this note, the papers [1] should be especially mentionned. Elements of the approach adopted here may be found in [2].

The authors wish to thank Ya.Sinai for very useful discussions and B.Gurevich for the permission to present some unpublished results.

II. Preliminaries

In this Section we fix our notation system. For details see, e.g., [3 , 4 , 5].

1. The space $M = R^\nu \times R^\nu$ with the usual topology is the phase space (p.s.) of the one-particle system. The points of M are denoted $(x,v), (x',v')$, etc . Denote: $B(x,a) = \{x' \in R^\nu : |x'-x| \leq a\}$, $M(a) = B(0,a) \times R^\nu$ and $M^c(a) = M \setminus M(a)$.

2. The collection of finite unordered subsets of M (including $\emptyset$) is denoted W^0 : this is the p.s. of the finite-particle system. The "points" of W^0 are denoted $\bar{w}, \bar{w}'$, etc . Given $\bar{w} \in W^0$, $|\bar{w}|$ denotes the cardinality of $\bar{w}$ viewed as the subset of M . W^0 is provided with a topology induced by that on M ; the "Lebesgue measure" over W^0 is defined by: $\lambda(d\bar{w}) = 1/|\bar{w}|! \prod_{(x,v)\in\bar{w}} (dx \times dv)$. Given $d \geq 0$, denote: $W^0_d = \{\bar{w} \in W^0 : \min_{(x,v),(x',v') \in \bar{w} : x \neq x'} |x-x'| \geq d\}$ (the p.s. of the hard-core particle system). A function $f : W^0 \to R^1$ is naturally identified with a sequence $\{f_n, n = 0,1,\dots\}$ of symmetric functions $f_n : M^n \to R^1$; this provides the definition of a C^k - function at a point $\bar{w} \in W^0$, $k = 0,1,\dots$.

3. The p.s. W of the infinite system is defined as the collection of all subsets $\underline{w} \subset M$ such that $\underline{w}_a = \underline{w} \cap M(a)$ is

finite for every $a > 0$. Clearly, $W^o \subset W$. The map $\underline{w} \in W \longmapsto \underline{w}_a \in W^o$ is denoted by π_a. W is provided with the natural topology defined by π_a, $a > 0$, (see e.g. 1) ; let $\mathfrak{B}$ denote the corresponding Borel σ -algebra. By $\mathfrak{B}(a)$ and $\mathfrak{B}^c(a)$ denote the sub- σ -algebras generated by $\underline{w}_a$ and $\underline{w}_a^c = \underline{w} \cap M^c(a)$, respectively. Given $d \geq 0$, denote: $W_d = \{ \underline{w} \in W : \pi_a(\underline{w}) \in W_d^o$, all $a > 0 \}$ (the p.s. of the infinite hard - core particle system). By $\mathcal{C}(W, a)$ we denote the space of bounded continuous functions $F: W \to R^1$ which are measurable w.r.t. $\mathfrak{B}(a)$.

4. A State $\mathbb{P}$ of the infinite particle system is a probability measure on $\mathfrak{B}$. Given a state $\mathbb{P}$, one defines the correlation measure $K_{\mathbb{P}}$ over W^o by

$$K_{\mathbb{P}}(A) = \int_W \mathbb{P}(d\underline{w}) \sum_{\bar{w} \in W^o : \bar{w} \subseteq \underline{w}} \chi_A(\bar{w})$$

where χ_A is the indicator of A. By $\mathbb{P}(F)$ (resp., $K_{\mathbb{P}}(f)$) we denote the expectation value of a function $F: W \to R^1$ (resp., $f: W^o \to R^1$) w.r.t. $\mathbb{P}$ (resp., $K_{\mathbb{P}}$).

In a wide situation there is (1-1)-correspondence between states and their correlation measures; for rigorous statements, see [4]. An important class of states is constituted by so-called Gibbs states. Let $\Phi : W^o \to R^1 \cup \{+\infty\}$ be a measurable function with $\Phi(\emptyset) = 0$. We say that a state $\mathbb{P}$ is a Gibbs state corresponding to the Gibbs potential Φ if the conditional probability $[\mathbb{P}(A|\mathfrak{B}^c(a))](\underline{w})$ for $A \in \mathfrak{B}(a)$ is proportional to $\int_{\pi_a A} \lambda(d\bar{w}) \exp[- \sum_{\bar{w}' \subseteq \bar{w} \cup \underline{w}_a^c : \bar{w}' \cap \bar{w} \neq \emptyset} \Phi(\bar{w}')]$. General sufficient conditions for a state $\mathbb{P}$ to be a Gibbs state are due to Kozlov [5].

5. We suppose for the simplicity that the particle interaction is described by a fixed pair potential (p.p.) $V : [0, +\infty) \to R^1 \cup \{+\infty\}$.

We shall impose throughout the paper the following restrictions on the p.p.: 1° the hard core: $V(r) = +\infty$ for $0 \leq r \leq d$ with $d \geq 0$, 2° the finite range: $V(r) = 0$ for $r \geq d_1 > d$, 3° the smoothness: $V \in C^1$ on the interval $(d, +\infty)$, and 4° $V(r) \nearrow +\infty$ as $r \searrow d$ for $d < r \leq \delta$ (the only exeption is: $V = 0$ for $r > d$, the case considered in Section V).

Let $\beta > 0$, $\mu \in R^1$ and $v_0 \in R^\nu$ be fixed. A Gibbs state $\mathbb{G}$ corresponding to

$$\Phi^{(0)}(\bar{w}) = \begin{cases} \beta\left(v^2/2 + v \cdot v_0 - \mu\right), & |\bar{w}| = 1, \ \bar{w} = \{(x,v)\} \\ \beta V(|x - x'|), & |\bar{w}| = 2, \ \bar{w} = \{(x,v),(x',v')\} \\ 0, & |\bar{w}| \geq 3 \end{cases} \tag{1}$$

is called an (β, μ, v_0) - equilibrium state (v_0 is the mean velocity).

III. Time evolution. The Gibbs postulate

To construct the time evolution of a state $\mathbb{P}$, one defines first the dynamics on the p.s. of the finite-particle system. Given $\bar{w} \in W_d^0$, consider the usual Hamilton system determining the motion of particles which have at $t=0$ the coordinates x and velocities v, $(x,v) \in \bar{w}$. It is convenient to label the functions $\underset{\sim}{x} : R^1 \to R^\nu$ and $\underset{\sim}{v} : R^1 \to R^\nu$ giving the one-particle trajectory in M by the initial data $(x,v) \in \bar{w}$. So, the equations of motion are

$$\begin{cases} \dot{\underset{\sim}{x}}_{(x,v)}(t) = \underset{\sim}{v}_{(x,v)}(t) \\ \dot{\underset{\sim}{v}}_{(x,v)}(t) = -\sum\limits_{(x',v')\in\bar{w}:\,x'\neq x} \left(\operatorname{grad} V(|q|) \Big|_{q=\underset{\sim}{x}_{(x,v)}(t)-\underset{\sim}{x}_{(x',v')}(t)} \right) \\ \underset{\sim}{x}_{(x,v)}(0) = x, \ \underset{\sim}{v}_{(x,v)}(0) = v. \end{cases} \tag{2}$$

The Hamiltonian of the system is

$$H(\bar{w}) = 1/2 \sum_{(x,v)\in\bar{w}} \Big[v^2 + \sum_{(x',v')\in\bar{w}:\,x'\neq x} V(|x-x'|) \Big] \tag{3}$$

Setting $\bar{w}(t) = \left\{ \left(\underset{\sim}{x}_{(x,v)}(t), \underset{\sim}{v}_{(x,v)}(t) \right)_{(x,v)\in\bar{w}} \right\}$, $t \in R^1$, one obtains the 1-parameter group of transformations $\bar{w} \in W_d^0 \mapsto \bar{w}(t) \in W_d^0$. If $\mathbb{P}$ is a state of the finite-particle hard-core system, i.e., $\mathbb{P}(W_d^0) = 1$, then the time evolution $T_t \mathbb{P}$ is given by

$$T_t \mathbb{P}(A) = \mathbb{P}\{\bar{w} \in W_d^0 : \bar{w}(-t) \in A\},\ A \in \mathfrak{B},\ A \subseteq W_d^0. \tag{4}$$

Given a general state $\mathbb{P}$ with $\mathbb{P}(W_d^0) = 1$, denote: $\mathbb{P}_a(A) = \mathbb{P}(\pi_a^{-1} A)$, $A \subseteq W_d^0$. Then $\mathbb{P}_a$ may be regarded as a state of the finite-particle system, and $T_t \mathbb{P}_a$ is defined by (4). We say that a family of states $\{\mathbb{P}^{(t)}, t \in R^1\}$ gives the <u>limit time evolution</u> corresponding to the p.p. V for the initial state $\mathbb{P}$ if for any $t \in R^1$, $a > 0$ and $F \in \mathcal{C}(W, a)$

$$\mathbb{P}^{(t)}(F) = \lim_{a' \to +\infty} T_t \mathbb{P}_{a'}(\pi_{a'}^{*} F) \tag{5}$$

where $(\pi^*_{a'} F) : W^o \to R^1$ is defined by: $(\pi^*_{a'} F)(\bar{w}) = F(\pi^{-1}_{a'} \bar{w})$ for $\bar{w} \in M(a')$ and $(\pi^*_{a'} F)(\bar{w}) = 0$, otherwise. In that case we denote: $\mathbb{P}^{(t)} = T_t \mathbb{P}$.

The existence of the limit time evolution (4) is closely connected with the problem of the construction of the limit dynamics on the p.s. W_d. We say that the limit dynamics exists for $\underline{w} \in W_d$ if for any $t \in R^1$ there exists the limit $\lim_{a \to +\infty} \underline{w}_a(t)$ where $\underline{w}_a(t) \in W^o_d$ is defined via the solution of (2) with the initial date $\underline{w}_a$. The problem of proving the existence of the limit time evolution for an initial state $\mathbb{P}$ with $\mathbb{P}(W_d) = 1$ is reduced to the problem of proving that the limit dynamics exists for $\mathbb{P}$ - a.a. $\underline{w} \in W_d$. Thus, one is interested to construct the limit dynamics on a subset $\hat{W} \subset W_d$ which is "as large as possible" and then to describe the states $\mathbb{P}$ concentrated on $\hat{W}$.

For $\nu = 1, 2$ and p.p. V satisfying the conditions 1°–4° above (with $d = 0$) and such that $|V'(r)| r \leq a + b V(r)$, $r > 0$, and $V(r) \geq c r^{-4}$, $0 < r < \delta_1 \leq d_1$ where a, b, c are positive constants, the limit dynamics is constructed in [6] on the set $\hat{W} = \{ \underline{w} \in W : \sup_{x \in R^\nu} \sup_{a > \ln_+ |x|} |B(x,a)|^{-1} \, |H(\underline{w} \cap (B(x,a) \times R^\nu))| < +\infty \}$

where $\ln_+ r = \max[1, \ln r]$, $r > 0$, and $|B(x,a)|$ denotes the volume of $B(x,a)$. There is a large set of states $\mathbb{P}$ which obey $\mathbb{P}(\hat{W}) = 1$; e.g., so is any Gibbs state corresponding to the Gibbs potential of the form (1) with V replaced by $\tilde{V} \geq c_1 + c_2 V$, $c_1, c_2 > 0$. The fact that the limit time evolution (4) exists for any (β, μ, v_0)-equilibrium state $\mathbb{G}$ (i.e., Gibbs state corresponding to a Gibbs potential $\Phi^{(0)}$ given by (1)) with arbitrary $\beta > 0$, $\mu \in R^1$, $v_0 \in R^\nu$ may be verified for any ν and a variety of p.p. V (see, e.g., [7]). As it has been expected, $T_t \mathbb{G} = \mathbb{G}$, i.e., every (β, μ, v_0)-equilibrium state is time-invariant.

We now pass to the discussion of the <u>Gibbs postulate</u>. The (β, μ, v_0) - equilibrium states defined in Section II. 5 are closely connected with three "classical" invariants of the motion: the total energy, total impuls and the number of particles. Given a state $\mathbb{P}$, suppose that the following limits exist and are constant for $\mathbb{P}$ - a.a. $\underline{w} \in W_d$:

$$\begin{cases} \bar{H}_{\mathbb{P}} = \lim\limits_{a \to +\infty} \dfrac{1}{|B(0,a)|} H(\bar{w}_a) \\ \bar{v}_{\mathbb{P}} = \lim\limits_{a \to +\infty} \dfrac{1}{|B(0,a)|} \sum\limits_{(x,v) \in \underline{w}_a} v \\ \bar{n}_{\mathbb{P}} = \lim\limits_{a \to +\infty} \dfrac{1}{|B(0,a)|} |w_a| \end{cases} \tag{5}$$

Our conjecture is: if the limit time evolution given by (4) exist for the initial state $\mathbb{P}$, then the limits (5) exist for the time-evolved states $T_t \mathbb{P}$, $t \in R^1$, and $\bar{H}_{T_t \mathbb{P}} = \bar{H}_{\mathbb{P}}$, $\bar{v}_{T_t \mathbb{P}} = \bar{v}_{\mathbb{P}}$, $\bar{n}_{T_t \mathbb{P}} = \bar{n}_{\mathbb{P}}$. Such a conjecture is now verified for the case we mentionned above (see [6]) where the limit time evolution is proved to exist. The main restriction on the state $\mathbb{P}$ which is needed here is a condition of vanishing its correlations at large distances (cluster property); an example of such a condition, the Rosenblatt mixing condition, is used in Section V.

Given a state $\mathbb{P}$, denote by $\mathcal{E}(\mathbb{P})$ the family of (β, μ, v_0)- - equilibrium states for which the limits (5) exist a.e. and give the values equal $\bar{H}_{\mathbb{P}}$, $\bar{v}_{\mathbb{P}}$ and $\bar{n}_{\mathbb{P}}$, respectively. In "good" cases $\mathcal{E}(\mathbb{P})$ consists of a unique state $\mathbb{G}_{\mathbb{P}}$ [e.g., so is the situation for $\nu = 1$, and for $\nu > 1$ with small values of $\bar{n}_{\mathbb{P}}$]. Suppose, for definiteness, that it is the case. The Gibbs postulate may be formulated as follows. <u>For a "large" class of p.p.</u> V <u>and initial states</u> $\mathbb{P}$ <u>the time-evolved states</u> $T_t \mathbb{P}$ <u>are converging</u>

(in the weak topology) as $t \to \pm\infty$ to the state $G_{\mathbb{P}}$. Among the conditions which one needs to impose on the initial state $\mathbb{P}$, the condition that its correlations vanish at large distances seems to play a crucial role.

The Gibbs postulate conjectured in such a form is based on a belief that, among possible "first integrals" of motion only those listed above can generate a "reasonable" invariant measure. Such a hypothesis is probably true for all "enough good" p.p. V . For "degenerate" cases considered in Section V there are many additional first integrals, and the set of limit states is more rich.

It is clear that the problem of rigorous proving the assertion stated above is very difficult. At the moment, one can prove the convergence to a limit state only for "simplest" time evolutions. In Section IV we discuss some recent results which may be considered as a support for the conjecture on the approach to equilibrium.

One can reformulate the Gibbs postulate in terms of finite-particle evolutions only: it is easy to check that the convergence conjecture above implies the existence of the (weak) limit

$$\lim_{a,|t|\to+\infty\,:\,|t|\le \varphi_{\mathbb{P}}(a)} T_t \mathbb{P}_a = G_{\mathbb{P}} \qquad \text{where} \quad \varphi_{\mathbb{P}}(a)\to+\infty \quad \text{as}$$

$a \to +\infty$. An interesting problem here is to investigate how the situation depends on boundary conditions (for instance: elastic reflection of particles on $\partial B(0,a)$) which may be inserted in the definition of finite-particle dynamics. Apparently, if $|t| \to +\infty$ "too rapidly", then $\lim_{a,|t|\to+\infty} T_t \mathbb{P}$ will not coinside with $G_{\mathbb{P}}$. The assertions of such type may be proved for the cases considered in Section V.

Concluding this Section, we briefly mention a "paradox" related to the behavior of the entropy of the time-evolved state $T_t \mathbb{P}$. Due to the "Liouville" character of the limit time evolution (3), the mean entropy per volume for $T_t \mathbb{P}$ is constant. However, taking the entropy of the restriction $T_t \mathbb{P}\restriction_{\mathfrak{B}(a)}$, one could expect the convergence (in general, non-monotonic) to the corresponding value for the limit state $\mathbb{G}_{\mathbb{P}}$. As above, we can prove that it is the case for the simplest evolutions considered in Section V.

IV. Time-invariant states

In this Section we consider the problem of describing states which are invariant w.r.t. the time evolution. We pass now to another version of constructing a time evolution which is, in a sense, more general and convenient for our purpose than the preceding one. Such a version is based on the so-called Bogoliubov hierarchy equations (B.h.e.) for the correlation function (or, more generally, for the correlation measure) of a time-evolved state. The conditions on the p.p. V which we impose in addition to those indicatad in Section II.5 are: $d>0$ and $V \in C^3$ on $(d,+\infty)$.

It is convenient to treat the B.h.e. as a unique equation. Denote by $C_0^1(W_d^0)$ the space of functions $f : W^0 \to R^1$ with $\operatorname{supp} f \subset \{\bar{w} \in \bar{W}_d^0 : \bar{w} \subset M(a)\}$ where $a = a(f)$ which are of class C^1 at every point $\bar{w} \in W^0$. A family $\{K^{(t)}, t \in R^1\}$ of measures over $\bar{W}_d^0$ is called a weak solution (see [8]) of the B.h.e. with the initial date $K_{\mathbb{P}}$ if (a) $K^{(0)} = K_{\mathbb{P}}$, (b) for any $f \in C_0^1(\bar{W}_d^0)$ the function $t \to K^{(t)}(f)$ is derivable in t, and

$$\frac{d}{dt} K^{(t)}(f) = K^{(t)}([f,H] + [f,\bar{U}]_*). \tag{6}$$

Here $\bar{U} : (x,\bar{w}) \in R^\nu \times \bar{W}^0 \mapsto R^1 \cup \{+\infty\}$ is given by: $\bar{U}(x,\bar{w}) =$

$= \sum_{(x',v')\in\bar{w}} V(|x-x'|)$, further, $[f,U]_*(\bar{w}) = \sum_{(x,v)\in\bar{w}} [f,\tilde{U}_x](\bar{w}\setminus\{(x,v)\})$ with $\tilde{U}_x = U(x,\cdot)$, and $[f,f']$ denotes the usual Poisson brackets: $[f,f'](\bar{w}) = \sum_{(x,v)\in\bar{w}} (\partial/\partial x\, f\, \partial/\partial v\, f' - \partial/\partial v\, f\, \partial/\partial x\, f')(\bar{w})$.

We do not dwell here on arguments leading to the definition of the time evolution via a weak solution of the B.h.e. We notice only that, given an initial state $\mathbb{P}$ for which the limit time evolution (4) exists and is continuous in t , the measure $K^{(t)} = K_{T_t\mathbb{P}}$ gives a weak solution of the B.h.e. with the initial date $K_{\mathbb{P}}$.

In terms of the B.h.e., time-invariant states $\mathbb{P}$ correspond to weak <u>stationary</u> solutions, i.e., their correlation measures satisfy the equation

$$K_{\mathbb{P}}([f,H]+[f,U]_*)=0, \quad f\in C_0^1(W_d^0). \tag{7}$$

The problem of describing solutions of (6) was studied in [9]. The authors consider a class $\mathcal{G}$ of all Gibbs states which correspond to Gibbs potentials Φ satisfying a number of conditions of a general type. The main result of [9] is that if $\mathbb{P}\in\mathcal{G}$ and $K_{\mathbb{P}}$ is a solution of (7) then $\mathbb{P}$ is a (β,μ,v_0) - equilibrium state, i.e. its Gibbs potential $\Phi_{\mathbb{P}}$ is of the form (1) with some $\beta>0$, $\mu\in R^1, v_0\in R^\nu$ [the fact that any (β,μ,v_0) - equilibrium state satisfies (7) is verified by simple arguments, see [9]].

It is natural to assume that, under some conditions, an initial Gibbs state $\mathbb{P}$ remains Gibbsian in the course of the limit time evolution and establish an equation describing the change of the Gibbs potential $\Phi^{(t)} = \Phi_{T_t\mathbb{P}}$ [the idea of [9] is based on a similar approach to a solution of (7)]. At the moment, such an equation is established [10] under some additional restrictions on the time evolution $\{T_t\mathbb{P},\ t\in R^1\}$ (the "clusterness" of the generating limit dynamics). The equation has the form

$$\frac{d}{dt}\Phi^{(t)} = [\Phi^{(t)}, H] + [\Phi^{(t)}, \bar{V}]_* \tag{8}$$

with the initial condition $\Phi^{(0)} = \Phi_{\mathbb{P}}$. Equation (8) may be considered as a "dual" to (6). Notice that (8) <u>may be solved</u> successively by passing from $|\bar{w}| = n$ to $|\bar{w}| = n+1$.

The use of the approach based on the equation (8) may be illustrated as follows. Suppose the p.p. V is repulsiv: $d/_{dr} V \leq 0$, and the initial Gibbs potential $\Phi_{\mathbb{P}}$ is such that $\Phi_{\mathbb{P}}(\bar{w}) = \beta(v^2/_2 + v \cdot v_0 - \mu)$ for $|\bar{w}| = 1$, $\bar{w} = \{(x, v)\}$ and $\lim \Phi_{\mathbb{P}}(\bar{w}) = 0$ whenever $|\bar{w}| = n > 1$ and $\max\limits_{(x,v),(x',v') \in \bar{w}} |x - x'| \to +\infty$. Then one can check [10] that the solution $\Phi^{(t)}$ of (8) approaches as $t \to \pm\infty$ the (β, μ, v_0) - equilibrium Gibbs potential (1). This does not mean, of course, that the corresponding state $T_t \mathbb{P}$ is weakly converging to a (β, μ, v_0) - equilibrium state, but in some sense clears up how such a convergence could hold.

V. Approach to limit states for simplest time evolutions

In this Section we deal with two types of the interparticle p.p. V. The first one is: $V(r) = 0$ for $r > d = 0$. Such an "interaction" generates the free motion of particles. The second type is the "pure hard core" p.p.: $V(r) = 0$ for $r > d > 0$. This type is considered only for $\nu = 1$. The convergence to limit states as $t \to \pm\infty$ has been studied in a number of papers (see [11, 12] and book [13]). The results presented here may be considered as an extension of the convergence theorem to a larger class of initial states which is natural from the point of view of Section III. For the proofs, see [14].

To start with, consider the case: $V(r)=0$ for $r>0$, the free motion. Let $\mathbb{P}$ be an initial state, and $K_{\mathbb{P}}$ be its correlation measure. By $K_{\mathbb{P}}^{(i)}$, $i=1,2,\ldots$, we denote the restriction of the measure $K_{\mathbb{P}}$ on the subsets of $\{\bar{w}\in W^{0}: |\bar{w}|=i\}$. In particular, it is convenient to consider $K_{\mathbb{P}}^{(1)}$ as a measure over M. A convenient sufficient condition for the limit time evolution (4) to exist may be given in terms of $K_{\mathbb{P}}^{(1)}$ only. Let

$\tau_t K_{\mathbb{P}}^{(1)}(A) = K_{\mathbb{P}}^{(1)}(\{(x,v)\in M : (x-tv,v)\in A\}),\ A\subseteq M,\ t\in R^1$.

Then $T_t\mathbb{P}$ exists if $\tau_t K_{\mathbb{P}}^{(1)}(M(a))<\infty$ for every $t\in R^1$ and $a>0$. In that case, $K_{T_t\mathbb{P}}^{(1)} = \tau_t K_{\mathbb{P}}^{(1)}$.

To state our convergence theorem we suppose some more. Namely, we suppose the following absolute continuity: there exists a measure $k^{(1)}$ over R^{ν} such that

$$K_{\mathbb{P}}^{(1)}(dx\times dv) \precsim k^{(1)}(dx)\times dv \tag{9,a}$$

with the Radon-Nicodym derivative

$$\frac{K_{\mathbb{P}}^{(1)}(dx\times dv)}{k^{(1)}(dx)\times dv} \leq f(v),\quad f\in L^1(R^{\nu},dv) \tag{9,b}$$

and

$$\sup_{y\in R^{\nu}} k^{(1)}(B(y,a)) < +\infty \text{ for any } a>0 \tag{9,c}$$

It is not hard to show that (9,a-c) imply the above bound: $\tau_t K_{\mathbb{P}}^{(1)}(M(a))<\infty$. Another condition we need is:

$$K^{(2)}_{\mathbb{P}}(d\bar{w}) \preceq (k^{(1)}(dx)\times dv)(k^{(1)}(dx')\times dv'),\ \bar{w}=\{(x,v),(x',v')\}\ ; \quad (10)$$

this means a "non-degeneracy" of the pair joint velocity distribution.

As it is said above, we also use the Rosenblatt mixing property [15]. Let $C_1, C_2 \subset R^\nu$ be two congruous cubes with the edges parallel to the coordinate axes, and $\mathfrak{W}(C_i)\subset\mathfrak{W}$ be the σ-algebra generated by $\underline{w}\cap(C_i\times R^\nu)$, $\underline{w}\in\bar{W}$, $i=1,2$. Let $\tilde{\alpha}_{\mathbb{P}}(C_1, C_2) = \sup_{A_i\subseteq W:\, A_i\in\mathfrak{W}(C_i),\, i=1,2} |\mathbb{P}(A_1\cap A_2)-\mathbb{P}(A_1)\mathbb{P}(A_2)|$, and $\alpha_{\mathbb{P}}(\ell, s) = \sup_{C_1,C_2:\, \operatorname{diam} C_i\le \ell,\, i=1,2;\ \operatorname{dist}(C_1,C_2)\ge s} \tilde{\alpha}_{\mathbb{P}}(C_1,C_2)$. We need the Rosenblatt mixing condition in the following form

$$\lim_{u\to+\infty} \alpha_{\mathbb{P}}(\nu^{-1/2}u, u) = 0 \quad (11)$$

To describe the limit states for the "free" evolution, fix $a>0$ and an absolutely continuous probability measure μ over R^ν. Consider the Gibbs state corresponding to the Gibbs potential $\Phi=\Phi_{a,\mu}$ of the form: $\Phi(\bar{w}) = -\ln\left(a\,\frac{\mu(dv)}{dv}\right)$ for $|\bar{w}|=1$, $\bar{w}=\{(x,v)\}$, and $\Phi(\bar{w})=0$ for $|\bar{w}|\ge 2$. Such a state is unique and is denoted $\mathbb{G}_{a,\mu}$. It is easy to check that $K^{(1)}_{\mathbb{G}_{a,\mu}}(dx\times dv) = a(dx\times\mu(dv))$, and $T_t\mathbb{G}_{a,\mu} = \mathbb{G}_{a,\mu}$.

The main result of [14] is as follows. Let a state $\mathbb{P}$ obey (9-11). Then the states $T_t\mathbb{P}$ defined by (4) (with $V(r)=0$ for $r>0$) converge as $t\to\pm\infty$ to a state $\mathbb{G}_{a,\mu}$ iff

$$\underline{\text{the measures}}\ \ \tau_t K^{(1)}_{\mathbb{P}}\ \ \underline{\text{converge to}}\ \ a(dx\times\mu(dv)). \quad (12)$$

In particular, if the state $\mathbb{P}$ is translationally-invariant, then $K^{(1)}_{\mathbb{P}}(dx\times dv) = a(dx\times\mu(dv))$, and (12) is automatically valid. For more general sufficient conditions for (12), see [14].

Now pass to the hard-core case. Let $V(r)=+\infty$ for $0\le r\le d$ and $V(r)=0$ for $r>d$ with $d>0$, and let $\nu=1$. We consider initial states $\mathbb{P}$ satisfying the conditions: (i) $\mathbb{P}(W_d)=1$, (ii) $\mathbb{P}$ is locally absolutely continuous and translationally invariant, (iii) $\sum_{m=0}^{\infty}\mathbb{P}_{m+n}(m)\,m<\infty$ for every $n=1,2,\dots$ where $\mathbb{P}_n(m)=\mathbb{P}(\{\underline{w}\in W_d : |\underline{w}_{nd}|=m\})$, (iv) $\mathbb{P}$ obeys (11). From (i) and (ii) it follows that $K^{(1)}_{\mathbb{P}}(dx\times dv)=\beta(dx\times\mu(dv))$ with $\beta<d^{-1}$, and relation (10) holds with $k^{(1)}(dx)=dx$.

To describe the limit states, fix a and μ as above, and consider the Gibbs potential $\Phi=\Phi_{d,a,\mu}$ of the form: $\Phi(\bar{w})=-\ln\left(a\frac{\mu(dv)}{dv}\right)$ for $|\bar{w}|=1$, $\bar{w}=\{(x,v)\}$, $\Phi(\bar{w})=+\infty$ for $\bar{w}\in\bar{W}^0\setminus\bar{W}^0_d$ and $\Phi(\bar{w})=0$ otherwise. The Gibbs state corresponding to such Φ is unique $[\nu=1!]$ and denoted as $\mathbb{G}_{d,a,\mu}$. The measure $K^{(1)}_{\mathbb{G}_{d,a,\mu}}$ is $\beta(dx\times\mu(dv))$ where $\beta=\frac{1}{1+ad}$. It is possible to show that the time-evolved state $T_t\mathbb{G}_{d,a,\mu}$ exists and coinsides with $\mathbb{G}_{d,a,\mu}$ itself.

A theorem proved in [14] asserts that, given a state $\mathbb{P}$ which satisfies (i-iv) above, the time evolution $\{T_t\mathbb{P}\}$ corresponding to pure hard-core interaction V exists, and the states $T_t\mathbb{P}$ approach the state $\mathbb{G}_{d,a,\mu}$ defined from the condition: $K^{(1)}_{\mathbb{P}}=K^{(1)}_{\mathbb{G}_{d,a,\mu}}$.

References

1. Lanford,O.E.,III, Classical Mechanics of One-Dimensional Systems of Infinitely Many Particles, I,II. Commun.Math.Phys.9(1969), 169-181; 11(1969), 257-292.

2. Gurevich, B.M., Sinai, Ya.G., Suhov Yu.M., On Invariant Measures for Dynamical Systems of One-Dimensional Statistical Mechanics. Uspekhi Matem.Nauk (Russian) 28:5 (1973), 45-82.

3. Lanford,O.E.,III, Time Evolution of Large Classical Systems. Lect. Notes in Phys. 38(1975), 1-97.

4. Lenard,A., States of Classical Statistical Mechanical System of Infinitely Many Particles,I,II. Arch.Rational Mech.Anal.59:3 (1975), 219-239; 241-256.

5. Kozlov,O.K., Gibbs'Description of Random Point Fields, Teorija Verojatn.Primen.(Russian) 21(1976), 348-365.

6. Dobrushin,R.L., Fritz,J., a) Non-Equilibrium Dynamics of One-Dimensional Infinite Particle Systems with a Hard-Core Interaction, Commun.Math.Phys.(to appear); b) Non-Equilibrium Dynamics of Two-Dimensional Infinite Particle System with a Singular Interaction, Commun.Math.Phys.(to appear).

7. Sinai, Ya.G., Construction of Dynamics for One-Dimensional Systems of Statistical Mechanics, Teor.Matem.Fizika (Russian) 11:2 (1972), 248-258; Presutti,E., Pulvirenti,E., Tirozzi,B., Time Evolution of Infinite Classical Systems with Singular, Long Range, Two Body Interactions. Commun.Math.Phys.47(1976), 81.

8. Gallavotti,G., Lanford,O.E.,III, Lebowitz,J.L., Thermodynamic Limit of Time-Dependent Correlation Functions for One Dimensional Systems. J.Math.Phys.,11(1972), 2898-2905; Sinai,Ya.G., Suhov,Yu.M., On the Existence Theorem for the Bogoliubov Hierarchy Equations. Teor.Matem.Fizika (Russian), 19:3(1974), 344-363.

9. Gurevich,B.M., Suhov,Yu.M., Stationary Solutions of the Bogoliubov Hierarchy Equations in Classical Statistical Mechanics,I,II. Commun.Math.Phys.49(1976), 63-96; 54(1977), 81-96; Part III. Commun. Math.Phys.(to appear); Part IV: in preparation.

10. Gurevich,B.M., Suhov,Yu.M., Time Evolution of Gibbs States, to appear.

11. Dobrushin,R.L., On the Poisson Law for the Particle Distribution in a Space. Ukr.Matem.Žurn.(Russian) 8:2 (1956), 127-134.

12. Volkovyssky,K.L., Sinai,Ya.G., a) Ergodic Properties of the Ideal Gas with Infinitely Many Degrees of Freedom. Funkz.Anal.Pril.(Russian) 5:4(1971), 19-21; b) Ergodic Properties of the Gas of One-Dimensional Hard Balls with Infinitely Many Degrees of Freedom.Funkz.Anal.Pril. (Russian) 6:1 (1972), 41-50; Aizenman,M., Goldstein,S.,Lebowitz,J.L., Ergodic Properties of an Infinite One Dimensional Hard-Rod System. Commun.Math.Phys.39(1974), 289-304.

13. Kerstan,J., Matthes,K., Mecke,J., Unbegrenzt Teilbare Punktprozesse. Berlin: Akademie-Ferlag 1974.
14. Dobrushin,R.L., Suhov Yu.M., Dynamical Systems of Statistical Mechanics, to appear in: Modern Problems of Mathematics (Russian). Moscow: VINITI-Edition 1978.
15. Rosenblatt,M., A Central Limit Theorem and a Strong Mixing Condition, Proc.Nat.Acad.Sci., USA 42:1 (1956), 43-47.

DYNAMICAL SYSTEMS WITH TURBULENT BEHAVIOR

by David RUELLE

IHES, 91440 Bures-sur-Yvette, France

1. Introduction.

Let the equation

$$\frac{dx}{dt} = X(x) \qquad \text{(continuous time)} \tag{1}$$

or

$$x_{t+1} = f(x_t) \qquad \text{(discrete time)} \tag{2}$$

describe the time evolution of some natural system. It is desirable that (1) or (2) define well-posed problems in the sense that the value of x at time t should depend continuously on the initial value x_o . In other words if an error δx_o is made on x_o, the error δx_t should be arbitrarily small for sufficiently small δx_o . For a bounded error δx_o , nothing prevents however δ_t from growing with t . If that happens to be the case, we say that we have sensitive dependence on initial condition. Actually what we have in mind is a situation where the sensitive dependence on x_o

(a) is not restricted to special choices of x_o

(b) remains after small perturbations of the evolution equation (1) or (2)

(c) is not due to x_t going to infinity *)

1.1. Example. Doubling of the circle. Consider the map $f: \varphi \mapsto 2\varphi \bmod 2\pi$ of the circle. We have exactly $\delta x_t = 2^t \delta x_o$ as long as $|\delta x_t| < \pi$. We have thus a sensitive dependence on initial condition, and one can check that it remains after small perturbation of the map f . Notice that the error $|\delta x_t|$ grows exponentially with t .

For continuous time, the significance of sensitive dependence on initial condition has first been appreciated in problems of fluid dynamics : turbulence and

---

*) Consider the differential equation

$$\frac{dx}{dt} = x$$

on $\mathbb{R}$. Then $x_t = x_o e^t$ and $\delta x_t = \delta x_o . e^t$ grows exponentially with t , but we shall not be interested in this example because the error δx_t grows only when $x_t \to \infty$.

weather prediction (see Lorenz [15], Ruelle and Takens [35]). It is clear for instance that if there is sensitive dependence on initial condition, it will not be possible to predict the weather accurately for long times. Another interesting and important problem is that of hamiltonian systems with many degrees of freedom (see Benettin, Galgani and Strelcyn [2]).

The evolution equation (2) for discrete time occurs naturally in ecology when the populations of different species in one year are given as a function of their populations the previous year. In some cases, irregular fluctuations are observed, which are believed to be associated with sensitive dependence on initial condition. See for instance May [18].

In the present review we shall address ourselves to the following problem : find the simplest cases where sensitive dependence on initial condition occurs. We shall in particular try to keep as low as possible the dimension of the space of the variable x in equation (1) or (2). We shall assume differentiability (of X or f) and distinguish three cases

I discrete time, f not necessarily invertible : <u>differentiable maps</u>
II discrete time, f invertible : <u>diffeomorphisms</u>
III continuous time : <u>flows</u>

To make a long story short, the smallest dimension for which sensitive dependence on initial condition develops in case I, II or III is respectively 1, 2 or 3. More details will be given below.

The material of the present review is organized as follows. In Section 2 we investigate the mathematical origin of sensitive dependence on initial condition, i.e. <u>hyperbolicity</u>. In section 3 we discuss sensitive dependence on initial condition in low dimension for differentiable maps, diffeomorphisms, and flows. In Section 4 we review some points of bifurcation theory. In Section 5 we discuss invariant measures describing the asymptotic behavior of general differentiable dynamical systems, we indicate some results and conjectures. The concluding Section 6 is devoted to general remarks on applications.

On the general subject of this review we refer to the following monographs : Marsden-Mc Cracken [17], Orsay turbulence seminar [40], Ruelle Duke Lectures and Duke conference [32], Lanford Bressannone Lectures (to appear), Berkeley turbulence seminar (to appear).

For the subject of turbulence proper, see in particular J.P. Gollub and H.L. Swinney [Onset of turbulence in a rotating fluid. Phys. Rev. Lett. <u>35</u>, 14, 927-930 (1975)] and J.B. Mc Laughlin and P.C. Martin [Transition to turbulence in a statically stressed fluid system. Phys. Rev. A <u>12</u>, 186-203 (1975)].

2. Hyperbolicity.

From now on we shall consider a time evolution $x \to f^t x$ on a compact n-dimensional manifold M. For each $x \in M$ there is a tangent space *) $T_x M$. We can choose an Euclidean metric on each $T_x M$, defining a Riemann metric on the manifold M. Since M is compact any two Riemann metrics are equivalent. Associated with any differentiable map $f:M \to M$ and for each $x \in M$, there is a tangent linear map $T_x f : T_x M \mapsto T_{f(x)} M$.

We shall assume that our time evolution (f^t) is of class C^r ($r=1,2,\ldots$ or ∞). For continuous t, f^t is the flow obtained by integrating a differential equation

$$\frac{d}{dt} f^t x = X(f^t x)$$

where X is a vector field on M, i.e. $X(x) \in T_x M$, and $x \to X(x)$ is assumed to be of class C^r (r times continuously differentiable). For discrete t, f^t is the t-th iterate of a C^r map $f:M \to M$. We can take t negative if f is a diffeomorphism (i.e. its inverse is of class C^1, hence C^r).

We shall follow the usual habit of calling "smooth" or "differentiable" something which is of class C^k for suitable (but unspecified) k.

Sensitive dependence on initial condition means that f^t stretches distances considerably for large t. Otherwise stated the linear maps $T_x f^t$ have large norm for large t. Notice that it is sufficient to have stretching in some direction; in other directions, $T_x f^n$ may be contracting. We shall refer loosely to this combination of stretching and contraction as <u>hyperbolicity</u>.

Notice that it is important to have hyperbolicity only asymptotically, i.e. on the set of limits of $f^t x$ for $t \to \infty$. This set of limit points is contained in the <u>non-wandering set</u> which we define now. A point $x \in M$ is wandering if it has a neighborhood U such that $U \cap f^t U = \emptyset$ for all sufficiently large t. The nonwandering set is the set of nonwandering points, it is closed and (f^t)-invariant.

Under the name of <u>Axiom A</u>, Smale [37] has formalized a notion of hyperbolicity, which has turned out to be extremely fruitful. The Axiom A diffeomorphisms and flows are now the best understood differentiable dynamical systems.

2.1. Definition

Axiom A for a diffeomorphism f consists of the following two conditions

*) M may be a sphere, torus, ... It is always possible to imagine that M is a n-dimensional submanifold of N-dimensional Euclidean space for some large N. In particular the tangent space $T_x M$ can then be identified with the usual geometric object. We shall assume that the manifold M and the Riemann metric are C^∞ (infinitely differentiable).

(Aa) Hyperbolicity (sensu stricto) of the nonwandering set Ω : for each $x \in \Omega$, T_xM is a direct sum $E_x^u \oplus E_x^s$ where the subspaces E_x^u , E_x^s depend continuously on x , $T_x f\, E_x^u = E_{f(x)}^u$, $T_x f\, E_x^s = E_{f(x)}^s$, and there exist $C > 0$ and $\lambda < 1$ such that, for all $x \in \Omega$, $n \geq 0$,

$$\|T_x f^n v\| \leq C\, \lambda^n \|v\| \qquad \text{if } v \in E_x^s$$

$$\|T_x f^{-n} v\| \leq C\, \lambda^n \|v\| \qquad \text{if } v \in E_x^u$$

(Ab) The periodic points are dense in the nonwandering set Ω .

For a flow, Axiom A is similar. Suppose for simplicity that the vector field X does not vanish. Then it is required that $T_xM = E_x^o \oplus E_x^u \oplus E_x^s$ for $x \in \Omega$, where E_x^o is one dimensional generated by $X(x)$, and E^u, E^s have properties as above. (Ab) again requires the density of periodic (i.e. closed) orbits in the nonwandering set Ω .

2.2. Asymptotic behavior in Axiom A dynamical systems.

For the study of Axiom A diffeomorphisms and flows we refer to the review of Smale [37], which is still very good reading, and the more recent monograph by Bowen [3], which is particularly interesting from our view point.

For an Axiom A diffeomorphism or flow, the nonwandering set Ω is the union of a finite number of compact invariant sets Ω_i such that for each Ω_i there is some x for which the orbit $\cup_t f^t x$ is dense in Ω_i . The Ω_i are called basic sets. A basic set Λ is an attractor if it has a neighborhood U such that $\cap_{t>0} f^t U = \Lambda$. The basin of the attractor is the $\cup_{t<0} f^t U$, i.e. the set of points x such that $f^t x \to \Lambda$ when $t \to +\infty$.

If Λ is an attractor for the Axiom A diffeomorphism f , any C^1-small perturbation f' of f has an attractor Λ' close to Λ , and there is a homeomorphism $h: \Lambda \mapsto \Lambda'$ close to the identity such that $f' = h \circ f \circ h^{-1}$. For an Axiom A flow (f^t) , we have a similar result where h sends orbits of f to orbits of f' , but the parametrization of orbits (by t) is in general not preserved. (These results are special cases of Smale's Ω-stability theorem, see Hirsch and Pugh [12], Pugh and Shub [27]).

Assuming that one has a C^2 diffeomorphism or flow one can show that the basins of the various attractors cover M up to a set of Lebesgue measure zero *) .

We define an "average" measure on the orbit of a point x by

$$\mu_{x,N} = \frac{1}{N} \sum_{n=0}^{N-1} \delta_{f^n x}$$

*) By Lebesque measure we mean the "volume" defined on M by any Riemann metric, the measure zero sets do not depend on the choice of the metric.

for a diffeomorphism or

$$\mu_{x,T} = \frac{1}{T} \int_0^T dt\ \delta_{f^t x}$$

for a flow. Then, for each attractor Λ there is a unique measure μ such that for almost all x in the basin of Λ with respect to Lebesque measure,

$$\text{vague} \lim_{N\to\infty} \mu_{x,N} = \mu \tag{3}$$

or

$$\text{vague} \lim_{T\to\infty} \mu_{x,T} = \mu \tag{4}$$

[vague lim means that for every continuous function φ on M, $\mu(\varphi) = \int \varphi(y)\ \mu(dy)$ is the limit of $\mu_{x,N}(\varphi)$ or $\mu_{x,T}(\varphi)$] .

The formula (3) or (4) shows that the asymptotic behavior of most points, for an Axiom A diffeomorphism or flow, is given by a finite number of measures associated with the attractors. One can show that the measure μ on the attractor Λ is the only measure which makes maximum the quantity

$$h(\mu) - \mu(\log J_+) \tag{5}$$

In this formula $h(\mu)$ is the measure-theoretic entropy (Kolmogorov-Sinai invariant) and $J_+(y)$ is the jacobian (with respect to a Riemann metric) of the map $E^u_x \mapsto E^u_{f'x}$ induced by $T_x f'$. The maximum of (5) is in fact 0 , and this variational principle for μ is related to the well-known variational principle of equilibrium statistical mechanics. For more details see Sinai [36], Ruelle [31], Bowen and Ruelle [4]).

2.3. <u>Sensitive dependence on initial condition in Axiom A dynamical systems</u>.

Let a point have an orbit asymptotic to an Axiom A attractor Λ . We shall have sensitive dependence on initial condition if stretching occurs on the attractor, i.e. if the dimension of the spaces E^u_x (with $x \in \Lambda$)' is strictly positive. In this case Λ is called a <u>strange attractor</u>. For a diffeomorphism, a non strange attractor is just an attracting periodic orbit. For a flow, it is an attractive periodic orbit, or an attracting fixed point.

2.4. <u>More general attractors</u>.

We shall say that a compact invariant set Λ (for a map f or flow (f^t)) is an attractor, if it has a neighborhood U such that $\bigcap_{t>0} f^t U = \Lambda$. Furthermore we want to impose an irreducibility condition (such that the union of two different attractors cannot be again considered to be an attractor). It is for instance reasonable to assume topological transitivity : there is $x \in \Lambda$ such that $\bigcup_t f^t x$ is dense in Λ .

2.5. Topological conditions on the map or flow.

The type of attractors which are possible, and therefore the occurence of sensitive dependence on initial condition, depend on global topological conditions on the map or flow. In particular, it will often be interesting to consider diffeomorphisms isotopic to the identity, (in particular, diffeomorphisms close to the identity).

2.6. Turbulent behavior.

Apart from the theory of Axiom A dynamical systems, and some remarkable results on quasi periodic systems (Arnold [1], Herman [10]) we do not have a good understanding of the asymptotic behavior of differentiable dynamical systems. Nevertheless two types of behavior may be recognized

(a) We may call turbulent a kind of time dependence for which the prototype is an orbit $f^t x$ asymptotic to a strange Axiom A attractor : there is sensitive dependence on initial condition, exponential decay of time correlations *), the average behavior is described by a measure μ with entropy $h(\mu) \neq 0$. Examples of turbulent behavior are known (see below) from dynamical systems which do not satisfy Axiom A.

(b) We may call non turbulent, or laminar, a kind of time dependence without sensitive dependence on initial condition, without decay of correlations, with average behavior described by a measure μ with entropy $h(\mu) = 0$. Examples are time dependence $f^t x$ asymptotic to a periodic orbit, or to a torus with quasi-periodic flow **) .

(c) That everything does not always fit neatly in one of the above types is shown in particular by the work of Newhouse discussed in Section 4.3.

For a certain type of behavior to be of interest, it should have a certain persistence under perturbations. For instance, a large class of Axiom A dynamical

---

*) Exponential decay of correlation has been proved for mixing Axiom A attractors for diffeomorphisms (see Sinai [36], Ruelle [31]) for flows the problem is open.

**) A quasi-periodic flow on the m-dimensional torus T^m , is a flow which is given, in suitable coordinates by

$$f^t \begin{pmatrix} x_1 \\ \vdots \\ x_m \end{pmatrix} = \begin{pmatrix} x_1 + a_1 t \pmod 1 \\ \vdots \\ x_m + a_m t \pmod 1 \end{pmatrix}$$

(the flow defined by a constant vector field).

systems (those with no cycles (or Ω-stable systems)) form an open set in the space of C^k diffeomorphisms or flows. Quasi-periodic motions "often" remain quasi-periodic under perturbation in some measure-theoretic sense [11].

3. Turbulent dynamical systems in low dimension.

Let f be a differentiable map of the compact manifold M into itself. If f is stretching and M one-dimensional, f cannot be one-to-one. Thus, if f is a diffeomorphism with sensitive dependence on initial condition, M has to have at least two dimensions corresponding to one stretching direction and one contracting direction. For a flow, the direction of flow is mapped into itself in an almost isometric manner by f^t ; therefore M has to have at least three dimensions corresponding to one flow direction, one stretching direction and one contracting direction. By this crude argument, the minimum dimension for which sensitive dependence on initial condition can occur for maps, diffeomorphisms and flows is thus respectively 1,2,3. A detailed study will show that this answer is indeed correct.

3.1. Differentiable maps in one dimension.

As shown by the Example 1.1., sensitive dependence on initial condition may occur for non-invertible maps in one dimension. However, the map of Example 1.1 cannot be obtained by continuous deformation of the identity, and the "turbulent behavior" is related to this topological feature.

Much interest has been devoted to maps of the interval $[0,1]$, like that of Fig.1.

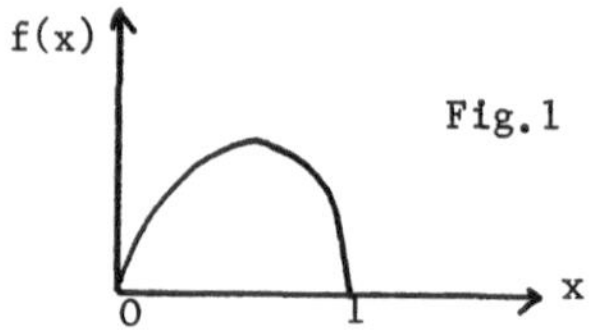

Fig.1

A very simple example is given by $f_R(x) = Rx(1-x)$, the so-called "logistic equation". Non trivial results on such maps have been obtained by Jacobson [13], Sharkovskii (see Li and Yorke [14], Stefan [39]) and by Milnor and Thurston [19] *).

To see how sensitive dependence on initial condition can develop, consider the "logistic equation" for $R = 4$. The change of variable $y = \frac{2}{\pi}$ arc sin $\sqrt{x}$ transforms f_4 into the map

$$\tilde{f} : y \to \begin{cases} 2y & \text{if } x \in [0,\frac{1}{2}] \\ 2(1-y) & \text{if } x \in [\frac{1}{2},1] \end{cases}$$

i.e. the broken linear transformation of Fig. 2.

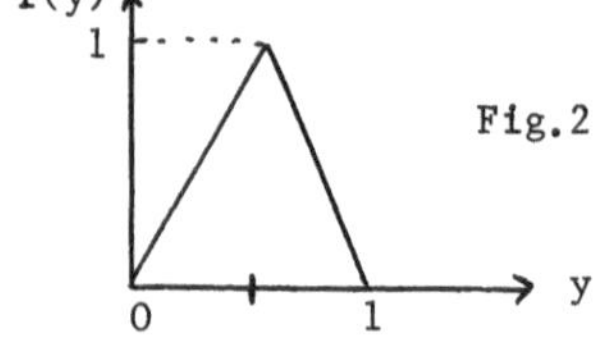

Fig.2

It is clear that this map has sensitive dependence on initial condition, and leaves Lebesgue measure invariant. Correspondingly, the map f_4 has sensitive dependence on initial condition, and leaves invariant the measure **)

---

*) For some interesting conjectures, see also Feigenbaum [6].

**) The Haar measure on the circle is invariant under doubling (Example 1.1). By projection on a diameter of the circle one recovers the invariance of the measure $\frac{1}{\pi}\frac{dx}{\sqrt{x(1-x)}}$ under f_4 . In this way the mysterious change of variable $x \mapsto y$ is easily understood.

$$\frac{1}{\pi} \cdot \frac{dx}{\sqrt{x(1-x)}}$$

For $R = 3.6785735.$, one can also prove that f_R has an invariant measure absolutely continuous with respect to the Lebesgue measure, and sensitive dependence with respect to initial condition (see Ruelle [33]). Sinai and Jakobson (private communication) have recently proved that this occurs for countably many values of R.

To assert that f_R produces turbulent behavior, one would like to know that the sensitive dependence on initial condition has a certain persistence. Here however we have only conjectures : it is likely that for a dense open set of values of $R \in [0,4]$ there is an attracting periodic orbit which attracts (Lebesgue -) almost all $x \in [0,1]$. It is however possible that f_R has an invariant measure absolutely continuous with respect to Lebesgue measure for R in a subset of non zero Lebesgue measure of $[0,4]$.

3.2. Remark. (Smale and Williams [38], May [18])

Suppose f is close to a map with sensitive dependence on initial condition. Although f itself may have an asymptotic behavior described by an attracting periodic orbit, this orbit is likely to be of high order and not very attractive. Then , especially in the presence of noise (say round-off errors in a machine computation), f will for all practical purposes appear turbulent. Contributing to this is also the fact (see Smale and Williams [28]) that a Cantor set may accompany the attracting periodic orbit, and although this Cantor set is not attracting, orbits tend to wander near it for a long time.

3.3. Diffeomorphisms in two dimensions.

Strange Axiom A attractors on two-dimensional manifolds have been known for a long time, for instance the toral automorphism

$$\begin{pmatrix} x \\ y \end{pmatrix} \to \begin{pmatrix} x + y & (\text{mod } 1) \\ x + 2y & (\text{mod } 1) \end{pmatrix}$$

on T^2 . It has been more difficult to construct a strange Axiom A attractor isotopic to the identity in a piece of $\mathbb{R}^2$, but Plykin [26] has provided such an example. Using Plykin's example one proves the following result.

Theorem. (Newhouse, Ruelle, Takens [22]).

Let M be a compact 2-dimensional manifold; then in every C^1 neighborhood of the identity there is an Axiom A diffeomorphism with a strange attractor. For a 2-torus, C^1 can be replaced by C^2 , and for a m-dimensional manifold - $m > 2$ - C^1 can be replaced by C^∞ .

Turbulent behavior occurs thus for diffeomorphisms in two dimensions (but not in one, because diffeomorphisms of the circle are clearly not "turbulent").

Unfortunately, theorems based on Plykin's example are in a sense misleading because Plykin's attractor is somewhat complicated and unlikely to occur in simple situations (for instance in diffeomorphisms with a simple analytic form). It is thus interesting to consider at this point a very simple diffeomorphism of $\mathbb{R}^2$ discussed numerically by Hénon [9], and which appears to have a "turbulent" attractor.

3.4. The Hénon attractor.

A polynomial map of second degree $\mathbb{R}^2 \to \mathbb{R}^2$ which has constant jacobian can be put by a linear change of coordinates in the form

$$f : \begin{pmatrix} x \\ y \end{pmatrix} \to \begin{pmatrix} y+1 - ax^2 \\ bx \end{pmatrix} \tag{6}$$

In particular the inverse of the map is in the same class again. Hénon has made a computer study of f for $a = 1.4$, $b = .3$, and finds what appears to be an attractor Λ of a new type (for pictures, see [9]) . The following facts appear to be true (numerically)

(a) Λ is the closure of the unstable manifold of one of the two fixed points of f .

(b) $\Lambda = \bigcap_{n\geq 0} f^n U$ where $fU \subset U$ and U is some explicitly computed quadrilateral.

We make a few comments

I. Λ cannot be an Axiom A attractor.

One reason is a result of Plykin [26], that if Λ is an Axiom A attractor in the plane with a neighborhood U such that $\Lambda = \bigcap_{n\geq 0} f^n U$, then U cannot be a disc with less than three holes. This would contradict (b) above.

Another reason is that the unstable manifold of the fixed point mentioned in (a) has arbitrarily sharp bends, preventing a continuous decomposition $T_x M = E^u_x \oplus E^s_x$.

II. (Due to S. Newhouse) Mañe [16] has shown that if U is a closed disc in the plane, and f a diffeomorphism such that $fU \subset$ interior U and

$$\Lambda = \bigcap_{n>0} f^n U \text{ is non wandering} \tag{7}$$

and (7) remains true for all small C^1 perturbations of f , then Λ is a single attracting fixed point. This shows that for the Hénon attractor, (7) cannot remain true under small perturbations.

III. In view of II one expects rapid changes in the structure of Λ as a is varied. In fact G. Parisi (private communication) has found that the "attractor" Λ is replaced by an attracting periodic orbit of order 7 for $a = 1.3$ and $b = .3$. This attracting periodic orbit exists only for a small range of values of a , and

and is somewhat difficult to see because points tend to it rather slowly.

IV. The Hénon map is in some sense very close to the "logistic equation" of 3.1. It is therefore not astonishing that for some values of a,b , an attracting set should occur which is not a periodic orbit. The interesting (and unsolved) question is whether the set of values of a,b for which a "turbulent" attractor occurs is of measure zero or not. In any case, Remark 3.2 applies.

V. (Due to S. Newhouse). The phenomenon of "infinitely many sinks" (see Section 4.3) does occur for the Hénon map for some values of a, b . Although such a statement cannot be really verified by computer studies, one would expect to see for some values of a,b , several attracting periodic orbits plus a residual "mess" consisting of unresolved periodic orbits and / or a turbulent attractor.

VI. A computer study gives the estimate

$$\lim_{n\to\infty} \frac{1}{n} \log \|T_x f^n\| \approx 0.4 \tag{8}$$

when x is in the basin of the Hénon attractor (S. Feit). This is consistent with the existence on the Hénon attractor of an asymptotic measure of the sort discussed in Section 2.2 for Axiom A diffeomorphisms. (See Section 5 below for a further discussion of this point). Interestingly, looking for negative values of the ℓ.h.s. of (8) is an efficient way of finding attracting periodic orbits numerically.

3.5. Flows in three dimensions.

One has the intuitive feeling that, for topological reasons, a flow in two dimensions cannot be turbulent (the orbits of different points stay locally parallel and are not mixed by the flow). This feeling is supported by the theorem that the topological entropy *) of a flow on a compact two dimensional manifold is always zero [see Young [44]]. There are however turbulent flows in three dimensions. Of special interest are the flows obtained by perturbation of a quasi-periodic flow; they occur in the discussion of fluid turbulence **), and correspond also to the weak coupling of several oscillators. For such flows we have the following result.

Theorem (Newhouse, Ruelle, Takens [22]).

Let $a = (a_1, \dots, a_m)$ be a constant vector field on the torus $T^m = \mathbb{R}^m/\mathbb{Z}^m$.

*) The topological entropy is the sup of the measure-theoretical entropies of invariant measures.

**) See Ruelle and Takens [35]. In that paper it was shown that "turbulence" may arise by small perturbations of a quasiperiodic flow on T^4 ; the next theorem improves that result in replacing T^4 by T^3 . Numerical evidence for turbulence on T^3 has been obtained by H. Sherman and J. Mc Laughlin [Power spectra of Nonlinear Coupled Waves, preprint].

<u>If</u> $m = 3$, <u>in every</u> C^2 <u>neighborhood of</u> a <u>there is a vector field satisfying Axiom A and having a strange attractor.</u>

<u>If</u> $m \geq 4$, <u>in every</u> C^∞ <u>neighborhood of</u> a <u>there is a vector field satisfying Axiom A and having a strange attractor.</u>

The proof of this theorem uses Plykin's example, and it is interesting to look for examples of turbulent attractors which need not satisfy Axiom A, but correspond to simple differential equations. Such an example has been introduced by Lorenz [15].

3.6. <u>The Lorenz attractor.</u>

We refer to the original paper and to earlier reviews (Ruelle in [40] and Lanford in [33]) for the general facts. A class of flows modelled after the Lorenz attractor has been defined by Guckenheimer [7] and Williams [43], and is now fairly well understood *). In particular it has been shown by Guckenheimer [8] that these flows have "codimension 2 Ω-stability". This means that for every attractor Λ' in the class there is a homeomorphism $h : \Lambda_{(u,v)} \mapsto \Lambda'$ where $\Lambda_{(u,v)}$ belongs to a reference family with two real parameters, such that h maps orbits to orbits (not preserving the parametrization by t in general). The situation is thus more complicated than that of Axiom A flows, but not very much more.

Apparently, asymptotic measures (cf. Section 2.2) can be defined as for Axiom A flows (Ia. Sinai, private communication).

---

*) Apart from the fact that it has not been proved that the Lorenz equations define a Lorenz attractor in the new sense !

4. Bifurcation theory.

If a differentiable dynamical system depends on parameters, its qualitative behavior may change as the parameters are changed. The study of these changes is called bifurcation theory. This is a more difficult problem than the understanding of individual systems, and therefore less well understood. Here we limit ourselves to the discussion of some special questions.

4.1. Hopf bifurcations and bifurcations of tori.

The relevance of this problem to turbulence has been discussed in Ruelle and Takens [35] *) and the Hopf bifurcations (for flows) leading to an attracting torus T^2 are well understood. For the more delicate transition from T^2 to T^3 see Chenciner and Iooss[5]. In fact one expects that transitions beyond T^2 will in general be "messy".

4.2. Quasi periodic flows.

The work of Arnold [1] on irrational rotations of the circle (or, equivalently, quasi periodic flows on T^2), has been remarkably completed recently by Herman [10], [11] . In particular, conditions are given such that, if (a_λ) is a one-parameter family of vector fields on T^2 , there is a set $S \subset [0,1]$, of Lebesgue measure close to 1 , such that the flow associated with ω_λ is quasi-periodic if $\lambda \in S$.

4.3. Infinitely many sinks.

Let Λ be a hyperbolic set (in the sense of Section 2.1) for a diffeomorphism f of the compact manifold M . Through each point x of Λ pass manifolds W_x^u and W_x^s tangent to E_x^n and E_x^s respectively (and of the same dimension) such that (W_x^n) and (W_x^s) are continuous f-invariant families. The manifolds W_x^n and W_x^s are called respectively unstable and stable. The occurence of a tangency between some W_x^u and W_y^s is important for bifurcation theory. In fact Newhouse [20], [21] has shown that if $\dim M = 2$, and f has such a tangency, then there is close to f an open set U of diffeomorphisms, each with a tangency (we use the C^r topology, $r \geq 2$) . Furthermore there is in U a residual set (= dense intersection of open sets) of diffeomorphisms which have infinitely many sinks (= attracting periodic orbits), or infinitely many sources as the case may be. This very interesting situation is not yet entirely analysed. It is however clear that it will easily lead to "messy" bifurcations, when a diffeomorphism f_λ crosses the region U as λ varies.

4.4. Nice bifurcations to turbulent attractors. (communicated by S. Newhouse).

In view of the above, one may wonder if one can go from an Axiom A diffeo-

*) Unnecessary conditions are stated in the discussion of the Hopf bifurcation for a diffeomorphism in [35]. For an improved version, see for instance Ruelle [30].

morphism f_o whose attractors are periodic points to an Axiom A diffeomorphism f_1 with a strange attractor, without going through a "messy" bifurcation. The answer is positive. The idea is to take a one-dimensional axiom A attractor (e.g. Plykin's attractor, or a solenoid) with a fixed point x. One perturbs f_1 so that x becomes attracting, and the rest of the attractor is split off as a non attracting Cantor set. The bifurcation thus obtained is simple in the sense that the topological type of f_λ changes for just one value of λ (at which f_λ no longer satisfies Axiom A). Also if the curve (f_λ) is slightly perturbed, there will be again one single bifurcation point at which the topological type changes.

4.5. Bifurcations in the Lorenz model.

The Lorenz equations (see [15]) contain - among others - a parameter r (proportional to the Rayleigh number). The bifurcations of the model for r not too large are reasonably well understood (at a numerical-heuristic rather than mathematical-rigorous level). We refer the reader to Kaplan and Yorke *) for a discussion of these bifurcations. Among other things, Kaplan and Yorke observe a phenomenon which they call preturbulence : something like turbulence is observed numerically while only periodic attractors exist. The explanation is that the solutions of the differential equations tend to remain for a long time close to some non attracting pieces of the nonwandering set (cf. Remark 3.2).

*) J.L. Kaplan and J.A. Yorke. Preturbulence : a regime observed in a fluid flow model of Lorenz (Preprint).

5. Invariant measures.

For simplicity we shall discuss only the case of a diffeomorphism f of the compact manifold M. If ρ is a f-invariant (Radon) measure on M, the limit *)

$$\lim_{n\to+\infty} \frac{1}{n} \log \|T_x f^n\| = \chi(x)$$

exists almost everywhere, defining the function χ, which is f-invariant and can be assumed constant if ρ is ergodic. This is a consequence of the noncommutative ergodic theorem of Oseledec [23]. More precisely, there is almost everywhere a splitting $T_x M = \bigoplus_i V_x^{(i)}$ such that

$$\lim_{n\to\pm\infty} \frac{1}{n} \log \|T_x f^n u\| = \chi_{(i)}(x) \quad \text{if} \quad u \in V_x^{(i)}$$

and

$$\max_i \chi_{(i)}(x) = \chi(x)$$

The above splitting of $T_x M$ depends measurably but not in general continuously on x. If ρ is ergodic, the <u>characteristic exponents</u> $\chi_{(i)}$ may be taken constant and, with the multiplicities $\dim V_x^{(i)}$, constitute the <u>spectrum</u> of (ρ, f).

The existence of stretching with respect to the measure ρ is expressed by the strict positivity of some $\chi_{(i)}$. In fact the sum $\chi_+(x)$ of the $\chi_{(i)}(x) \dim V_x^{(i)}$ over the positive $\chi_{(i)}$ is the maximum expansion coefficient for a multivector $u_1 \wedge u_2 \wedge \ldots \wedge u_r$. We have also

$$\chi_+(x) = \lim_{n\to+\infty} \frac{1}{n} \log \|T_x^{\wedge} f^n\|$$

where $T_x^{\wedge} f$ is the action associated with $T_x f$ on the exterior algebra of $T_x M$.

<u>Theorem.</u> [34] <u>For every f-invariant measure</u> ρ,

$$h(\rho) \leq \int \rho(dx)\, \chi_+(x)$$

<u>where</u> $h(\rho)$ <u>is the entropy of</u> μ.

<u>Question 1.</u> <u>Do we have</u>

$$\max_\rho \left[h(\rho) - \int \rho(dx)\, \chi_>(x)\right] = 0 \quad ?$$

This is known to be true if f satisfies Axiom A [31], or is the time one map of an Axiom A flow [4], or has a smooth invariant measure [24].

<u>Question 2.</u> <u>Take</u> ρ <u>ergodic and assume that all</u> $\chi_{(i)}$ <u>are</u> $\neq 0$. <u>Can one define for</u> ρ <u>almost all</u> x, <u>a smooth local stable manifold tangent to</u>

*) Norms are taken with respect to some arbitrary Riemann metric.

$\sum_{\chi_{(i)} > 0} V_x^{(i)}$ at x, <u>such that these form an invariant family ? Similarly for unstable manifolds.</u>

The answer is positive for Axiom A [37] and for an ergodic component of a smooth invariant measure [25].

<u>Question 3.</u> <u>Assuming a positive answer to Question 2, suppose that</u> ρ <u>has conditional measures on the unstable manifolds, which are absolutely continuous (with respect to the measure defined by the Riemann metric), can one conclude that</u>

$$h(\rho) = \int \rho(dx)\, \chi_{>}(x) \qquad ?$$

The answer is positive for Axiom A .

These questions pertain to the problem of trying to identify the limits when n tend to ∞ of the measures

$$f^n m \ , \ \frac{1}{n} \sum_{k=0}^{n-1} f^k m$$

for m smooth on M, and also for m-almost all x, of

$$\frac{1}{n} \sum_{k=0}^{n-1} \delta_{f^k x}$$

where δ_x is the unit mass at x. The problem has been solved when f satisfies Axiom A (see Section 2.2), and (apart from the limits of $f^n m$) when f has a smooth invariant measure [24].

Let us remark that the Hénon attractor apparently has an asymptotic measure ρ with one characteristic exponent $\simeq 0.4$, and the other (determined by the constant Jacobian - 0.3) $\simeq -1.6$. One may conjecture that $h(\rho) \simeq 0.4$.

6. Concluding remarks.

From a mathematical viewpoint, the subject of the present review consists of partial answers to very difficult problems, and it may appear premature to ask general questions about the turbulent behavior of differentiable dynamical systems. Such questions, however, are imposed by the applications, and it is important to realize that these applications have potentially a conceptual interest as great as the underlying mathematics. For instance, we have briefly mentioned the problems of population dynamics in ecology. Such problems, and related problems of evolution, are to a certain extent understood at an intuitive and verbal level, but in general a conceptual understanding is lacking. It is thus very desirable to build a mathematical framework in which one could deal with such problems as, e.g. why temperate forests are species poor, while tropical rain forests are species rich with considerable spatial (and presumably temporal) fluctuations (see for instance [29], [42], [28]).

It is perhaps appropriate to suggest that mathematical physicists have in principle the aptitudes and interests which would allow them to attack successfully such problems.

To conclude let me record the historical importance of Thom's ideas [41] in convincing scientists of the usefulness of qualitative dynamics in the natural sciences. That some physicists had similar ideas is shown by the following quotation from the Feynman Lectures in Physics.

"The next great era of awakening of human intellect may well produce a method of understanding the qualitative content of equations. Today we cannot. Today we cannot see that the water flow equations contain such things as the barber pole structure of turbulence that one sees between rotating cylinders. Today we cannot see whether Schrödinger's equation contains frogs, musical composers, or morality - or whether it does not. We cannot say whether something beyond it like God is needed, or not. And so we can all hold strong opinions either way."

Acknowledgements.

It is a pleasure to record my indebtedness to Sheldon Newhouse for much help in the preparation of this report.

References.

[1] V.I. Arnold. Small denominators. I mappings of the circumference onto itself. Amer. Math. Soc. Transl. (Ser.2) 46, 213-284 (1965).

[2] G. Benettin, L. Galgani, and J.-M. Strelcyn. Kolmogorov entropy and numerical experiments. Phys. Rev. A.14, 2338-2345 (1976).

[3] R. Bowen. Equilibrium states and the ergodic theory of Anosov diffeomorphisms. Lecture Notes in Mathematics N° 470. Springer, Berlin, 1975.

[4] R. Bowen and D. Ruelle. The ergodic theory of Axiom A flows. Inventiones math. 29, 181-202 (1975).

[5] A. Chenciner and G. Iooss. Bifurcation of a torus T^2 into a torus T^3. Preprint.

[6] M.J. Feigenbaum. Quantative universality for a class of nonlinear transformations. Preprint.

[7] J. Guckenheimer. A strange, strange attractor. Preprint.

[8] J. Guckenheimer. Structural stability of Lorenz attractors. Preprint.

[9] M. Hénon. A two-dimensional mapping with a strange attractor. Commun. Math. Phys. 50, 69-77 (1976).

[10] M.R. Herman. Conjugaison C^∞ des diffeomorphismes du cercle pour presque tout nombre de rotations. CRAS to appear.

[11] M.R. Herman. Mesure de Lebesgue et nombre de rotation. Preprint.

[12] M.W. Hirsch and C.C. Pugh. Stable manifolds and hyperbolic sets. Proc. Sympos. Pure Math. 14, 133-163 AMS. Providence, R.I., 1970.

[13] M.W. Jakobson. On smooth mappings of the circle into itself. Mat. Sbornik 85 (127) N°2 (6), 163-188 (1971).

[14] T.Y. Li and J. Yorke. Period three implies chaos. SIAM J. Applied Math. To appear.

[15] E.N. Lorenz. Deterministic nonperiodic flow. J. atmos. Sci. 20, 130-141 (1963).

[16] R. Mañe. The stability conjecture on two-dimensional manifolds. Preprint.

[17] J.E. Marsden and M. Mc Cracken. The Hopf bifurcation and its applications. Applied Math. Sci. 19. Springer, New York, 1976.

[18] R.M. May. Simple mathematical models with very complicated dynamics. Nature 261, 459-467 (1976).

[19] J.W. Milnor and W. Thurston. Unpublished.

[20] S. Newhouse. Diffeomorphisms with infinitely many sinks. Topology 13, 9-18 (1974).

[21] S. Newhouse. The abundance of wild hyperbolic sets and non-smooth stable sets for diffeomorphisms. Preprint.

[22] S. Newhouse, D. Ruelle, and F. Takens. Occurence of strange axiom A attractors near quasi periodic flows on T^m, $m \geq 3$. Preprint.

[23] V.I. Oseledec. A multiplicative ergodic theorem. Ljapunov characteristic numbers for dynamical systems. Trudy Moskov. Mat. Obšč. 19, 179-210 (1968). English translation. Trans. Moscow Math. Soc. 19, 197-231 (1968).

[24] Ja. B. Pesin. Ljapunov characteristic exponents and ergodic properties of smooth dynamical systems with an invariant measure. Dokl. Akad. Nauk SSSR 226 N°4, 774-777 (1976). English translation. Soviet Math. Dokl. 17 N°1, 196-199 (1976).

[25] Ja. B. Pesin. Invariant manifold families which correspond to the nonvanishing characteristic exponents. Izv. Akad. Nauk SSSR, Ser. Mat. 40 N°6, 1332-1379 (1976).

[26] R.V. Plykin. Sources and currents of A-diffeomorphisms of surfaces. Mat. Sbornik 94 N°2(6), 243-264 (1974).

[27] C.C. Pugh and M. Shub. The Ω-stability theorem for flows. Inventiones Math. 11, 150-158 (1970).

[28] P.J. Regal. Ecology and evolution of flowering plant dominance. Science 196, 622-629 (1977).

[29] P.W. Richards. The tropical rain forest. Cambridge University Press, Cambridge, 1966.

[30] D. Ruelle. Bifurcation in the presence of a symmetry group. Archive Rat. Mech. Anal. 51, 136-152 (1973).

[31] D. Ruelle. A measure associated with Axiom A attractors. Amer. J. Math. 98, 619-654 (1976).

[32] D. Ruelle. Statistical mechanics and dynamical systems, and papers from the 1976 Duke turbulence conference. Duke University Mathematics Series 3, Durham, 1977.

[33] D. Ruelle. Applications conservant une mesure absolument continue par rapport à dx sur [0,1] . Commun. Math. Phys. 55, 47-51 (1977).

[34] D. Ruelle. An inequality for the entropy of differentiable maps. Preprint.

[35] D. Ruelle and F. Takens. On the nature of turbulence. Commun. Math. Phys. 20, 167-192 (1971); 23, 343-344 (1971).

[36] Ja. G. Sinai. Gibbs measures in ergodic theory. Russian Math. Surveys 166, 21-69 (1972).

[37] S. Smale. Differentiable dynamical systems. Bull. Amer. Math. Soc. 73, 747-817 (1967).

[38] S. Smale and R. Williams. The qualitative analysis of a difference equation of population growth. Preprint.

[39] P. Štefan. A theorem of Šarkowskii on the existence of periodic orbits of continuous endomorphisms of the real line. Commun. Math. Phys. 54,237-248 (1977).

[40] R. Temam (editor). Turbulence and Navier-Stokes equations. Lecture Notes in Mathematics N° 565. Springer, Berlin, 1976.

[41] R. Thom. Stabilité structurelle et morphogénèse. W.A. Benjamin, Reading, Mass., 1972.

[42] T.C. Whitmore. Tropical rain forests of the far east. Clarendon Press, Oxford, 1975.

[43] R.F. Williams. The structure of Lorenz attractors. Preprint.

[44] L.S. Young. Entropy of continuous flows on compact 2-manifolds. Preprint.

GENERIC PROPERTIES OF NAVIER-STOKES EQUATIONS

R. TEMAM

Laboratoire d'Analyse Numérique - Bâtiment 425
Université de Paris-Sud
91405 - Orsay, France

1 - INTRODUCTION.

In this article we intend to review and describe some properties of the set of steady-state solutions of the Navier-Stokes equations of viscous incompressible fluids.

We first recall the equations. Let Ω be the region of the space $\mathbb{R}^3$ filled by the fluid (we assume that Ω is an open bounded set of $\mathbb{R}^3$). We denote by $u = \{u_1, u_2, u_3\}$, the velocity of the particle of fluid at point x at time t ($u = u(x,t)$, $x \in \Omega$, $t > 0$), and $p = p(x,t)$ denotes the pressure at point x at time t . Then

$$\frac{\partial u}{\partial t} - \nu\, \Delta u + (u.\nabla)u + \nabla p = f \quad \text{in} \quad \Omega \times [0,\infty[\ . \tag{1.1}$$

$$\nabla u = 0 \quad \text{in} \quad \Omega \times [0,\infty[\ ; \tag{1.2}$$

here f represents volumic forces and $\nu > 0$ is the inverse of the Reynolds number, $\nu = Re^{-1}$. This system is completed by the initial condition for the velocity

$$u(x,0) = u_0(x) \quad \text{in} \quad \Omega \ , \tag{1.3}$$

(u_0 given), and, assuming that the boundary Γ of Ω is materialized and solid, we have the boundary condition

$$u(x,t) = \phi(x,t) \ , \ \text{on} \quad \Gamma \times [0,\infty[\ ; \tag{1.4}$$

ϕ is the (given) velocity of $\Gamma(\phi(x,t)$ velocity of point x at time $t)$ and (1.4) is the non slip condition.

In part of the article, we will assume for simplicity that Γ is at rest, i.e., $\phi = 0$.

If we assume that f (and ϕ , if $\phi \neq 0$) is independent of time, we may look for the solutions of (1.1), (1.2), (1,4) which are independent of time : we get the steady state Navier-Stokes equations :

$$(1.5) \qquad -\nu\,\Delta u + (u.\nabla)u + \nabla p = f \quad \text{in} \quad \Omega\ , \ (f = f(x))\ ,$$

$$(1.6) \qquad \nabla u = 0 \quad \text{in} \quad \Omega$$

$$(1.7) \qquad u = 0 \quad \text{on} \quad \Gamma \quad (^1)$$

u, p solutions of (1.5)-(1.7) are called steady state solutions of the Navier-Stokes equations.

In the mathematical theory of Navier-Stokes equations, there is an essential simplification which is due to Leray [8], [9], [10] and which allows the decoupling of the equations (1.1)-(1.4) into two parts

- An infinite dimensional differential equations involving only u :

$$(1.8) \qquad \frac{du}{dt} = M(u,t)\ , \quad t > 0$$

$$(1.9) \qquad u\big|_{t=o} = u_o\ .$$

- When u is known, the determination of p through (1.1). This is a trivial step, not to be considered in the sequel.

Later on we will recall the principle of the reduction of (1.1)-(1.4) to (1.8)-(1.9), and we will describe (1.8) in more details ; let us say already that $u(t)$ is a point of an infinite dimensional Hilbert space H , and for each t , $v \longmapsto M(v,t)$ is a mapping say from H into itself.

If f is independent of t , then M is independent of t , $v \longmapsto M(v)$ is a vector field on H , and (1.8)-(1.9) become :

$$(1.10) \qquad \frac{du}{dt} = M(u)$$

$$(1.11) \qquad u(0) = u_o\ .$$

In this case, a steady state solution of the Navier-Stokes equation is just an equilibrium point for the differential equation (1.10), i.e. a solution of

$$(1.12) \qquad M(u) = 0\ .$$

(^{1}) Or $u(x) = \phi(x)$ on Γ if $\phi \neq 0$.

The research of steady state solutions of the Navier-Stokes equations is not a simple problem since (1.12) is an equation in an infinite dimensional space H . Our motivation for studying the set S of solutions of (1.12) is related (although in a very far way) to Turbulence : S is an invariant manifold for (1.10), and actually the simplest one. We recall that an invariant manifold $\sum$ is a subset of H such that

$$u_0 \in \sum \Longrightarrow u(t) \in \sum , \forall t ,$$

where u is the solution of (1.10)-(1.11). The invariant manifolds play an essential role in two recent approaches to Turbulence ;

- the well-known approach of Ruelle and Takens, [12],
- the approach of C. Foias [3] , based on the concept of statistical solutions of Navier-Stokes equations.

The study of (1.12) is understood as a preliminary effort in the study of invariant manifolds for Navier-Stokes. Most of the result which will be described are based on joint papers with C. Foias [5], [6], [7].

2 - THE FUNCTIONAL FORM OF THE EQUATIONS.

The basic space H is a subspace of $\mathbb{L}^2(\Omega) = L^2(\Omega)^3$, the space of square-integrable vector functions on Ω ; $\mathbb{L}^2(\Omega)$ admits a decomposition into two orthogonal spaces

$$\mathbb{L}^2(\Omega) = H \oplus G ,$$

where G is the space of gradients of scalar functions $p \in L^2(\Omega)$ such that $\nabla p \in \mathbb{L}^2(\Omega)$. Its orthogonal complement H is shown to be

(2.1) $$H = \{u \in \mathbb{L}^2(\Omega) , \nabla u = 0 , u.n = 0 \text{ on } \Gamma\} ,$$

n the unit normal on Γ , pointing outward with respect to Ω.

Let P denote the orthogonal projector in $\mathbb{L}^2(\Omega)$ onto H . We denote by A the unbounded operator whose domain $D(A) \subset H$ is the set of vector functions u in $\mathbb{L}^2(\Omega)$ together with their first and secund derivatives, and which vanish on Γ . For $u \in D(A)$, we set

(2.2) $$Au = - P \Delta u \ (\in H) .$$

It is known (cf. for instance R. Temam [16]), that A is an isomorphism from $D(A)$ onto H , that A is self-adjoint, strictly positive and A^{-1} is compact.

For $u, v \in D(A)$ we write,

(2.3)
$$\begin{cases} B(u,v) = P((u.\nabla)v) \\ B(u) = B(u,u) \end{cases}$$

which makes sense (cf. [16]) ; moreover, B is a bilinear compact operator from $DA \times D(A)$ into H , and

$$(B(u,v),v)_H = 0 \quad , \quad \forall\, u,v \in D(A) \ .$$

The form (1.8) of the time-dependent Navier-Stokes equations is just obtained by applying the operator P to both sides of (1.1). We get

(2.4)
$$\frac{du}{dt}(t) + \nu\, Au(t) + B(u(t)) = f(t) \ ,$$

i.e.

(2.5)
$$M(u,t) = -\nu\, Au - B(u) + f(t) \ .$$

We assume now that f is independent of t ; we then obtain (1.10) with $M(u) = -\nu\, Au - Bu + f$:

(2.6)
$$\frac{du}{dt} + \nu\, Au + B(u) = f$$

(2.7)
$$u(0) = u_o \ .$$

An equilibrium point of (2.6) is an u in $D(A)$ such that

(2.8)
$$\nu\, Au + B(u) = f \ .$$

3 - GENERAL PROPERTIES.

For $\nu > 0$ fixed and f fixed in H , we call $S(f,\nu)$ the set of $u \in D(A)$ which satisfies (2.8).

The classical properties of $S(f,\nu)$ are the following one (cf. for instance [16]).

(i) $S(f,\nu) \neq \emptyset$: existence theorem proved by approximation or by the Leray-Schauder degree theory

(ii) $S(f,\nu)$ is reduceed to one point if

$$(2.9) \qquad \nu^2 > c_1 |f|_H ,$$

where c_1 is a constant depending only on Ω and $|f|_H$ is the norm of f in H.

This is not a very interesting case for Turbulence, since it corresponds to a situation where the fluid is very viscous, or the forces which are applied are very small (laminarflow). When (2.9) holds, any solution $u(t)$ of (2.6) converges, for $t \longrightarrow \infty$, to $S(f,\nu)$ which is then a stable attractor.

(iii) $S(f,\nu)$ is compact in $D(A)$ (and H).

Other properties of $S(f,\nu)$ are the following ones (cf. Foias-Temam [5], [6], [7]) :

(iv) $S(f,\nu)$ is homeomorphic to a compact set of $\mathbb{R}^m$, for a sufficiently large m :

$$m \geqslant m_1(f,\nu) ;$$

m_1 depends on $|f|_H$ and ν, is an increasing function of its first argument and a decreasing function of ν. Consequently,

$$m_1(|f|_H,\nu) \leqslant m_1(R,\nu_0)$$

if

$$|f|_H < R < +\infty , \quad \nu > \nu_0 > 0 ,$$

and we can use the same m for all these f and ν.

(v) $S(f,\nu)$ is a compact real analytic set. More precisely, it is the image by an analytic mapping of a $\mathcal{C}$-real analytic set of $\mathbb{R}^m$ (same m as in (iv)).

Let us sketch the proof of (iv) : since A^{-1} is a compact self adjoint operator in H it possesses a sequence of eigenfunctions $\{w\}_{i\ i=1}^{\infty}$ which constitute an orthonormal basis of H :

$$Aw_i = \lambda_i w_i , \ i \geqslant 1 , \quad 0 < \lambda_1 \leqslant \lambda_2 \leqslant \dots , \ \lambda_i \longrightarrow +\infty \text{ as}$$

$i \longrightarrow +\infty$. Let P_m be the orthogonal projector in H onto the space spanned by $w_1,\dots,w_m$. It is shown that for m sufficiently large, P_m is one to one ; because of (iii), $P_m S(f,\nu)$ is compact and P_m is a homeomorphism between $S(f,\nu)$ and $P_m S(f,\nu)$.

The proof of (v) is given in [7].

4 - GENERIC PROPERTIES.

We now establish some generic properties of $S(f,\nu)$, i.e. properties which are generic with respect to the data, f,ν . A typical result is the following one :

Theorem 4.1. For every $\nu > 0$ fixed, there exists an open dense subset $\mathcal{O}$ of H , such that $S(f,\nu)$ is finite for every $f \in \mathcal{O}$. Furthermore card $S(f,\nu)$ is odd and constant on each connected component of $\mathcal{O}$.

For the proof of this theorem, we distinguish two parts :

a) all the properties of $S(f,\nu)$, except that card $S(f,\nu)$ is odd

b) the oddness of card $S(f,\nu)$.

The point a) is just a particular case of the following Lemma.

Lemma 4.1. Let X and Y be two Banach spaces and let N be a $\mathcal{C}^1$, (non linear) Fredholm mapping from X into Y , N proper of index 0.

Then the set R_N of regular values of N is a dense open set of Y , card $N^{-1}(y)$ is finite, $\forall\, y \in R_N$, and this number is constant on a connected component of R_N .

We recall that N Fredholm of index 0 , means that $N'(x)$ is a linear Fredholm operator of index 0 , $\forall\, x \in X$. A point $y \in Y$ is regular if $N'(x)$ is an isomorphism from X onto Y for every x satisfying $N(x) = y$. All the results contained in this Lemma are easy consequences of the implicit function theorem and the properness of N . The only non trivial result is the density of R_N , which follows from the infinite dimensional version of the Sard's theorem, due to Smale (cf. [15]).

The properties a) are obtained by applying this Lemma with

$$X = D(A) \ , \ Y = H \ , \ N(u) = \nu\, Au + B(u) \ ;$$

the reader is referred to [6], [13], [17].

The proof of the oddness of the number of points of $S(f,\nu)$ is, on the other hand, a straightforward application of the topological degree theory of Elworthy and Tromba [1] , [2] . These authors define for a nonlinear Fredholm mapping N between two Banach spaces X, Y, the topological degree of N at a regular point y . This number is actually independent of y , and is related to the set $N^{-1}(y)$. In the present case, we consider $X = D(A)$, $Y = H$, $N(u) = \nu\, Au + B(u)$, $\forall\, u \in D(A)$. Using (ii), we easily compute the degree of N at a point $y = f$ whose norm in H is a sufficiently small ((2.9) satisfied) ; this degree is one, and by the definition of the degree, $S(f,\nu)$ has an odd number of points for every regular f .

Discussion of a)

We would like to show that, in some sense, the fact that $S(f,\nu)$ is finite for all the f in a dense subset of H cannot be improved ($\mathcal{O} \neq H$). Actually in proving that the Navier-Stokes operator $N(u) = \nu Au + B(u)$ satisfies the assumption of Lemma 4.1, we only use the following properties of H, A, B :

(α) H is a Hilbert space

($\alpha\alpha$) A is an unbounded linear operator in H with domain $D(A) \subset H$; A is closed self-adjoint, strictly positive and A^{-1} is compact.

($\alpha\alpha\alpha$) B is a bilinear compact operator from $D(A) \times D(A)$ into H, and

$$(B(u,u),u)_H = 0 \,, \qquad \forall\, u \in D(A) \,.$$

Then the application of Lemma 4.1, conducts to an abstract analogue of Theorem 4.1, for a mapping N, $N(u) = \nu Au + B(u)$ from $D(A)$ into H, where H, A, B satisfy (α), ($\alpha\alpha$), ($\alpha\alpha\alpha$). Our remark is that at this level of generality (i.e. without using more properties of the Navier-Stokes equations), Theorem 4.1 cannot be improved, i.e. we cannot assert that $\mathcal{O} = H$ or $\mathcal{O} \neq H$.

This is shown by two examples which enter into the framework of this abstract version of Theorem 4.1, and for which $\mathcal{O} = H$ in one case, $\mathcal{O} \neq H$ in the other case.

Example 1.

The first example is related to the Burgers equation. In one dimension, Ω = the interval $(0,1)$, we consider the following differential equation :

$$-\nu \frac{d^2u}{dx^2} + u\frac{du}{dx} = f \quad \text{on} \quad (0,1) \tag{4.1}$$

$$u(0) = u(1) = 0 \,. \tag{4.2}$$

It is easy to write this problem in the form of a functionnal equation $N(u) = \nu Au + B(u) = f$. We set $H = L^2(\Omega)$, $A = -\frac{d^2}{dx^2}$ with domain, the set of u in $L^2(0,1)$, whose first and secund derivatives belong to $L^2(0,1)$, and such that $u(0) = u(1) = 0$. We set also $B(u,v) = u\frac{dv}{dx}$.

We skip the verification of the assumptions of Lemma 4.1, but we observe that $N'(u).v = 0$ is equivalent to

$$\begin{cases} -\nu \dfrac{d^2v}{dx^2} + u\dfrac{dv}{dx} + v\dfrac{du}{dx} = 0 \\[2ex] v(0) = v(1) = 0 \end{cases}$$

and, by elementary calculus, $v \equiv 0$, so that $N'(u)$ is an isomorphism, $\forall u \in D(A)$. Whence the set of regular values of N is the whole space $L^2(0,1)$. The number of solutions of (4.1)-(4.2) is independent of f , but this number is trivially 1 for $f = 0$ ($f = 0$ implies $u = 0$) ; $\mathcal{O} = H$ for this example.

Example 2.

This example is due to Gh. Minea [11], and is a finite dimensional one, $H = \mathbb{R}^3$.

We set A = the identity, and $B(u,v)$ is defined by

$$B(u,v) = \begin{cases} B_1 u.v \\ B_2 u.v \\ B_3 u.v \end{cases}$$

where the B_i are the following 3×3 matrices

$$B_1 = \begin{pmatrix} 0 & 0 & 0 \\ 0 & \delta & 0 \\ 0 & 0 & \delta \end{pmatrix} , \quad B_2 = \begin{pmatrix} 0 & -\delta & 0 \\ 0 & 0 & 0 \\ 0 & 0 & 0 \end{pmatrix} , \quad B_3 = \begin{pmatrix} 0 & 0 & -\delta \\ 0 & 0 & 0 \\ 0 & 0 & 0 \end{pmatrix}$$

The verification of (α)-$(\alpha\alpha\alpha)$ is trivial. Now an elementary calculation shows that there are one or three solutions for $\nu Au + B(u) = f$, if $f = \{f_1, f_2, f_3\}$ with $|f_2| + |f_3| \neq 0$. When $|f_2| + |f_3| = 0$, we find a complete circle in $S(f,\nu)$:

$$u_1 = \frac{1}{\delta} , \quad u_2^2 + u_3^2 = (f_1 - \frac{1}{\delta}) \frac{1}{\delta} \qquad (\text{if } f_1 > \frac{1}{\delta})$$

and beside that an isolated solution

$$u_2 = u_3 = 0 , \qquad u_1 = f_1 .$$

5 - OTHER PROPERTIES AND REMARKS.

i) For generic values of f , the set of bifurcation point for (2.8) is made of isolated values of ν which can only accumulate at $\nu = 0$. More precisely let

$$S_1(f) = \bigcup_{\nu > 0} S(f,\nu) .$$

This set is a real analytic set ; for generic values of f , this set is made of a real analytic manifold of dimension 1 , whose projection on the ν axis is the

whole interval $]0,+\infty[$. The set of singular points of this continuum (which includes the set of all bifurcating values of ν) is discrete and finite on every interval $]\nu_0,+\infty[$, $\nu_0 > 0$. Besides that $S(f,\nu)$ contains isolated points or manifolds whose projection on the ν axis is discrete and finite on every interval, $]\nu_0,+\infty[$, $\nu_0 > 0$.

ii) If the boundary data ϕ is different from zero, then we have also similar generic properties with respect to f, ν, ϕ . For instance for every $\nu > 0$ and $f \in \mathcal{C}^{\alpha}(\bar{\Omega})^n$ fixed, there exists a dense open set of values of ϕ , and $S(f,\nu,\phi)$ is finite for every ϕ in this set. Cf. [14].

iii) Let us consider the Taylor problem, i.e. the motion of a fluid between two infinite coaxial rotating cylindars, let us assume that one of the cylindars is at rest and let us denote by ω the angular velocity of the secund cylindar. We assume that body forces f are applied : for generic values of f and in particular for arbitrarily small forces f , the set of steady solutions is finite for almost all values of ω .

iv) Similar results are valid for time periodic solutions of Navier-Stokes equations ; cf. [17], [14] .

REFERENCES.

[1] K.D. Elworthy and A.J. Tromba - Degree theory on Banach manifolds Proc. Symp. Pure Math., Vol.18, A.M.S. (1970) p.86-94.

[2] K.D. Elworthy and A.J. Tromba - Differential structures and Fredholm maps on Banach manifolds Proc. Symp. Pure Math., Vol.15, A.M.S. (1970) p.45-94.

[3] C. Foias - Solutions statistiques des équations de Navier-Stokes. Cours au Collège de France, 1974.

[4] C. Foias - Statistical study of Navier-Stokes equations Rend. Sem. Mat. Univ. Padova, 48 (1972) p.219-348 et 49 (1973) p.9-123.

[5] C. Foias and R. Temam - On the stationary statistical solutions of the Navier-Stokes equations Publ. Math. d'Orsay, n° 120-75-28 (1975).

[6] C. Foias and R. Temam - Structure of the set of stationary solutions of the Navier-Stokes equations Comm. Pure Appl. Math., 30 (1977), p.149-164.

[7] C. Foias et R. Temam - Remarques sur les équations de Navier-Stokes stationnaires et les phénomènes successifs de bifurcation Annali Scuola Norm. Sup. Pisa Volume dédié à J. Leray, à paraître.

[8] J. Leray - Etude de diverses équations intégrales non linéaires et de quelques problèmes que pose l'hydrodynamique J. Math. Pures et Appl. 13 (1933), p.1-82.

[9] J. Leray - Essai sur les mouvements plans d'un liquide visqueux que limitent des parois
J. Math. Pures et Appl., 13 (1934) p.331-418.

[10] J. Leray - Sur le mouvement d'un liquide visqueux emplissant l'espace
Acta Math., 63 (1934), p.193-248.

[11] Gh. Minea - Remarques sur l'unicité de la solution stationnaire d'une équation de type Navier-Stokes
Revue Roumaine Math. Pures et Appl., Tome XXI n°8 (1976), p.1071-1075.

[12] D. Ruelle and F. Takens - On the nature of Turbulence
Comm. Math. Phys., 20 (1971), p.167-192, and 23 (1971), p.343-344.

[13] J.C. Saut - Exposé dans le Séminaire d'Equations aux Dérivées Partielles non linéaires
Publication mathématique d'Orsay, en cours de parution.

[14] J.C. Saut et R. Temam - Propriétés de l'ensemble des solutions stationnaires des équations de Navier-Stokes : généricité par rapport aux données aux limites
C.R. Ac. Sc. Paris, à paraître (1977).

[15] S. Smale - An infinite dimensional version of Sard's Theorem
Amer. J. Math. 87 (1965), p.861-866.

[16] R. Temam - Navier-Stokes equations, Theory and Numerical Analysis
North Holland and Elsevier, Amsterdam-New York, 1977.

[17] R. Temam - Une propriété générique de l'ensemble des solutions stationnaires ou périodiques des équations de Navier-Stokes
Actes du Colloque franco Japonais, Tokyo, Septembre 1976.

A LIMIT THEOREM FOR TURBULENT DIFFUSION

H. Kesten

Cornell University, Ithaca, New York 14853

G. C. Papanicolaou

Courant Institute of Mathematical Sciences, New York University, N. Y. 10012

Introduction

We shall present a theorem motivated by analytical and numerical results given in [1] and elsewhere. This theorem does not cover many interesting physical situations. The answer can also be obtained rather easily by formal perturbation theory. Our primary concern is in establishing precise conditions for the validity of such calculations. Complete proofs, some other theorems and examples are given in a paper [2] by the authors. References to related work are also given in [2].

Statement of the Theorem

Let (Ω,F,P) be a probability space and let $F(x,\omega): R^d\times\Omega \to R^d$ be jointly measurable relative to $F \times B(R^d)$ (B = σ-algebra of Borel sets in R^d). We assume that for P almost all ω the random field $F(x) = (F_i(x,\omega))$ is three times continuously differentiable in $x = (x_1,x_2,\ldots,x_d)$. We also assume that $F(x)$ is strictly stationary, i.e., for each $h \in R^d$ and points $y_1,y_2,\ldots,y_n$ in R^d the joint distribution of

$$F(y_1+h),\ F(y_2+h),\ \ldots,\ F(y_n+h)$$

is the same as that of

$$F(y_1),\ F(y_2),\ \ldots,\ F(y_n)\ .$$

We define a process $x(t) = x(t,\omega)$ with values in R^d, $t \geq 0$, $\omega \in \Omega$, as the solution of the differential equation

$$(1)\qquad \frac{dx(t,\omega)}{dt} = v + \varepsilon F(x(t,\omega),\omega)\ , \qquad x(0,\omega) = x_0 \in R^d\ .$$

Here $v \in R^d$, $v \neq 0$, is a fixed vector, $\varepsilon \in (0,1]$ is a parameter which we shall let tend to zero and x_0 is the (nonrandom) initial position. Equation (2.1) has a unique solution for almost all ω in view of the hypotheses on F.

It is convenient to write (1) in component form along v and perpendicular to v as follows.

$$(2)\qquad \frac{d\tilde{x}_1(t,\omega)}{dt} = |v| + \varepsilon F^{(1)}(\tilde{x}_1(t,\omega),\ \tilde{x}_2(t,\omega),\omega)\ , \qquad \tilde{x}_1(0,\omega) = x_0^{(1)} \in R\ ,$$

(3) $$\frac{d\tilde{x}_2(t,\omega)}{dt} = \varepsilon F^{(2)}(\tilde{x}_1(t,\omega),\tilde{x}_2(t,\omega),\omega) , \qquad \tilde{x}_2(0,\omega) = x_0^{(2)} \in R^d ,$$

where $x_0^{(2)}$ is perpendicular to v, $x_0 = x_0^{(2)} + x_0^{(1)} v/|v|$ and

(4) $$\tilde{x}_1(t,\omega) = \frac{(x(t,\omega),v)}{|v|} = \sum_{i=1}^{d} \frac{1}{|v|} x_i(t,\omega)v_i ,$$

(5) $$\tilde{x}_2(t,\omega) = x(t,\omega) - \tilde{x}_1(t,\omega) \frac{v}{|v|} ,$$

with the definitions

(6) $$F^{(1)}(x_1,x_2,\omega) = \frac{1}{|v|} \left(F(x_1 \frac{v}{|v|} + x_2,\omega),v\right) ,$$

(7) $$F^{(2)}(x_1,x_2,\omega) = F\left(x_1 \frac{v}{|v|} + x_2,\omega\right) - F^{(1)}(x_1,x_2,\omega) \frac{v}{|v|} ,$$

Note that $(\tilde{x}_1(t,\omega), \tilde{x}_2(t,\omega))$ is a process with values in R^{d+1} but $\tilde{x}_2(t,\omega)$ is always perpendicular to v and we have

(8) $$x(t,\omega) = \tilde{x}_1(t,\omega) \frac{v}{|v|} + \tilde{x}_2(t,\omega) .$$

We are interested in the asymptotic behavior of $x(t,\omega)$ as $\varepsilon \to 0$ and $t \to \infty$ with $\varepsilon^2 t$ = constant i.e. in the so-called weak coupliminglimit or diffusion limit, etc., in which the usual second order perturbation theory becomes an exact lmiit. For this purpose we need additional hypotheses about the random field $F(x,\omega)$ as follows.

If $\beta = (\beta_1,\dots,\beta_d)$, $\beta_i \geq 0$ integers, is a multiindex and $|\beta| = \beta_1+\dots+\beta_d$, we denote by D^β partial derivatives as usual. We shall assume that for each $0 \leq M < \infty$ and each β with $0 \leq \beta \leq 3$, there is a large integer $p < \infty$ (the precise value of p is not important) such that

(9) $$\int_\Omega \left(\sup_{|x|\leq M} |D^\beta F(x,\omega)|\right)^p P(d\omega) = E\left\{\left(\sup_{|x|\leq M} |D^\beta F(x)|\right)^p\right\} < \infty , \quad 0 \leq |\beta| \leq 3 .$$

Since F is stationary, if (2.9) holds for some M, say M = 1, then it holds for all other $M < \infty$.

We assume that F has mean zero

(10) $$\int_\Omega F(x,\omega) \, P(d\omega) = E\{F(x)\} = 0 .$$

Let $\Lambda \in B(R^d)$ and let F_Λ denote the minimal σ-algebra of subsets of Ω in F generated by sets of the form

$$\left\{\omega \in \Omega \mid F(x,\omega) \in A, \; A \in B(R^d), \; x \in \Lambda\right\} .$$

With $v \in R^d$, $v \neq 0$ fixed and $0 \leq M \leq 0$ we shall use the notation

$$F_s^t(v,M) = F_\Lambda \;, \quad -\infty \leq s \leq t \leq +\infty \;, \tag{11}$$

where

$$\Lambda = \left\{ x \in R^d \mid s \leq \frac{(x,v)}{|v|} \leq t, \; \left| x - \frac{(x,v)v}{|v|^2} \right| \leq M \right\} . \tag{12}$$

When M and v are fixed we write F_s^t for short.

We shall assume that P is strongly mixing on the σ-algebras F_s^t as follows. With $v \neq 0$ and M fixed define

$$\alpha(t,v,M) = \sup_{\substack{A \in F_{t+s}^{\infty} \\ B \in F_{-\infty}^{s}}} |P(AB) - P(A)P(B)| \;, \qquad t \geq 0 \;. \tag{13}$$

If $\alpha(t) = \alpha(t,v,M) \to 0$ as $t \to \infty$, we say that P is strongly mixing. For the result that follows we need to assume in addition a rate of mixing, that is, for some large integer $p < \infty$ (not necessarily the same p as in (9)) we have

$$\int_0^\infty [\alpha(t,v,M)]^{1/p} \, dt < \infty \;. \tag{14}$$

Let us now return to (2) and (3) and the process $(\tilde{x}_1(t,\omega), \tilde{x}_2(t,\omega))$. To state our result define

$$x_1^\varepsilon(t,\omega) = \tilde{x}_1\left(\frac{t}{\varepsilon^2},\omega\right) - \frac{|v|t}{\varepsilon^2} \;, \tag{15}$$

$$x_2^\varepsilon(t,\omega) = \tilde{x}_2\left(\frac{t}{\varepsilon^2},\omega\right) \;, \tag{16}$$

$$x^\varepsilon(t,\omega) = x\left(\frac{t}{\varepsilon^2},\omega\right) - \frac{vt}{\varepsilon^2} \;, \tag{17}$$

so that, from (8), we have

$$x^\varepsilon(t,\omega) = x_1^\varepsilon(t,\omega)\frac{v}{|v|} + x_2^\varepsilon(t,\omega) \;. \tag{18}$$

Let Q^ε denote the probability measure induced by $(x_1^\varepsilon(t), x_2^\varepsilon(t))$ on $C([0,\infty), R^{d+1})$.

<u>Theorem</u>. <u>If</u> $v \neq 0$ <u>and</u> (9), (10) <u>and</u> (14) <u>hold for a certain large</u> $p < \infty$, then Q^ε <u>converges weakly to the measure</u> Q <u>corresponding to the diffusion</u> $(x_1(t), x_2(t))$ <u>in</u> R^{d+1} <u>with the infinitessimal generator</u>

(19) $$L = \frac{1}{2} a^{(1))} \frac{\partial^2}{\partial x_1^2} + \sum_{j=1}^{d} a_j^{(12)} \frac{\partial^2}{\partial x_1 \partial x_{2j}} + \frac{1}{2} \sum_{i,j=1}^{d} a_{ij}^{(22)} \frac{\partial^2}{\partial x_{2i} \partial x_{2j}}$$

$$+ b^{(1)} \frac{\partial}{\partial x_1} + \sum_{j=1}^{d} b_j^{(2)} \frac{\partial}{\partial x_{2j}} ,$$

(20) $$Q\left(x_1(0) = x_0^{(1)} , x_2(0) = x_0^{(2)}\right) = 1 .$$

Here

(21) $$a^{(11)} = \frac{1}{|v|} \int_{-\infty}^{\infty} E\left\{F^{(1)}(0,x_2) \; F^{(1)}(t,x_2)\right\} dt ,$$

(23) $$a_j^{(12)} = \frac{1}{|v|} \int_{-\infty}^{\infty} E\left\{F^{(1)}(0,x_2) \; F_j^{(2)}(t,x_2)\right\} dt ,$$

(23) $$a_{ij}^{(22)} = \frac{1}{|v|} \int_{-\infty}^{\infty} E\left\{F_i^{(2)}(0,x_2) \; F_j^{(2)}(t,x_2)\right\} dt ,$$

(24) $$b^{(1)} = \frac{1}{|v|} \sum_{j=1}^{d} \int_{0}^{\infty} E\left\{F_j^{(2)}(0,x_2) \frac{\partial F^{(1)}(t,x_2)}{\partial \partial x_{2j}}\right\} dt$$

$$- \frac{1}{|v|} E\left\{(F^{(1)}(0,x_2))^2\right\} ,$$

(25) $$b_j^{(2)} = \frac{1}{|v|} \sum_{i=1}^{d} \int_{0}^{\infty} E\left\{F_i^{(2)}(0,x_2) \frac{\partial F_j^{(2)}(t,x_2)}{\partial x_{2i}}\right\} dt .$$

Examples of random fields satisfying all the above conditions are given in [2].

References

[1] R. H. Kraichnan, Diffusion by a Random Velocity Field, The Physics of Fluids 13 (1970) pp. 22-31.

[2] H. Kesten and G. C. Papanicolaou, Limit Theorems for Turbulent Diffusion, to appear.

MANY PARTICLE SCATTERING AMPLITUDES

M.C.Polivanov

Steclov Mathematical Institute
Academy of Science of the USSR
Moscow, USSR.

1. Introduction.

I would report here some results obtained during the last year and concerning the subject of many particle scattering amplitudes. These are obtained by a group including beside this author A.A.Logunov, B.V.Medvedev, M.A.Mestvirishvili, L.A.Muzafarov, V.P.Pavlov and A.D.Sukhanov. Part of these results is published in a series of papers [1,2,3,4].

In the recent time many-particle scattering become a subject of intencive study and different semi-phenomenological models were presented such as multiperipherical, multireggeon, parton, statistical etc.

But the real understanding and estimation of these models is hardened due to the lacking of basic information on the analytic structure of the many-particle amplitudes. Thus e.g. the crossing-symmetry for these amplitudes which is always assumed, basing on two-particle intuition and was never proved here to exist even in the simplest situations. The same can be said about the so-called "generalized optical theorem" conjectured in [5] and relating the distribution function of inclusive process

$$a+b \rightarrow c + \text{all other} \qquad \text{(I)}$$

to the absorbtive part of the 3-particle "forward" scattering

$$a+b+c \rightarrow a'+b'+c' \qquad \text{(II)}$$

Meanwhile it is not at all clear without detailed study of analytic properties even how to define the absorbtive part of the amplitude (II).

Our investigations were aimed in the first place to clear up just these two points: crossing-symmetry and generalized optical theorem.

2. General Framework.

Our general setting is the usual Bogolubov system of axioms for the causal S-matrix [6,7]. The central object here is the S-matrix considered as a functional of the asymptotic fields:

$$S[\varphi] = \sum_{n=1}^{\infty} \frac{1}{n!} \int dx_1 \dots dx_n \, \mathcal{S}_n(x_1 \dots x_n) : \varphi(x_1) \dots \varphi(x_n) : \qquad (1)$$

The S-matrix obeys the usual conditions of unitarity, Poincare-invariance, spectrality and the causality condition

$$\frac{\delta}{\delta\varphi(y)} \left\{ \frac{\delta S}{\delta\varphi(x)} S^+ \right\} = 0 \quad \text{if} \quad x \gtrsim y . \qquad (2)$$

The sign in the causality condition shows that the asymptotic fields are chosen to be out-fields.

The main technical trick, which allows to reduce arbitrary matrix elements of the S-matrix to the vacuum expectation values and then to use causality in order to find support restrictions, is the reduction formula of the form

$$[a^{\pm}(p), S] = \mp (2\pi)^{-3/2} \int dx \, e^{\mp ipx} \frac{\delta S}{\delta\varphi(x)} \qquad (3)$$

$$\equiv \mp \Gamma_p^{\mp} \frac{\delta S}{\delta\varphi(x)} ,$$

where $a^{\pm}(p)$ are creation and annihilation operators of the asymptotic out-states. We have introduced here a shorthand $\Gamma_p^{\mp}$ for respective integral, which is widely used below. In order to get reasonable Heisenberg operators like current

$$J(x) = i \frac{\delta S}{\delta\varphi(x)} S^+, \qquad (4)$$

T-products of currents

$$\mathcal{J}_n(x_1, \dots, x_n) = \frac{\delta^n S}{\delta\varphi(x_1) \dots \delta\varphi(x_n)} S^+ \qquad (5)$$

or retarded products

$$R(x;x_1,\ldots,x_n) = \frac{\delta J(x)}{\delta\varphi(x_1)\ldots\delta\varphi(x_n)} \qquad (6)$$

we have at a suitable moment use the stability condition: $S|\alpha\rangle = |\alpha\rangle$, allowing to insert S^+ . But the last holds only if $|\alpha\rangle$ is a vacuum or one-particle state. This restricts the applicability of the usual reduction technique if we consider $\langle m|S|n\rangle$ with $m,n>2$.

3. Modified Reduction Technique.

To this end we introduce the following modification of reduction technique. Let us consider both in- and out-bases. Then, evidently,

$$\langle m|S|n\rangle = \langle m|\tilde{n}\rangle \quad , \text{ where } \langle m| \text{ - is out-state}, \quad \langle\tilde{n}| \text{ - is in-state} \qquad (7)$$

and

$$\tilde{\varphi}(x) = S\varphi(x)S^+ = \varphi(x) + \int d\zeta\, D(x-\zeta)\, J(\zeta), \qquad (8)$$

or, for creation-annihilation operators,

$$\tilde{a}^\pm(p) = S a_p^\pm S^+ = a_p^\pm - i\,\Gamma_p^\mp J(x). \qquad (9)$$

Now, if we continue to consider the S-matrix and all other Heisenberg operators, such as $J(x)$, $\mathcal{S}_n(x_1,\ldots,x_n)$, $R(x;x_1,\ldots,x_n)$, as functionals of the out-fields and introduce two kinds of variational derivatives with respect to the out-field and to the in-field denoted by

$$\delta_k^+ \equiv \frac{\delta}{\delta\varphi(x_k)} \quad , \quad \delta_k^- \equiv \frac{\delta}{\delta\tilde{\varphi}(x_k)} \qquad (10)$$

then we introduce in addition to the reduction formula (3) another one:

$$[\,\tilde{a}^\pm(p), S\,] = \mp\,\Gamma_p^\mp \frac{\delta S}{\delta\tilde{\varphi}(x)} \quad . \qquad (11)$$

For the general Heisenberg operator:

$$[a^-(p_k), F] = \Gamma_k^+ \delta_k^+ F, \tag{12$^+$}$$

$$[F, \tilde{a}^+(p_k)] = \Gamma_k^- \delta_k^- F. \tag{12$^-$}$$

Evidently δ_k^- can be expressed in terms of δ_k^+ :

$$\delta_k^- F = \delta_k^+ F - i[F, J_k]. \tag{13}$$

Applying the last formula to the current operator $J_\ell \equiv J(x_\ell)$ we have

$$\delta_k^- J_\ell = \delta_k^+ J_\ell - i[J_\ell, J_k] = \delta_\ell^+ J_k . \tag{14}$$

The last equality here is a result of well-known relation

$$\delta_k^+ J_\ell - \delta_\ell^+ J_k = i[J_\ell, J_k]. \tag{15}$$

From Eq.(14) we see that causality condition (4) has now two forms

$$\delta_k^+ J_\ell = 0 \quad , \text{ if } \quad x_k \lesssim x_\ell \tag{4$^+$}$$

$$\delta_k^- J_\ell = 0 \quad , \text{ if } \quad x_k \gtrsim x_\ell \tag{4$^-$}$$

The reduction of the matrix element $\langle m | \tilde{n} \rangle$ based on the formulas (12) leads to the operator of the form

$$\delta_{i_1}^{\#} \ldots \delta_{i_{m+n-1}}^{\#} J_{i_{m+n}} \quad , \quad \# = + \text{ or } - , \tag{16}$$

which is due to $(4^{\pm})$ a generalized retarded operator.

The algebra of operators δ_i^+ , δ_j^- together with commutativity of $\delta_i^{\#}$ of the same sign and Eq.(16), which may be written as a commutator

$$[\delta_k^+, \delta_\ell^-] F = -i[F, \delta_k^+ J_\ell] \tag{17}$$

is a realization of the algebra of Steinmann arrows. Vacuum expectation values of the operators (16) are Steinmann monomials. All the results of the theory of generalized retarded functions such as Steinmann identities and Ruelle commutators [9] may be reproduced inside this realization. In our experience the existence of the simple and transparent realization in terms of δ^+ , δ^- - symbols proves to be extremely helpful in the case of higher order ($n \geq 6$) functions, when combinatorics becomes really hard and the general algebraic relations are not always enough to guess the result.

Note that the hints to this realization in fact go back to an old paper of Logunov [8] and to the textbook of Bogolubov, Logunov and Todorov [7].

4. Three-particle scattering. Kinematics.

The theory of generalized retarded functions with its rich algebraic and analytic structure is a fascinating realm, but in general it is too complicated, if we like to obtain some immediate answers to the question posed in the introduction.

Thus we switched to the detailed investigation of the simplest non-trivial many-particle amplitude - the process of $3 \to 3$ " forward" scattering. Moreover just the knowledge of the analytic properties of this amplitude shed a light on the inclusive process of the type (I).

We begin with description of kinematics of the process (II). We have first six momenta p_j , p_j' ($j = a, b, c$), which become pairwise equal for the forward scattering. After using all additional conditions, like $p_i^2 = m_i^2$, $\Sigma p_j = \Sigma p_j'$ etc., we leave for the forward case with three independent kinematical variables. We choose them in the following way, which is best suited also for the inclusive process (I). Let us fix a laboratory system of the particle a : $p_a = (m, \underline{0})$ and introduce the two 4 -vectors:

$$\Delta = \frac{1}{2}(p_c - \alpha p_b) \qquad (18)$$

$$Q = \frac{1}{2}(p_c + \alpha p_b) - \eta \Delta$$

where

$$\alpha = \frac{p_c^0}{p_b^0}, \qquad \eta = \frac{m_c^2 - \alpha^2 m_b^2}{-4 m_\Delta^2}, \qquad m_\Delta^2 = \underline{\Delta}^2 \qquad (19)$$

These vectors are chosen so that

$$\Delta^{\circ} = 0, \quad Q\Delta = -\underline{Q}\underline{\Delta} = 0. \tag{20}$$

The following three scalars are independent variables:

$$\omega = Q^{\circ}, \qquad \alpha, \quad m_{\Delta}^{2} = -\Delta^{2} = \underline{\Delta}^{2}. \tag{21}$$

In this variables

$$\begin{aligned} p_a &= (m, \underline{Q}) \\ p_b &= \frac{1}{\alpha}(\omega, |\underline{Q}|\underline{e}_1 - (1-\eta)m_{\Delta}\underline{e}_2) \\ p_c &= (\omega, |\underline{Q}|\underline{e}_1 + (1+\eta)m_{\Delta}\underline{e}_2) \end{aligned} \tag{22}$$

where

$$|\underline{Q}| = \sqrt{\omega^2 - Q^2}, \quad \underline{e}_1 = \underline{Q}/|\underline{Q}|, \quad e_2 = \underline{\Delta}/|\underline{\Delta}|, \quad \underline{e}_1\underline{e}_2 = 0.$$

It is easy to see that

$$Q^2 = m_c^2 + m_{\Delta}^2(1+\eta)^2$$

is independent of ω. The variable ω is proportional to the missing mass variable $(p_a + p_b - p_c)^2$ of the process (I). This is a variable of analytic continuation of the amplitude. The other variables would be fixed inside physically admissible region. This region for ω and α is:

$$\begin{aligned} &\omega \geq \sqrt{m_c^2 + m_{\Delta}^2(1+\eta)^2} \\ &0 < \alpha < \infty \end{aligned} \tag{23}$$

The admissible region of α for the process (I) is

$$0 < \alpha < 1.$$

5. Reduction of the amplitude.

In the first step of the reduction we split out one of the creation in-operators and use the Eq.(9) in order to write

$$\langle p_a' p_b' p_c' | \widetilde{p_a p_b p_c} \rangle = \langle p_a' p_b' p_c' | a_b^+ | \widetilde{p_a p_c} \rangle - i\Gamma_b^- \langle p_a' p_b' p_c' | J_b | \widetilde{p_a p_c} \rangle \quad (24)$$

The first term contributes to the disconnected part, and this contributions would be further omitted without mentioning. Further commuting J_b with a_j^- from the left and $\tilde{a}_j^+$ from the right using Eqs (12) we come to the following result:

$$\langle p_a' p_b' p_c' | \widetilde{p_a p_b p_c} \rangle =$$

$$= i\Gamma_{b'} \langle p_{a'} p_{c'} | J_b a_{b'}^- | \widetilde{p_a p_c} \rangle - i\Gamma_b^- \Gamma_{b'}^+ \langle p_a' p_c' | \tilde{a}_c^+ \delta_{b'}^+ J_b | p_a \rangle \quad (25)$$

$$- i\Gamma_b^- \Gamma_{b'}^+ \Gamma_c^- \Gamma_{c'}^+ \langle p_a' | \delta_{c'}^+ \delta_c^- \delta_{b'}^+ J_b | p_a \rangle .$$

The first two terms in the r.h.s. of (25) have nasty analytic properties and no straightforward interpretation. But we may prove that these terms do not contribute to the physical amplitude if we fix kinematic variables inside some admissible region. In this case the whole amplitude (connected) is equal to the last term and we come to the following representation of the scattering amplitude

$$\langle a'b'c' | S | abc \rangle'_{conn} = \frac{i\,\delta(p_a + p_b + p_c - p_a' - p_b' - p_c')}{(2\pi)^2 \sqrt{\prod_1^6 2p_i^0}} \mathcal{R}(p_a, p_b, p_c, p_a', p_b', p_c') \quad (26)$$

(a dash $\langle \cdots \rangle'$ is a sign of the special kinematic region), where for the "forward" scattering

$$\mathcal{R}(p_a, \cdots) = -2p_a^0 \int \exp\{i(p_c(y_c' - y_c) + ip_b y_b')\} \cdot$$

$$\cdot \langle a' | r(y_c, 0, y_{c'}, y_{b'}) | a \rangle \, dy_c \, dy_{c'} \, dy_{b'} \quad (27)$$

Here $r(\ldots)$ is the following generalized retarded operator:

$$r(y_c, y_b, y_{c'}, y_{b'}) = \delta_{c'}^+ \delta_c^- \delta_b^- J_{b'} \quad (28)$$

Following a different way of reduction we come to the expression (26) with $\mathcal{R}$ changed to ("forward" case):

$$\mathcal{A}(p_a,p_b,p_c,p_a',p_b',p_c') = -2p_a^0 \int \exp\{ip_c(y_c'-y_c)+ip_b y_b'\} \cdot \langle a'|\mathfrak{A}(y_c,0,y_{c'},y_{b'})|a\rangle \tag{29}$$

with another generalized retarded operator

$$\mathfrak{A}(y_c,y_b,y_{c'},y_{b'}) = \delta_c^+\,\delta_{c'}^-\,\delta_{b'}^-\,J_b\,. \tag{30}$$

Generalized retarded functions $\mathcal{R}$ and $\mathcal{A}$ play the role of retarded and advanced amplitudes in the usual case of $2\to 2$ amplitude. The dependence on the ω is concentrated in the exponentials in Eqs (27) and (29); each of them in our kinematics have the form

$$ip_c(y_c'-y_c)+ip_b y_b' = i(\omega z^0-|\underline{Q}|(\underline{e}_1\underline{z})) - im_0\left[(1+\eta)(\underline{e}_1\cdot(\underline{y}_c'-\underline{y}_c)-\tfrac{1}{\alpha}(1-\eta)(\underline{e}_2\,\underline{y}_{b'})\right], \tag{31}$$

where

$$z = y_c'-y_c+\frac{1}{\alpha}y_b'\,. \tag{32}$$

The study of support properties of $\tau(\dots)$ and $\mathfrak{A}(\dots)$ shows that $\mathcal{R}$ can be continued analytically to the upper half-plane in ω and $\mathfrak{A}$ – to the lower half-plane.

6. Absorbtive part and the Structure of Analiticity Domain.

Constructing a difference of $\mathcal{R}$ and $\mathcal{A}$ we define an absorbtive part of the amplitude:

$$A(p_a,\dots,p_c') = (2i)^{-1}\{\mathcal{R}(p_a,\dots,p_c')-\mathcal{A}(p_a,\dots,p_c')\}$$

Absorbtive part can be expressed in terms of the commutators of currents and variational derivatives thereof. Using the Eq.(13) in order to change all δ^+-variations in the generalized retarded operators into δ^--variations we come to the following expression:

$$A(p_a,\dots,p_c') = -p_a^0 \int dy_b' \, dy_c \, dy_c' \exp\{ip_c(y_c'-y_c)+ip_b y_b'\} \cdot \sum_{\nu=1}^{6} \langle a' | [O_1^\nu, O_2^\nu] | a \rangle$$

$$= p_a^0 \sum_{\nu=1}^{6} A_\nu(p_a,\dots,p_c) .$$

Explicit form of the operators O_i^ν and the graphical representation of different terms see on the Table 1. Thus we see that tvelve different chanuels contribute to the absorbtive part of our amplitude. The contribution of the inclusive processes giving rise to the generalized optical theorem is contained in terms A_1. The other terms also admit direct physical interpretation in terms of scattering amplitudes or form-factors (black blobs on the Table I).

The use of spectrality shows that for the "forward" three-particle scattering under some conditions on the masses of particles a, b, c and on the free kinematic variables α and m_Δ^2 there is a gap in the real axis of ω, which allows an introduction of a single analytic function with cuts beginning at physical thresholds and poles corresponding to one-particle intermediate states.

The following results are obtained. For the very special case, where the masses at two particles are equal to zero (e.g. $\gamma+\gamma+N \to \gamma+\gamma+N$) we can use the rather trivial tricks in order to avoid so called "unphysical region" of complex momenta of particles and to write down a real dispersion relations for this amplitude. In this case we see at once that the conjectures of crossing-symmetry and generalized optical theorem are verified.[3]

In the case, where only one particle has zero mass we come to the situation of "singularity island" near the origin, resembling classical results of Bros, Epstein and Glaser [4].

In the general case of non-zero masses of all the three particles we see only that in the neighborhood of the two cuts there exist finite regions of analiticity which can be connected only by a path lying in the complex-masses space. In the physical situation of real masses this two neighborhoods remain disjoint.

References.

1. V.P.Pavlov, M.C.Polivanov, A.D.Sukhanov, Proceedings of the X International School on High-Energy Physics, Dubna, 1976.
2. B.V.Medvedev, V.P.Pavlov, M.C.Polivanov, A.D.Sukhanov, Proceedings of the International Seminar on Deep Inelastic Processes on High Energies, Serpukhov, 1977, to be published.
3. A.A.Logunov, M.A.Mestverishvili, B.V.Medvedev, V.P.Pavlov, M.C. Polivanov, A.D.Sukhanov, Theor. Math. Phys., to be published.
4. L.A.Musafarov, V.P.Pavlov, Theor. Math. Phys., to be published.
5. H.Mueller, Phys. Rev., D2, 2963 (1970).
6. N.N.Bogolubov, B.V.Medvedev, M.C.Polivanov, Voprosy teorii dispersionnykh sootnoshenii, Moscow, 1958.
7. N.N.Bogolubov, A.A.Logunov, I.T.Todorov, Osnovy axiomaticheskogo podkhoda v kvantovoi teorii polia, Moscow, 1969.
8. A.A.Logunov, Voprosy teorii dispersionnykh sootnoshenii dlia neuprugikh processov, Dubna, 1959.
9. H.Epstein, in: Axiomatic field Theory, Boulder School, 1965.

TABLE 1.

ν	O_1^ν	O_2^ν	A_ν
1	$\delta^-_{c'} J_b$	$\delta^-_c J_{b'}$	
2	$\delta^-_c J_b$	$\delta^-_{c'} J_{b'}$	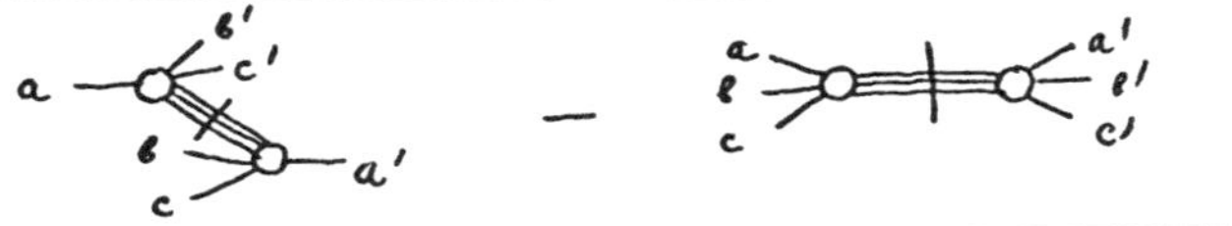
3	J_b	$\delta^-_c \delta^-_{c'} J_{b'}$	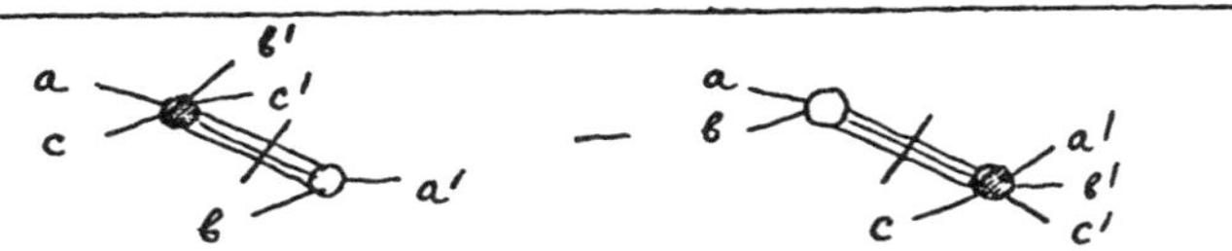
4	$\delta^-_c \delta^-_{c'} J_b$	$J_{b'}$	
5	$\delta^-_b \delta^-_c J_{b'}$	$J_{c'}$	
6	J_c	$\delta^-_{b'} \delta^-_{c'} J_b$	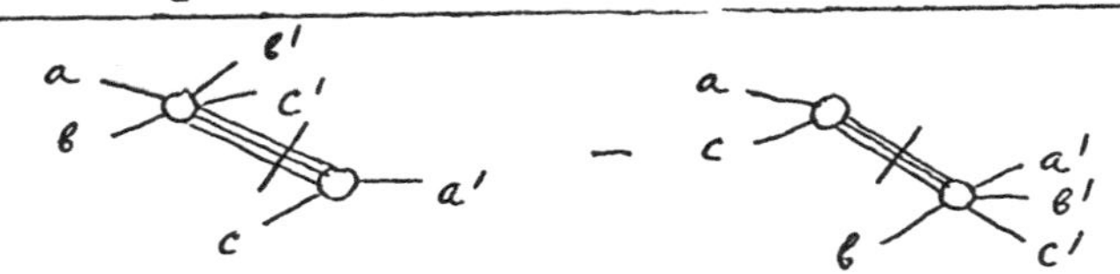

A REMARK ON EQUATIONS OF MOTION IN ASYMPTOTICALLY FREE THEORIES

I.V.Tutin, O.I.Zavialov
Steklov Mathematical Institute, Moscow

1.We'll argue here that in typical asymptotically free theories the system of equations for the renormalized Green functions (in short, equations of motion) does not define the solution uniquely.

This remark might be important in the following context. The asymptotic freedom means in particular that the exact values of some renormalization constants are zero. This leads to additional symmetry of the equations of motion which is lacking in perturbation theory The example is given by the $\frac{\lambda}{4!}\varphi^4$ - theory with $\lambda > 0$. The equation of motion here takes the form

$$(\Box + m^2)\varphi(x) = \lim_{\xi\to 0}\left[\frac{\lambda}{6} Z(\xi^2)\varphi(x+\xi)\varphi(x)\varphi(x-\xi) - \Delta(\xi^2)\varphi(x) + \sigma(\xi^2)(\xi\partial)^2\varphi(x) + q(\xi^2)\Box\varphi(x)\right] \quad (1)$$

where $Z(\xi^2), \sigma(\xi^2), q(\xi^2), \Delta(\xi^2)$ are c-number functions closely related to the renormalization constants. Due to the asymptotic freedom one finds:

$$\lim_{\xi\to 0} Z(\xi^2) = 0; \quad \lim_{\xi\to 0} Z(\xi^2)\varphi(x+\xi)\varphi(x) = \lim_{\xi\to 0} Z(\xi^2)\varphi(x+\xi)\varphi(x-\xi) = 0$$

(the last relation is valid up to vacuum diagrams). Indeed the renormalization-group methods [1] show that when $\xi\to 0$

$$Z(\xi^2) \sim (\ln\xi^2)^{-1}; \quad \varphi(x+\xi)\varphi(x) \sim \varphi(x+\xi)\varphi(x-\xi) \sim (\ln\xi^2)^{1/3} \quad (2)$$

Let φ be the solution of (1). Then the field $\varphi_\tau(x) = \varphi(x) + \tau$ satisfies "almost the same "equation since the bilinear terms like $\lambda/6\, Z(\xi^2)\,\tau\,\varphi_\tau(x+\xi)\,\varphi_\tau(x-\xi)$ arising due to field-translation vanish in the limit $\xi = 0$. So,the "translated" equation differs from the initial one only by some constant in the r.-h.-side. This property was first noticed by Brandt [2] and was called "partial τ-invariance". Suppose some momenta cut-off Λ is introduced into the theory. If the field $\varphi(x)$ corresponds to the renormalized Lagrangian $\mathcal{L}_\Lambda$ then the field $\varphi_c(x)$ corresponding to the Lagrangian $\mathcal{L}_c = \mathcal{L}_\Lambda + C_\Lambda\varphi$ will satisfy the "translated" equation in the limit $\Lambda\to\infty$ (with the appropriate choice of C_Λ). Identifying the fields φ_τ and φ_c ,made in [2] , has lead to the conclusion that for $n>2$ the proper parts of n-points Green functions vanish at zero momenta and that the effective potential of the theory is positive.

Our remark shows however that this important conclusion is not plausible. In the limit $\Lambda=\infty$ the number of solutions of the equation of motion increases simultaneously with arising of the additional symmetry. The Lagrangian generating the field φ_{τ} (contrary to the field φ_c) contains besides the vertices from $\mathcal{L}_c$ also the vertices of the type $\mathcal{L}'=Z_\Lambda(0)\varphi^3$. No doubt, these vertices give a vanishing contribution $3Z_\Lambda(0)\varphi^2$ to the equation of motion (when $\Lambda\to\infty$). Nevertheless the product φ^3 is more singular than φ^2 and so the contribution of the vertice $Z_\Lambda(0)\varphi^3$ to the exact Green functions might well remain finite.

2.Let us illustrate these heuristic arguments by a simple non-relativistic model with a Lagrangian

$$\mathcal{L}=\mathcal{L}_0+\mathcal{L}_1=[:i\varphi^+\dot{\varphi}-\varphi^+\omega\varphi:]+[\tfrac{\lambda}{2}:(\varphi^+\varphi)^2]. \tag{3}$$

Here $\varphi(x)$ is the field in $(n+1)$ - space-time dimensions, $\omega(\bar{p})=$ $=\alpha(\bar{p}^2)^{n/2}+\varkappa$ with constants α and $\varkappa$ being positive. The regularized free propagator has the form

$$\langle T\varphi_0(x)\varphi_0^+(y)\rangle=\frac{i}{(2\pi)^{n+1}}\int dp\, e^{-ip(x-y)}\frac{e^{-\frac{\bar{p}^2}{\Lambda^2}}}{p_0-\omega(\bar{p})+i\epsilon}\equiv\mathcal{D}_\Lambda(x-y) \tag{4}$$

where $px=p_0x_0+\bar{p}\bar{x}$ and $\langle T\varphi_0\varphi_c\rangle=\langle T\varphi_c^+\varphi_c^+\rangle=0.$
The function $\mathcal{D}_\Lambda(x)$ satisfies the equation

$$\Box_\Lambda\mathcal{D}_\Lambda(x)=\delta(x),\quad \Box_\Lambda\equiv[\partial_0+i\omega(i\vec{\partial})]\,e^{-\frac{\vec{\partial}^2}{\Lambda^2}} \tag{5}$$

and posesses the property $\mathcal{D}_\Lambda(x)=0$ if $x_0<0$. This property simplifies the set of non-vanishing diagrams and allows to calculate 2-point and 4-point Green functions exactly (for example 2-point functions coincide with the free ones).Equation of motion for the field, corresponding to the interaction $\mathcal{L}_1$ is

$$\Box_\Lambda\varphi(x)=\lim_{\xi\to0}i\lambda Z_\Lambda(\xi)\,\varphi^+(x+\xi)\,\varphi(x)\,\varphi(x-\xi) \tag{6}$$

where

$$Z_\Lambda(\xi)=\left[1+i\lambda\int dx\,\mathcal{D}_\Lambda(x)\,\mathcal{D}_\Lambda(x-\xi)\right]^{-1} \tag{7}$$

It is easy to calculate that

$$Z_\Lambda(0)=\left[1+\frac{\lambda S^{(n-1)}}{2(2\pi)^n\alpha}\ln\frac{\Lambda}{\varkappa}+Q(\Lambda)\right]^{-1} \tag{8}$$

where Q is a bounded function and $S^{(n-1)}$ is the measure of $(n-1)$-dimensional unit sphere. One finds also that when $\xi \to 0$

$$Z_\Lambda(\xi) \sim (\ln \xi^2)^{-1}. \tag{9}$$

So, the situation is similar to that of $\lambda\varphi^4$ -theory. Let us now change a bit the initial interaction, adding to it bilinear terms. Choose

$$\mathcal{L}_2 = \mathcal{L}_1 + \mathcal{L}' = :\frac{\lambda}{2}(\varphi^+\varphi)^2: + \frac{1}{2}\mu Z_\Lambda(0)(:\varphi^+\varphi^+ + \varphi\varphi:). \tag{10}$$

Equation of motion now takes the form

$$\Box_\Lambda \varphi(x) = \lim_{\xi\to 0} i\lambda Z_\Lambda(\xi)\varphi^+(x+\xi)\varphi(x)\varphi(x-\xi) + i\mu Z_\Lambda(0)\varphi^+(x) \tag{11}$$

and differs from (6) only by the last term. But due to (8) $Z_\Lambda(0) \to 0$ when $\Lambda \to \infty$ and this term drops out of (11). It means that the limiting Green functions (and the limit can be easily seen to exist) $G^{(1)}$ and $G^{(2)}$ which correspond to interactions $\mathcal{L}_1$ and $\mathcal{L}_2$ respectively, satisfy the same set of renormalized equations. Direct calculations in first two orders in μ and exactly in λ show however that $G^{(1)}$ and $G^{(2)}$ are different. In particular for $G_{11}(x|y) \equiv$ $\equiv \langle T\varphi(x)\varphi^+(y)\rangle$, $G_{20}(x_1,x_2) \equiv \langle T\varphi(x_1)\varphi(x_2)\rangle$ and $G_{02}(y_1,y_2) \equiv \langle T\varphi^+(y_1)\varphi^+(y_2)\rangle$ one finds

$$\begin{aligned} G_{11}^{(2)}(x|y) &= D_\infty(x-y) = G_{11}^{(1)}(x|y), \\ G_{20}^{(2)}(x_1,x_2) &= i\mu\int dx\, D_\infty(x_1-x)D_\infty(x_2-x) \neq G_{20}^{(1)} = 0, \\ G_{02}^{(2)}(y_1,y_2) &= i\mu\int dy\, D_\infty(y-y_1)D_\infty(y-y_2) \neq G_{02}^{(1)} = 0. \end{aligned} \tag{12}$$

3. The mechanism leading to the discussed degeneracy is simple - the vertices in Lagrangian are more singular than the corresponding terms in equations of motion. Indeed when $\Lambda \to \infty$ the vertices φ^2 and $(\varphi^+)^2$ from the Lagrangian $\mathcal{L}_2$ give rise to singularities of the type $(\ln \Lambda)$. These singularities are compensated by the zero of the factor $Z_\Lambda(0)$. So their contribution to Green functions proves to be finite though it doesn't change the equation of motion. The mechanism is sure to be present in other models, in particular in $\lambda\varphi^4$- model, $\lambda > 0$. Let us consider once more the term $Z_\Lambda(0)\varphi^3$

arising in this model after field-translation. The renormalization group methods show that the vertice φ^3 introduces the singularities of the type $(\ln \Lambda)$. Therefore the product $Z_\Lambda(0)\,\varphi^3$ leads to finite contribution. So, the asymptotically free $\lambda\varphi^4$-theory is probably still unstable.

References:

1.K.Symanzik. Lett.Nuov.Cim. 6,77 (1973)
Comm.Math.Phys.34,7 (1973)

2.R.Brandt. Phys.Rev. D14, 3381 (1976) and the literature cited there.

SHORT-DISTANCE EXPANSION FOR PRODUCTS OF CURRENT-LIKE OPERATORS

S.A.Anikin, O.I.Zavialov
Steklov Mathematical Institute, Moscow.

In ref.'s [1-4] Wilson expansions for the products of Heisenberg local fields $A(x)$ have been verified in every order of perturbation theory. For the needs of high-energy physics it is important to know also the short-distance behaviour for products of more complicated operators, say, currents. Such operators are linear combinations of Zimmermann [5] normal products $N^{(a)}[A^n(x)]$. The derivation of Wilson expansions for these objects is given below.

Let $\mathcal{B}_{\{\mu\}}(x)$ be a monomial of the asymptotic free field $\varphi(x)$:

$$\mathcal{B}_{\{\mu\}}(x) = :\varphi_{(\mu_1)}(x)\ldots\varphi_{(\mu_m)}(x):$$

$\{\mu\}$ being the multi-index $\{\mu\}=\{(\mu_1)\ldots(\mu_m)\}$ with

$$(\mu_i)=(\mu_{i0},\mu_{i1},\mu_{i2},\mu_{i3});\ |\mu_i|=\mu_{i0}+\mu_{i1}+\mu_{i2}+\mu_{i3};\ (\mu_i)!=\mu_{i0}!\cdot$$

$$\cdot\mu_{i1}!\,\mu_{i2}!\,\mu_{i3}!\ ;\qquad \varphi_{(\mu_i)}(x)=(\partial_0)^{\mu_{i0}}\ldots(\partial_3)^{\mu_{i3}}\varphi(x).$$

The composite field corresponding to $\mathcal{B}_{\{\mu\}}(x)$ will be denoted $B^{(a)}_{\{\mu\}}(x)$:

$$B^{(a)}_{\{\mu\}}(x) \equiv N^{(a)}[A_{(\mu_1)}(x)\ldots A_{(\mu_m)}(x)]$$

where $a \geq \dim \mathcal{B}_{\{\mu\}}(x)$. It is defined by

$$B^{(a)}_{\{\mu\}}(x) = S^{+}\otimes\left[R^{(a)}\mathcal{B}_{\{\mu\}}(x)\,E_0(\mathcal{J})\right]\equiv S^{+}\otimes \check{B}^{(a)}_{\{\mu\}}(x) \qquad (1)$$

where $\mathcal{J}=i\int\mathcal{L}(x)dx$; $\mathcal{L}(x)$ being the unrenormalized Lagrangian, $E_0(\mathcal{J})\equiv\exp[\mathcal{J}]$, $S=R^{(a)}E_0(\mathcal{J})$, the cross in the circle indicating the ordinary multiplication, the products without this mark understood as time-ordered. $R^{(a)}$ is the Bogolubov-Parasiuk renormalization based on the subtraction operators $M^{(a)}$ (for graphs containing the vertice x) and $M^{(4)}$ (for graphs which do not contain this vertice). In φ^4 -theory this means that on any Feynman diagram Γ in the perturbation series for $\mathcal{B}_{\{\mu\}}E_0(\mathcal{J})$ or $E_0(\mathcal{J})$ the renormalization $R^{(a)}$ acts according to the "forest formulae"

$$R^{(a)}\Gamma = \vdots(1-M_1)\ldots(1-M_k)\vdots\,\Gamma \qquad (2)$$

where M_i transforms the coefficient function of the i-th proper divergent subgraph Γ_i into the polynomial equal to the sum of ω_i first terms of its Mc.Lorian series in the external momenta and the symbol $\vdots\,\vdots$ cancels the contribution of overlapping subgraphs. For the subgraph Γ_i with ℓ_i external legs the divergence index ω_i is $\omega_i = a - \ell_i$ if the vertice x is contained in Γ_i (that is $M_i = M^{(a)}$) and $\omega_i = 4 - \ell_i$ if $x \notin \Gamma_i$ (that is $M_i = M^{(4)}$). The chronological product $(B^{(a_1)}_{\{\mu_1\}}(x+\xi_1) B^{(a_2)}_{\{\mu_2\}}(x+\xi_2))^{(a)}$ of two composite fields, which is our main object, is defined via

$$(B^{(a_1)}_{\{\mu\}_1}(x+\xi_1)\, B^{(a_2)}_{\{\mu\}_2}(x+\xi_2))^{(a)} = S^+ \otimes [R^{(a,a_1,a_2)} b_{\{\mu\}_1}(x+\xi_1) \cdot \tag{3}$$
$$\cdot\, b_{\{\mu\}_2}(x+\xi_2)\, E_0(\mathcal{J})] \equiv S^+ \otimes (\check{B}^{(a_1)}_{\{\mu\}_1}(x+\xi_1)\, \check{B}^{(a_2)}_{\{\mu\}_2}(x+\xi_2))^{(a)}$$

where the renormalization $R^{(a,a_1,a_2)}$ is based on subtraction operators $M^{(a)}, M^{(a_1)}$, and $M^{(a_2)}, M^{(4)}, (a \geqslant a_1 + a_2 - 4)$ that is in the forest formulae (2) $M_i = M^{(a)}$ if $x+\xi_1 \in \Gamma_i, x+\xi_2 \in \Gamma_i$; $M_i = M^{(a_1)}$ if $x+\xi_1 \in \Gamma_i$, $x+\xi_2 \notin \Gamma_i$; $M_i = M^{(a_2)}$ if $x+\xi_1 \notin \Gamma_i, x+\xi_2 \in \Gamma_i$ and $M_i = M^{(4)}$ if $x+\xi_1 \notin \Gamma_i$, $x+\xi_2 \notin \Gamma_i$.

Next we introduce several structure formula which solve the combinatories of renormalization and which we use later. These formula have been obtained in [6] (see also [7,8]):

$$\check{B}^{(a)}_{\{\mu\}}(x) = S \otimes B^{(a)}_{\{\mu\}}(x) = E_0(\mathcal{J}_{ren}) \frac{1}{1 + M^{(a)} E_1(\mathcal{J}_{ren})}\, b_{\{\mu\}}(x) \equiv \tag{4}$$
$$\equiv E_0(\mathcal{J}_{ren})\, {}'B^{(a)}_{\{\mu\}}(x)\,; \quad {}'B^{(a)}_{\{\mu\}} = [1 + M^{(a)} E_1(\mathcal{J}_{ren})]^{-1}\, b_{\{\mu\}}(x);$$

$$(\check{B}^{(a_1)}_{\{\mu\}_1}(x+\xi_1)\, \check{B}^{(a_2)}_{\{\mu\}_2}(x+\xi_2))^{(a)} = E_0(\mathcal{J}_{ren}) \frac{1}{1 + M^{(a)} E_1(\mathcal{J}_{ren})} \cdot \tag{5}$$
$$\cdot\, [1 - M^{(a)}]\, {}'B^{(a_1)}_{\{\mu\}_1}(x+\xi_1)\, {}'B^{(a_2)}_{\{\mu\}_2}(x+\xi_2).$$

We use here the short notations $E_1(\mathcal{J}) = \exp[\mathcal{J}] - 1$, $E_2(\mathcal{J}) = \exp[\mathcal{J}] - 1 - \mathcal{J}$; $\mathcal{J}_{ren} = i \int \mathcal{L}_{ren}(x)\, dx$ where $\mathcal{L}_{ren}(x)$ is the renormalized Lagrangian. It was shown in [6] that

$$\mathcal{J}_{ren} = \mathcal{J} - M^{(4)} E_2(\mathcal{J} - M^{(4)} E_2(\mathcal{J} - \ldots)\ldots). \tag{6}$$

Relations (4)-(6) need some comments. First, some intermediate regularization is supposed to be present in the theory. This regularization will be removed only on the last step. Second, (4)-(6) are the relations between formal power series. If one expands (4)-(6) in powers of the coupling constant (say, in powers of $\mathcal{J}$) the operators M will never occur in the denominator and the typical term in the right-hand side, let us say, of eq.(4) will look like $M\mathcal{L}\ldots\mathcal{L}M\mathcal{L}\ldots$ $\ldots\mathcal{L}M\mathcal{L}\ldots\mathcal{L}\,\mathcal{B}(x)$. Now $\mathcal{L}\ldots\mathcal{L}\,\mathcal{B}(x)$, being the time-ordered product of the vertice $\mathcal{B}(x)$ and the Lagrangian vertices $\mathcal{L}$, can be represented via Feynman rules in the normal form $\mathcal{L}\ldots\mathcal{L}\,\mathcal{B}(x) \longrightarrow F(x)$ where

$$F(x) = \sum \frac{1}{e!} \int F_e(x|y_1\ldots y_e) : \mathcal{L}(y_1)\ldots\mathcal{L}(y_e) : dy_1\ldots dy_e \,. \qquad (7)$$

Here F_e are certain sums of Feynman diagrams on which the operator M is perfectly defined. So $M\mathcal{L}\ldots\mathcal{L}\,\mathcal{B}(x)$ is again the generalized vertice and we can proceed in the same way to the next M thus obtaining the whole set of renormalized diagrams representing the left-hand side.

Let us write explicitely the result of action of M on any functional of the form (7):

$$M^{(a)}F(x) = \sum_{\{\lambda\}:\ \Sigma|\lambda_i|+e\leq a} \mathcal{B}_{\{\lambda\}}(x)\frac{(-i)^{\Sigma|\lambda_j|}}{e!(\lambda_1)!\ldots(\lambda_e)!}\left\langle F(0)\,\tilde{\mathcal{L}}^{(\lambda_1)}_{(0)}\ldots\tilde{\mathcal{L}}^{(\lambda_e)}_{(0)}\right\rangle^{prop}. \qquad (8)$$

The tildes in (8) signify Fourier transformation with respect to field coordinates and the supercripts (λ_i) in the vacuum-expectation value denote that the derivative of the respective order (λ_i) is taken in the corresponding momentum variable.

Let us now construct the auxiliary operator $\mathfrak{M}^{(a)}_x$ defined on normal products: $\mathcal{L}(y_1)\ldots\ldots\mathcal{L}(y_e)$:and by linearity on arbitrary Wick monomials:

$$\mathfrak{M}^{(a)}_x : \mathcal{L}(y_1)\ldots\mathcal{L}(y_e) : = \sum_{e+\Sigma|\lambda_i|\leq a} \frac{(y_1-x)^{(\lambda_1)}\ldots(y_e-x)^{(\lambda_e)}}{(\lambda_1)!\ldots(\lambda_e)!}\,\mathcal{B}_{\{\lambda\}}(x)\,. \qquad (9)$$

Since any polylocal Heisenberg operator $F(x_1,\ldots,x_n)$ of the form

$$F(x_1,\ldots,x_n) = \sum_e \frac{1}{e!}\int dy_1\ldots dy_e\, F_e(x_1\ldots x_n|y_1\ldots y_e) : \mathcal{L}(y_1)\ldots\mathcal{L}(y_e) :$$

is a linear combination of normal products: $\mathcal{L}(y_1)\ldots\mathcal{L}(y_e)$: the operator $\mathfrak{M}^{(a)}_x$ can be readily applied to $F(x_1,\ldots,x_n)$. One gets

$$\mathfrak{M}^{(a)}_x F^{x\text{-prop}}(x+\xi_1,\ldots,x+\xi_n) = \sum_{e+\Sigma|\lambda_i|\leq a}\frac{(-i)^{\Sigma|\lambda_i|}}{e!\,(\lambda_1)!\ldots(\lambda_e)!}\,\cdot \qquad (10)$$
$$\cdot\,\mathcal{B}_{\{\lambda\}}(x)\left\langle F(\xi_1\ldots\xi_n)\,\tilde{\varphi}^{(\lambda_1)}_{(0)}\ldots\tilde{\varphi}^{(\lambda_e)}_{(0)}\right\rangle^{x\text{-prop}}.$$

The x-proper part entering (10) is the set of such diagrams which become proper after the vertices $x+\xi_1,\ldots,x+\xi_n$ collapse in the point x. Comparing (10) and (8) one finds that

$$\lim_{\xi_i\to 0}\mathfrak{M}^{(a)}_x F^{x\text{-prop}}(x+\xi_1,\ldots,x+\xi_n) = M^{(a)}F(x,\ldots,x)\,. \qquad (11)$$

In particular on one-point functionals both operators coinside:

$$\mathcal{M}_x^{(a)} F^{prop}(x) = M^{(a)} F(x). \tag{12}$$

Now we turn directly to short-distance expansions. Since $M^{(a)}$ transforms any two-point functional $F(x+\xi_1, x+\xi_2)$ into something proportional to $\delta^{(q)}(\xi_1-\xi_2)$, the eq. (5) reads

$$(\check{B}^{(a_1)}_{\{\mu\}_1}(x+\xi_1)\check{B}^{(a_2)}_{\{\mu\}_2}(x+\xi_2))^{(a)} = E_0(\mathcal{J}_{ren})\,{}'B^{(a_1)}_{\{\mu\}_1}(x+\xi_1)\,{}'B^{(a_2)}_{\{\mu\}_2}(x+\xi_2) \tag{13}$$

if $(\xi_1-\xi_2)^2<0$. On the other hand elementary transformations lead to

$$E_0(\mathcal{J}_{ren})\,{}'B^{(a_1)}_{\{\mu\}_1}(x+\xi_1)\,{}'B^{(a_2)}_{\{\mu\}_2}(x+\xi_2) = E_0(\mathcal{J}_{ren})\frac{1}{1+\mathcal{M}^{(a)}E_1(\mathcal{J}_{ren})}\,\cdot \tag{14}$$

$$\cdot\,\mathcal{M}^{(a)}E_0(\mathcal{J}_{ren})\,{}'B^{(a_1)}_{\{\mu\}_1}(x+\xi_1)\,{}'B^{(a_2)}_{\{\mu\}_2}(x+\xi_2) + Q(x,\underline{\xi})$$

where $\mathcal{M}^{(a)}$ coinsides with $\mathcal{M}_x^{(a)}$ on the x-proper parts of the corresponding functionals and is defined to be zero on the parts which are not x-proper. The functional Q takes the form

$$Q(x,\underline{\xi}) = E_0(\mathcal{J}_{ren})\frac{1}{1+\mathcal{M}^{(a)}E_1(\mathcal{J}_{ren})}[1-\mathcal{M}^{(a)}]\,{}'B^{(a_1)}_{\{\mu\}_1}(x+\xi_1)\,{}'B^{(a_2)}_{\{\mu\}_2}(x+\xi_2). \tag{15}$$

Choose $a \geq a_1+a_2$. The crucial point is that $Q(x,\underline{\xi})$, tends to zero like $\rho^{a-a_1-a_2+1}(\ln\rho)^m$ as $\rho\to 0$, $\xi_1=\rho\xi_1^0$, $\xi_2=\rho\xi_2^0$. Indeed all the asymptotic fields $\mathcal{L}$ occuring in normal products expansion of ${}'B^{(a_1)}_{\{\mu\}_1}(x+\xi_1)\,{}'B^{(a_2)}_{\{\mu\}_2}(x+\xi_2)$ have either $x+\xi_1$ or $x+\xi_2$ as their arguments. According to (9) the operator $\mathcal{M}^{(a)}$ transforms them into their Teylor series in the powers of ξ. That means that $(1-\mathcal{M}^{(a)})\,{}'B^{(a_1)}_{\{\mu\}_1}(x+\xi_1)\,{}'B^{(a_2)}_{\{\mu\}_2}(x+\xi_2)$ has a zero of order $a-a_1-a_2+1$ in the limit $\xi\to 0$. All the other parts of the right-hand side of (15) introduce at most logarithmic singularities in this limit. The proof is based on the fact that to every diagram contributing to (15) there corresponds the forest formulae (2) where M_i are now either M or $\mathcal{M}$ (compare (15) and (5)). So, for any of such diagrams one can derive the parametric representation which exibits the necessary property. The detailed proof is lengthy and will be published elsewhere.

Let us return to (14). According to (10) we have

$$\mathcal{M}^{(a)}E_0(\mathcal{J}_{ren})\,{}'B^{(a_1)}_{\{\mu\}_1}(x+\xi_1)\,{}'B^{(a_2)}_{\{\mu\}_2}(x+\xi_2) = \sum_{\ell+\Sigma|\lambda_i|\leq a} K^{\{\lambda\}}(\xi_1,\xi_2)\,\ell_{\{\lambda\}}(x) \tag{16}$$

where for $(\xi_1-\xi_2)^2<0\ ,\ \xi_1^0>\xi_2^0$

$$K^{\{\lambda\}}_{(\xi_1,\xi_2)}=\frac{(-i)^{\Sigma|\lambda_i|}}{\ell!(\lambda_1)!\ldots(\lambda_\ell)!}\left\langle B^{(a_1)}_{\{\mu\}_1}(\xi_1)\,B^{(a_2)}_{\{\mu\}_2}(\xi_2)\,\tilde{\varphi}^{(\lambda_1)}_{(0)}\ldots\tilde{\varphi}^{(\lambda_\ell)}_{(0)}\right\rangle^{T\text{-prop}} \tag{17}$$

Now we take into account that up to C-number coefficients $K^{\{\lambda\}}(\xi_1,\xi_2)$ every term $b_{\{\lambda\}}(x)$ in (16) gives rise to one-point functional in (14). Thus due to (12) the operator $\mathcal{M}^{(a)}$ in the denominator of (14) can be changed into $M^{(a)}$. Consequently the factor $E_0(s_{ren})[1+\mathcal{M}^{(a)}E_1(s_{ren})]^{-1}$ transforms every $b_{\{\lambda\}}(x)$ into $\breve{B}^{(a)}_{\{\lambda\}}(x)$ (see eq.(4)). Multiplying both sides of (14) by S^+ from the left we obtain in the left-hand side the product of composite fields and in the right-hand side - the linear combination of composite fields with singular coefficients. In other words we get the desired Wilson expansion:

$$B^{(a_1)}_{\{\mu_1\}}(x+\xi_1)\,B^{(a_2)}_{\{\mu_2\}}(x+\xi_2)=\sum_{\{\lambda\}:\,\ell+\Sigma|\lambda_i|\leq a}K^{\{\lambda\}}_{(\xi_1,\xi_2)}\,B^{(a)}_{\{\lambda\}}(x)^+ \tag{18}$$

$$+\,O\left(\xi^{a-a_1-a_2+1}(\ln\xi)^m\right)$$

where the coefficients $K^{\{\lambda\}}(\xi_1,\xi_2)$ are given by (17).

References:

1.W.Zimmermann.Ann.of Phys. 77, 570 (1973).

2.R.Brandt.Ann.of Phys.52, 122 (1969) .

3.T.Clark. Nucl.Phys. B81, 263 (1974).

4. S.A.Anikin, O.I.Zavialov.Theor.Math.Phys.(Soviet)27,N3 (1976).

5.W.Zimmermann.Brandeis Lectures. MIT Press 1970.

6.O.I.Zavialov,S.A.Anikin.Theor.Math.Phys.(Soviet)26, N2 (1976); Dubna preprint D2-9259 (1975).

7.O.I.Zavialov.Proc.XVIII Int.Conf.High Energy Phys.Tbilisi 1976 v.II T59.

8.S.A.Anikin,M.C.Polivanov,O.I.Zavialov.Fortschr.der Phys., September 1977 (in print).

THE USE OF EXTERIOR FORMS IN FIELD THEORY

W. Thirring

Institut für Theoretische Physik
Universität Wien

Cartan's calculus formalizes the quantities appearing in Maxwell's and Einstein's equations and is therefore particularly suited for expressing them. Unfortunately, it is not yet as commonly used as it deserves because the practitioner frequently don't know it and the mathematicians seldomly do explicit calculations. I would like to summarize here the main points of this formalism. For details see [1].

(1) Basis

The central objects in field theory are differential forms (≡ antisymmetric covariant tensor fields). The ones of rank p, p = 0,1, ..., m = dimension of the manifold M (= p-forms) are a module E_p over E_o, the real-valued functions. The exterior product $\wedge : E_p \times E_q \rightarrow E_{p+q}$ makes

$$\bigcup_{p=0}^{m} E_p$$

to a graded algebra. If e^j is a local basis in E_1 one can construct a ($\binom{m}{p}$-dimensional) basis in E_p with $\wedge$:

$$e^{j_1 \cdots j_p} := e^{j_1} \wedge e^{j_2} \wedge \ldots \wedge e^{j_p} .$$

Thus every p-form ω can be written

$$\omega = \sum_{(j)} \omega_{j_1 \cdots j_p} e^{j_1 \cdots j_p}, \quad \omega_{(j)} \in E_o. \tag{1.1}$$

(2) Exterior Derivative

The differential processes of vector calculus are synthesized in the operation $d: E_p \rightarrow E_{p+1}$ with the properties

(i) $d(\omega_1+\omega_2) = d\omega_1 + d\omega_2, \; \omega_{1,2} \in E_p$

(ii) $d(\omega_1\wedge\omega_2) = (d\omega_1)\wedge\omega_2 + (-)^p \omega_1\wedge d\omega_2, \; \omega_1 \in E_p, \; \omega_2 \in E_q$ (2.1)

(iii) $d\, d\omega = 0, \; \omega \in E_p$.

Functions x^j over a neighbourhood are called independent if the $dx^j \in E_1$ are independent. Using them as a basis $e^j = dx^j$ the exterior derivative becomes simply

$$d\omega = d\left(\sum_{(j)} \omega_{j_1 \ldots j_p} e^{j_1 \ldots j_p}\right) = \sum_{(j)} \frac{\partial \omega_{j_1 \ldots j_p}}{\partial x^j} e^{j\, j_1 \ldots j_p}.$$

(3) Pseudo-Riemannian Structure

A scalar product $\langle | \rangle$ in E_1

$$\langle e^i | e^j \rangle = g^{ij} = g^{ji}, \quad \text{Det } g^{ij}(x) \neq 0 \;\forall\; x \in M$$

also gives an isomorphism $*: E_p \leftrightarrow E_{m-p}$. Because of linearity one has to specify it only for a basis:

$$*e^{j_1 \ldots j_p} = \varepsilon^{j_1 \ldots j_m} g_{j_{p+1},k_1} \cdots g_{j_m,k_{m-p}} e^{k_1 \ldots k_{m-p}} \frac{\sqrt{g}}{(m-p)!}, \tag{3.1}$$

$g = |\text{Det } g_{ik}|, \quad g_{ik}\, g^{kj} = \delta_i{}^j, \quad \varepsilon$ = totally antisymmetric tensor.

It allows to define the co-derivative $\delta: E_p \to E_{p-1}$:

$$\delta = * \circ d \circ * \; (-)^{m(p+1)+s}$$

$$(-)^s = \text{signature of } g. \tag{3.2}$$

(4) Integration

A p-form ω defines a measure over p-dimensional submanifolds N_p. The corresponding integral is simply denoted by $\int_{N_p} \omega$. It is the inverse operation to d in the sense that partial integration generalizes to Stokes' theorem

$$\int_{N_{p+1}} d\omega = \int_{\partial N_{p+1}} \omega, \tag{4.1}$$

∂N = boundary of N, ω with compact support.

$\nu \in E_{m-p}$ defines

$$\nu(\omega) = \int_M \nu \wedge \omega, \quad \omega \in E_p,$$

a linear functional over E_p. Correspondingly we shall also admit distribution-type p-forms (currents) [2], understanding that due care has to be exercised when multiplying them. Of greater importance will be linear maps $E_p \to E_{q|\bar{x}}$ where the latter denotes the antisymmetric covariant tensors at the point $\bar{x} \in M$ [3]. In particular one can define a "δ-function" $\delta_{\bar{x}} \in E_{p|\bar{x}} \otimes E_{m-p}$ which reproduces the value at a point $\bar{x}$ of a p-form ω

$$\int \delta_{\bar{x}} \wedge \omega = \omega_{|\bar{x}}. \tag{4.2}$$

$\delta_{\bar{x}}$ has its support in $\bar{x}$ and there it can be written in a natural basis $\bar{e}$ for $E_{p|\bar{x}}$ and e of E_{m-p}

$$\delta_{\bar{x}} = \bar{e}^{i_1 \dots i_p} \otimes {*}e_{i_1 \dots i_p} \frac{(-)^{p(m-p)}}{p!} \delta(x_1-\bar{x}_1)\, \delta(x_2-\bar{x}_2) \dots \delta(x_m-\bar{x}_m)\ . \tag{4.3}$$

(5) Differential Equations

The differential equations encountered in classical field theory are of such a structure that exterior and co-derivative of p-forms are specified. The appropriate tool for solving such an equation is the "Green-function" $G_{\bar{x}}$. Like $\delta_{\bar{x}}$ it is actually a tensor-valued current $\in E_{p|\bar{x}} \otimes E_{m-p}$ and satisfies the equation

$$(\delta d + d\delta) G_{\bar{x}} = - \delta_{\bar{x}}\ . \tag{5.1}$$

As a consequence of Stokes' theorem and the rules for derivatives one obtains for $\bar{x} \in N \subset M$, Dim N = Dim M

$$\omega_{|\bar{x}} = (-)^{p+m} \int_N [dG_{\bar{x}} \wedge \delta\omega - \delta G_{\bar{x}} \wedge d\omega] - \int_{\partial N} [\delta G_{\bar{x}} \wedge \omega - (-)^{s} {*}dG \wedge {*}\omega]\ . \tag{5.2}$$

Since we have not yet demonstrated the existence of $G_{\bar{x}}$ so far this result is only formal. By explicit construction of $G_{\bar{x}}$ in certain cases one sees that (5.2) is valid provided ∂N is nowhere tangent to the light cone of $\bar{x}$. If this is the case, (5.2) solves the problem as it expresses ω at $\bar{x}$ by $d\omega$ and $\delta\omega$ in N and ω and ${*}\omega$ at ∂N. For some manifolds like $\mathbb{R}^m$, $g = -dx^0 \otimes dx^0 + dx^1 \otimes dx^1 + \dots + dx^{m-1} \otimes dx^{m-1}$ there is a $G_{\bar{x}}$ with the additional property that its support is contained in the past light cone of $\bar{x}$. In this case (5.2) solves the initial-value problem: If $N = \{x^i \in \mathbb{R}^m : t_0 \leq x^0 \leq t_1\}$ only the part of ∂N with $x^0 = t_0$ contributes and ω is determined by $d\omega$, $\delta\omega$ and the values of ω and ${*}\omega$ for $x^0 = t_0$.

(6) Electrodynamics

Maxwell's Equations

They refer to the case $m = 4$, $p = 2$, and for the field form $F \in E_2$ they read

$$dF = 0, \quad \delta F = J = \text{charge-current} \in E_1 \tag{6.1}$$

or in integral form

$$\int_{\partial N_3} F = 0, \quad \int_{N_3} {*J} = - \int_{\partial N_3} {*F} . \tag{6.2}$$

Since $\delta \circ \delta = 0$, they imply current conservation: $\delta J = 0$, in integral form

$$\int_{\partial N_4} {*J} = 0 .$$

In some manifolds like $\mathbb{R}^4$, $\delta J = 0$ implies the existence of $F \in E_2$ such that $\delta F = J$. In general the latter equation is stronger than $\delta J = 0$. In particular, if $N_3 = \{x^i \in \mathbb{R}^4 : x^0 = t\}$ is compact and without boundary (like in $\mathbb{R} \times T^3$) then the total charge

$$Q(t) = \int_{N_3} {*J}$$

is not only constant in t but zero:

$$\int_{N_3} {*J} = \int_{\partial N_3} {*F} = \int_{\emptyset} {*F} = 0 . \tag{6.3}$$

The usual procedure for solving (6.1) is to conclude from $dF = 0$ that $F = dA$, go to the Lorentz-gauge $\delta A = 0$ and then solve $(\delta d + d\delta)A = J$. This is not so good because one wants to express F in terms of the initial values of F and $*F$ and not in terms of A and its derivatives. (5.2) gives in $\mathbb{R}^4$ directly the solution of the Cauchy problem if one uses

$$G^{ret}_{\bar{x}} = - \frac{1}{4\pi} \bar{e}_{\alpha\beta} * e^{\alpha\beta} \, \delta((\bar{x}-x)^2) \, \Theta(\bar{x}^0 - x^0) : \tag{6.4}$$

$$F|_{\bar{x}} = \int_N dG^{ret}_{\bar{x}} \wedge J - \int_{\partial N} [\delta G^{ret}_{\bar{x}} \wedge F + {*dG^{ret}_{\bar{x}}} \wedge {*F}] . \tag{6.5}$$

If $N = \{x \in \mathbb{R}^4 : x^0 > t\}$ and the limit $t \to -\infty$ exists, these equations express F in terms of an incoming field and the usual Lienard-Wiechert potentials. In some situations other Green-functions are called for. For instance, on a metallic surface B the boundary condition is simply $F_{|B}$: = restriction of F to B = 0. In such a situation $\partial N = B \cup S$ in (5.2) will have a space-like part S corresponding to an initial surface and B. To have only S contribute to $\int_{\partial N}$ we need a Green-function such that

$$*dG_{\bar{x}\,|B} = 0 \;. \tag{6.6}$$

With this Green-function (5.2) solves the initial-value problem

$$F_{\bar{x}} = \int_N dG_{\bar{x}} \wedge J - \int_S [\delta G_{\bar{x}} \wedge F + *dG_{\bar{x}} \wedge *F] \;. \tag{6.7}$$

If $B = \{x \in \mathbb{R}^4 : x^1 = 0\}$ (metallic mirror) a Green-function satisfying (6.6) is simply $(1 + R)G^{ret}_{\bar{x}}$ where R is the reflection $x^1 \to -x^1$. Similarly the $G_{\bar{x}}$ for a rectangular wave-guide is constructed by a sequence of reflections. In this way the causal structure of G becomes obvious, whereas in usual procedure [4] after piecing a plane wave together remain the questions

(a) Has one found all solutions?

(b) Does nothing propagate faster than light?

(7) Gravitation

Here the e^i play the role of potentials and the connection-forms $\omega^i{}_k \in E_1$:

$$de^i = \omega^i{}_k \wedge e^k, \qquad dg_{ik} = g_{ij}\,\omega^j{}_k + g_{kj}\,\omega^j{}_i \tag{7.1}$$

correspond to the field strength. However, by a change of the basis:

$$e^i \to A^i{}_k\, e^k, \qquad A^i{}_k \in E_o \;, \tag{7.2}$$

the $\omega^j{}_k$ transform inhomogeneously. Only in the curvature-forms $R^i{}_k \in E_2$:

$$R^i{}_k = d\omega^i{}_k + \omega^i{}_j \wedge \omega^j{}_k \tag{7.3}$$

these inhomogeneous terms drop out. Thus, if one wants to construct a Lagrange-function $L \in E_4$ which is invariant under (7.2) the simplest possibility after $*\mathbb{1}$ is (a constant apart)

$$L = R^{ik} \wedge {}^*e_{ik} + L_{\text{matter}} \, . \tag{7.4}$$

Variation of e^i gives Einstein's equations in a form advocated by Trautman [5]

$$R^i{}_k \wedge {}^*e^{km} = \frac{\delta L_{\text{matter}}}{\delta e_m} =: {}^*T^m \; \varepsilon \; E_3 \, . \tag{7.5}$$

However, it seems desirable to have an equation like the inhomogeneous Maxwell equation where the coderivative of a 2-form is a current. This can easily be done by writing in the left side of (7.5) the contribution $(d\omega^i{}_k) \wedge {}^*e_i{}^{km}$ as $d(\omega^i{}_k \wedge {}^*e_i{}^{km}) + \omega^i{}_k \wedge d{}^*e_i{}^{km}$. Retaining only the first term on the left hand side we obtain (in units $8\pi\kappa = 1$)

$$d(\omega^i{}_k \wedge {}^*e_i{}^{km}) = {}^*T^m + {}^*t^m$$

$${}^*t^m = - \omega_{bg} \wedge (\omega^m{}_r \wedge {}^*e^{rbg} + \omega^g{}_r \wedge {}^*e^{mbr}) \, . \tag{7.6}$$

The interpretation of (7.6) is obviously that the currents of the energy (m = 0) and momentum (m = 1,2,3) have a contribution T from matter and one t from gravitation. (7.6) has as a consequence not only the conservation law

$$\delta(T^m + t^m) = 0 \tag{7.7}$$

but the stronger statement

$$\int_{N_3} ({}^*T^m + {}^*t^m) = \int_{\partial N_3} \omega^i{}_k \wedge {}^*e_i{}^{km} \, . \tag{7.8}$$

In particular, if we have a space-like compact N_3 without boundary then

$$\int_{N_3} ({}^*T^m + {}^*t^m) = 0 \, ,$$

that is, the total energy and momentum in a closed universe are zero.

References

[1] Y. Choquet-Bruhat, C. DeWitt-Morette, M. Dillard-Bleick: Analysis Manifolds and Physics, North Holland, 1977
H. Flanders, Differential Forms, Academic Press 1963
E. Hlawka, Acta Phys. Austr. Suppl. 7, Springer 1970
H. Holman and H. Rummler, Alternierende Differentialformen, Bibliographisches Institut 1972
M. Spivak, Calculus on Manifolds, Benjamin 1965
W. Thirring, Klassische dynamische Systeme, Springer 1977

[2] G. de Rham, Variétés Différentiables, Hermann, Paris, 1955

[3] W. Thirring, Klassische Feldtheorie, Springer 1978

[4] J.D. Jackson, Classical Electrodynamics, J. Wiley & Sons, 1967

[5] A. Trautman, Theory of Gravitation, in: The Physicist's Conception of Nature, J. Mehra Ed., Reidel (1973)

SHORT COMMUNICATIONS

ON EXTENSIONS OF FLOWS IN THE PRESENCE OF SETS OF SINGULARITIES

Michael Aizenman*
Departments of Mathematics and Physics
Princeton University
Princeton, N.J. 08540, USA

Abstract

Measure preserving (m.p.), flows in $\mathbb{R}^n$ may have intersecting trajectories. This kind of singularity may prevent the uniqueness or, in other cases, even the existence of a m.p. flow with a given velocity field $\underline{v}$, without contradicting the "integrability condition ": $\underline{\nabla}\underline{v} = 0$ in a weak sense. Examples are constructed to prove that, in $\mathbb{R}^n$ $n \geq 3$, these phenomena can not be ruled out by the proposed condition, $\underline{v} \in L^2$, or by any condition on the moments of $\underline{v}$. Thus the question of a useful criterion, especially for the study of non stationery flows (e.g. those described by Navier-Stokes equations) remains open. On the positive side, a criterion is given which ensures that a measure preserving flow, with $\underline{v} \in L^p$ $p > 1$, has no flux through compact sets, e.g. sets of possible singularities, whose "dimension" is low enough. (The "dimension", as used here, is not necessarily integral.) The upper bound on the "dimension" increases with p via an inequality which is optimal (possibly not strictly), as shown by the examples.

I. Introduction

Measure preserving flows in Euclidian spaces are encountered in various subjects. An example is the "time evolution" flow in the phase space of a classical system. Another important class of examples are the stationary flows of an incompressible fluid.

It is often very useful to describe a flow which is absolutely continuous in time by its "infinitesimal generator", which is usually taken to be represented by the field of velocity of that flow.

The following discussion bears on the fundamental questions about the relation between flows and velocity fields. It will be shown that certain assumptions on the velocity field, which are reasonable and may seem sufficient, do not rule out various examples of different flows with the same velocity fields and other examples of velocity fields which do not correspond to any measure preserving flows.

In many situations of interest the flow is not stationary. Following Euler (not Lagrange) one usually studies such a flow by looking at the time evolution of its velocity field. This may often be described by a differential equation, e.g. by the Navier-Stokes equation. At present, there is no useful condition on the initial data which guarantees that the resulting velocity field does specify uniquely an existing flow. The examples which are constructed point at the insuficiency of some suggested criteria.

In some situations one may have additional information about the flow which reduces the questions of existence and uniqueness to a "small" subset of the space. We give suficient conditions for a measure preserving flow to avoid sets of correspondingly low "dimension" (in a sense which leads also to non-integral dimensions). The criterion may be used to obtain positive answers in the former situations. The inequality which is required is optimal, (possibly not strictly) as shown by the same class of counter-examples which was mentioned above.

II. Setup

Any measure refered to in this article is, unless specified to the contrary, the corresponding Lebesgue measure.

Definition 1: A measure preserving (m.p.) flow in $\mathbb{R}^n$ is a collection, T_t, of measurable mappings $T_t: \mathbb{R}^n \to \mathbb{R}^n$ which satisfy:

1) $\forall\, t_1, t_2 \in \mathbb{R},\ x \in \mathbb{R}^n$

$$T_{t_1} T_{t_2} \underline{x} = T_{t_1+t_2} \underline{x}$$

2) $\forall\, t \in \mathbb{R}$ and any measurable A $\mathbb{R}^n$

$$|T_t^{-1} A| = |A|.$$

Here $|A|$ is the measure of A.

Definition 2: $\underline{v}: \mathbb{R}^n \to \mathbb{R}^n$ is the velocity field of a flow T_t if for almost every (a.e.) $\underline{x} \in \mathbb{R}^n$.

$$T_t\underline{x} = \underline{x} + \int_0^t du\, \underline{v}(T_u\underline{x}) \qquad \forall\, t \in \mathbb{R} \tag{1}$$

We shall discuss only flows which do have some velocity field. Notice that all the trajectories of any such flow, except for an invariant set of measure zero, are continuous as functions of "time". Further, if T_t is a measure preserving flow with velocity $\underline{v}$ then for any $f \in C^1$, such that $\underline{v}\underline{\nabla} f \in L^1$,

$$\int \underline{v}\, \underline{\nabla}\, f(\underline{x}) d^n\underline{x} = \frac{d}{dt}\bigg|_{t=0} \int f(T_t\underline{x})\, d^n\underline{x} = 0. \tag{2}$$

Thus, a necessary requirement for T_t to preserve (Lebesque) measure is

$\underline{\nabla}\, \underline{v} = 0$ weakly, in some proper sense. (3)

III. Basic questions:

Two questions which are fundamental for the use of the infinitesimal description of flows are:

Q. 1: Let $\underline{v}$ satisfy (3). Under what conditions does there exist a measure preserving flow whose velocity is $\underline{v}$?

Q. 2: Under what conditions is a m.p. flow uniquely characterized by $\underline{v}$?

A condition which is well known to be sufficient for both existence and uniqueness (in compact regions) is Lipschitz continuity of $\underline{v}$, [1]. This, however, is very restrictive, especially in situations in which $\underline{v}$ itself evolves in time, e.g. via Navier-Stokes equations. For the latter case there is no known useful restriction on the initial conditions which guarantees Lipschitz continuity of $\underline{v}$ at any later time.

As a starting point for a better understanding of the questions let us consider an example first discussed by E. Nelson [2].

Example 1 (the "small bang"):

In $\mathbb{R}^2$, a general way to generate vector fields which satisfy (3) is by:

$$\underline{v}(x) = J\underline{\nabla}f(x)$$

$$\text{with} \quad J \text{ - the orthogonal rotation } J = \begin{pmatrix} 0 & 1 \\ -1 & 0 \end{pmatrix} \tag{4}$$

and f any function which is regular "enough." In particular, let Θ be the polar angle and

$$f(\underline{x}) = \begin{cases} \Theta(\underline{x}) & \pi \geq \Theta \geq \Theta \\ 2\pi - \Theta(\underline{x}) & 2\pi \geq \Theta > \pi \end{cases} \tag{5}$$

Substituting (5) in (4) one gets the vector field

$$\underline{v}_N(\underline{x}) = \frac{\hat{\underline{x}}}{|\underline{x}|} \operatorname{sgn} x_2 \qquad \left(\hat{\underline{x}} = \frac{x}{|x|} \right) \tag{6}$$

Figure 1: Two different m.p. flows with the same velocity field - $\underline{v}_N$.

The corresponding flow is uniquely defined on the complement of the line $\{x_2 = 0\}$ and describes a radial motion; towards the origin ($\underline{x} = 0$) in the lower-half-plane and away from it in the upper-half-plane. This, partially defined, flow has many different measure preserving extensions parametrized by the transformation which the angle undergoes at the moment of passage through the origin (examples are indicated in fig. 1). All these extensions have $\underline{v}_N$ as their velocity field.

As one would expect, the velocity field $\underline{v}_N$, which has a non-vanishing flux through a point, is diverging at that point. $|v|$ as a function of the distance to the point is proportional to the inverse of the area available to the flux. As a consequence the second moment of $\underline{v}$ is infinite i.e. $v \notin L^2_{loc.}$, though $v \in L^{2-\varepsilon}_{loc.}$ for any $\varepsilon > 0$.

This example and additional considerations led to the following question [2] which, has its answer been positive, would have given a usefull answer to Q. 1 and to Q. 2.

Q. 3: If $\underline{v}$: $R^n \longrightarrow R^n$ is of compact support and

a) $\underline{\nabla}\,\underline{v} = 0$ weakly, in $\mathcal{S}'$,
b) $\underline{v} \in L^2$

does there exist a unique m.p. flow with velocity $\underline{v}$?

Before answering Q.3 let us raise a related question (which will not be answered here). D. Ruelle mentioned in his talk some limitations on possible dynamics imposed by the dimension of the space.

In dimension 1: there is only one, up to scaling, m.p. flow on $\mathbb{R}^1$.

In dimension 2: any diffeomorphism on a compact, two dimensional, manifold has zero topological entropy [3].

The latter does not generalize to the class of flows whose velocity is smooth on the complement of a 1-point set. This can be easily seen by constructing a variant of the "small bang" example.

Q. 4: Does the above generalize to the class of two dimensional m.p. flows with L^2-velocities?

Or: has a two dimensional flow to have infinite kinetic energy to brake the bonds of dimensionality?

IV. Some negative answers

The first part of the results reported here are negative answers to the question Q. 3 [4].

A.1: Let $n \geq 3$. $\exists$ $\underline{v}$: $\mathbb{R}^n \longrightarrow \mathbb{R}^n$, of compact support, such that

a) $\underline{\nabla}\,\underline{v} = 0$ in $\mathcal{S}'$
b) $\underline{v} \in L^\infty$

and which is the velocity field of various different m.p. flows.

A.2: Let $n \geq 3$. $\exists$ $\underline{v}$: $\mathbb{R}^n \longrightarrow \mathbb{R}^n$, which satisfies the conditions mentioned in A.1, which is not the velocity field of any m.p. flow.

The construction of the counter-examples is motivated by the following consideration. The non-uniqueness in example 1 is caused by the convergence of trajectories. This in turn results in the divergence of $\underline{v}$. Intuitively, in $\mathbb{R}^3$, a flow which has a non vanishing flux through a line may be less singular than a flow which converges to a point. It would, however, still have the same singularity as $\underline{v}$ (which was two dimensional). Is it possible that the convergence of trajectories onto an even larger set is consistent with lower singularity for $\underline{v}$?

In the rest of this section we describe a construction of various m.p. flows in $\mathbb{R}^3$ which pass through sets of "dimensions" even larger than 1 on which their orbits intersect. As expected: the higher the "dimension" of the set of singularity the lower will be the singularity of $\underline{v}$ (see [4] for a more detailed argument).

We start by describing a construction of various subsets of R^3 of dimension which may be any number in [0,2].

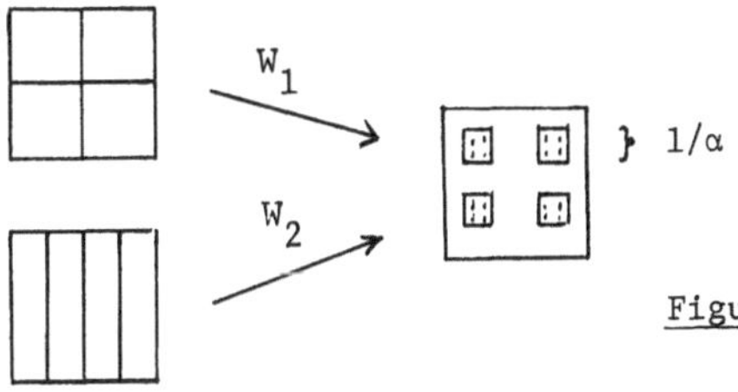

Figure 2

Construction 1: Let W_1 be a 1-1 mapping of the unit square, $(0,1]^2$, into itself under which the square is partitioned into k smaller squares, each of which is uniformly contracted by a factor α. In the next step perform a mapping of each of the images of the k squares into itself by properly scaling W_1. An infinite itteration of that procedure gives a 1-1 mapping of the unit square which we denote by $L(W_1)$, with the range $RL(W_1)$.

As easily seen, the immage of $RL(W_1)$ under the uniform dialation of R^2 by α consists of k disjoint translates of it (permitting rotations). Thus, it is very reasonable to define the dimension of $RL(W_1)$, ν, by $k = \gamma^\nu$, i.e.

$$\nu = \frac{\ln k}{\ln \gamma} \tag{7}$$

In fact, there are more general definitions of dimension (e.g. Hausdorff dimension, [5]) which in this case agree with (7).

We shall also use the following variant of construction 1.

Construction 2: Let W_2 be a mapping of $(0,1]^2$ with the same immage as W, which partitions the square to k equal strips (rather than squares) which are then dialated uniformly in each direction. Perform the next transformation as in construction 1, paying attention to keep the same fibers unpartitioned. $L(W_2)$ - the result of an infinite itteration of that procedure, is a mapp with the same range as $L(W_1)$. The important difference is that $L(W_2)$ is not 1-1: the pre-image of any point in $RL(W_2)$ is a whole fiber ($\{\underline{x} \in (0,1]^2 | \, x_1 = \text{const}\}$, for the transformation in fig. 2).

Let now W be a mapping of the type discussed in either of the previous constructions. We shall construct a partially defined flow "through" the unit cube, $(0,1]^3$, which "enters" it through the upper face, $\{x_3=1\}$, and "leaves" through the set $\{\underline{x} \in (0,1]^3 | x_3=0, \ (x_1, \ x_2) \in RL(W)\}$. The meaning of partially defined (p.d.)

flows and of related concepts should be clear from the context. The definitions are delayed to the next section.

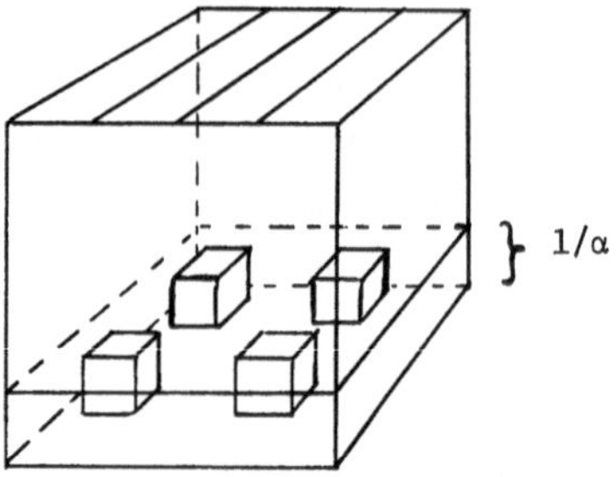

Figure 3

Construction 3:

i) Let $\hat{T}_t$ be a m.p. flow which in

$\{\underline{x} \in (0,1]^3 \mid 1 \geq x_3 > 1/\alpha\}$

satisfies:

1) $\hat{T}_t$ is uniquely characterized by its velocity field $\underline{\hat{v}}$.
2) $\hat{T}_t$ has no flux through the side faces $\{x_1 = 0,1\}$, $\{x_2 = 0,1\}$ (i.e. $\underline{\hat{v}}$ is transversal there).
3) $\underline{\hat{v}}$ is continuous at the upper face, $\{x_3 = 1\}$, where it is identically $(0,0,-1)$.
4) $\underline{\hat{v}}$ is continuous from above at the lower face, $\{x_3 = 1/\alpha\}$, where it is $\frac{\alpha^2}{k}$ $(0,0,-1)$ on the set$\{(x_1,x_2)$ Range W, $x_3 = 1/\alpha\}$ and $(0,0,0)$ otherwise
5) For some $\bar{\tau} > 0$:

$$\hat{T}_{\bar{\tau}}(x_1,x_2,1) = (W(x_1,x_2),\ 1/\alpha) \quad \forall\, (x_1,x_2) \in (0,1]^2$$

ii) We now define $\underline{v}$ in $(0,1]^3$ by:

1) $\underline{v} = \underline{\hat{v}}$ in $\{\underline{x} \in (0,1]^3 \mid 1 \geq x_3 > 0\}$
2) $\underline{v} = 0$ in $\{\underline{x} \in (0,1]^3 \mid (x_1,x_2) \notin \text{Range } W,\ 1/\alpha \geq x_3 > 0\}$
3) The remaining region

$\{\underline{x} \in (0,1]^3 \mid (x_1,x_2) \in \text{Ran } W,\ 1/\alpha \geq x_3 > 0\}$

is a union of k cubes. Define $\hat{v}_1$ in the "upper" part of each of them by

$$\underline{v}(\underline{x}) = \frac{\alpha^2}{k}\ \underline{v}(\underline{x}'), \tag{8}$$

where $\underline{x}'$ is the point in the unit cube which corresponds to $\underline{x}$ under the proper rotation and dialation which takes the small cube onto $(0,1]^3$.

4) Itterate ad infinitum the steps 2) and 3), properly scaled, in each of the smaller cubes.

iii) Clearly, the resulting vector field corresponds to a p.d. flow T_t which is uniquely defined in $(0,1]^3$. It has the properties:

1) For $\tau = \frac{\bar{\tau}}{1 - k/\alpha^3}$

$$T\,(x_1,x_2,1) = (\,L(W)(x_1,x_2),\ 0) \quad \forall\,(x_1,x_2) \in (0,1]^2$$

2) T_t has a uniform flux into the unit cube through the face $\{x_3=1\}$ described by the velocity $(0,0,-1)$.

3) T_t has no flux through the "side" faces, $\{x_{1,2}=0,1\}$, and through the complement of $RL(W) \times \{0\}$ on the face $\{x_3=0\}$.

4) T_t is measure preserving in the domain of its definition.

We are now ready to reach for the main examples. Let W_1, W_2 be mappings of the type discussed in constructions 1 and, correspondingly, 2. By means of construction 3, utilizing W_1 and then W_2, we obtain vector fields $\underline{v}_1$, $\underline{v}_2$ defined in $(0,1]^3$. In what follows Z denotes the reflection $Z:(x,y,z) \longmapsto (x,y,-z)$.

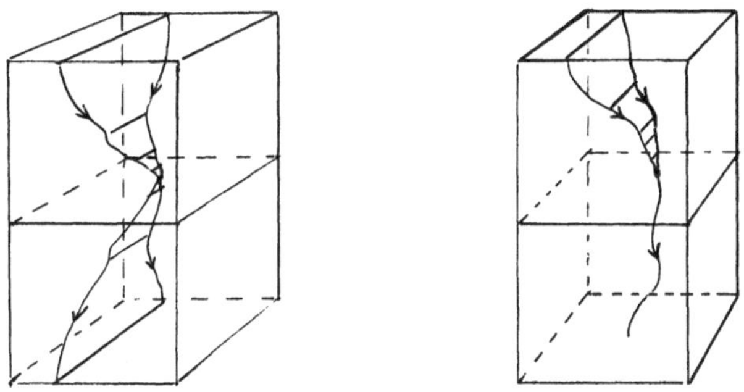

Figure 4 a) b)

Example 2 (non uniqueness): Define $\underline{u}_1$ in the set

$$D = (0,1]^3 \cup Z\,(0,1]^3$$

by:

$$\underline{u}_1(\underline{x}) = \underline{v}_2(\underline{x})$$

$$\underline{u}_1(Z\underline{x}) = -Z\underline{v}_2(\underline{x}), \qquad \forall\ \underline{x} \in (0,1]^3.$$

$\underline{u}_1$ is very simple on the boundry of D. It is $(0,0,-1)$ on the faces $\{x_3=\pm 1\}$ and tangential on the "sides". Such a vector field can be easily extended, so that in the complement of D it would define uniquely a flow with a uniformly bounded velocity field whose support is compact.

In D, $\underline{u}_1$ is the velocity field of various m.p. flows, all of which locally agree with $\hat{T}_t$ inside $(0,1]^3$ and with its reflection $(-Z)$ in $Z(0,1]^3$. Under these flows fibers are contracted to points and then expand again (see fig. 4a). The different extensions are parametrized by the collection of (m.p.) transformations which the fibers undergo at the instant of contraction to a point.

Example 3 (non existence): Define $\underline{u}_2$ in D by:

$$\underline{u}_2(\underline{x}) = \underline{v}_2(\underline{x})$$

$$\underline{u}_2(Z\underline{x}) = -Z\underline{v}_1(\underline{x}), \qquad \forall\ \underline{x} \in (0,1]^3 .$$

In the complement of D extend $\underline{u}_2$ as in the previous example.

Had $\underline{u}_2$ been the velocity field of a flow, it would contract fibers to points, as in the previous example, but in the region of expansion ($x_3<0$) it would remain 1-1 (see fig. 4b).Thus $\underline{u}_2$ is not the velocity field of any m.p. invertible flow. Nevertheless, the "net flux" of $\underline{u}_2$ into any set is zero and it satisfies condition (3).

$\underline{v}_1$, $\underline{v}_2$ have simple scaling properties. Using these, one gets an expression for the moments of $\underline{u}_1$, $\underline{u}_2$ which shows that

$$\underline{u}_{1,2} \in L^p \text{ if and only if } (p-1)(2-\nu)<1. \tag{9}$$

Here ν is the "dimension", as given by (7), of the set $RL(W_1) = RL(W_2)$.

As we expected, one may obtain counter-examples with less divergent velocities by having "larger" sets of singularities. The degree of divergence is measured here by the moments and the "size" of sets by the "dimension" (which is defined more precisely in the next section).

Perhaps the most surprising examples are obtained with $\nu = 2$. These result from $W_{1,2}$: $(0,1]^3 \longrightarrow (0,1]^3$ which are onto. The scaling factor in (8) is then 1 and one gets uniformly bounded $\underline{u}_1$, u_2. In other words, the condition in (9) extends to cover $\nu = 2$, $p = \infty$.

Similar effects occur in $\mathbb{R}^n$, $n > 3$. Examples are constructed in the same way, replacing $\mathbb{R}^2$ by hyperplanes. The inequality in (9) is then replaced by

$$(p-1)(n-\nu-1)<1. \tag{10}$$

V. A positive result

One would like to have a simple condition on $\underline{v}$ which will guarantee that the phenomena which was exhibited (non-uniqueness and non-existence) do not occur. As it follows from the examples, the criterion would have to request more than some bounds on the moments of $\underline{v}$. However, the finitness of some moment (i.e. $\int |\underline{v}|^p\, d^n\underline{x} < \infty$) is a usefull data. We shall now describe how it leads to the required conclusion, when suplemented by the additional information that $\underline{v}$ has no such singularity on the complement of a set of low dimension.

Definition 3: Let $S_-(\cdot) \leq 0 \leq S_+(\cdot)$ be two measurable functions which vanish only on sets of zero measure. A partially defined (p.d.) m.p. flow in R^n, defined up to (S_-, S_+), is a one parameter family of mappings, T_t, such that

1) each T_t takes $\{\underline{x} \in \mathbb{R}^n | S_-(\underline{x}) < t < S_t(\underline{x})\}$ into $\mathbb{R}_n$
2) $S_{\substack{+\\(-)}}(T_t \underline{x}) = S_{\substack{+\\(-)}}(\underline{x}) - t$
3) $T_{t_2} T_{t_1} \underline{x} = T_{t_1+t_2} \underline{x}$ whenever t_1, $t_1 + t_2$ $(S_-(x), S_+(\underline{x}))$
4) T_t are measure preserving.

Thus, $S_{\substack{+\\(-)}}(\cdot)$ serve as stopping times up to which the flow is locally defined.

Definition 4: Let $A \subset \mathbb{R}^n$ be a compact set, T_t a p.d. flow whose orbits are continuous (almost surely). T_t avoids A if the set

$$\{\underline{x} \in \mathbb{R}^n \setminus A | \inf_{S_-(\underline{x}) < t < S_+(\underline{x})} \text{dist.}(T_t \underline{x}, A) = 0 \}$$

has zero (Lebesgue) measure.

In the situations which we now have in mind, A is a set on whose complement the flow is uniquely defined by $\underline{v}$. We shall address the question

Q. 5: Let a m.p., p.d., flow be defined on the complement of a compact set $A \subset R^n$, i.e. $S_+(x) < \infty \Rightarrow T_{S_+(\underline{x})} \underline{x} \in A$ (defined by continuity). Under what conditions (on $\underline{v}$ and A) does the flow avoid A? (In which case the flow has a unique global "extention" - itself).

The answer which will be provided uses a notion of dimensionality. To introduce it we denote, for $A \subset \mathbb{R}^n$,

$$S_r(A) = \{\underline{x} \in \mathbb{R}^n | \text{ dist } (\underline{x}, A) = r\}$$

$|S_r(A)|$ - the surface measure of $S_r(A)$ (given by the n-1 dimensional Hausdorff measure, [5]). The dimension of "simple" sets (points, line segments etc.) is reflected in the power law by which $|S_r(A)|$ decreases as $r \to 0$. We now generalize this relation.

Definition 5: For any closed $A \subset \mathbb{R}^n$:

1) $\nu(A) = \inf \{\delta | \sup_{r<1} |S_r(A)| \; r^{n-1-\delta} < \infty \}$
2) Let C(A) denote the collection of countable closed coverings of A. The "dimension" of A is

$$\nu(A) = \inf \{ \sup_i \{\nu(A_i)\} \mid \{A_i\} \in C(A)\}.$$

With this terminology we answer Q. 5 by the following result [6].

Theorem: Let T_t be a measure preserving p.d. flow in $\mathbb{R}^n$ with the velocity field $\underline{v}$, such that

1) $\underline{v} \in L^p_{loc.}$ with $p > 1$.

Then T_t avoids any compact set $A \subset \mathbb{R}^n$ whose "dimension" satisfies

2) $(p-1)\ (n-1-\nu(A)) > 1$.

Remark: The same holds for transformations which preserve any other measure whose density (with respect to Lebesque measure) is uniformly bounded.

Notice that the examples constructed in section IV shown that the inequality can not be strictly weakened.

*Supported by U. S. National Science Foundation grant No. MCS 75-21684 A01.

References:

[1] E. Nelson, Topics in Dynamics-I: Flows, Princeton University Press (1969).

[2] E. Nelson, "Les ecoulements incompressibles d'energie finie", Colloques Internationaux du Centre National de la Recherche Scientifique, 117 p. 159, 1962.

[3] L.S. Young, "Entropy of continuous flows on compact 2-manifolds" (preprint).

[4] M. Aizenman, "On vector fields as generators of flows; a counterexample to Nelson's conjecture," Annals of Mathematics (to appear).

[5] H. Federer, Geometric Measure Theory, Springer-Verlag (1969).

[6] M. Aizenman, "A sufficient condition for the avoidance of sets by measure preserving flows in $\mathbb{R}^n$" (preprint).

ON THE UNIQUENESS OF THE EQUILIBRIUM STATE FOR PLANE ROTATORS

J. Bricmont
J.R. Fontaine
Institut de Physique Thèorique
Universitè de Louvain
Louvain-la-Neuve, Belgium

L. Landau
Department of Mathematics
Bedford College
University of London
Regent's Park NWI, England

Abstract: We study the classical statistical mechanics of the plane rotator, and show that there is a unique translation invariant equilibrium state in zero external field, if there is no spontaneous magnetization. Moreover, this state is then extremal in the equilibrium states. In particular there is a unique phase for the two dimensional rotator, and a unique phase for the three dimensional rotator above the critical temperature. It is also shown that there is a unique equilibrium state in a sufficiently large external field.

1. - The Model and Results.

At each point i of the lattice Z^d is associated a spin variable $\underset{\sim}{\sigma}_i \in R^2$. The a priori probability distribution for $\underset{\sim}{\sigma}_i$ is assumed rotation invariant:

$$d\nu(\underset{\sim}{\sigma}_i) = d\lambda(r_i)\, d\phi_i$$

The measure $d\lambda$ must satisfy certain conditions [2], which are satisfied for example by the fixed length ($d\lambda = \delta(r-b)\,dr$) and uniform ($d\lambda = \chi(r\in[0,b])\,d\nu$) distributions. The interaction between spins is given by a traslation invariant, ferromagnetic pair interaction:

$$H = -\Sigma_{ij} J_{ij}\, \underset{\sim}{\sigma}_i \cdot \underset{\sim}{\sigma}_j - \underset{\sim}{h} \cdot \Sigma\, \underset{\sim}{\sigma}_i$$

where $J_{ij} = J(i-j) \geq 0$ and $\underset{\sim}{h}$ denotes the external field.

An equilibrium state is an infinite volume limit of finite region Gibbs states with some boundary condition, or equivalently a probability measure on the infinite volume configuration space satisfying the DLR equations.

A phase is an equilibrium state which is invariant under the lattice translations.

Definition. The spontaneous magnetization M is defined by

$$M = \lim_{\|\underset{\sim}{h}\|\to 0} \rho_{\underset{\sim}{h}}(\underset{\sim}{n}\cdot\underset{\sim}{\sigma}_i)$$

where $\underset{\sim}{n}$ is a unit vector in the direction of the external field $\underset{\sim}{h}$ and $\rho_{\underset{\sim}{h}}$ is any phase in the external field $\underset{\sim}{h}$.

We remark that m is independent of the choice of $\rho_{\underset{\sim}{h}}$ [2]. Our basic results are [2]:

Theorem 1. If the spontaneous magnetization $M = 0$ then there is a unique phase in zero external field. Moreover, this phase is then extremal in the equilibrium states. This phase is also the unique quasi-periodic state.

Theorem 2. Let $d\lambda(z) = \delta(z-b)\,dz$. Then for a large enough external field the equilibrium state is unique.

Corollary

If the lattice dimension d=2, there is a unique phase in zero external field. If d=3, there is a unique phase for $T > T_c$, where Tc is the critical temperature for spontaneous magnetization.

2. Method of Proof.

To show uniqueness of the phase (Th. 1) we use the equivalence between phases and tangents to the graph of the pressure [5,6]. We consider the perturbation to the Hamiltonian H:

$$H_\lambda = H - \lambda \sum \tau_i f$$

where we sum over the lattice translates of f. If we canshow that the pressure P_λ is differentiable at λ =0 it follows that all invariant equilibrium states take the same value on f.
More precisely:

Lemma. Let there exist a sequence of positive numbers $\{\lambda_n\}_{n\in N}$ and another one of negative numbers $\{\lambda'_n\}_{n\in N}$, both converging to zero, and phases $\rho_{\lambda'_n}$, ρ_{λ_n} of $H_{\lambda'_n}$, H_{λ_n} such that $\lim_{n\to\infty} \rho_{\lambda'_n}(f) = \lim_{n\to\infty} \rho_{\lambda_n}(f)$. Then all phases take the same value on f.

In order to apply the lemma, one applies Ginibre's inequalities [4] and generalized Griffith's inequalities [1,2,3] to show that the limit $\lambda \searrow 0$ coincides with the limit $\lambda' \nearrow 0$.

To show uniqueness of the equilibrium state we use the fact that in a sufficiently large external field, the generalized Griffith's inequalities are satisfied by states with arbitrary boundary conditions. This enables us to bound all equilibrium states by two states ρ_M and ρ_m. From the Lee-Yang theorem and an inductive procedure we show the identity of ρ_M with ρ_m and this gives uniqueness of the equilibrium state.

3. References.

(1) Bricmont, J.: Correlation inequalities for two-component fields. Ann. Soc. Sc. Brux. 90 (1976) 245-252.
(2) Bricmont, J., Fontaine, J. R., Landau, L.J.: On the Uniqueness of the Equilibrium State for Plane Rotators. To appear in Comm. math. Phys.
(3) Dunlop, F.: Correlation inequalities for multicomponent rotators. Comm. math. Phip. 49 (1976) 247-256.
(4) Ginibre, J.: General Formulation of Griffith's Inequalities. Comm. math. Phys. 16 (1970) 310-328.
(5) Israel, R.: Thesis, Princeton University.
(6) Ruelle, D.: Statistical Mechanics (New York, Benjamin 1969).

A GEOMETRIC APPROACH TO THE SOLUTION OF CONFORMAL INVARIANT FIELD EQUATIONS

E.A. TAGIROV
Laboratory of Theoretical Physics, Joint Institute for Nuclear Research
141980 Dubna, USSR
I.T. TODOROV
Institute of Nuclear Research and Nuclear Energy, Bulgarian Academy of Sciences
Sofia 1113, Bulgaria

Summary.

The analogue of a massless scalar field equation with a dimensionless self-coupling in a (curved)n-dimensional space-time V_n is

$$\left(\Box + \frac{n-2}{4(n-1)} R\right)\varphi + \lambda \varphi^{\frac{n+2}{n-2}} = 0 \qquad (1)$$

where $\Box$ is the covariant d'Alembert operator and R is the scalar curvature of V_n. (The necessity of the 1/6R term - for n = 4 - for the conformal invariance and the correct physical interpretation of the corresponding free field equation - with λ = 0 - has been made clear in (1,2)). Eq. (1) is left invariant under the conformal mapping

$$V_n \rightarrow \bar{V}_n, \qquad g_{\mu\nu}(x) \rightarrow \bar{g}_{\mu\nu}(x) = \Omega^2(x) g_{\mu\nu}(x); \qquad (2)$$

if φ is a solution of (1) in V_n, then

$$\bar{\varphi}(x) = \Omega^{1-\frac{n}{2}} \varphi(x) \qquad (3)$$

is a solution of the corresponding equation in $\bar{V}_n$. The transformation law for the scalar curvature can be written in the form

$$\left[\Box + \frac{R}{2(n-1)}\right]\Omega - \frac{\bar{R}}{2(n-1)}\Omega^3 + \frac{n-4}{2\Omega} g^{\alpha\beta}\partial_\alpha\Omega\,\partial_\beta\Omega = 0 \qquad (4)$$

We observe that for n = 4 and R = const Eq.(4) coincides with Eq.(1).This observation allows to reduce the problem of classifying the solutions of the nonlinear wave equation

$$\Box\varphi + \lambda\varphi^3 = 0 \qquad (5)$$

in Minkowski space, according to their invariance groups, to the known classification of conformally flat spaces of constant scalar curvature by their isometry groups (3). If the group G of V_n has k parameters ($k \leqslant 10$) then the $SO(4,2)/Z_2$ conformal transformations of $\Omega(x)$ give rise to a 15-k family of solutions.

The solutions corresponding to the local factors Ω of the de Sitter and the anti de Sitter spaces are

$$\varphi_{\rho\varepsilon}(x-a) = \sqrt{\frac{-2\varepsilon}{\lambda}}\,\frac{2\rho}{\rho^2-\varepsilon(x-a)^2}\ ;\ \rho>0,\ \mathrm{sign}\,\lambda=\varepsilon \tag{6}$$

where ε is + or -1, respectively. The ε = -1 solutions have been studied in (4), while the ε = 1 ones correspond to Lipatov's pseudoparticle solutions (5), used for the saturation of functional integral representations of Euclidean Green functions. The solutions corresponding to the static cylindric universe $\tilde{M}$, which plays the role of the universal covering of the conformal compactification of Minkowski space,

$$\varphi_{\ell}(x-a,u) = \frac{2}{\ell\sqrt{-\lambda}}\left\{\left[1-\left(\frac{x-a}{\ell}\right)^2\right]^2+\left[\frac{2u(x-a)}{\ell}\right]^2\right\}^{-\frac{1}{2}};\quad \begin{matrix} u=(u^0,\underline{u}),\\ u^2=1, \end{matrix} \tag{7}$$

have also been known before (6). They are remarkable for being everywhere regular (and bounded) and for carrying a finite energy. A number of previously unknown solutions have also been found in this manner.

The Corrigan-Fairlie-'t Hooft-Wilczek ansatz

$$A_\mu(x) = q_{\sigma\mu}\,\partial_\sigma \ln\varphi(x) \tag{8}$$

where the matrices $q_{\sigma\mu}$ satisfy the commutation relations

$$[q_{\sigma\mu}, q_{\tau\nu}] = \delta_{\sigma\tau}q_{\mu\nu}+\delta_{\mu\nu}q_{\sigma\tau}-\delta_{\sigma\nu}q_{\mu\tau}-\delta_{\mu\tau}q_{\sigma\nu}$$

of the Lie algebra SO(4), gives to any (pure imaginary time) solution of (5) a solution of the (Euclidean) Yang-Mills equations (8,9). In particular, the Euclidean counterpart of the solution (6) for ε = 1 corresponds to the Yang-Mills instanton solution first found in (7). The same type of ansatz was also used to study Minkowski space Yang-Mills equations (8,9). That makes our analysis also relevant for this problem.

References.

1) R.Penrose, Conformal treatment of infinity, in: Relativity, Groups and Topology, ed. C.M.De Witt and B.De Witt, Les Houches Summer School, 1963(Gordon and Breach N.Y.; 1964) pp. 565-584.

2) N.A.Chernikov, E.A.Tagirov, Quantum theory of scalar field in de Sitter space--time, Ann. Inst. H.Poincarè 9, 109 (1968).

3) A.Z.Petrov, New Methods in General Relativity (Nauka, M., 1965); see also A.Z.Petrov, Einstein Spaces (Pergamon Press, Oxford, 1969) (especially Chapter 6, pp. 257-275).

4) S.Fubini, A new approach to conformal invariant field theories, Nuovo Cimento 34A, 521 (1976).

5. L.N.Lipatov, Divergence of the perturbation series and pseudoparticles, Zh. Eksp. Teor. Fiz., Pisma 25, 116 (1977).

6) L.Castell , Exact solutions of the $\lambda\varphi^4$ theory, Phys. Rev. D6, 536 (1972).

7) A.A.Belavin, A.M.Polyakov, A.S.Schwartz, Yu.S.Tyupkin, Pseudoparticle solutions of the Yang-Mills equations, Phys. Letters 59B, 85 (1975).

8) J.Cervero, L.Jacobs, C.Nohl, Elliptic solutions of classical Yang-Mills theory, Phys. Letters 69B, 351 (1977).

9) W.Bernreuther, A note on classical solutions of the Yang-Mills equations in Minkowski space, C.T.P. Publ. 626, MIT (1977).

STABILITY, DETAILED BALANCE AND KMS CONDITION FOR QUANTUM SYSTEMS

Alberto Frigerio (°), Vittorio Gorini
Istituto di Fisica dell'Università, Milano, Italy
Istituto Nazionale di Fisica Nucleare, Sezione di Milano, Italy

Maurizio Verri (*)
Istituto di Matematica del Politecnico, Milano, Italy
Istituto Nazionale di Fisica Nucleare, Sezione di Milano, Italy

Abstract. We give characterizations of the states of thermodynamic equilibrium of infinitely extended and finite quantum systems in terms of stability under interactions and of a quantum condition of detailed balance.

1. Among the fundamental problems in statistical mechanics is the characterization of the states of thermodynamic equilibrium of large (macroscopic) systems. A large system in a state of thermodynamic equilibrium can be regarded as a heat bath for small systems in thermal contact with it.

Therefore, it should be possible to characterize an equilibrium state ω of a large system R by the requirement that it acts as a heat bath for any small system S weakly coupled to it, in the sense that it drives the small system to a terminal state ρ which is independent of the where and the how of the coupling, for a sufficiently large class of couplings [1, 2]. Here we give a characterization of this property in the framework of the rigorous theory of the weak coupling limit [3] and within the algebraic approach to quantum statistical mechanics. In this way, we are able to show that this property implies the KMS condition on R, that the terminal state of S is the canonical state ρ_β at the same temperature of the bath, and that the reduced dynamics of S satisfies a fully quantum mechanical condition

---

(°) A fellowship from the Italian Ministry of Public Education is acknowledged.

(*) A fellowship from the Italian National Science Council (C.N.R.) is acknowledged.

of detailed balance with respect to ρ_β [1, 4]. Thus we obtain an independent justification of the KMS condition as the correct characterization of a state of thermodynamic equilibrium for finite as well as for infinitely extended quantum systems.

If one is not interested in the details of the reduced dynamics of S, but in the justification of the KMS condition alone, one can characterize thermal equilibrium between S and R by the requirement that the state $\rho \otimes \omega$ is stable under interactions between the two systems. From this the KMS condition can be derived, with the use of clustering properties which are milder than those introduced so far in the literature.

2. We describe R by a triple $(\mathfrak{A},\{\alpha_t\},\omega)$, $\mathfrak{A}$ a C^*-algebra, $\{\alpha_t\}_{t\in\mathbb{R}}$ a strongly continuous group of *-authomorphisms of $\mathfrak{A}$, and ω a state on $\mathfrak{A}$ such that $\omega\circ\alpha_t=\omega$ (it would be sufficient to require $\{\alpha_t\}$ to be a weakly*-continuous group of *-authomorphisms of the von Neumann algebra $\pi_\omega(\mathfrak{A})''$, leaving ω invariant). The spatially confined quantum system S is described by an algebra of observables $\mathcal{B}(\mathcal{H})$, $\mathcal{H}$ a separable Hilbert space, with a Hamiltonian H such that $\exp(-\beta H)$ is trace class for all $\beta>0$.

Assume that there exists a suitable self-adjoint subset $\mathcal{R}$ of $\mathfrak{A}$, stable under $\{\alpha_t\}$, such that for all couplings of the form

$$\lambda V = \lambda \sum_{i=1}^{n} F_i \otimes \varphi_i$$

$F_i=F_i^*\in\mathcal{B}(\mathcal{H})$, $\varphi_i=\varphi_i^*\in\mathcal{R}$, $\omega(\varphi_i)=0$, Davies' weak coupling limit technique [3] can be applied, leading to a description of the reduced dynamics of S in the interaction picture by a dynamical semigroup [5]. The explicit form of the generator of the latter is [3, 1]:

$$L(\sigma)=\sum_{\substack{r,r',s,s'\\ \varepsilon_r-\varepsilon_{r'}=\varepsilon_s-\varepsilon_{s'}}}\ \sum_{i,j=1}^{n}(F_j)_{rr'}(F_i)_{s's}\Big(-i\,s_{ij}(\varepsilon_r-\varepsilon_{r'})\,\delta_{r's'}\,[P_{rs},\sigma] + \hat{h}_{ij}(\varepsilon_r-\varepsilon_{r'})\Big[P_{s's}\,\sigma\,P_{rr'}-\tfrac{1}{2}\{P_{rr'}P_{ss'},\sigma\}\Big]\Big), \tag{1}$$

where $\sigma\in\mathcal{T}(\mathcal{H})$, the space of trace class operators on $\mathcal{H}$, and

(i) $P_{rr'}=|r\rangle\langle r'|$, $\{|r\rangle\}$ a complete orthonormal set (c.o.n.s.) of eigenvectors of H, $H|r\rangle=\varepsilon_r|r\rangle$,

(ii) $(F_j)_{rr'} = \langle r | F_j | r' \rangle$,

(iii) $\hat{h}_{ij}(\lambda) = \int_{-\infty}^{\infty} dt \; e^{-i\lambda t} \, h_{ij}(t), \quad h_{ij}(t) = \omega(\varphi_j \, \alpha_t(\varphi_i)),$

(iv) $s_{ij}(\lambda) = i \int_0^{\infty} dt \; e^{-i\lambda t} \, h_{ij}(t) - \frac{i}{2} \hat{h}_{ij}(\lambda)$.

The series (1) converges in the trace norm.

Remark that L depends on the reservoir observables φ_j only through their two-point correlation functions in the state ω. However, the application of the weak coupling limit technique requires conditions on the whole set of n-point correlation functions for all n , which can only be checked when R is quasifree [3].

In the above framework, we characterize $R = (\mathfrak{A}, \{\alpha_t\}, \omega)$ as a heat bath by the condition that there exists a state $\rho \in \mathcal{T}(\mathcal{H})$ which is stationary under the free dynamics of S and under all the dynamical semigroups obtained in the weak coupling limit for all admissible couplings, i.e.

(i) $[H, \rho] = 0$ and

(ii) $L(\rho) = 0$ for all couplings.

It is possible to give conditions on the effectiveness of the coupling in order that ρ is approached as $t \to \infty$ by all initial states [6]. However, these are not directly relevant to our problem.

3. The proof of the KMS condition runs as follows.

(a) Let the spectrum of H be nondegenerate. Then, by (i), $\rho = \sum_r \rho_r P_{rr}$. A straightforward computation [1], using the arbitrariness of the coupling, shows that (ii) implies

$$\rho_r \hat{h}_{ji}(-\lambda) = \rho_s \hat{h}_{ij}(\lambda) \quad \text{if} \quad \varepsilon_s - \varepsilon_r = \lambda, \tag{2}$$

(b) If ρ is faithful, it follows from (2) that there is a function

$$\mu(\lambda) = \hat{h}_{ji}(-\lambda) / \hat{h}_{ij}(\lambda) = \rho_s / \rho_r \quad \text{if} \quad \varepsilon_s - \varepsilon_r = \lambda, \tag{3}$$

which is positive and multiplicative. Then, since by our hypotheses the functions $\hat{h}_{ij}(\lambda)$ are continuous [3], we have

$$\mu(\lambda) = \exp(-\beta\lambda) \quad \text{for some real} \quad \beta. \tag{4}$$

Combining (3) with (4), it follows that ω satisfies the KMS condition on $\mathcal{R}$ at inverse temperature β, and that $\rho = \rho_\beta = (\mathrm{Tr}\, e^{-\beta H})^{-1} e^{-\beta H}$

If $\mathcal{H}$ is infinite-dimensional, this forces β to be positive.

(c) If we do not assume ρ to be faithful, but require its functional form $\rho = \rho(H)$ not to depend on H, we can easily see that (2) leads to the following alternative:

(c_1) either $\rho_s = \delta_{rs}$, with ε_r the smallest eigenvalue of H, and the support of all $\hat{h}_{ij}(\lambda)$ is contained in the positive half-line;[1]

(c_2) or ρ is faithful.

In the latter case, the KMS condition follows as in (b). In the former case, ω is a ground state, at least when restricted to $\mathcal{R}$.

If R is a quasifree fermion reservoir and ω is a quasifree state, $\mathcal{R}$ can be chosen to be the set of finite linear combinations of creation and annihilation operators with compact support in energy space, and the KMS condition (or the ground state property) can be extended to the whole algebra by means of the expansion formulas for the n-point correlation functions. In general, a sufficient condition in order to be able to perform the extension of the KMS condition to $\mathfrak{A}$ is that $\mathcal{R}$ can be chosen such that $\mathcal{R} \cup \{\mathbb{1}\}$ is either a uniformly dense subset or a strongly dense $*$-subalgebra of $\mathfrak{A}$ [1].

In view of the difficulty of checking Davies' conditions for a sufficiently large $\mathcal{R}$ when the reservoir is not quasifree, it is worthwhile to remark that the relation (2) can be derived from a stability condition on the state $\rho \otimes \omega$ and the assumption that ω is L^1-clustering in time for the observables of a uniformly dense self-adjoint subset $\mathcal{S}$ of $\mathfrak{A}$.

In this case one finds, as in [7],

$$\int_{-\infty}^{\infty} dt\, \rho\otimes\omega\,(V\alpha_t^\circ(W)) = \int_{-\infty}^{\infty} dt\, \rho\otimes\omega\,(\alpha_t^\circ(W)\,V) \tag{5}$$

[1] If $\mathcal{H}$ is finite-dimensional, ε_r might also be the largest eigenvalue of H, and the support of $\hat{h}_{ij}(\lambda)$ would be contained in the negative half-line.

for all V,W in $\mathcal{B}(\mathcal{H})\otimes\mathcal{S}$, where $\alpha_t^{o}(B\otimes A)= e^{iHt}Be^{-iHt}\otimes\alpha_t(A)$. Taking $V=|r\rangle\langle s|\otimes\varphi_i$ and $W=|s\rangle\langle r|\otimes\varphi_j$,(5) yields (2), and we reach the same conclusions as before. Notice that we do not need any asymptotic abelianess in norm [8] , nor clustering properties of higher order correlation functions [7,8] .

4. We remark that (2) is the proper quantum generalization of the property of detailed balance. Indeed, if the spectrum of H is nondegenerate, the subspace of density matrices which are diagonal in the energy representation is invariant under the dynamical semigroup which describes the reduced dynamics of S in the interaction picture, and the evolution in this space is determined by a Pauli master equation with transition rates

$$W_{rs}= \sum_{i,j=1}^{n} (F_j)_{sr}(F_i)_{rs}\hat{h}_{ij}(\varepsilon_s-\varepsilon_r) . \tag{6}$$

Then, (2) yields [3,9]

$$W_{rs}\,\rho_s = W_{sr}\,\rho_r , \tag{7}$$

which is the familiar detailed balance condition for a classical discrete Markov process [10]. On the other hand, the insertion of (2) in the full quantum generator L provides conditions also on the transition rates among off-diagonal matrix elements, and these, together with (7), express detailed balance in its full quantum mechanical form.
It is worthwhile to observe that quantum detailed balance can be expressed in a purely algebraic form, which makes no reference to the reservoir. Indeed, by passing to the dual generator L^* on $\mathcal{B}(\mathcal{H})$, one sees that (2) amounts to the decomposition $L^*= L_h^* + L_s^*$, where L_h^* is a Hamiltonian generator, and where

$$\langle L_h^*(A), B\rangle =-\langle A, L_h^*(B)\rangle , \quad \langle L_s^*(A), B\rangle =\langle A, L_s^*(B)\rangle \tag{8}$$

for all A, B in $\mathcal{B}(\mathcal{H})$, with $\langle A, B\rangle=\rho(A^*B)$, ρ being the stationary state [4, 1] .

5. We have shown here that a heat bath is KMS via the property of detailed balance. However, we believe that this property is important in its own right, and not merely instrumental in the derivation of the KMS

condition. Indeed, we have shown in [1] that detailed balance (in the form (8) implies the KMS condition even when the state ρ is not assumed a priori to be independent of the coupling. Conversely, if one assumes ω to be KMS at inverse temperature β, then (2) holds with $\rho = \rho_\beta$, $R = (\mathfrak{A}, \{\alpha_t\}, \omega)$ is a heat bath and the reduced dynamics of S satisfies detailed balance with respect to ρ_β [3,9]. Furthermore, quantum detailed balance has important applications to the derivation of the Onsager relations and to entropy production, as discussed in [4,9,11].

References.

1. A.Kossakowski, A.Frigerio, V.Gorini and M.Verri: Quantum Detailed Balance and KMS Condition, Commun. Math. Phys., to appear.
2. G.L.Sewell: Ann. Phys. (N.Y.) 85, 336 (1974), and Lecture notes at the Scuola di Perfezionamento in Fisica, University of Milan (unpublished).
3. E.B.Davies: Commun. Math. Phys. 39, 91 (1974).
4. R.Alicki: Rep. Math. Phys. 10, 249 (1976).
5. A.Kossakowski: Rep Math. Phys. 3, 247 (1972); V.Gorini, A.Kossakowski and E.C.G.Sudarshan: J.Math Phys. 17, 821 (1976); G. Lindblad: Commun. Math. Phys. 48, 119 (1976).
6. H.Spohn: An Algebraic Condition for the Approach to Equilibrium of an Open N-level System, Lett. Math. Phys., to appear; D.E.Evans: Irreducible Quantum Dynamical Semigroups, Commun. Math. Phys. 54, 293 (1977); A.Frigerio: Quantum Dynamical Semigroups and Approach to Equilibrium, preprint University of Milan, 1977.
7. R.Haag, D.Kastler and E.Trych-Pohlmeyer:Commun. Math. Phys.38, 173 (1974).
8. O.Bratteli, D.Kastler: Commun. Math. Phys. 46, 37 (1976).
9. K.Hepp: Z.Phys. B20, 53 (1975); Lecture Notes in Physics 39,138 (1975).
10. S.R.De Groot, P.Mazur: Nonequilibrium Thermodynamics, North Holland Publishing Company, Amsterdam 1962.
11. H.Spohn, J.Lebowitz: Irreversible Thermodynamics for Quantum Systems Weakly Coupled to Thermal Reservoirs, Adv. Phys. Chem., to appear.

STOCHASTICITY AND IRREVERSIBILITY IN INFINITE MECHANICAL SYSTEMS

Gérard G. EMCH

Dpts of Mathematics and of Physics, University of Rochester (USA)

The concepts of Kolmogorov entropy and Kolmogorov flows are extended to quantum dynamical systems described in the language of von Neumann algebras. This generalization carries over to the quantum realm the result that the entropy of non-singular K-flows is strictly positive ; in particular, this entropy is again infinite for the quantum generalization of the flow of Brownian motion.

I. INTRODUCTION. Our aim is to extend the classical theory of Kolmogorov and Sinai [1,2,3,4] to situations encountered in Quantum Statistical Mechanics. Specifically a classical flow $\{\Omega,\mu,T(\mathbb{R})\}$ can be viewed algebraically as a triple $\{\mathfrak{N},\phi,\alpha(\mathbb{R})\}$ where : $\mathfrak{N}$ is a von Neumann algebra acting on a separable Hilbert space $\mathfrak{H}$ [namely $\mathfrak{N} = \mathcal{L}^\infty(\Omega,\mu)$, the elements of which are regarded as bounded multiplication-operators on $\mathfrak{H} = \mathcal{L}^2(\Omega,\mu)$] ; ϕ is a faithful normal state on $\mathfrak{N}$ [namely $\phi : f \in \mathcal{L}^\infty(\Omega,\mu) \mapsto \langle\phi;f\rangle = \int_\Omega d\mu(\omega)f(\omega) \in \mathbb{C}$] ; and $\alpha(\mathbb{R})$ is a strong-op. continuous, one-parameter group of automorphisms of $\mathfrak{N}$ such that $\phi\circ\alpha(t) = \phi$ for all t in $\mathbb{R}$ [namely $\alpha(t)[f](\omega) = f(T(t)[\omega])$]. A generalized flow $\{\mathfrak{N},\phi,\alpha(\mathbb{R})\}$ is defined by omitting in the above description the condition that $\mathfrak{N}$ be abelian [and thus renouncing to the particular features written above in square-brakets] . As an example, consider the quasi-free generalized flow defined from a triple $\{\mathfrak{K}, C, U(\mathbb{R})\}$ where $\mathfrak{K}$ is a separable Hilbert space ; C is a self-adjoint operator on $\mathfrak{K}$ with $0 < C < I$; and $U(\mathbb{R})$ is a continuous one-parameter group of unitary operators acting on $\mathfrak{K}$, such that $[U(t), C] = 0$ for all t in $\mathbb{R}$. $\{\mathfrak{N}, \phi, \alpha(\mathbb{R})\}$ is then constructed as follows. To $\hat{\phi} : k \in \mathfrak{K} \mapsto \exp\{-\|k\|^2/4\}$ and $\nu : (k_1,k_2) \in \mathfrak{K}\times\mathfrak{K} \mapsto \exp\{i\,\mathrm{Im}\,(Ck_1,k_2)/2\}$ corresponds a non-Fock, factor representation $W(\mathfrak{K})$ of the CCR on $\{\mathfrak{K},\nu\}$. We identify $\mathfrak{N}$ as $W(\mathfrak{K})''$; ϕ as the normal extension of $\hat{\phi}$ to $\mathfrak{N}$; and $\alpha(t)$ as the normal extension to $\mathfrak{N}$ of $\alpha(t) : W(k) \in W(\mathfrak{K}) \mapsto W(U(t)k) \in W(\mathfrak{K})$.

For a classical flow $\{\Omega,\mu,T(\mathbb{R})\}$ the concept of dynamical entropy can be approached in two ways which are conceptually different, but nevertheless mathematically equivalent. Whereas in both approaches the dynamical entropy $H_\mu(T)$ is defined as the sup of $H_\mu(\zeta,T)$ over all (finite) measurable partitions ζ of Ω, the difference between the two approaches comes in the definition of $H_\mu(\zeta,T)$, the entropy of the partition ζ under the evolution $T(\mathbb{R})$.

In the first approach, one defines :

$$H_\mu(\zeta,T) \equiv \lim_{n\to\infty} H_\mu(\zeta \vee \ldots \vee T^n[\zeta])/(n+1)$$

where : $T = T(1)$; $\zeta \vee \ldots \vee T^n[\zeta]$ is the smallest measurable partition of Ω which refines all $T^k[\zeta]$ $(0 \leq k \leq n)$; and, for any (finite) partition ζ of Ω into mutually disjoint measurable subsets ζ_k:

$$H_\mu(\zeta) = \Sigma_k\, h[\mu(\zeta_k)]$$
$$h : x\epsilon[0,1] \mapsto -x\log x \;\epsilon\; \mathbb{R}^+.$$

In the second approach, one defines :

$$H_\mu(\zeta,T) \equiv \lim_{n\to\infty} H_\mu(\zeta|T^{-1}[\zeta] \vee \ldots \vee T^{-n}[\zeta])$$
$$= \lim_{n\to\infty} H_\mu(T^n[\zeta]|\; \zeta \vee T[\zeta] \vee \ldots \vee T^{n-1}[\zeta]),$$

where $H_\mu(\zeta|\zeta_1)$ is the entropy of ζ conditioned by ζ_1 with respect to μ.

When it comes to generalize the dynamical entropy $H_\mu(T)$ to a generalized flow $\{\mathcal{N},\phi,\alpha(\mathbb{R})\}$, where $\mathcal{N}$ is not abelian, two difficulties have to be mastered.

The <u>first difficulty</u> is that, with ζ denoting an arbitrary (finite) partition of the identity into mutually orthogonal projectors F_k in the non-abelian von Neumann algebra $\mathcal{N}$, the measurement of ζ can (!) perturb the state ϕ on $\mathcal{N}$, thus introducing a stochastic element which does <u>not</u> pertain to the time-evolution.

The <u>second difficulty</u> is that ζ and $\alpha(t)[\zeta]$ do in general not commute, so that the question arises as to what object should take the place of the minimal refinement $\vee$.

Connes and Størmer [5] succeeded in extending the first definition of $H_\mu(\zeta,T)$ in such a manner that it becomes useful for the classification of Bernoulli shifts on the hyperfinite II_1-factor. This generalization however misses one physically important feature, namely that the second definition of $H_\mu(\zeta,T)$ has an immediate interpretation as the information gained by repeating a measurement of ζ in the course of time ; in that sense $H_\mu(T)$ is a measure of the "stochasticity" present in the "deterministic" evolution $T(\mathbb{R})$. In contrast, we concentrate on this interpretation, and thus propose [6,7] instead an extension of the second definition, keeping in mind that even a quantum measurement only involves partitions of the identity into mutually orthogonal projections in $\mathcal{N}$.

II. ADMISSIBLE PARTITIONS. To handle the first of the two difficulties mentioned in the introduction, we recall two facts.

First fact : According to von Neumann [8] , if ζ is a partition of the identity on $\mathfrak{H}$ into mutually orthogonal projectors F_k in $\mathcal{N}$, the effect of the measurement of ζ is to change the state ϕ into a new state $\zeta[\phi]$ given by :

$$\zeta[\phi] = \Sigma_k \lambda_k \phi_k \quad \text{where}$$
$$\lambda_k = \langle\phi; F_k\rangle; \ \langle\phi_k; \bullet\rangle = \lambda_k^{-1} \langle\phi; F_k \bullet F_k\rangle.$$

Second fact : According to the Tomita-Takesaki theory [9] , there exists a unique continuous, one-parameter group $\sigma(\mathbb{R})$ of automorphisms of $\mathcal{N}$ with respect to which ϕ satisfies the KMS boundary condition. Moreover the algebra $\mathcal{N}_\phi$ of fixed points of $\mathcal{N}$ under $\sigma(\mathbb{R})$ coincides with :

$$\{N\in\mathcal{N} \mid \langle \phi; NM-MN \rangle = 0 \text{ for all } M \text{ in } \mathcal{N}\}.$$

Note in particular that if $\mathcal{N}$ is abelian $\mathcal{N}_\phi = \mathcal{N}$ and $\sigma(t) = id$ for all t in $\mathbb{R}$. Returning to our general case, we have :

Lemma 1 : The following conditions on a partition ζ in $\mathcal{N}$ are equivalent : (i) $\zeta[\phi] = \phi$; (ii) $(\alpha(t)[\zeta])[\phi] = \phi$ for all t in $\mathbb{R}$; (iii) $\zeta \subset \mathcal{N}_\phi$ (iv) $\{\alpha(t)[\zeta] \mid t\in\mathbb{R}\}'' \subseteq \mathcal{N}_\phi$.

We call admissible a partition $\zeta\subset\mathcal{N}$ which satisfies one (and thus all) of the conditions of this lemma. We denote by Z the set of all admissible partitions ; by ζ'' the von Neumann algebra generated by ζ ; and by M the set of all von Neumann subalgebras of $\mathfrak{N}_\phi$. Clearly every partition $\zeta \subset \mathfrak{N}$ is admissible if and only if ϕ is a trace on $\mathfrak{N}$, which is in particular the case when $\mathfrak{N}$ is abelian.

III. CONDITIONAL ENTROPY. As in the classical theory [10] the entropy of a partition $\zeta = \{F_k\} \subset \mathfrak{N}$ with respect to a state ψ on $\mathfrak{N}$ is to be defined as

$$H_\psi(\zeta) = \Sigma_k\, h\,[\langle \psi ; F_k \rangle] \quad \text{with}$$

$$h : x \in [0,1] \mapsto -x\log x \in \mathbb{R}^+ .$$

Lemma 2 : Let $\zeta = \{G_j\}$ and $\zeta_1 = \{F_k\}$ be two (finite) admissible partitions, $\mathfrak{X} = \zeta''$ and $\mathfrak{X}_1 = \zeta_1''$. Then :

$$H_\phi(\zeta|\zeta_1) \equiv \Sigma_k\, \lambda_k\, H_{\phi_k}(\zeta)$$

$$= \Sigma_j \langle \phi \, ; \, h\,[\mathcal{E}(G_j|\mathfrak{X}_1)] \rangle$$

$$= \mathrm{Sup}_{x \in S} \{\Sigma_k (\langle \phi ; h\,[\mathcal{E}(x_k|\mathfrak{X})] \rangle - \langle \phi ; h\,[\mathcal{E}(x_k|\mathfrak{X}_1)] \rangle)\}$$

where : λ_k and ϕ_k are defined as in section II ;
$\mathcal{E}(.|\mathfrak{X})$ is the unique faithful normal conditional expectation from $\mathfrak{N}$ onto $\mathfrak{X}$ with $\phi \circ \mathcal{E}(.|\mathfrak{X}) = \phi$; S is the set of all finite families $x = \{x_k\}$ of positive elements in $\mathfrak{N}_\phi$ with $\Sigma_k\, x_k = I$.

Clearly if $\mathfrak{N}$ is abelian $H_\phi(\zeta|\zeta_1)$ reduces to the usual entropy of ζ conditioned by ζ_1 with respect to ϕ.

Lemma 3 : The mapping

$$H_\phi : (\zeta ; \mathfrak{M}) \in Z \times M \mapsto H_\phi(\zeta|\mathfrak{M}) \in \mathbb{R}$$

defined by $H_\phi(\zeta|\mathfrak{M}) \equiv \mathrm{Inf}_{\zeta_1 \subset \mathfrak{M}}\, H_\phi(\zeta|\zeta_1)$ satisfies :

(i) $H_\phi(\zeta|\mathfrak{M}) \geq 0$;

(ii) $H_\phi(\zeta|\mathfrak{M}) = 0$ if and only if $\zeta \subset \mathfrak{M}$;

(iii) $\zeta \subseteq \zeta_o$ implies $H_\phi(\zeta|\mathfrak{M}) \leq H_\phi(\zeta_o|\mathfrak{M})$ for all $\mathfrak{M}$ in M ;

(iv) $\mathfrak{M}_1 \subseteq \mathfrak{M}_2$ implies $H_\phi(\zeta|\mathfrak{M}_1) \geq H_\phi(\zeta|\mathfrak{M}_2)$ for all ζ in Z.

Remarks : 1) Lemmata 2 and 3 justify calling $H_\phi(\zeta|\mathfrak{M})$ the entropy of the partition ζ conditioned by the von Neumann algebra $\mathfrak{M}$ with respect to the state ϕ. 2) Since our definition does not require $\mathfrak{M}$ to be abelian, it allows to bypass the second difficulty mentioned in the introduction. Indeed with $\mathfrak{M}_{n-1}[\zeta] \equiv \{\alpha(m)[\zeta] \mid o \leq m \leq n-1\}''$, the expression $H_\phi(\alpha(n)[\zeta] \mid \mathfrak{M}_{n-1}[\zeta])$ is now well-defined for every admissible partition ζ. 3) A less strict adherence to quantum measurement theory could have suggested to introduce a "conditional entropy" defined by :

$$\hat{H}_\phi(\zeta \mid \mathfrak{M}) \equiv \Sigma_j <\phi;\ h\,[\mathcal{E}(G_j \mid \mathfrak{M})]>.$$

<u>IV. GENERALIZED KOLMOGOROV ENTROPY</u>. Because of Lemma 3, the following limit exists for every $\zeta \subset \mathfrak{N}_\phi$:

$$H_\phi(\xi,\alpha) = \lim_{n\to\infty} H_\phi(\alpha(n)[\xi] \mid \mathfrak{M}_{n-1}[\xi]) .$$

We take it as our definition of the <u>entropy of the partition</u> $\zeta \subset \mathfrak{N}_\phi$ <u>under the time evolution</u> $\alpha(\mathbb{R})$. We use this expression to define the <u>Kolmogorov entropy of the generalized flow</u> $\{\mathfrak{N},\phi,\alpha(\mathbb{R})\}$ as :

$$H_\phi(\alpha) = \mathrm{Sup}_{\zeta\varepsilon Z, H_\phi(\zeta)<\infty}\, H_\phi(\zeta,\alpha).$$

From our previous remarks it is clear that $H_\phi(\alpha)$ reduces to the usual Kolmogorov entropy when $\mathfrak{N}$ is abelian.

<u>V. GENERALIZED K-FLOWS</u> A generalized flow $\{\mathfrak{N},\phi,\alpha(\mathbb{R})\}$ is said to be a <u>generalized K-flow</u> if there exists a von Neumann algebra $\mathcal{A} \subset \mathfrak{N}$ such that :

(i) $\mathcal{A} \subsetneq \alpha(t)[\mathcal{A}]$ for all t in $\mathbb{R}^+$;
(ii) $V_t\, \alpha(t)[\mathcal{A}] = \mathfrak{N}$; (iii) $\bigcap_t \alpha(t)[\mathcal{A}] = \mathbb{C}\,I$;
(iv) $\mathcal{A} = \sigma(t)[\mathcal{A}]$ for all t in $\mathbb{R}$.

A generalized K-flow is said to be <u>singular</u> if $\mathfrak{N}_\phi = \mathbb{C}\,I$.

<u>Lemma 4</u> : <u>Let</u> $\{\mathfrak{N},\phi,\alpha(\mathbb{R})\}$ <u>be a generalized K-flow, and</u> $\alpha_\phi(t)$ <u>be the restriction of</u> $\alpha(t)$ <u>to</u> $\mathfrak{N}_\phi$. <u>Then</u> $\{\mathfrak{N}_\phi,\phi,\alpha_\phi(\mathbb{R})\}$ <u>is again a generalized K-flow</u>.

<u>Theorem</u> : <u>The entropy of every non-singular generalized K-flow is strictly positive</u>.

If $\mathfrak{N}$ is abelian we recall that $\mathfrak{N}_\phi = \mathfrak{N}$ so that this theorem does indeed generalize the classical result according to which the entropy of a non-trivial (i.e. $\mathcal{N} \neq \mathbb{C}$ I) classical K-flow never vanishes. We should also mention that several of the properties of classical K-flows (e.g. Lebesgue spectrum) extend to our generalized K-flows [7].

VI. STATISTICAL MECHANICS. We remark in conclusion that the flow of Brownian motion [11] has a quantum analog, namely the Ford-Kac-Mazur model for the diffusion of a quantum harmonic oscillator coupled with a thermal bath [7,12]. The total dynamical system, constituted by the harmonic oscillator plus its bath, turns out to be a generalized K-flow where $\mathcal{N}$ is a type III_λ-factor. Its generalized Kolmogorov entropy can be computed to be infinite [7]. So is in fact the usual Kolmogorov entropy of the classical K-flow of Brownian motion, the latter providing in a similar manner a deterministic model for the diffusion equation of a classical particle in a harmonic potential. The parallel between the classical Kolmogorov theory and its non-commutative, algebraic extension is therefore continued into the realm of non-equilibrium statistical mechanics. The quantum generalization of the flow of Brownian motion has been further extended to the concept of quasi-free generalized K-flows [13] ; these appear as the quantum stochastic processes (or minimal dilation [14]) naturally associated to a class of Markov semi-groups of completely positive maps on von Neumann algebras ; the latter occur in the study of certain irreversible processes, of which the diffusion of a quantum particle in an harmonic well is a particular case.

1. A.N. Kolmogorov, A New Metric Invariant of Transcient Dynamical Systems and Automorphisms of Lebesgue Spaces, Dokl. Akad. Nauk 119, 861 (1958).
2. A.N. Kolmogorov, On the Entropy per Time Unit as a Metric Invariant of Automorphisms, Dokl. Akad. Nauk 124, 754 (1959).
3. Ya. Sinai, On the Concept of Entropy for Dynamical Systems, Dokl. Akad. Nauk 124, 768 (1959).
4. Ya. Sinai, Dynamical Systems with countably-multiple Lebesgue Spectrum, Izvestia Mat. Nauk 25, 899 (1961).
5. A. Connes and E. Størmer, Entropy for Automorphisms of Type II_1 von Neumann Algebras, Acta Mathematica 134, 289 (1975).
6. G.G. Emch, Positivity of the K-Entropy on Non-Abelian K-Flows, z. Wahrscheinlichkeitstheorie verw. Gebiete 29, 241 (1974).
7. G.G. Emch, Generalized K-Flows, Commun. math. Phys. 49, 191 (1976).

8. J. von Neumann, Grundlagen der Quantenmechanik, Springer, Berlin (1932).

9. M. Takesaki, Tomita's theory of Modular Hilbert Algebras, Springer Lecture Notes in Mathematics No. 128 (1970).

10. A.J. Khinchin, Mathematical Foundations of Information Theory, Uspechi Mat. Nauk 7 (1953) and 11 (1956).

11. T. Hida, Stationary Stochastic Processes, Princeton University Press, (1970).

12. G.G. Emch, Non-Equilibrium Statistical Mechanics, Acta Physica Austriaca, Suppl. XV, 79 (1976) ; and : A Dilation Problem in Non-Equilibrium Statistical Mechanics, preprint, Rochester, (1977).

13. G.G. Emch, S. Albeverio and J.P. Eckmann, Quasi-Free Generalized K-Flows, preprint, Geneva, (1977).

14. G.G. Emch, Minimal Dilations of CP-Flows, to appear in Proc. U.S. Japan Conference on Operator Algebras, R.V. Kadison, ed.

WHY THE KMS STATES?

W.Pusz
Department of Mathematical Methods in Physics
Warsaw University, 00-682 Hora 74 Warsaw
Poland

After the papers of Kubo [2] Martin and Schwinger [3] it was only recognized [1] that the KMS condition can be used to characterizing the equilibria at the infinite systems and that the KMS states are straightforward generalizations of Gibbs states describing the equilibria for finite systems. But up to now there is a need for physical motivations to justify the use just the KMS states as equilibrium states. A review of the situation was given in lectures of H.Araki, R.Haag and D.Kastler.

Here we would like to give another motivation based rather upon the second principle of thermodynamics than purely dynamical considerations. In what follows we sketch some results of [4].

Suppose that $\mathfrak{A}$ is a C^*-algebra at observables associated with the physical system and that α_t is one parameter automorphism group describing the time evolution of it. Consider the class of influences upon the system which rely on changing the external conditions - by the external condition we have in mind any external field. This means that the system is thermally isolated so in general it is not closed one but still its dynamics can be described by one parameter family of automorphisms.

If $\quad \mathbb{R}^1 \ni t \longmapsto h_t = h_t^* \in \mathfrak{A}$

describes the perturbation of the dynamics then for perturbed dynamics α_t^h we get

$$\begin{cases} \dfrac{d\alpha_t^h(A)}{dt} = i\alpha_t^h\left(\dfrac{\delta}{i}(A) + [h_t, A]\right) \\ \alpha_0^h = id \end{cases}$$

where δ is the generator of the unperturbed dynamics α.

In the case when $\alpha_t(A) = e^{itH}Ae^{-itH}$ then $\delta(A) = i[H,A]$ thus h_t is really the additional term in Hamiltonian H.

Suppose the system is in the state ω then one can compute the work performed upon the system in the period $[0,T]$ by the external forces:

$$L^h(\omega) = \int_0^T \omega_t\left(\frac{dh_t}{dt}\right)dt = \int_0^T \omega\left(\alpha_t^h\left(\frac{dh_t}{dt}\right)\right)dt$$

As the characteristic properties at a state ω describing the equilibrium state of the system we adopt

Definition 1 ω is passive for α iff $L^h(\omega) \geq 0$ for any cyclic change of the external conditions h_t i.e.

$$\mathbb{R}^1 \ni t \to h_t = h_t^* \in \mathfrak{A}$$

$$h_t = 0 \quad \forall t \notin [0,T]$$

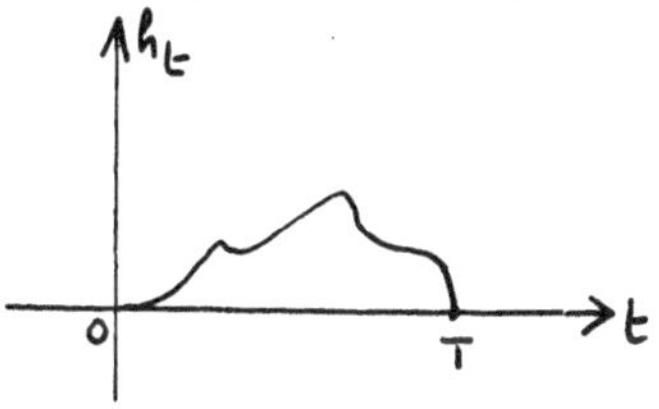

This condition of passivity means that there is no possibility to extract work from the system by such influences - so this essentialy refers to the second thermodynamical law. As an example of such a situation one can have in mind compressing and decompressing the gas in the cylinder to the previous volume without exanging heat with surroundings.

One can prove that passive states are stationary states for unperturbed dynamics α. Moreover it can be seen that all α-KMS states for some inverse temperature $\beta \geq 0$ and ground states are passive. By the linearity at the passivity condition one can see that in general this does not characterize the state with definite temperature because any mixture of KMS states for different temperatures is passive; to avoid such a situation one has to adopt stronger assumptions. The most natural assumption in this context characterizing the system with definite temperature is

<u>Definition</u> 2 ω is <u>completely passive</u> for α iff the product state $\overset{n}{\otimes}\omega$ is a passive state for the dynamics $\overset{n}{\otimes}\alpha$ at the composed system $\overset{n}{\otimes}\mathfrak{A}$ for all n = 1, 2, 3.......

One can also consider states ω describing "pure phases". As mathematical description of such a situation we adopt the weakly clustering assumption. This condition we need for some amenable group G connecting with the dynamical group α but here we state it in very simple version.

<u>Definition</u> 3 An α invariant state ω is weakly clustering for α iff

$$\frac{1}{2T}\int_{-T}^{T}\omega(A\alpha_t B)\,dt \underset{T\to\infty}{\longrightarrow} \omega(A)\,\omega(B)$$

Now we can state

<u>Theorem</u>

Suppose that either

1° ω is completely passive

or

2° ω is passive and weakly clustering

then ω is α-KMS state for some inverse temperature $\beta \geq 0$ or ω is a ground state for α.

Moreover if 2° holds and ω is non-tracial state then ω is type I primary state (pure state) whenever it is a ground state and type III_1 primary state for KMS case with $\beta > 0$.

To give an idea why this theorem holds we sketch the proof.

Let $\{\pi, U, \mathcal{H}, \Omega\}$ be GNS representation associated with the state ω (as was mentioned before ω is α-invariant). We know that Ω is cyclic and Ω: $U_t\Omega = \exp itH\,\Omega$ where H is the Hamiltonian describing the evolution in the representation. Suppose that <u>Ω is separating</u> for $\mathcal{M} \doteq \pi(\mathfrak{A})''$ then one can associate with Ω a modular operator Δ by Tomita-Takesaki theory. It follows easily from this theory that H and Δ strongly commute thus they possess joint spectrum $S = S_p(H, \log\Delta)$ and this set is symmetric. It can be proven that it follows from passivity that

$$S = S_p(H, \log\Delta) \subset \{(E,p)\in\mathbb{R}^2 \mid Ep \leq 0\} \equiv V$$

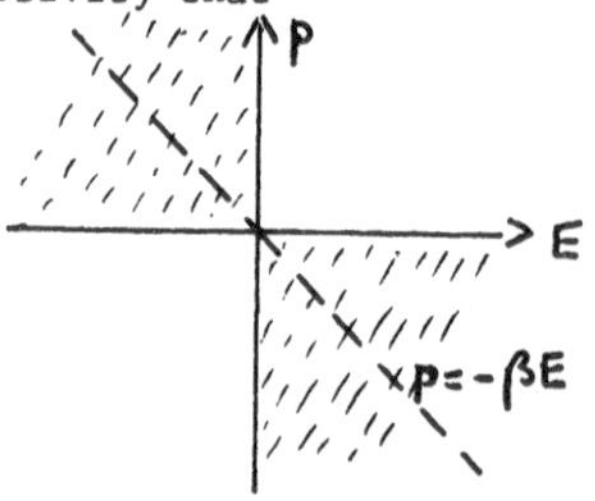

From the additional assumptions of complete passivity or of weakly clustering one can deduce that S is a group contained in V but the only possibility in this case is that S is contained in a straight line thus there exists $\beta \geq 0$ such that $\Delta = e^{-\beta H}$ and this means that ω is α-KMS for inverse temperature β.

If Ω is not separating one can prove that ω is a ground state for α (i.e. $H \geq 0$). For details see [4].

References.

1 R.Haag, N.Hugenholtz, M.Winnink: "On the equilibrium states in quantum statistical mechanics" Commun. Math. Phys. 5, 215 (1967).

2 R.Kubo: "Statistical-mechanical theory of irreversible processes I. General theory and simple applications to magnetic and conduction problems" J. Phys. Soc. Japan 12, 570 (1957).

3 P.C.Martin, J.Schwinger: "Theory of many particle systems J" Phys. Rev. 115, 1342 (1959).

4 W.Pusz, S.L.Woronowicz: "Passive states and KMS states for general quantum system" to appear.

A COMMENT TO THE TALK BY E.SEILER

G.Gallavotti, F.Guerra, S.Miracle-Solé

The formula given by Wegner may be rigorously proven as follows (for instance).

Let Σ_L be a square surface with side L lying on a coordinate plane of the lattice Z^3 on which an Ising model is resting.

Put for every pair of nearest neighbours $f = (i,j)$:

$$\sigma_f = \sigma_i \sigma_j$$

and let

$$\beta_f = \beta^* \quad \text{if} \quad f \cap \Sigma_L = \phi$$

$$\beta_f = -\beta^* \quad \text{if} \quad f \cap \Sigma_L \neq \phi$$

We wish to show that:

$$\lim_{\Lambda \uparrow R^3} \frac{Z_\Lambda}{Z^0_\Lambda} = \frac{\sum_\sigma \exp \sum_{f \subset \Lambda} \beta^*_f \sigma_f}{\sum_\sigma \exp \sum_{f \subset \Lambda} \beta^* \sigma_f} = \exp(O(\beta^4))\, L$$

This is an immediate consequence of the Gruber-Kunz cluster expansion which in our case can be performed as follows:

(if $\mathcal{B}_\Lambda$ = set of subsets of "bonds", i.e. sets of n.n. pairs):

$$Z_\Lambda = \sum_{\underline{\sigma}} \prod_f \left(1 + (e^{+\beta^*_f \sigma_f} - 1\right) = \sum_{F \subset \mathcal{B}_\Lambda} \sum_{\underline{\sigma}} \prod_f \left(e^{\beta^*_f \sigma_f} - 1\right) =$$

$$= 2^{|\Lambda|} \sum_{\substack{x_1 \dots x_k \\ x_i \cap x_j = \phi \quad i \neq j}} \prod_{i=1}^{k} \zeta(x_i)$$

where $x_1 \dots x_k$ are the "connected components" of the family F of bonds and

$$\zeta(X) = \sum_{\underline{\sigma}_X} 2^{-|X|} \prod_{f \in X} \left(e^{\beta^*_f \sigma_f} - 1\right)$$

The logarithm of the sum expressing Z_Λ can be expanded into a series of products of the ζ's and it is well known that only "connected diagrams" appear in this expansion. More precisely let N_Λ = set of the families $\Gamma = (x_1, \dots, x_5)$ of connected components such that if $x_i, x_j \in \Gamma$ then there is a chain $i_1 = i, i_2, \dots, i_p = j$ for which $x_{i_k} \cap x_{i_{k+1}} \neq \emptyset$, $k = 1, \dots, p-1$.

The elements of Γ may contain several times the same x: so we may represent Γ as $(x_1^{n_1}, \dots, x_p^{n_p})$ where $n_1, \dots, n_p > 0$ are the (integer) multiplicities and $x_i \neq x_j$. Let $|\Gamma| = \sum_i n_i$. Gruber and Kunz results read in this case

i) $$Z_\Lambda = 2^{|\Lambda|} \exp \sum_{\Gamma \subset N_\Lambda} \varphi^T(\Gamma)\, \zeta(\Gamma)$$

provided the series converges absolutely, where if $\Gamma = (x_1^{n_1}, \dots, x_p^{n_p})$:

ii) $\zeta(\Gamma) = \prod_1^r \zeta(x_i)^{n_i}$

and the numbers $\varphi^T(\Gamma)$ are certain combinatorial constants (independent on Λ if $\Gamma \in N_\Lambda$) verifying

iii) $|\varphi^T(\Gamma)| \le |\Gamma|^{-1}$

An expansion similar to the i) above can be made for the Z_Λ^0 and calling $\zeta^0(x)$ the corresponding ζ's

$$Z_\Lambda \cdot Z_\Lambda^{0\,-1} \doteq \langle \exp -2\beta^* \sum_{f \subset \Sigma_L} \sigma_f \rangle_0 = \exp \sum_{\Gamma \cap \Sigma_L \neq \emptyset} \varphi^T(\Gamma)\,(\zeta(\Gamma) - \zeta^0(\Gamma))$$

since $\zeta(\Gamma) = \zeta^0(\Gamma)$ if $\Gamma \cap \Sigma_L = \emptyset$ (provided the series in i) for Z_Λ and the analogous for Z_Λ^0 converge).

It is obvious, however, that

$$|\zeta(\Gamma)| \le |e^{\beta^*} - 1|^{\sum n_i |x_i|} \doteq \xi(\Gamma) \quad ; \quad |\zeta^0(\Gamma)| \le \xi(\Gamma)$$

which implies the convergence of the series (because of ii)) if β^* is small and also (since Γ is connected):

$$\sum_{\Gamma \ni 0} |\varphi^T(\Gamma)|\, \xi(\Gamma) \le O(\beta^*)^\ell$$

if β^* is small.

Finally remark that the Γ's for which $\zeta(\Gamma) \neq \zeta_0(\Gamma)$ are those which contain at least one x containing a closed circuit intersecting Σ_L an odd number of times (recall that x is a set of bonds and therefore it may contain "closed circuits").

The convergence of the series then implies that the lowest order dominates

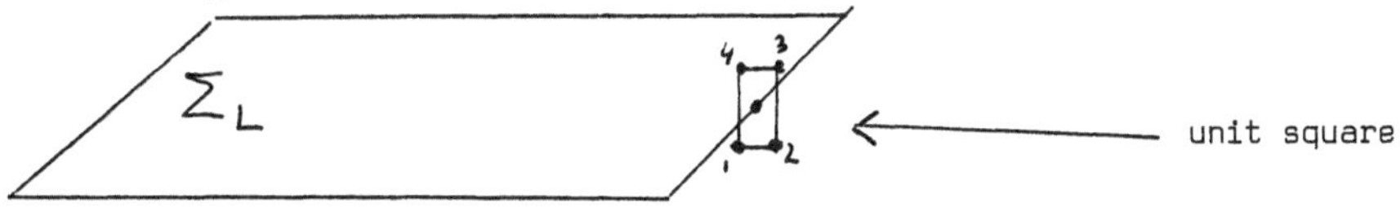

which corresponds to the sum

$$\exp \sum_{\gamma \supset \Sigma_L} \zeta(\gamma)\, \varphi^T(\gamma)$$

where γ is a set of four pairs of n.n. forming a square as 1 2 3 4 in the picture.

The $\varphi^T(\gamma)$ for such a configuration is +1 and

$$\zeta(f) = \sum_{\sigma_1\sigma_2\sigma_3\sigma_4} 2^{-4} (e^{-\beta^*\sigma_1\sigma_4} - 1)(e^{-\beta^*\sigma_4\sigma_3} - 1)(e^{-\beta^*\sigma_3\sigma_2} - 1)(e^{-\beta^*\sigma_2\sigma_1} - 1) = -\beta^{*4} + O(\beta^{*6})$$

while $\zeta_0(f) = +\beta^{*4} + O(\beta^{*6})$

hence

$$Z_\Lambda Z_\Lambda^{0\,-1} = \exp\left(-2\beta^{*4}(4L) + O(\beta^{*6})L\right)$$

since also the contribution from the more complicated Γ's must be at least of order β^{*6}.

It seems clear to us that the formula that Wegner gives without proof in his work has been obtained by an expansion of the above type which is very familiar expansion technique for the physicists.

References:

1) E.Wegner; J. Math. Phys., 12, 2259, 1971.

2) C.Gruber, A.Kunz: Comm. Math. Phys. 22, 133, 1971.

Selected Issues from

Lecture Notes in Mathematics

Lecture Notes in Physics

Vol. 44: R. A. Breuer, Gravitational Perturbation Theory and Synchrotron Radiation. VI, 196 pages. 1975.

Vol. 45: Dynamical Concepts on Scaling Violation and the New Resonances in e^+e^- Annihilation. Edited by B. Humpert. VII, 248 pages. 1976.

Vol. 46: E. J. Flaherty, Hermitian and Kählerian Geometry in Relativity. VIII, 365 pages. 1976.

Vol. 47: Padé Approximants Method and Its Applications to Mechanics. Edited by H. Cabannes. XV, 267 pages. 1976.

Vol. 48: Interplanetary Dust and Zodiacal Light. Proceedings 1975. Edited by H. Elsässer and H. Fechtig. XII, 496 pages. 1976.

Vol. 49: W. G. Harter and C. W. Patterson, A Unitary Calculus for Electronic Orbitals. XII, 144 pages. 1976.

Vol. 50: Group Theoretical Methods in Physics. 4th International Colloquium. Nijmegen 1975. Edited by A. Janner, T. Janssen, and M. Boon. XIII, 629 pages. 1976.

Vol. 51: W. Nörenberg und H. A. Weidenmüller. Introduction to the Theory of Heavy-Ion Collisions. IX, 273 pages. 1976.

Vol. 52: M. Mladjenović, Development of Magnetic β-Ray Spectroscopy. X, 282 pages. 1976.

Vol. 53: D. J. Simms and N. M. J. Woodhouse, Lectures on Geometric Quantization. V, 166 pages. 1976.

Vol. 54: Critical Phenomena. Sitges International School on Statistical Mechanics, June 1976. Edited by J. Brey and R. B. Jones. XI, 383 pages. 1976.

Vol. 55: Nuclear Optical Model Potential. Proceedings 1976. Edited by S. Boffi and G. Passatore. VI, 221 pages. 1976.

Vol. 56: Current Induced Reactions. International Summer Institute, Hamburg 1975. Edited by J. G. Körner, G. Kramer, and D. Schildknecht. V, 553 pages. 1976.

Vol. 57: Physics of Highly Excited States in Solids. Proceedings 1975. Edited by M. Ueta and Y. Nishina. IX, 391 pages. 1976.

Vol. 58: Computing Methods in Applied Sciences. Proceedings 1975. Edited by R. Glowinski and J. L. Lions. VIII, 593 pages. 1976.

Vol. 59: Proceedings of the Fifth International Conference on Numerical Methods in Fluid Dynamics. 1976. Edited by A. I. van de Vooren and P. J. Zandbergen. VII, 459 pages. 1976.

Vol. 60: C. Gruber, A. Hintermann, and D. Merlini, Group Analysis of Classical Lattice Systems. XIV, 326 pages. 1977.

Vol. 61: International School on Electro and Photonuclear Reactions I. Edited by C. Schaerf. VIII, 650 pages. 1977.

Vol. 62: International School on Electro and Photonuclear Reactions II. Edited by C. Schaerf. VIII, 301 pages. 1977.

Vol. 63: V. K. Dobrev et al., Harmonic Analysis on the n-Dimensional Lorentz Group and Its Application to Conformal Quantum Field Theory. X, 280 pages. 1977.

Vol. 64: Waves on Water of Variable Depth. Edited by D. G. Provis and R. Radok. 231 pages. 1977.

Vol. 65: Organic Conductors and Semiconductors. Proceedings 1976. Edited by L. Pál, G. Grüner, A. Jánossy and J. Sólyom. 654 pages. 1977.

Vol. 66: A. H. Völkel, Fields, Particles and Currents. VI, 354 pages. 1977.

Vol. 67: W. Drechsler and M. E. Mayer, Fiber Bundle Techniques in Gauge Theories. X, 248 pages. 1977.

Vol. 68: Y. V. Venkatesh, Energy Methods in Time-Varying System Stability and Instability Analyses. XII, 256 pages. 1977.

Vol. 69: K. Rohlfs, Lectures on Density Wave Theory. VI, 184 pages. 1977.

Vol. 70: Wave Propagation and Underwater Acoustics. Edited by J. Keller and J. Papadakis. VIII. 287 pages. 1977.

Vol. 71: Problems of Stellar Convection. Proceedings 1976. Edited by E. A. Spiegel and J. P. Zahn. VIII, 363 pages. 1977.

Vol. 72: Les instabilités hydrodynamiques en convection libre forcée et mixte. Edité par J. C. Legros et J. K. Platten. X, 202 pages. 1978.

Vol. 73: Invariant Wave Equations. Proceedings 1977. Edited by G. Velo and A. S. Wightman. VI, 416 pages. 1978.

Vol. 74: P. Collet and J.-P. Eckmann, A Renormalization Group Analysis of the Hierarchical Model in Statistical Mechanics. IV, 199 pages. 1978.

Vol. 75: Structure and Mechanisms of Turbulence I. Proceedings 1977. Edited by H. Fiedler. XX, 295 pages. 1978.

Vol. 76: Structure and Mechanisms of Turbulence II. Proceedings 1977. Edited by H. Fiedler. XX, 406 pages. 1978.

Vol. 77: Topics in Quantum Field Theory and Gauge Theories. Proceedings, Salamanca 1977. Edited by J. A. de Azcárraga. X, 378 pages 1978.

Vol. 78: Böhm, The Rigged Hilbert Space and Quantum Mechanics. IX, 70 pages. 1978.

Vol. 79: Group Theoretical Methods in Physics. Proceedings, 1977. Edited by P. Kramer and A. Rieckers. XVIII, 546 pages. 1978.

Vol. 80: Mathematical Problems in Theoretical Physics. Proceedings, 1977. Edited by G. Dell'Antonio, S. Doplicher and G. Jona-Lasinio. VI, 438 pages. 1978.

Lecture Notes in Physics